Lecture Notes in Mathematics

A collection of informal reports and seminars
Edited by A. Dold, Heidelberg and B. Eckmann, Zürich

Tulane University Ring and Operator Theory Year, 1970–1971
Volume I

246

Lectures on Rings and Modules

Springer-Verlag
Berlin · Heidelberg · New York 1972

AMS Subject Classifications (1970): 16 A 06, 16 A 08, 16 A 10, 16 A 12, 16 A 14, 16 A 20, 16 A 26, 16 A 30, 16 A 36, 16 A 40, 16 A 42, 16 A 46, 16 A 50, 16 A 56, 16 A 60, 16 A 62, 16 A 64, 16 A 66

ISBN 3-540-05760-9 Springer-Verlag Berlin · Heidelberg · New York
ISBN 0-387-05760-9 Springer-Verlag New York · Heidelberg · Berlin

Offsetdruck: Julius Beltz, Hemsbach/Bergstr.

ACKNOWLEDGEMENTS

The special year in the Theory of Rings and Operator Algebras was the third program of a similar kind held at Tulane University, following one in the Theory of C*-Algebras and one in Category Theory. These programs were made possible solely through a grant from the Ford Foundation. We wish to express our sincere gratitude to the Ford Foundation for its generous support and cooperation.

A few participants were supported by other agencies: Hans H. Storrer held a fellowship of the Schweizerischer Nationalfonds zur Föderung der Wissenschaftlichen Forschung while he spent the year at Tulane University. Silviu Teleman was able to spend several months at Tulane University, thanks to the exchange program between the National Academy of Sciences of the U. S. A. and the Academy of Sciences of the Socialist Republic of Rumania.

Tulane University provided the facilities for the program and administered the grant.

Most of these notes were typed efficiently and patiently by the secretarial staff of the Mathematics Department of Tulane University.

Karl Heinrich Hofmann

PREFACE

From September 1970 through May 1971 Tulane University organized a special year long program in the theory of non-commutative rings and operator algebras. Visitors from various institutions of the U.S.A. and abroad contributed series of lectures in which they covered recent advances in their own field of specialty. These notes contain these lectures to the extent that they have not appeared elsewhere.

The contributions are collected in three volumes; the division is roughly the following: Volume I represents the purely algebraic side of the program with lectures in the structure theory of rings and modules; in Volume II we collected contributions to the analysis section of the program with lectures on operator algebras and applications to the theory of operators on Hilbert space and to harmonic analysis; Volume III contains applications of topology to ring theory in the form of selected lectures in the theory of representation of rings by sections in sheaves.

LIST OF PARTICIPANTS IN RING THEORY YEAR

Horst Behncke, University of Heidelberg, Heidelberg, Germany

Gerhard Betsch, University of Tübingen, Tübingen, Germany, and The University of Arizona, Tucson, Arizona

Ann Boyle, University of New Hampshire, Durham, New Hampshire

P. M. Cohn, University of London, London, England

V. Dlab, Carleton University, Ottawa, Canada

John Ernest, University of California, Santa Barbara, California

Carl Faith, Rutgers University, Princeton, New Jersey

Alfred W. Goldie, University of Leeds, Leeds, England

G. Hauptfleisch, Rand Afrikaans University, Johannesburg, South Africa

Herbert Heyer, University of Tübingen, Tübingen, Germany

Klaus Keimel, University of Tours, Tours, France

Kwangil Koh, North Carolina State University, Raleigh, North Carolina

J. Lambek, McGill University, Montreal, Quebec, Canada

Horst Leptin, University of Heidelberg, Heidelberg, Germany

Gerhard Michler, University of Tübingen, Tübingen, Germany

R. S. Pierce, University of Hawaii, Honolulu, Hawaii

J. R. Ringrose, University of Newcastle, Newcastle, England

Teishirô Saitô, Tohoku University, Sendai, Japan

Hans H. Storrer, Research Institute for Mathematics, Swiss Federal Institute of Technology, Zurich, Switzerland

Masamichi Takesaki, University of California, Los Angeles, California

Silviu Teleman, Mathematical Research Institute of the Rumanian Academy of Sciences, Bucarest, Rumania

Contents

P.M.COHN

Skew Fields of Fractions, and the Prime Spectrum of a General Ring

VLASTIMIL DLAB and CLAUS MICHAEL RINGEL

Balanced Rings

CARL FAITH

Modules Finite over Endomorphism Ring

L. FUCHS

The Cancellation Property for Modules

ALFRED W. GOLDIE

The Structure of Noetherian Rings

SKEW FIELDS OF FRACTIONS, AND THE PRIME SPECTRUM OF A GENERAL RING

by

P. M. Cohn

University of London
London, England

It is well known that a commutative ring has a field of fractions if and only if it is an integral domain; moreover, the field of fractions, when it exists, is unique up to isomorphism and its elements have the form ba^{-1}, where a,b lie in the ring and $a \neq 0$. This construction is of great importance in algebraic geometry, where it leads to the notion of localization at a prime ideal. In the case of a polynomial ring, or more generally, an affine ring, the prime ideals correspond to algebraic subvarieties, and the set of all prime ideals carries a natural topology, the Zariski topology. Moreover, much of this carries over to arbitrary commutative rings.

When one tries to do the same for general rings, one is hampered by the fact that not every ring has a field of fractions, or even a homomorphism into a (skew) field. One will therefore have to begin by looking for conditions under which a field of fractions exists. This will limit the class of rings we can consider, but the least we must ask is that it should contain the analogue of polynomial rings, i.e. the free algebras. We shall see that fields of fractions exist in that case, but are far from unique and we shall need the notion of a universal field of fractions. The first few lectures will be taken up with the development of this notion; after that we shall look at an analogue of the prime spectrum for general rings, suggested by these ideas.

1. The category of R-fields

Most of what we do will be either trivial or well known (or both) for commutative rings, so our rings are generally non-commutative. Likewise fields will be "not necessarily commutative division rings". All rings have a 1, which acts unitally and is preserved by homomorphisms, subrings etc. If R is any ring, the set of non-zero elements of R is denoted by R^*. In case R^* is non-empty and closed under multiplication, R is called an _integral domain_.

Let R be a ring. By an _R-ring_ one understands a ring A with a homomorphism $R \to A$. (This notion is to be distinguished carefully from that of an R-algebra; this is a ring with a homomorphism of R into the center of A). Of course an R-ring homomorphism is a map $\phi : A \to B$ of R-rings which is a homomorphism of rings, such that the triangle shown commutes.

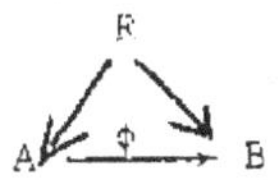

By an _R-field_ we shall mean an R-ring K which is a field, and such that K is the least field containing the image of R. If, moreover, the canonical mapping $R \to K$ is injective, the R-field is called a _field of fractions_ for R. Any map f between R-fields is necessarily an isomorphism. For ker f is a proper ideal in K, hence 0,

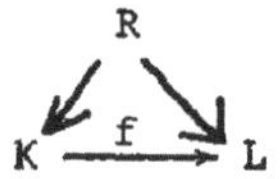

and im f is a subfield of L containing the image of R, hence im f = L. We therefore need to generalize the notion of homomorphism. Let us define a specialization of R-fields K,L as an R-ring homomorphism f from an R-subring K_0 of K to L such that $x \in K_0$, $xf \neq 0$ implies $x^{-1} \in K_0$.

If we have a specialization $f: K \to L$, the domain K_0 of f is a local ring, i.e. the non-units form an ideal. For $x \in K_0$ is a unit precisely if $xf \neq 0$, so the set of non-units is ker f. Moreover, $L \supseteq \text{im } f \cong K_0/\ker f$, and this is a field containing the image of R. Since L is the least field with this property, we have L = im f, thus every specialization is surjective.

Given any ring R, we define a category $\mathfrak{F}_R$ as follows: the objects are R-fields and the maps specializations. Of course we must check that the composition of maps is defined: Given specializations $f: K \to L$, $g: L \to M$, let K_0, L_0 be the domains of f,g respectively and put $K_1 = \{x \in K_0 \mid xf \in L_0\}$, $f_1 = f|K_1$. We claim that $f_1 g: K_1 \to M$ is a specialization. If the canonical map $R \to K$ is denoted by μ_K, we have $R\mu_K f = R\mu_L$, hence $R\mu_K \subseteq K_1$, so that K_1 is an R-ring. Moreover, if $x \in K_1$ and $xf_1 g \neq 0$, then $xf = xf_1 \neq 0$, so $x^{-1} \in K_0$ and $(x^{-1})f_1 = (xf_1)^{-1} \in L_0$ hence $x^{-1} \in K_1$. This shows that $f_1 g$ is indeed a specialization.

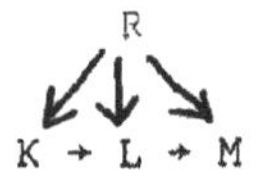

An initial object in the category $\mathfrak{F}_R$ of R-fields is called a <u>universal</u> <u>R-field</u>. Thus an R-field U is universal if, for any R-field K, there exists a unique specialization from U to K. Clearly the universal R-field is unique up to isomorphism, if it exists at all.

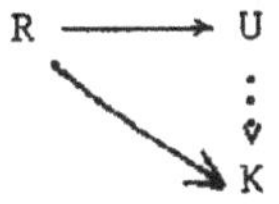

Moreover, if R has a universal R-field, U say, then R has a field of fractions if and only if U is a field of fractions. In that case we call it a <u>universal</u> <u>field</u> <u>of</u> <u>fractions</u>. We note however that R need not have a universal R-field, or indeed any R-field at all.

We now give some examples to illustrate the definitions.

1. Let R be a commutative ring. The R-fields correspond precisely to the prime ideals of R. Given an R-field K, the kernel of the natural map $\mu_K : R \to K$ is a prime ideal, and conversely, if g is a prime ideal of R, then $R \to (R/g)$ (where (A) is the field of fractions of A) gives us an R-field. R has a universal R-field iff the nil radical is prime (i.e. the set of prime ideals has a least element). There is a universal field of fractions iff R is an integral domain (i.e. the zero ideal is prime).

2. Let R be a (left or right) Noetherian ring. Any R-field is obtained as a field of fractions of R/A , for a suitable ideal A . Now R/A is again Noetherian, and an integral domain, hence (by Goldie's theorem) it is a (left or right) ore domain and the field of fractions of R/A is

unique up to isomorphism; its elements have the form $a^{-1}b$ (or ba^{-1}) where $a,b \in R$ and $a \neq 0$.

3. In general there may be no R-fields at all. E.g. let $R = k_2$, the ring of all 2×2 matrices over a field k. This ring is simple, so if there is an R-field K, the map $R \to K$ must be injective, which is absurd, because R has zero-divisors. One might argue that rings like k_2 require a separate theory, which allows total matrix rings over (skew) fields. There is some justification in this, but the basic objection remains, since there are integral domains that cannot be embedded in fields (Malcev [12]), and even integral domains that have no homomorphisms into fields ([5]; recall that homomorphisms preserve 1, and in a field $1 \neq 0$).

4. Let F be the free group on a set X and M the free metabelian group on X (i.e. satisfying the law $((x,y), (z,t)) = 1$). Then F and M can both be ordered and hence the group algebras kF, kM can be embedded in fields (cf. [4], [14]), but both contain the semigroup algebra of the free semigroup on X, in other words the free associative algebra $k\langle x\rangle$. This means that $k\langle x\rangle$ has two essentially different fields of fractions. Actually it is not hard to get infinitely many different fields of fractions of $k\langle x\rangle$. Fix $n > 1$ and let R be the k-algebra generated by a and b subject to the relation $ba = ab^n$. Then R is a right ore domain and so has a field of fractions; the subalgebra generated by $x = a$ and $y = ab$ is free on x and y,

and the field of fractions obtained for $k\langle x,y\rangle$ in this way is different for each n (see [10]).

2. The universal Σ-inverting ring

Let R be a ring and K a field of fractions for R. If R is commutative, every element u of K can be written $a^{-1}b$ $(a,b \in R)$. Thus u is obtained by solving

$$au - b = 0. \qquad (1)$$

In the general case, even when we have a field of fractions, its elements cannot usually be obtained in this simple way. Even when we have adjoined inverses for all elements of R^* there may still be elements without an inverse, so that we need to perform repeated inversions, as in $(ab^{-1}c + de^{-1}f)^{-1}$. However, we shall be able to manage with a single inversion if we replace $\underline{a}$ in (1) by a matrix. In other words, we can take care of repeated inversions by taking the inverse of a matrix. In the commutative case this gives nothing new: to invert a matrix A, we need only invert the element $\det A$, which in the commutative case is a polynomial in the elements of A. In general this is not so, here $\det A$ is a <u>rational</u> function of the elements of A (modulo commutators, at least; see [9]), and it is to avoid these rational functions that we need to operate with the matrices themselves.

The first step will be to adjoin the inverses of certain matrices to R. Since our aim is to construct fields, we shall confine ourselves to square matrices, the only ones that can be inverted over a field.

Given a set Σ of square matrices over a ring R, and a homomorphism $f:R \to R'$, we shall say f is <u>Σ-inverting</u> if Af is invertible over R' for all $A \in \Sigma$.

To facilitate the construction of Σ-inverting homomorphisms, we make the obvious but important observation that there always is a universal Σ-inverting homomorphism:

<u>Theorem 1</u>. Given any ring R and any set Σ of matrices over R, there is a ring R_Σ and a Σ-inverting homomorphism $\lambda:R \to R_\Sigma$ which is universal in the sense that any Σ-inverting homomorphism can be factored uniquely by λ.

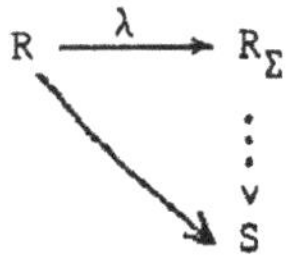

Proof. To obtain a presentation for R_Σ let us take a presentation for R and for each $n \times n$ matrix $A \in \Sigma$ adjoin n^2 indeterminates, written as a matrix A', with defining relations $AA' = A'A = I$. The natural mapping $R \to R_\Sigma$ is a homomorphism, and it is clear how any Σ-inverting homomorphism $f:R \to S$ can be factored by $\lambda:R_\Sigma$ is generated by elements correspond either to elements of R (whose image is prescribed) or to components of a matrix A', and in that case they must be mapped to the corresponding components of $(Af)^{-1}$.

Although this result provides us with a ring in which the matrices of Σ become invertible, there is no way of telling when the map λ is injective, or even whether $R_\Sigma \neq 0$. All we can say is that if there exists a Σ-inverting injection, then (by the usual argument) the universal Σ-inverting map is also injective. In these lectures we take the point of view that we know (by some means or other) that λ is injective and make deductions from this fact. Later we shall meet some non-trivial cases where this assumption is realized.

3. The Σ-rational closure of a mapping

Let R be any ring and Σ a set of square matrices over R. For any Σ-inverting homomorphism $f:R \to S$ we can define the Σ-rational closure as the set $R_\Sigma(S)$ of all components of matrices $(A^f)^{-1}$ where $A \in \Sigma$. We shall be interested in conditions under which this rational closure is a ring, so let us define a set Σ of matrices to be admissible if (i) $1 \in \Sigma$ (the 1×1 matrix consisting of the unit-element), (ii) Σ admits elementary (row or column) operations and (iii) if $A,B \in \Sigma$, then $\begin{pmatrix} A & C \\ 0 & B \end{pmatrix} \in \Sigma$ for any matrix C of the appropriate size (and the zero-matrix 0).

Let $f:R \to S$ be any homomorphism; the set of all matrices inverted by f is always admissible. For clearly 1^f is always invertible, and invertibility is unaffected by elementary transformations. Moreover, if A,B are invertible and C is any matrix of suitable size, then

$$\begin{pmatrix} A & C \\ 0 & B \end{pmatrix} \quad \text{has the inverse} \quad \begin{pmatrix} A^{-1} & -A^{-1}CB^{-1} \\ 0 & B^{-1} \end{pmatrix}.$$

Moreover, the matrices inverted will all be square if we postulate that S has the invariant basis property, for that amounts to saying: every invertible matrix is square.

We now show that for an admissible set Σ, the Σ-rational closure under any Σ-inverting map is a subring. This is a straightforward generalization of the Schutzenberger-Nivat criteria for rationality of power series (corresponding to conditions c,b respectively in the next theorem; see [15]).

Theorem 2. Let R be any ring and Σ an admissible set of matrices over R. If $f:R \to S$ is a Σ-inverting homomorphism then the Σ-rational closure $\bar{R}$ is a subring of S containing R^f and for any $x \in S$, the following conditions are equivalent:

a) $x \in \bar{R}$

b) $x = u_1$ is the first component of the solution of a matrix equation

(1) $$Au + a = 0,$$

where $A \in \Sigma^f$ and a is a column in R^f,

c) $x = u_1$ is the first component of the solution of a matrix equation

(2) $$Au - e_1 = 0,$$

where $A \in \Sigma^f$ and $e_1 = (1,0,\ldots,0)^T$ (T means: transpose).

Proof. Clearly $c \Rightarrow b$, but if b holds, we have

$$\begin{pmatrix} 0 & 1 \\ A & a \end{pmatrix}\begin{pmatrix} u \\ 1 \end{pmatrix} - \begin{pmatrix} 1 \\ 0 \end{pmatrix} = 0,$$

which has the form (2): the matrix lies in Σ^f, because Σ is admissible. Thus $b \Leftrightarrow c$. Now $a \Leftrightarrow c$ by definition of $\bar{R}$ and the fact that Σ admits elementary transformations.

To show $\bar{R}$ is a subring we use (b). If $c \in R^f$, it satisfies

$$1.\quad u - c = 0$$

hence $\bar{R} \supseteq R^f$. It remains to show that $\bar{R}$ admits subtraction and multiplication. If u_1, v_1 are the first components of the solutions of $Au + a = 0$, $Bv + b = 0$, respectively, then $u_1 - v_1$ is the first component of the solution of

$$(3)\qquad \begin{pmatrix} 1 & -1\ 0 & 1\ 0 \\ 0 & A & 0 \\ 0 & 0 & B \end{pmatrix} \begin{pmatrix} u_1 - v_1 \\ u \\ v \end{pmatrix} + \begin{pmatrix} 0 \\ a \\ b \end{pmatrix} = 0,$$

while $u_1 v_1$ is the first component of the solution of

$$(4)\qquad \begin{pmatrix} A & a\ 0 \\ 0 & B \end{pmatrix} \begin{pmatrix} uv_1 \\ v \end{pmatrix} + \begin{pmatrix} 0 \\ b \end{pmatrix} = 0.$$

In both cases the matrix lies in Σ^f, by admissibility of Σ.

The matrices in (3) and (4) are linear in the elements of A and B. This shows that if R is generated, as ring, by a set X, over a coefficient domain k, then the matrices occurring in (1), (2) can be taken to be linear in X.

The next question is whether the rational closure necessarily contains the inverses of those of its elements that are invertible in S. This question can be answered affirmatively if every matrix in S is either a zero-divisor or invertible, which holds e.g. when S is a field. To prove this result, let us define the (i,j)-minor of a matrix A as the matrix obtained by omitting the ith row and j^{th} column from A. We also note that there is no loss of generality in taking R to be a subring of S.

Lemma. Given rings R,S where $R \subseteq S$, let $u_1 \in S$ be rational over R, say it is the solution of (2), and denote the (1,1)-minor of A by A_{11}. Then

(i) u_1 is a left zero-divisor in S iff A_{11} is a left zero-divisor over S,

(ii) if A_{11} is invertible in S, then u_1 has a left inverse in S.

Proof. Write $A = (a_1,\ldots,a_n)$, $A_{11} = (a_2',\ldots,a_n')$ (so a_i' is obtained by removing the first element from a_i). Then A_{11} is a left zero-divisor iff there exist $v_2,\ldots,v_n \in S$, not all 0, satisfying

$$a_2'v_2 + \ldots + a_n'v_n = 0. \tag{5}$$

This equation is equivalent to

$$a_2v_2 + \ldots + a_nv_n + e_1x = 0 \tag{6}$$

for some $x \in S$, which on putting $v = (0,v_2,\ldots,v_n)^T$, may be written

$$Av + e_1x = 0.$$

Here $x \neq 0$, for otherwise $Av = 0$, and since A is invertible we find $v = 0$, a contradiction. Now $Av = -e_1x = Aux$, by (2), hence $v = ux$, and so $u_1x = 0$, i.e. u_1

is a left zero-divisor. Conversely, if $u_1x = 0$, then $Aux + e_1x = 0$, when written out, gives

$$a_2v_2 + \dots + a_nv_n + e_1x = 0,$$

on putting $v_i = u_ix$. But this is just (6), and not all the v_i vanish because $x \neq 0$. Hence we obtain (5); therefore A_{11} is a left zero-divisor.

Next assume A_{11} invertible, then $B = (e_1, a_2, \dots, a_n)$ is also invertible. Put $v = (1, u_2, \dots, u_n)^T$, then we have

$$Bv + a_1u_1 = 0.$$

Now the equation $Bv + a_1 = 0$ has a unique solution w in S, and $Bv = -a_1u_1 = Bwu_1$, hence $v = wu_1$ i.e. $w_1u_1 = 1$. This shows that u_1 has a left inverse as claimed.

4. The description of R-fields by localizing sets

We now show how to relate R-fields to admissible sets of matrices. In the commutative case each R-field is completely specified by its kernel, i.e. the set of all elements mapping to 0, and these sets are characterized as the prime ideals of R. In general, these sets are not enough (because a given integral domain can have non-isomorphic fields of fractions). Instead we shall characterize the R-fields by the sets of matrices that are made invertible; they are characterized below as the saturated localizing sets of matrices (to be defined).

Theorem 3. Let R be any ring. Then

i) if Σ is a set of matrices such that the universal Σ-inverting ring R_Σ is a local ring, then the residue-class field of R is an R-field,

ii) if K is an R-field and Σ is the set of all matrices over R which are inverted in K, then Σ is admissible and R_Σ is a local ring with residue-class field isomorphic to K.

Remark. A local ring was defined as a ring in which the non-units form an ideal. By convention we also assume $1 \neq 0$, i.e. we do not admit $\{0\}$ as local ring.

Proof of theorem. Let Σ be such that R_Σ is a local ring, with residue class field K. By composing the natural maps we get a homomorphism $R \to R_\Sigma \to K$, and here K is a field and an R-ring. Let K' be the least subfield generated by the image of R. Any matrix A in Σ maps to an invertible matrix A_1 in K; this matrix is a non-zero-divisor over K' and hence has an inverse already in K'. Thus K' contains inverses of all matrices in Σ and hence contains a generating set of K, so $K = K'$, i.e. K is an R-field.

Now let K be an R-field and denote by Σ the set of matrices over R that map to invertible matrices over K. Then Σ is admissible, since K has the invariant basis property. Further, by definition of R_Σ as the universal Σ-inverting ring we have a commutative triangle

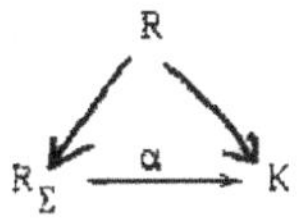

and we have to show that R_Σ is a local ring with residue class field K. This will follow if we prove that any element of R_Σ not in $\ker\alpha$ is invertible in R_Σ.

Let $u_1 \in R_\Sigma$ satisfy the equation

$$Au + e_1 = 0,$$

where $A \in \Sigma$, and assume $u_1 \notin \ker\alpha$. Write again A_{11} for the (1,1)-minor of A. Then u_1^{α} is non-zero, hence invertible, and so A_{11}^{α} is not a left zero-divisor, by the lemma in §3, and so it is invertible. Hence $A_{11} \in \Sigma$ and by another application of the lemma we find that u_1 has a left inverse in R_Σ, say $xu_1 = 1$. Clearly $x^{\alpha} \neq 0$, so x has a left inverse y, and $u_1 = yxu_1 = y$, i.e. $xu_1 = u_1x = 1$. This shows u_1 has an inverse, and the proof is complete.

This result shows that R-fields are completely determined by the admissible sets of matrices in R. Let us call a set Σ of matrices in R <u>localizing</u> if it is admissible and R_Σ is a local ring. Theorem 3 then says that R-fields correspond to localizing sets of matrices, but to get a (1,1)-correspondence we must restrict ourselves to "saturated" sets. Let Σ be any set of matrices over R such that $R_\Sigma \neq 0$ and denote by Σ' the set of all matrices of R that are invertible in R_Σ. Then

$$\Sigma' \supseteq \Sigma,$$

but equality need not hold. This is analogous to the fact that in a commutative ring the set inverted in R_M is generally larger than M. Let us call Σ <u>saturated</u> if it is the precise set of matrices inverted in R_Σ. We supplement the description of R-fields given by Th. 3 in two respects: i) we give a criterion for Σ to be localizing and ii) we translate the relation between specializations to the corresponding matrix sets. First we need a lemma describing local rings.

<u>Lemma</u>. Let R be a non-zero ring in which for each $x \in R$, either x has a left inverse or $1-x$ has a right inverse. Then R is a local ring.

Proof. R is a local ring if the non-units form an ideal, or equivalently, for any non-unit x, $1-x$ is a unit. So it will be enough to show that R has no one-sided inverses.

Assume $xy = 1$, $yx \neq 1$, then $yx(yx-1) = y(xy-1)x = 0$ hence yx has no left inverse, therefore $1-yx$ has a right inverse, z say, and $x = x(1-yx)z = (1-xy)xz = 0$, a contradiction. Thus R is a local ring, as claimed.

<u>Theorem 4</u>. Let R be any ring and Σ an admissible set of matrices such that $R_\Sigma \neq 0$. If Σ satisfies the condition L: For any $A \in \Sigma$, either $A_{11} \in \Sigma$ or $A+e_{11} \in \Sigma$ (where A_{11} is the (1,1)-minor of A and e_{11} the usual matrix unit),

then Σ is localizing. Conversely, any saturated localizing set Σ satisfies L.

Proof. Any element $x \in R_\Sigma$ is obtained as solution $x = u_1$ of an equation

$$Au + e_1 = 0 \tag{7}$$

with $A \in \Sigma$. If u_1 has no left inverse in R_Σ, then A_{11} is not invertible in R_Σ. Hence $A_{11} \notin \Sigma$, by L, $A+e_{11} \in \Sigma$ and we can solve the equation

$$(A+e_{11})v + e_1 = 0. \tag{8}$$

Now (8) can be written

$$Av + e_1(1+v_1) = 0$$

and combining this with (7) we find $Av = -e_1(1+v_1) = Au(1+v_1)$ hence $v = u\,(1+v_1)$ i.e. $v_1 = u_1(1+v_1)$ and so

$$(1-u_1)(1+v_1) = 1.$$

Thus for any $x \in R_\Sigma$ either x has a left-inverse or $1-x$ has a right inverse. By the lemma R_Σ is a local ring.

Conversely assume R_Σ is a local ring, with residue class field K. A matrix A over R maps to an invertible matrix in K iff it maps to an invertible matrix in R_Σ, i.e. iff $A \in \Sigma$. Suppose $A \in \Sigma$, but $A_{11} \notin \Sigma$, and consider the equation

$$Au + e_1 = 0 \tag{9}$$

over K. Since A_{11} is not invertible in K, u_1 has no left inverse in K and so is 0. This means that the inverse A^{-1} of A has (1,1)-element 0. Therefore

$$(A+e_{11})A^{-1} = \begin{pmatrix} 1 & c \\ 0 & I \end{pmatrix}$$

where $c = (c_2,\ldots,c_n)$ is the first row of A^{-1} (after the (1,1)-element). By elementary cloumn operations we can transform the matrix on the right to I, hence $A+e_{11}$ has a right inverse over K and therefore an inverse. But this means that $A+e_{11} \in \Sigma$ as asserted.

Let K be any R-field. Then the natural mapping $R \xrightarrow{f} K$ is a ring epimorphism. For if $\alpha,\beta : K \to S$ are homomorphisms, their images are subfields of S and the subset of K on which both agree is a subfield containing Rf, i.e. the whole of K. On the other hand, the mapping f is not usually flat unless R is an ore domain.

Given two R-fields, K and L, if there is a specialization $\alpha : K \to L$, it is uniquely determined, because, as we have just seen, the map $R \to K$ is an epimorphism, and hence so is the map $R \to \mathrm{dom}\,\alpha$. We shall now describe the condition for a specialization in terms of the sets of matrices that are inverted. Recall that by Th. 3 every R-field has the form K_Σ of a residue-class field with respect to R_Σ where Σ is a localizing set of matrices.

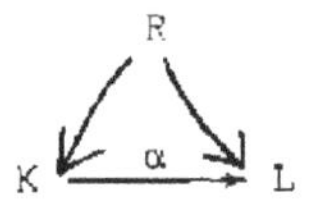

<u>Theorem 5</u>. Given two localizing sets of matrices Σ_1, Σ_2 of a ring R, there is a specialization $K_{\Sigma_1} \to K_{\Sigma_2}$ iff $\Sigma_1 \supseteq \Sigma_2$.

Proof. Let R_i $(i=1,2)$ be the universal Σ_i-inverting ring: by hypothesis, R_i is a local ring with maximal ideal m_i and residue class field K_i. Assume $\Sigma_1 \supseteq \Sigma_2$ and let R_{12} be the subring of R_1 generated by the inverses of matrices in Σ_2. Then R_{12} is a homomorphic image of R_2, the universal Σ_2-inverting ring, and the natural homomorphism $R_1 \to K_1$ maps R_{12} to $R'_{12} = R_{12}/(R_{12} \cap m_1)$. Now R_{12} is a local ring (being the image of R_2) and $R_{12} \cap m_1$ is a proper ideal, therefore the natural homomorphism $R_2 \to K_2$ can be taken via R'_{12}, giving a homomorphism from a local R-subring of K_1 into K_2; any $x \in R'_{12}$ which maps to a non-zero element of K_2 must come from an invertible element of R_2 and hence be itself invertible.

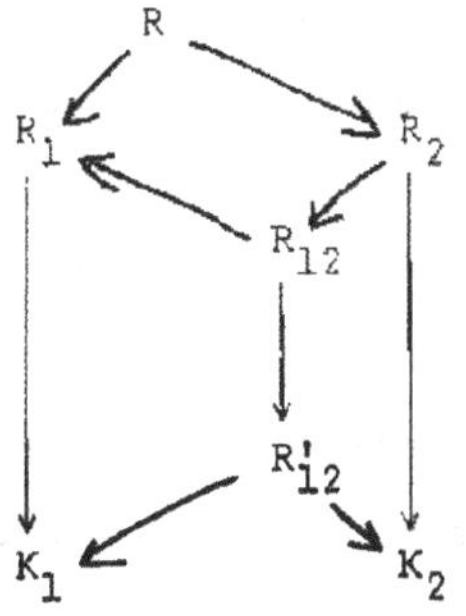

Conversely let $\alpha: K_1 \to K_2$ be a specialization of R-fields, and let $f_i: R \to K_i$ be the canonical homomorphism. Take $A \in \Sigma_2$ then A^{f_2} has an inverse which is the image of a matrix B in K_1: $(A^{f_2})(B^{\alpha}) = I$, hence $A^{f_1}B = I+C$ where $C^{\alpha} = 0$. Hence $I+C$ has an inverse and therefore so does A^{f_1}, i.e. $A \in \Sigma_1$, thus $\Sigma_1 \supseteq \Sigma_2$ as claimed.

From this theorem we immediately obtain conditions for a universal R-field to exist:

Corollary. A ring R has a universal R-field iff the collection of localizing sets of matrices has a greatest element.

5. Sufficient conditions for the existence of universal fields of fractions

We now come to the problem of constructing R-fields when they exist. We already know that they will be of the form R_Σ/m_Σ, for some set Σ of matrices; our main problem is to find the appropriate conditions to put on the family Σ. If we actually want a field of fractions, we must be sure to exclude zero-divisors from Σ, but this condition on Σ is not necessary for getting R-fields, and in any case not sufficient. To illustrate the situation, take an integral domain R that is neither left nor right Ore. This means that there exist $a,b,c,d \neq 0$ in R such that $Ra \cap Rb = 0$, $cR \cap dR = 0$. It follows easily that the matrix

$$A = \begin{pmatrix} a \\ b \end{pmatrix} (c\ d) = \begin{pmatrix} ac & ad \\ bc & bd \end{pmatrix}$$

is a non-zero divisor in R_2. But its image in any field has rank less than 2 and so cannot have an inverse. This suggests the following

Definition. A matrix A over a ring R is said to be full if it is square, say $n\times n$, and not of the form $A = PQ$, where P is $n\times r$, Q is $r\times n$ and $r < n$.

This definition is taken to mean for $n = 1$, that $a \in R$ is full if it is $\neq 0$. Thus any non-zero ring with zero-divisors provides us with examples of full zero-divisors. For $n > 1$, take the ring $k[x,y,z,t]/(xt-yz)$ (coordinate-ring of the quadric), then the matrix $\binom{xy}{zt}$ has zero determinant and so is a zero-divisor, but it is full.

We shall often be interested in conditions under which the set Φ of all full matrices is admissible. $1 \in R$ is full precisely if $1 \neq 0$, i.e. if R is not the zero-ring. Secondly, Φ always admits elementary operations, but the third condition for admissibility (if $A,B \in \Phi$, then $\binom{AC}{DB} \in \Phi$) is not generally true, and is rather hard to verify in actual cases. E.g. let $F(n)$ be the condition "I_n is full". Then there exist rings in which $F(\nu)$ for $\nu = 1,\ldots,n-1$ does not imply $F(n)$. (See [5]). A ring in which $F(n)$ holds for all n is said to have the <u>strong invariant basis property</u> (clearly it implies the invariant basis property, and in [5] it is shown to be stronger than the latter).

In the next result we give conditions for the rational closure under a Φ-inverting map to be a field:

<u>Theorem 6</u>. Let R,S be rings and $f:R \to S$ a Φ-inverting homomorphism, and assume that either

i) $S \neq 0$ and the set Φ of full matrices in R is admissible, or

ii) S has the strong invariant basis property, then f is injective, and the Φ-rational closure is the universal field of fractions of R.

Proof. In either case $S \neq 0$. Let $a \in R$, $a \neq 0$, then a is full, hence af has an inverse, and so $af \neq 0$, i.e. f is injective, so we may take R to be embedded in S, without loss of generality. Next assume i) and denote by $\bar{R}$ the Φ-rational closure of R in S. We claim that $\bar{R}$ is a field.

Let $x \in \bar{R}$, say $x = u_1$ is the first component of the solution of

$$Au + a = 0, \tag{1}$$

where A is a full matrix over R. Write $A = (a_1,\ldots,a_n)$ and put $A_1 = (a,a_2,\ldots,a_n)$, $v = (1,u_2,\ldots,u_n)^T$, then (1) can be written

$$a_1u_1 + \ldots + a_nu_n + a = 0,$$

i.e.

$$A_1v + a_1u_1 = 0. \tag{2}$$

We shall show that if $u_1 \neq 0$, then A_1 is full. Assuming this for the moment, we can solve

$$A_1w + a_1 = 0$$

in S; in fact the solution lies in $\bar{R}$. Hence $A_1v = -a_1u_1$ $= A_1wu_1$, so $v = wu_1$ and equating first components we find $w_1u_1 = 1$. Thus every non-zero element of $\bar{R}$ has a left inverse, which proves that $\bar{R}$ is a field.

It remains to show that A_1 is full. If this is not so, we can write $A_1 = BC$ where B is $n\times r$, C is $r\times n$ and $r < n$. Let C_1 be the $r \times (n-1)$ matrix consisting of the last $n-1$ columns of C, then

$$A = (a_1,a_2,\ldots,a_n) = (a_1,BC_1) = (a_1,B)\begin{pmatrix}1 & 0\\ 0 & C_1\end{pmatrix}.$$

Here $(a_1\ B)$ is $n \times (r+1)$ and $r+1 \leq n$. Since A is full, we have $r+1 = n$ and $(a_1\ B)$ is full. Now (2) can be written

$$(a_1\ B)\binom{u_1}{Cv} = 0,$$

and since every full matrix over R is invertible over S, we find $u_1 = 0$, a contradiction, which shows A_1 to be full.

We have now shown that $\bar{R}$ is a field, i.e. a field of fractions of R, and by the Cor. to Th. 5, it is the universal field of fractions.

In case ii), let Φ' be the set of matrices inverted. By hypothesis, $\Phi' \supseteq \Phi$, and by the strong invariant basis property in S every invertible matrix is full, so $\Phi' = \Phi$. Now Φ' is admissible and so we are reduced to case i). This completes the proof.

In practice the way one applies this theorem is to construct the universal Φ-inverting ring and apply the hypotheses. Thus we obtain the

Corollary. A ring R has a field of fractions K in which every full matrix is invertible iff

i) the set Φ of full matrices over R is admissible,

and

ii) the universal Φ-inverting ring R_Φ is non-zero.

Moreover, when this is so, $K \cong R_\Phi$ is the universal field of fractions of R.

We also note the following: if R is a ring with a Φ-inverting mapping (into some other ring $\neq 0$), so that the universal field of fractions K exists, then any automorphism α of R maps any full matrix into a matrix which is again full (an 'honest' mapping), and hence α extends to an automorphism $\bar{\alpha}$ of K. Clearly $\bar{\alpha}$ is uniquely determined by α.

Before giving cases to which Th. 6 can be applied, let us give some examples. We have already noted that in the commutative case the field of fractions exists iff the ring is an integral domain and in that case it is necessarily the universal field of fractions.

1. A ring R with no universal R-field. $R = \mathbb{Z}_6$ (integers mod 6), has two R-fields $\mathbb{Z}_2$ and $\mathbb{Z}_3$, and no others.

2. A ring R with universal R-field but no field of fractions: $\mathbb{Z}_4$.

3. A ring R with a field of fractions but no universal R-field. Here we need to look at non-commutative examples, by the remark made earlier. The following example is due to G. M. Bergman:

Let R be a ring with two non-isomorphic fileds of fractions K,L say. Since every field of fractions is given by the matrices which become invertible, there must be a matrix A over R which is invertible in L but not in K (interchange K,L if necessary). Let $\bar{R}$ be the ring of all polynomials in a commuting indeterminate t : $a_0 + a_1 t + \ldots + a_n t^n$ where $a_0 \in R$, $a_i \in K$ $(i > 0)$. Then $\bar{R} \subseteq K[t] \subseteq K(t)$, so $\bar{R}$

has a field of fractions. If it has a universal field of fractions U say, then every matrix over $\bar{R}$ invertible in <u>any</u> $\bar{R}$-field must be invertible in U. Now $\bar{R}$ has L as $\bar{R}$-field: $\bar{R} \to R \to L$, where the map $\bar{R} \to R$ is given by putting $t = 0$, and $R \to L$ is the given embedding. The matrix A over $\bar{R}$ has an inverse in L and so must be invertible in U. However, in K and hence in K(t), A is not invertible and hence a zero-divisor. In fact we can write $A = PQ$, where P,Q are $n\times r$, $r\times n$ matrices over K, and $r < n$ (since A is singular, its columns are linearly dependent on $r < n$ columns, which we may take as our matrix P). In $\bar{R}$ we have $At^2 = Pt.Qt$, showing that At^2 is not full in $\bar{R}$, and so has no inverse in U. But U is a field of fractions, so $t \neq 0$ in U and A,t are invertible, hence so is At^2, a contradiction.

Let us return for a moment to the problem of embedding rings in skew fields. We have the necessary condition:

(3) $$\forall\ x,y \qquad xy = 0 \Rightarrow x = 0 \text{ or } y = 0.$$

It is a result of universal algebra (see e.g. [4] p. 235) that the condition for embeddability of an integral domain in a field can be expressed in the form of <u>quasi-identities</u>, i.e. conditions of the form

(4) $$A_1 \,\&\, A_2 \,\&\, \ldots \,\&\, A_n \Rightarrow B,$$

where $A_1, A_2, \ldots, A_n$, B are equations and all variables occurring have a universal quantifier. This is analogous to the situation in groups: the conditions for embedding semi-groups in groups can all be expressed in the form of quasi-identities. In fact such a set has been determined explicitly by Malcev [13]; here there is no exception like (3), basically because groups unlike fields are closed under direct products.

Some instances of conditions (4) have been found by A. A. Klein [11]. In a field, any nilpotent $n\times n$ matrix C satisfies $C^n = 0$, and this can be put into the form (4). In this way we obtain necessary conditions for embeddability of rings in fields. But as Bergman has observed, Klein's conditions are not sufficient. To see this, consider a ring R with 27 generators over a field, written for convenience in the form of 3×3 matrices X,Y,P with defining relations (in matrix form)

(5) $XY = YX = I$ (i.e. X is invertible), $P^2 = P$,

$$X^{-1}PX = I - P.$$

Bergman has proved that this ring R satisfies all of Klein's conditions. But it cannot be embedded in a field; in fact it does not even possess a homomorphism into a field. For if it did, the image of P could be diagonalized, with 0's and 1's on the main diagonal, and necessarily equal numbers of each, by (5). But this is impossible because there are 3 diagonal elements.

6. Firs: a class of rings with a universal field of fractions

In the preceding sections we have investigated the structure of R-fields under the hypothesis that they exist. As we have seen, this is not true for every ring, and for a given ring it may be a difficult problem to decide whether it has any homomorphisms into a field. We shall now describe a family of rings for which the universal Φ-inverting mapping is injective. As we saw, this guarantees the existence of the field of fractions. It will be necessary here to quote results without proof. The proofs can be found in [2,3,6,7].

By a <u>free right ideal ring</u>, right <u>fir</u> for short, we understand a ring R in which every right ideal is free, as module over R, and any two bases of a free module have the same number of elements (invariant basis property). <u>Left firs</u> are defined correspondingly, and a <u>fir</u> is a left and right fir.

In the commutative case, firs are just principal ideal domains. Every principal ideal domain, commutative or not, is a fir, but the class of firs is very much wider than that. It includes e.g. i) free algebras over a field (on any free generating set), ii) group algebras of free groups and iii) free products of (skew) fields.

Every finitely presented module M over a fir R has a resolution

$$0 \to R^r \to R^s \to M \to 0.$$

The number $s-r = \chi(M)$ depends only on M and not on the presentation used; it is called the <u>characteristic</u> of M. It is easily seen that over a commutative fir the characteristic is non-negative, and finitely generated torsion modules may be defined by the condition $\chi(M) = 0$. Analogously we define for an arbitrary fir, a <u>torsion module</u> as a module M such that i) $\chi(M) = 0$ and ii) $\chi(M') \geq 0$ for all submodules M' of M. The torsion modules over R form a full subcategory $\mathcal{T}_R$ of the category $\mathcal{M}_R$ of all right R-modules and homomorphisms, and one has <u>Theorem A [6]</u>. The category $\mathcal{T}_R$ of torsion modules over a fir R is an abelian category in which all objects have finite composition length.

Any module M with a presentation

$$R^r \to R^s \to M \to 0$$

can be defined by an $r \times s$ matrix (or $s \times r$, depending on how you multiply). In particular, torsion modules are defined by certain square matrices and we may ask which matrices occur. The answer is simple: The module defined by a square matrix A is torsion iff A is full [6]. We can therefore use the Jordan-Hôlder theorem in the category $\mathcal{T}_R$ to obtain a unique factorization theorem for full matrices over a fir. We shall not repeat all the definitions and results (see [6]), but recall that an <u>atom</u> in a ring is a non-unit which cannot be written as a product of two non-units. Given $a \in R$, the ring $\mathrm{End}_R(R/aR)$ is called the <u>eigenring</u> of a. A simple application of Schur's lemma shows that the eigenring of an atom in a fir is a field. More generally, this holds for the eigenring of a full atom in a total matrix ring over a fir.

Our aim is to show that for a fir, the natural mapping $R \to R_\Phi$ is injective. If this fails, it already fails in a finite subset of Φ, namely those involved in the resulting relation, so we have to show that $R \to R_\Sigma$ is injective, where Σ is any finite set of full matrices. We may take all these matrices to be of the same size, say $n \times n$, by bordering the undersize ones with 0's and 1's. Thus, passing to the total matrix ring $T = R_n$, we have a finite set S of elements of T which are full, as matrices over R, and we must show $T \to T_S$ is injective. By splitting each element into its atomic factors we may take S to consist of atoms. Now for any $p,q \in S$, $\mathrm{Hom}(T/pT, T/qT)$ is 0 or a field. In the second case $T/pT \cong T/qT$ and we can omit q (unless $q = p$), for inverting p will already invert q. Now our problem is solved by the following

Theorem B [7]. Let R be any ring and S a finite subset of R. Then the universal S-inverting homomorphism $R \to R_S$ is injective provided that S consists of non-zero-divisors, and for any $p,q \in S$,

$$\mathrm{Hom}_R(R/pR,\, R/qR) = \begin{cases} 0 & \text{if } p \neq q, \\ \text{a field} & \text{if } p = q. \end{cases}$$

In fact the theorem even holds for infinite S, by the same localization principle that was used before.

With the help of this theorem it then follows that every fir has a universal field of fractions in which every full matrix can be inverted. We consider one application of this result.

Algebraic geometry is essentially the study of solutions of algebraic equations, i.e. equations of the form $f(x_1,\ldots,x_n) = 0$. Here f could equally well be a rational function, but then one could write $f = f_1/f_2$ where f_1, f_2 are polynomials without a common factor and the problem can essentially be reduced to the study of solutions of $f_1 = 0$. In the non-commutative case this simplification cannot be made, and we also have to specify which elements are to commute with our variables. An example will make this clear:

Let E be a field with center C. A general polynomial in non-commuting indeterminates $x_1,\ldots,x_n$ with coefficients in E is an element ϕ of the free algebra $E\langle x_1,\ldots,x_n\rangle$. This means that x_i commutes with all elements of E and so can only be specialized to elements of C. We want to be able to specialize x_i to any element of E, and this will generally only commute with elements of C, so we have to form polynomials in $x_1,\ldots,x_n$ with coefficients in E which do not satisfy any relations save that each x_i commutes with the elements of C. The resulting algebra may briefly be expressed as a free product over C: $E * C\langle x_1,\ldots,x_n\rangle$. We observe that this algebra may also be described as the tensor E-ring on the direct sum of n copies of the E-bimodule

$E \otimes_C E$. This form of writing the algebra shows that it is a fir [ ], and for this it is not necessary that C should be the center of E, but merely a subfield (not necessarily central). Let us generalize the set-up slightly by taking a subfield D of E, and writing $D \cap C = k$. Then k will be a central subfield of D and it makes sense to substitute elements of E for $x_1,\dots,x_n$ in the elements ϕ of $R = D *_k k\langle x_1,\dots,x_n\rangle$. In this way we get a mapping

$$R \times E^n \to E,$$

the evaluation mapping, and the zero-sets of sets of elements of R are subsets of E^n, forming the family of closed sets in the analogue of the Zariski-topology on E^n. This situation has been studied by Procesi [16] who has shown that if E is finite-dimensional over C, say $(E:C) = t$, so that $E \cong C^t$ as vector space, then $E^n \cong C^{tn}$ is a homeomorphism, taking the usual Zariski-topology in C^{tn}.

To study zero-sets of rational functions we take the universal field of fractions of $R = D * k\langle x_1,\dots,x_n\rangle$, K say, and for each $\phi \in K$ define a mapping from a subset of E^n to E. This is again the evaluation mapping, but it may not be defined on the whole of E^n (e.g. x^{-1} is not defined for $x = 0$). We thus consider the points of E^n where ϕ is defined and vanishes, or more simply, the points of E^n where ϕ^{-1} is undefined. On writing down the equation $Au + a = 0$, in which ϕ^{-1} is a component of the solution, we see that instead of zero-sets of rational functions we may consider the sets where a given family of matrices becomes singular (recall that a matrix over a skew field is either invertible or a two-sided zero-divisor, i.e. singular).

The zero-sets obtained in this way again form a topology on E^n, generally finer than the topology considered before (because we have zero-sets of rational functions as well as of polynomials). It seems natural to take this as the Zariski-topology on E^n. The study of zero-sets on E^n is likely to show many parallels with commutative theory, as well as some differences. E.g. one can show that when E has an infinite center, then E^n is irreducible, in the sense that it cannot be written as the union of two proper closed subspaces [1]. On the other hand, it seems most unlikely that E^n is Noetherian (i.e. that its open sets satisfy ACC).

Besides the zero-sets in E^n it is natural to consider, for a given subset F of E^n, the set of all square matrices over K which become singular on F. They form the analogue of a radical ideal in commutative ring theory. We shall call such sets of matrices ideal sets; they will be studied more closely in the next section.

7. Ideal sets

Algebraic geometry shows how to define a contravariant functor $R \to \text{Spec } R$ from commutative rings to topological spaces, which for affine rings yields the corresponding algebraic variety. The points of Spec R are just the prime ideals of R, i.e. the kernels of homomorphisms of R into fields. To obtain an analogue in the non-commutative case one may consider the set of all kernels of homomorphisms of R into skew fields. But now it is no longer possible to recognize the field of fractions from its kernel alone, because as we saw, a given ring may have several non-isomorphic fields of fractions. Thus we must find a more precise description of these homomorphisms, and one way of doing this is to take the set of all square matrices in R that map to singular matrices. Since over a field, a matrix is either singular or invertible, prescribing the singular matrices amounts to prescribing the invertible matrices, and this is enough to determine the field, by Th. 3.

We shall introduce the notion of an *ideal set* in arbitrary rings, which abstracts the properties of the "set of matrices mapping to singular matrices over a field". Moreover, in rings with a universal field of fractions inverting all full matrices, suitable ideal sets can in fact be interpreted in this way.

Let A,B be two matrices of the same size, over R, which differ only in the i-th row. Then we shall say that the *determinantal sum* of A and B with respect to the i-th row exists and is the matrix C, whose rows are the same as those of A (and B) except the i-th, which is the sum of the i-th rows of A and B. We write

$$C = A \nabla B,$$

indicating in words the row whose elements are added. Observe that when R is commutative, so that determinants are defined, we have $\det C = \det A + \det B$. The determinantal sum with respect to a column is defined in a similar fashion.

Let R be any ring. A _pre-ideal set_ in R is a set $\mathcal{P}$ of square matrices over R satisfying the following four conditions.

1. $\mathcal{P}$ includes all non-full matrices,
2. $\mathcal{P}$ admits elementary row or column operations,
3. If $A, B \in \mathcal{P}$ and their determinantal sum C (with respect to some row or column) exists, then $C \in \mathcal{P}$,
4. If $A \in \mathcal{P}$, then $\begin{pmatrix} A & 0 \\ 0 & B \end{pmatrix} \in \mathcal{P}$ for all square matrices B.

If, further,

5. $\begin{pmatrix} A & 0 \\ 0 & 1 \end{pmatrix} \in \mathcal{P} \Rightarrow A \in \mathcal{P}$,

we call $\mathcal{P}$ an _ideal set_, _proper_ in case $1 \notin \mathcal{P}$. We shall generally denote ideal sets by greek upper case letters Π, Ω etc. and use latin script letters for preideal sets.

We note some elementary consequences of the definitions:

a) If Π is an ideal set, any matrix with a zero row or column lies in Π. For if $A = (A,0)$ say, then $A = A_1 \, (I \; 0)$ and this shows that A is not full.

b) Let Π be an ideal set. Then for any square A, B and any C of the appropriate size,

$$\begin{pmatrix} A & C \\ 0 & B \end{pmatrix} \in \Pi \Leftrightarrow \begin{pmatrix} A & 0 \\ 0 & B \end{pmatrix} \in \Pi .$$

For, given A, B, C, write $A = \begin{pmatrix} a \\ A_1 \end{pmatrix}$ $C = \begin{pmatrix} c \\ C_1 \end{pmatrix}$ where a, c is the first row in each case, then

$$(1) \qquad \begin{pmatrix} A & C \\ 0 & B \end{pmatrix} = \begin{pmatrix} a & c \\ A_1 & C_1 \\ 0 & B \end{pmatrix} = \begin{pmatrix} a & 0 \\ A_1 & C_1 \\ 0 & B \end{pmatrix} \nabla \begin{pmatrix} 0 & c \\ A_1 & C_1 \\ 0 & B \end{pmatrix}$$

where the determinantal sum is with respect to the first row. Now

$$\begin{pmatrix} 0 & c \\ A_1 & C_1 \\ 0 & B \end{pmatrix} = \begin{pmatrix} 0 & c \\ I & C_1 \\ 0 & B \end{pmatrix} \begin{pmatrix} A_1 & 0 \\ 0 & I \end{pmatrix} ;$$

thus the second matrix on the right of (1) is not full, and lies in Π. Hence

$$\begin{pmatrix} A & C \\ 0 & B \end{pmatrix} \in \Pi \Leftrightarrow \begin{pmatrix} a & 0 \\ A_1 & C_1 \\ 0 & B \end{pmatrix} \in \Pi .$$

By elementary transformations we see that the same argument applies to the other rows of C. So we can make them all 0 and the result follows.

c) In an ideal set Π, if $A \in \Pi$ then for any square matrix B of the same size, AB and $BA \in \Pi$.

Let $A \in \Pi$ and let B be any matrix of the same size. Then $\begin{pmatrix} A & 0 \\ I & B \end{pmatrix} \in \Pi$, and by elementary transformations this becomes in turn $\begin{pmatrix} 0 & -AB \\ I & B \end{pmatrix} \rightarrow \begin{pmatrix} AB & 0 \\ B & I \end{pmatrix} \rightarrow \begin{pmatrix} AB & 0 \\ 0 & I \end{pmatrix}$. Hence $AB \in \Pi$, and similarly $BA \in \Pi$.

Let Π be an ideal set and write $\Pi = \bigcup_n \Pi_n$, where Π_n is the set of all $n \times n$ matrices in Π. From c) above it is clear that Π_1 is an ideal in R. In fact, one may think of Π_1 as a "first approximation" to Π, in a way which may be made more precise by thinking of ideal sets as the sets of matrices which become singular under a homomorphism into a field. This would suggest that in a commutative ring R one has the relation

(2) $$A \in \Pi \Longleftrightarrow \det A \in \Pi_1.$$

But a look at some examples shows that neither of these implications holds; it raises the question for which commutative rings either implication (2) (or both) are true.

Let Π_λ be any family of ideal sets, then it is clear that $\Pi = \bigcap \Pi_\lambda$ is again an ideal set. We can therefore speak of the least ideal set containing a given set $\mathfrak{X}$ of square matrices, viz. the intersection of all ideal-sets containing $\mathfrak{X}$. We use this fact to clarify the relation between ideal sets and preideal sets.

Proposition. Let $\mathfrak{P}$ be any preideal set. Then the least ideal set containing $\mathfrak{P}$ is given by

(3) $$\Pi = \{A \mid \begin{pmatrix} A & 0 \\ 0 & I \end{pmatrix} \in \mathfrak{P} \text{ for some unit-matrix } I\}.$$

Proof. Let Π be defined by (3), then $\Pi \supseteq \mathfrak{P}$, and any ideal set containing $\mathfrak{P}$ must also contain Π, so it only remains to be shown that Π is itself an ideal set. Properties 1.-4.

are clear; if $\begin{pmatrix} A & 0 & 0 \\ 0 & 1 & 0 \\ 0 & 0 & I_n \end{pmatrix} \in \mathcal{P}$, this means that $\begin{pmatrix} A & 0 \\ 0 & I_{n+1} \end{pmatrix} \in \mathcal{P}$, hence $A \in \Pi$, so Π also satisfies 5. and therefore is an ideal set as claimed.

The least ideal set containing $\mathcal{X}$ is also called the ideal set <u>generated</u> by $\mathcal{X}$. From the above Prop. we have the immediate

<u>Corollary</u>. A preideal set generates a proper ideal set iff it contains no unit-matrix I (of any size).

An ideal set Π is called <u>prime</u> if it is proper and

$$\begin{pmatrix} A & 0 \\ 0 & B \end{pmatrix} \in \Pi \Rightarrow A \in \Pi \text{ or } B \in \Pi.$$

An ideal set Π is called <u>semiprime</u> if

$$\begin{pmatrix} A & 0 \\ 0 & A \end{pmatrix} \in \Pi \Rightarrow A \in \Pi.$$

These concepts have another description using the following notion of a product of ideal sets. Given two ideal sets Π_1, Π_2, their <u>product</u>, denoted by $\Pi_1\Pi_2$ is defined as the ideal set generated by all $\begin{pmatrix} A_1 & 0 \\ 0 & A_2 \end{pmatrix}$ with $A_i \in \Pi_i$. We observe that this product is commutative and is contained in their intersection: $\Pi_1\Pi_2 = \Pi_2\Pi_1 \subseteq \Pi_1 \cap \Pi_2$. The following lemma is useful in constructing product ideal sets:

<u>Lemma</u>. Let $\mathcal{P}_1, \mathcal{P}_2$ be any sets of square matrices, $\mathcal{P}$ the set of matrices $\begin{pmatrix} A_1 & 0 \\ 0 & A_2 \end{pmatrix}$ $A_i \in \mathcal{P}_i$ and Π_1, Π_2, Π the ideal sets generated by $\mathcal{P}_1, \mathcal{P}_2, \mathcal{P}$ respectively, then $\Pi = \Pi_1\Pi_2$.

Proof. Clearly $\mathfrak{P} \subseteq \Pi_1\Pi_2$, hence $\Pi \subseteq \Pi_1\Pi_2$. To prove equality it is enough to show that $\begin{pmatrix} A_1 & 0 \\ 0 & A_2 \end{pmatrix} \in \Pi$ for any $A_i \in \Pi_i$. Now any $A_i \in \Pi_i$ is obtained from the matrices in $\mathfrak{P}_i$ and the non-full matrices by operations corresponding to properties 2-5 of ideal sets. All these operations, carried out on $\begin{pmatrix} A_1 & 0 \\ 0 & P \end{pmatrix}$ (for some $P \in \mathfrak{P}_2$) instead of A_1, will still give a matrix in Π. Similarly for the second matrix on the diagonal, thus $\begin{pmatrix} A_1 & 0 \\ 0 & A_2 \end{pmatrix} \in \Pi$ and so $\Pi = \Pi_1\Pi_2$.

<u>Theorem 8</u>. Let Π be an ideal set in any ring R. Then

i) Π is prime iff Π is proper and $\Pi_1\Pi_2 \subseteq \Pi \Rightarrow \Pi_1 \subseteq \Pi$ or $\Pi_2 \subseteq \Pi$, for any ideal sets Π_1, Π_2,

ii) Π is semiprime iff $\Omega^2 \subseteq \Pi \Rightarrow \Omega \subseteq \Pi$ for any ideal set Ω.

Proof. Let Π be prime and $\Pi \supseteq \Pi_1\Pi_2$ but $\Pi \not\supseteq \Pi_i$ $(i=1,2)$. Then there exists $A_i \in \Pi_i$ but $A_i \notin \Pi$. Since Π is prime, we have $\begin{pmatrix} A_1 & 0 \\ 0 & A_2 \end{pmatrix} \notin \Pi$, but $\begin{pmatrix} A_1 & 0 \\ 0 & A_2 \end{pmatrix} \in \Pi_1\Pi_2 \subseteq \Pi$, a contradiction.

Conversely, assume that Π satisfies the conditions stated, and let $\begin{pmatrix} A_1 & 0 \\ 0 & A_2 \end{pmatrix} \in \Pi$. Let (A_i) be the ideal set generated by A_i, then by the lemma, $(A_1)(A_2)$ is generated by $\begin{pmatrix} A_1 & 0 \\ 0 & A_2 \end{pmatrix}$, and hence $(A_1)(A_2) \subseteq \Pi$. Therefore $(A_i) \subseteq \Pi$ for $i = 1$ or 2, say $i = 1$, and so $A_1 \in \Pi$; thus Π is prime.

Similarly, if Π is semiprime and $\Omega^2 \subseteq \Pi$, let $A \in \Omega$, then $\begin{pmatrix} A & 0 \\ 0 & A \end{pmatrix} \in \Omega^2 \subseteq \Pi$, hence $A \in \Pi$, so $\Omega \subseteq \Pi$ as claimed. Conversely, if Π satisfies the given condition and

$\begin{pmatrix} A & 0 \\ 0 & A \end{pmatrix} \in \Pi$, then the square of (A), viz. $(A)^2$, is generated by $\begin{pmatrix} A & 0 \\ 0 & A \end{pmatrix}$, which lies in Π, therefore $(A)^2 \subseteq \Pi$. By hypothesis, $(A) \subseteq \Pi$, whence $A \in \Pi$.

We next elucidate the relation between prime and semiprime ideal sets. Let Π be any ideal set and define

$$\sqrt{\Pi} = \{A \mid \begin{pmatrix} A & & 0 \\ & A & \\ 0 & & \ddots & \\ & & & A \end{pmatrix} \in \Pi\}.$$

Clearly $\sqrt{\Pi} \supseteq \Pi$; we assert that $\sqrt{\Pi}$ is the least semiprime ideal set containing Π. Any semiprime ideal set containing Π must contain $\sqrt{\Pi}$. For on writing $\operatorname{diag}(A^{(r)})$ for the diagonal sum of r copies of A, we have if $A \in \sqrt{\Pi}$, then $\operatorname{diag}(A^{(r)}) \in \Pi$ for some r, and taking k so large that $2^k \geq r$, $\operatorname{diag}(A^{(2^k)}) \in \Pi$. Hence, by an easy induction, any semiprime ideal set containing Π must contain A. It remains to show that $\sqrt{\Pi}$ itself is a semiprime ideal set. 1,2,4,5 are clear, to prove 3, (determinantal sum), we note that

$\operatorname{diag}((A \nabla B)^n)$ is a determinantal sum of terms $\operatorname{diag}(C_1,\ldots,C_n)$

where each C_i is A or B. Hence if $\operatorname{diag}(A^{(r)})$ and $\operatorname{diag}(B^{(s)})$ lie in $\sqrt{\Pi}$ then $\operatorname{diag}((A \nabla B)^{r+s-1}) \in \sqrt{\Pi}$. Thus $\sqrt{\Pi}$ is an ideal set; if $\operatorname{diag}(A^{(2)}) \in \sqrt{\Pi}$, say $\operatorname{diag}(A^{(2r)}) \in \Pi$, then $A \in \sqrt{\Pi}$, so $\sqrt{\Pi}$ is indeed semiprime.

To relate semiprime to prime ideal sets we first show that the familiar method of constructing prime ideals as maximal ideals still works for sets.

Lemma. Let Σ be any set of matrices containing 1 and closed under diagonal sums, and Π a maximal ideal set disjoint from Σ. Then Π is prime.

Proof. Let $\Pi_1\Pi_2 \subseteq \Pi$ but $\Pi_i \not\subseteq \Pi$, then $\Pi_i \cap \Sigma \neq \emptyset$. Take $A_i \in \Pi_i \cap \Sigma$ then $\begin{pmatrix} A_1 & 0 \\ 0 & A_2 \end{pmatrix} \in \Sigma$ (since Σ is admissible), but $\begin{pmatrix} A_1 & 0 \\ 0 & A_2 \end{pmatrix} \in \Pi_1\Pi_2 \subseteq \Pi$, showing that $\Pi \cap \Sigma \neq \emptyset$, a contradiction. Moreover, $1 \notin \Pi$; thus Π is prime as claimed.

Of course maximal ideal sets disjoint from a given set Σ always exist; the collection of ideal sets is clearly inductive, so that Zorn's lemma applies. All we need to be sure about is that it is not empty. For example, given any ideal set Ω and any admissible set of matrices Σ such that $\Omega \cap \Sigma = \emptyset$, there always exists a maximal ideal set containing Ω and disjoint from Σ.

Theorem 9. An ideal set is semiprime iff it is an intersection of prime ideal sets.

Proof. Let $\Pi = \bigcap \Pi_\lambda$ where Π_λ is prime. If $\begin{pmatrix} A & 0 \\ 0 & A \end{pmatrix} \in \Pi$, then $\begin{pmatrix} A & 0 \\ 0 & A \end{pmatrix} \in \Pi_\lambda$ for all λ, hence $A \in \Pi_\lambda \subseteq \Pi$, so Π is semiprime.

Conversely, let Π be semiprime. It will be enough to find for each $A \notin \Pi$ a prime ideal set Π_A containing Π but not A, for then $\Pi = \bigcap \{\Pi_A \mid A \notin \Pi\}$.

Let $A \notin \Pi$ be given and consider the set Σ of all elementary transforms of triangular sums of A and 1, i.e. matrices with A and 1 down the main diagonal, and zeros below these blocks. This is an admissible set of matrices disjoint from Π. For any matrix in Σ is elementarily equivalent to one of the form

$$\begin{pmatrix} A & & & * \\ & I & & \\ & & A & \\ & & & \cdot \\ 0 & & & & \cdot \\ & & & & & \cdot \end{pmatrix},$$

and if this lies in Π then so does A, against the hypothesis. Let Π_A be a maximal ideal set disjoint from Σ and containing Π. By the lemma, Π_A is prime, and $A \notin \Pi_A$, so $\bigcap \Pi_A = \Pi$. This completes the proof.

So far we have operated entirely on a formal level; we needed no assumption on our rings, and did not interpret the results in any way. But ideal sets were supposed to generalize the notion of an ideal, and perhaps the most basic use of an ideal in a ring is to form quotient rings. Now this is not possible with arbitrary ideal sets, for the following reason: to form the quotient ring with respect to an ideal A means "putting all the elements in A equal to 0". To form the quotient ring with respect to an ideal set Π means: "making all the matrices in Π singular (whatever that may mean)". The singularity of a matrix is defined over a field (even skew), but not over a general ring. One possible definition (which reduces to the usual one for skew fields, but not for all commutative rings) is to take "singular" to mean "not full". Then forming a quotient ring with respect to an ideal set Π would mean: Finding a ring-epimorphism making all the elements in Π non-full. It is not clear how to do this in general, but at least for prime ideal sets there is no problem: the situation is explained in the next result, which establishes the link between prime ideal sets and admissible sets, in analogy with the connection between prime ideals and multiplicative sets in the commutative case.

Theorem 10. Let R be any ring. Then

i) an ideal set Π is prime iff its complement in the set of all square matrices is admissible. When this is so, Σ is a localizing set, provided that $R_\Sigma \neq 0$.

ii) The complement of a saturated admissible set of full matrices is an ideal set iff it is closed under determinantal sums. When this is so, the ideal set is prime.

Proof. i) Denote the complement of Π by Σ, then always admits elementary transformations. $1 \in \Sigma \Longleftrightarrow 1 \notin \Pi$ and the third condition for Σ holds iff $A, B \notin \Pi \Rightarrow$ $\begin{pmatrix} A & C \\ 0 & B \end{pmatrix} \notin \Pi$ for all C, i.e. $\begin{pmatrix} A & 0 \\ 0 & B \end{pmatrix} \notin \Pi$. But this is just the condition for Π to be prime.

Now assume Π prime and let $A \in \Sigma$, $A_{11} \notin \Sigma$ (where A_{11} is the (1,1)-minor of A), then $A_{11} \in \Pi$ and hence $A' = \begin{pmatrix} 1 & a \\ 0 & A_{11} \end{pmatrix} \in \Pi$ where $a = (a_{12}, \ldots, a_{1n})$ is the first row of A, without its first element. Now

$$A + e_{11} = A \nabla A' \tag{4}$$

(formed with respect to the first column). If $A + e_{11} \notin \Sigma$ then $A + e_{11} \in \Pi$ and hence, by (4), $A \in \Pi$, a contradiction. This proves that $A + e_{11} \in \Sigma$, and so Σ is a localizing set, by Theorem 4.

ii) Let Σ be a saturated admissible set of full matrices, and Π its complement. Then Π certainly satisfies conditions 1,2. To prove 4,5 we must show

a) $\begin{pmatrix} A & 0 \\ 0 & B \end{pmatrix} \in \Sigma \Rightarrow A \in \Sigma$ and b) $A \in \Sigma \Rightarrow \begin{pmatrix} A & 0 \\ 0 & 1 \end{pmatrix} \in \Sigma$. Of these b) is clear, and a) follows because Σ is saturated: if $\begin{pmatrix} A & 0 \\ 0 & B \end{pmatrix}$ is invertible in R_Σ, then so is A, and hence $A \in \Sigma$. Thus Π is an ideal set precisely when it satisfies 3. If it does, it is a prime ideal set, by the first part.

Now let R be a ring with a universal field of fractions which is Φ-inverting (where Φ is the set of all full matrices over R). Then for the complement Σ of any prime ideal set Π we have $R_\Sigma \neq 0$ (because $R_\Phi \neq 0$ and $\Sigma \subseteq \Phi$). Hence R_Σ is a local ring, called the <u>localization</u> at the prime Π.

Here we have had to postulate $R_\Phi \neq 0$ to ensure that $R_\Sigma \neq 0$. But it seems plausible that this assumption is superfluous. This would be the case if we could prove the following

<u>Conjecture</u>. Let R be any ring and Σ the complement of a prime ideal set; then $R_\Sigma \neq 0$. [This conjecture has already been proved, see Appendix.]

If R has <u>any</u> proper ideal set Π, then the set Σ of all unit matrices is closed under diagonal sums and disjoint from Π. So we can find a maximal ideal set containing Π and disjoint from Σ, and by the lemma, this will be a prime ideal set. Thus prime ideal sets exist, provided any proper ideal sets exist at all. Let Π_0 be the ideal set generated by ϕ; clearly this is the least ideal set in R, and the above argument shows that R has prime ideal sets iff Π_0 is proper. Now if the above conjecture were true, each prime ideal set in R would lead to a local ring and hence an R-field. Conversely, every R-field leads to a prime ideal set, viz. the matrices which become singular over the field. Thus we have the following

Consequence of the conjecture. A ring R has an R-field iff the least ideal set in R is proper.

In fact, the condition given for the existence of R-fields is necessary, and an affirmative answer to the conjecture would prove it sufficient. Thus a way of testing the conjecture would be to take rings without R-fields and try to show that the least ideal set is improper. E.g. rings without strong basis property have no R-field: the unit matrix (of a certain size) is not full. A sufficient condition for the least ideal set to be proper is that the following hold:

i) R has the strong invariant basis property (i.e. all unit-matrices are full), and

ii) the determinantal sum of two non-full matrices is non-full.

For then the non-full matrices form a preideal set, which clearly generates a proper ideal set.

8. The prime spectrum of a ring

We now consider a general ring R and its set X of all prime ideal sets. As for prime ideals in commutative rings, we can put a topology on X which enables us to construct a sheaf, but so far we have not been able to identify R as the ring of global sections.

Let R be any ring, and denote by $X = X(R)$ the collection of all prime ideal sets in R. Of course we shall need to restrict R if X is to be non-empty, and even when this is so, we need to know for which $x \in X$, the localization R_x is non-zero. Let us call a prime x good if $R_x \neq 0$, and bad otherwise. E.g. if $R_\Phi \neq 0$, then R has primes and every one of them is good. For each $x \in X$ we denote by R_x the localization, its unique maximal ideal by m_x, and the residue class field by $k(x) = R_x/m_x$. Any matrix A over R has a value $A(x)$, say, in $k(x)$, and $A \in x$ iff $A(x)$ is singular.

With every set ξ of square matrices we associate a subset of X, defined by

$$V(\xi) = \{x \in X \mid x \supseteq \xi\}.$$

We write $V(A)$ in place of $V(\{A\})$. The properties of V are given in

Theorem 11. i) $V(0) = X$, $V(1) = \emptyset$.

ii) $\xi \subseteq \xi' \Rightarrow V(\xi) \supseteq V(\xi')$.

iii) $V(\bigcup_\lambda \xi_\lambda) = \bigcap_\lambda V(\xi_\lambda)$.

iv) If Π is any ideal set, $V(\sqrt{\Pi}) = V(\Pi)$.

v) Let $\xi\xi' = \{\begin{pmatrix} A & 0 \\ 0 & A' \end{pmatrix} \mid A \in \xi,\ A' \in \xi'\}$,

then $V(\xi\xi') = V(\xi) \cup V(\xi')$.

Proof. i) - iii) are clear, iv) follows because $x \supseteq \Pi \Longleftrightarrow x \supseteq \sqrt{\Pi}$, and to prove v), note that $x \in V(\xi) \cup V(\xi') \Longleftrightarrow x \supseteq \xi$ or $x \supseteq \xi' \Longleftrightarrow x \supseteq \xi\xi'$.

This theorem shows that the $V(\xi)$ satisfy the axioms for closed sets. The resulting topology is called the spectral topology on X.

With every subset Y of X we associate a set of matrices $j(Y)$, defined by

$$j(Y) = \bigcap_{y \in Y} y.$$

We observe that $j(Y)$ is always a semiprime ideal set (by Theorem 9); $j(\{x\}) = x$. Its properties are given in

Theorem 12. i) $j(\phi)$ = set of all square matrices, $j(X)$ = least semiprime ideal set,

ii) $Y \subseteq Y' \Rightarrow j(Y) \supseteq j(Y')$,

iii) $j(\bigcup Y_\lambda) = \bigcap_\lambda j(Y_\lambda)$,

iv) $j(V(\xi))$ = radical of ideal set generated by ξ,

v) $V(j(Y)) = \bar{Y}$.

Again i) - iii) are clear. iv). Let Π be the ideal set generated by ξ, then $V(\xi) = V(\Pi) = V(\sqrt{\Pi})$, hence $j(V(\xi)) = j(V(\sqrt{\Pi})) = \sqrt{\Pi}$, the last equality holding because $\sqrt{\Pi}$ is semiprime. v) Clearly $V(j(Y))$ is closed and contains Y, so $V(j(Y)) \supseteq \bar{Y}$. Now let Z be closed and contain Y, say $Z = V(\Pi) \supseteq Y$; this means $y \supseteq \Pi$ for any $y \in Y$, hence $\Pi \subseteq j(Y)$ and so by Theorem 11, ii) $Z = V(\Pi) \supseteq V(j(Y))$. This proves $\bar{Y} = V(j(Y))$.

We thus have a bijection between the closed sets of X and the semiprime ideal sets in R. A given x is a closed point of X iff x is a maximal ideal set; in general, the $y \in \overline{\{x\}}$ correspond to specializations of x. In fact this means $y \supseteq x$, hence (by Theorem 5), $k(x) \to k(y)$ is a specialization, in the sense defined earlier.

For any square matrix A, let us define its "support" in X as

$$D(A) = \{x \in X \mid A \notin x\}.$$

Thus $D(A)$ consists of the primes at which A is non-singular. As complement of $V(A)$, $D(A)$ is an open set.

<u>Theorem 13</u>. i) The open sets $D(A)$ form a base of open sets for X, ii) Each $D(A)$ is quasi-compact. In particular, $X = D(1)$ is quasicompact.

Proof. i) If $\mathcal{U}$ is open, $\mathcal{U} = X \setminus V(\xi)$ for some ξ. Now $V(\xi) = \bigcap_{A \in \xi} V(A)$ by Theorem 11, hence $\mathcal{U} = X \setminus V(\xi) = \bigcup_{A \in \xi} D(A)$.

ii) By i) it is enough to show that every covering of $D(A)$ by sets $D(A_\lambda)$ has a finite subcovering. Assume $D(A) \subseteq \bigcup D(A_\lambda)$ and let Π be the ideal set generated by the A_λ, then

$$V(A) = \bigcap V(A_\lambda) = V(\Pi) = V(\sqrt{\Pi}),$$

hence (by Theorem 12, iv)) $\sqrt{\Pi}$ is the radical of the ideal set generated by A. Thus $A \in \sqrt{\Pi}$, $\operatorname{diag}(A^{(r)}) \in \Pi$, and hence $\operatorname{diag}(A^{(r)})$ already lies in the ideal set generated by finitely many of the A_λ; it follows that $D(A) \subseteq D(A_{\lambda_1}) \cup \ldots \cup D(A_{\lambda_n})$.

The sets $D(A)$ are often called the basic open sets. We note incidentally that the set of good primes is closed, or equivalently, the set of bad primes is open. For if x is bad, then $R_x = 0$, and this means that 1 belongs to the ideal of relations defining R_x. It therefore lies in the ideal generated by finitely many such relations, arising from the inversion of matrices $A_1,\ldots,A_n$. Let $A = \operatorname{diag}(A_1,\ldots,A_n)$, then the universal A-inverting ring R_A is 0, hence $R_y = 0$ for any $y \not\ni A$, i.e. any $y \in D(A)$. Thus x has a whole neighborhood of bad primes as claimed.

We now construct a presheaf on X; we must find a system of rings $R(A)$ corresponding to the open sets $D(A)$, such that for $D(A) \supseteq D(A')$ there is a homomorphism $R(A) \to R(A')$ which combines in the correct fashion (so as to give a contravariant functor from basic open sets and inclusions to rings and homomorphisms). One's first impulse is to take R_A, the universal A-inverting ring; but this is not quite the right choice. Let us say for brevity, a set of matrices is localizing if it is the complement of a prime ideal set (this slightly restricts the previous usage of the term). Now define $R(A)$ as R_{Σ_A}, the universal Σ_A-inverting ring, where Σ_A is the intersection of all localizing sets containing A. Clearly we always have a mapping $R_A \to R(A)$, but this need not be an isomorphism.

Let us observe that the correspondence $R \longmapsto X(R)$ is a contravariant functor: given any homomorphism of rings $\phi: R \to S$, we obtain a map $\phi^* : X(S) \to X(R)$ which is easily seen to be continuous. In the particular case of the rings $R(A)$ we can say more:

Lemma. The canonical homomorphism $f: R \to R(A)$ (for any square matrix A) induces a homeomorphism

$$\phi^* : X(R(A)) \to D(A).$$

For the primes of $R(A)$ are just the primes of R not containing A, i.e. $D(A)$. Thus each $x \in D(A)$ comes from one prime of $R(A)$ and one only. Thus ϕ^* is an order-isomorphism, and hence a homeomorphism.

It is now clear how to obtain our sheaf. We have a presheaf, consisting of the rings $R(A)$ over the basic open sets $D(A)$. Clearly if $D(A) \supseteq D(A')$ then $A \in x \Rightarrow A' \in x$, hence for any localizing set Σ, $A' \in \Sigma \Rightarrow A \in \Sigma$, so that $\Sigma_{A'} \supseteq \Sigma_A$ and we have a natural homomorphism $R_{\Sigma_A} \to R_{\Sigma_{A'}}$, i.e. $R(A) \to R(A')$. It is clear that these maps combine in the right way, so we have a presheaf of rings over $X = X(R)$. The corresponding sheaf is denoted by $\tilde{R}$. The first step is to determine the stalk at x: this turns out to be the localization at x, as one would hope:

Lemma. $R_x = \varinjlim_{A \notin x} R(A)$.

Proof. For any $A \nmid x$ we have a homomorphism $R(A) \to R_x$, and this is such as to make the accompanying triangle (for $D(A) \supseteq D(A'))$ commute. Therefore we have a homomorphism

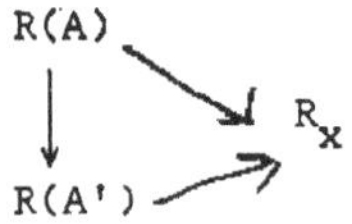

$$(1) \qquad \varinjlim R(A) \to R_x .$$

Now let $c \in R_x$, then c is obtained by inverting some matrix $A \nmid x$ (e.g. c might be the (1,1)-element of A^{-1}), hence c is the image of an element $c_A \in R(A)$. If $c_A, c'_A \in R(A)$ both map to c, then they are identified due to the inversion of a matrix A' such that $D(A) \supseteq D(A') \ni x$, hence in $R(A')$, we have $c'_{A'} = c_{A'}$. This shows that (1) is in fact an isomorphism, as claimed.

Let $\Gamma(X,\tilde{R})$ be the ring of global sections of our sheaf $\tilde{R}$, then each $a \in R$ defines a global section $\hat{a}$, and it is clear that the mapping $a \longmapsto \hat{a}$ is a homomorphism

$$(2) \qquad R \to \Gamma(X,\tilde{R}).$$

We ask: When is this map an isomorphism? Let I_x be the kernel of the canonical homomorphism $R \to R_x$, then clearly $\bigcap_{x \in X} I_x$ is the kernel of the mapping (2), thus we see:

The mapping (2) is injective iff for each $a \in R$, $a \neq 0$, there is a homomorphism of R into a local ring not annihilating a. In fact the I_x are characterized as the kernels of mappings into local rings.

The question when (2) is surjective is of more interest. Let us first indicate one necessary condition.

Proposition. Let A be a matrix over R such that the ideal set generated by A is improper. Then the image of A under the mapping (2) is invertible, hence each component of the matrix A^{-1} in $\Gamma(X,\tilde{R})$ defines a global section, and for (2) to be surjective, any matrix over R generating an improper ideal set must be invertible in R.

This follows immediately: the ideal set (A) generated by A is improper iff it is not contained in any prime, i.e. if $A \notin x$ for all x, i.e. $A(x)$ is non-singular for all x.

Let us say that R is I-closed (I for 'inverse') if $R \neq 0$ and every matrix generating an improper ideal set is invertible over R. If R is I-closed, then 0 (as $|x|$ matrix) is not invertible over R, so the ideal set generated is proper. This shows that the least ideal set of R is proper and hence $X(R) \neq \emptyset$.

In the sheaf $\tilde{R}$ we saw that the stalk over x was R_x; by definition I_x is the kernel of the map $R \to R_x$, so R/I_x is a subring of R_x, and we obtain a subsheaf of $\tilde{R}$ by cutting down the stalk to R/I_x. Let us denote this subsheaf by $\tilde{R}'$. Its stalks are obtained from R by mapping to 0 the elements that are mapped to 0 as a consequence of inverting certain matrices (but without actually inverting the matrices).

Another representation, closely related to the representation of R in $\tilde{R}$, is obtained by mapping each R_x to $k(x)$. In any ring R let us define a <u>stem ideal</u> as an ideal P such that R/P is embeddable in a (skew) field, and denote by $\mathcal{U}$ the space of stem ideals of R with the hull-kernel topology. Given $x \in X$, let $\phi(x)$ be the kernel of the natural mapping $R \to k(x)$; clearly $\phi(x) \in \mathcal{U}$, so ϕ is a mapping of X into $\mathcal{U}$. It is in fact a surjection, by Theorem 3, and it is not hard to see that ϕ is continuous, but the topology on X is not in general obtainable by transporting that of $\mathcal{U}$. To see this, let us find the quotient topology induced on $\mathcal{U}$ by X. In the quotient topology, $V \subseteq \mathcal{U}$ is open iff $\phi^{-1}(V)$ is open, i.e. if for each $u \in V$ and each R-field $K \supseteq R/u$, there is a matrix A invertible in K such that $u' \in V$ for all R/u' that are contained in an R-field inverting A. If in this definition we restrict A to be a 1×1 matrix, we get the usual hull-kernel topology on $\mathcal{U}$. The present topology is therefore finer.

When $\mathcal{U}$ is given the quotient topology in this way, ϕ is of course continuous, and so $\mathcal{U}$ is again quasi-compact, but there are no grounds for believing that the topology on X is the one obtained by pull-back from $\mathcal{U}$. This would be so, if the following condition were satisfied:

Given a matrix A, and $x \in D(A)$, there exists a matrix B such that $x \in D(B)$ and for any y such that $\phi(y) = \phi(y')$ and $y' \in D(B)$, it follows that $y \in D(A)$.

Or put differently: given A, and $x \in X$ such that A is invertible in $k(x)$, there exists B, invertible in $k(x)$, such that A is invertible in any $k(y)$ such that the image of R in $k(y)$ can be embedded in a field inverting B.

It does not seem easy to check this condition in particular rings. Of course even when it does not hold, we could use the pull-back to define a finer topology on X; but this would no longer be quasi-compact.

To show that the mapping (2) on p. 49 is surjective, we must show the following: If $D(A_i)$ is a covering of X and $s_i \in R(A_i)$ such that s_i, s_j have the same image for all B with $D(B) \subseteq D(A_i) \cap D(A_j)$, then there exists $s \in R$ with image s_i in $R(A_i)$.

The usual way to prove such assertions is by partitions of unity. Given a covering of X by sets $D(A_i)$, by quasi-compactness we can find a finite subcovering: $X \subseteq D(A_1) \cup \ldots \cup D(A_n)$. Define $\Pi_i = \bigcap \{x \in X \mid x \notin D(A_i)\}$, thus Π_i is the intersection of all primes containing A_i. Denote by Π the ideal set generated by $\Pi_1, \ldots, \Pi_n$, then Π is improper. For if Π were proper, it would be contained in some x, thus $x \supseteq \Pi_i$ for all i and so $x \notin D(A_i)$ all i, a contradiction. Hence Π is indeed improper. At this point one obtains in the commutative theory a partition of unity. Here one merely obtains matrices $A_i' \in \Pi_i$ such that the ideal set generated by $A_1', \ldots, A_n'$ is improper, and it is difficult to see how to evaluate this information.

Alternatively, one may try to form partitions of unity over the stem ideal space, but this also seems problematic, quite apart from the fact that difficulties arise if the topology on X is not obtainable by pull-back from $\mathcal{U}$.

At any rate, it seems not unreasonable to end with another Conjecture. The necessary condition found earlier for the mapping $R \to \Gamma(X,\tilde{R})$ to be surjective, viz. that R be I-closed, is also sufficient.

The material in sections 1-5 is in course of publication [7,8], and only enough has been included to make the rest intelligible. Section 6 summarizes material that has for the most part been published [3,6,7]. On the other hand the material of section 7-8 is new as far as I am aware. I would like to acknowledge the help of various members of the audience for their helpful reaction; in particular Karl Hofmann for several discussions on representations of rings by sections, which enabled me to see the problems much more clearly than before.

Appendix: All primes are good.

This appendix is devoted to proving the conjecture made on p.42. Thus we shall prove the following

Theorem. Let R be any ring and Π a prime ideal set. Then there exists an R-field K such that Π is the precise class of matrices mapped to singular matrices in the homomorphism $R \to K$.

Let Π be as in the theorem and denote its complement by Σ. Then by Theorem 3, R_Σ is a local ring with residue class field K, hence $R_\Sigma \neq 0$ and this establishes the conjecture on p.42 that all primes are good (in the terminology introduced on p. 44).

In outline the proof goes as follows: We construct the elements of K as classes of solutions of equations of the form $Au + a = 0$. To verify that we have a field, we use the following lemma from [17], which expresses all the field operations in terms of division and the operation $x\theta = 1 - x$.

Lemma 1. Let G be a group with a mapping θ of the set $G_1 = \{x \in G \mid x \neq 1\}$ into itself such that

(i) $(yxy^{-1})\theta = y(x\theta)y^{-1}$ $\quad (x \in G_1,\ y \in G)$,

(ii) $x\theta^2 = x$,

(iii) $(xy^{-1})\theta = [(x\theta)(y\theta)^{-1}]\theta.(y^{-1})\theta$ $\quad (x,y \in G_1,\ x \neq y)$,

(iv) $e = (x^{-1})\theta.x.(x\theta)^{-1}$ is independent of x.

Then there exists a unique field K, whose multiplicative group coincides with G, and on G, $x\theta = 1-x$ $(x \in G_1)$, $e = -1$.

Secondly we shall need the following method of constructing a group as a quotient of a semigroup. In Lemma 2, the semigroup M is not assumed to have a unit element.

Lemma 2. Let M be a non-empty semigroup and M' a subsemigroup of M such that

(i) $xy \in M' \Rightarrow yx \in M'$ and

(ii) for each $x \in M$ there exists an $\bar{x} \in M$ such that $\bar{\bar{x}} = x$, $x\bar{x} \in M'$.

Then the relation defined on M by

$$(1) \qquad x \sim y \quad \text{whenever} \quad xz\bar{y} \in M' \quad \text{for some} \quad z \in M'$$

is a congruence on M whose quotient is a group.

Proof. We first show that (1) is an equivalence. Let $a \in M$, then $b = a\bar{a} \in M'$, hence for any $x \in M$, $b\bar{x}x \in M'$, so by (i), $xb\bar{x} \in M'$, i.e. '$\sim$' is reflexive. Next, if $x \sim y$, $y \sim z$, say $xu\bar{y}$, $yv\bar{z} \in M'$, where $u,v \in M'$, then by (ii) $\bar{y}y \in M'$ and so $u\bar{y}yv \in M'$. Now $x.u\bar{y}yv.\bar{z} \in M'$, therefore $x \sim z$, i.e. '$\sim$' is transitive. To establish symmetry, assume $x \sim y$, say $xu\bar{y} \in M'$, $u \in M'$. Then $\bar{y}xu \in M'$ and $u\bar{y}x \in M'$, hence $\bar{y}xu.u\bar{y}x \in M'$. Further, $\bar{y}xu.u \in M'$, hence $xu^2\bar{y} \in M'$ and so $y\bar{y}.xu^2\bar{y}.x\bar{x} \in M'$. But $\bar{y}xu.u\bar{y}x \in M'$, therefore $y \sim x$ and the symmetry of '$\sim$' is proved. [Here we have the rather unusual situation where symmetry is harder to prove than transitivity.]

Next we observe that

$$(2) \qquad x \sim y \Rightarrow \bar{x} \sim \bar{y}.$$

For if $xu\bar{y} \in M'$ $(u \in M')$ then $\bar{x}xu\bar{y}y \in M'$, hence $\bar{x} \sim \bar{y}$.

Given any $x,y \in M$ we have $y\bar{y}\bar{x}x \in M'$, hence $xy\bar{y}\bar{x} \in M'$ and so, for any $u \in M'$, $xy\bar{y}\bar{x}u \in M'$, and $\bar{y}\bar{x}uxy \in M'$. Thus

$$(3) \qquad \bar{y}\bar{x} \sim \overline{xy}.$$

Now let $x \sim x_1$, $y \sim y_1$, say $xu\overline{x_1}$, $yv\overline{y_1} \in M'$ $(u,v \in M')$. Then $\bar{x}x.yv\overline{y_1} \in M'$, hence $xyv\overline{y_1}\bar{x} \in M'$ and $xu\overline{x_1} \in M'$, so

$$(4) \qquad xy.v\bar{y}_1\bar{x}.xu\bar{x}_1.x_1y_1.\overline{x_1y_1} \in M'.$$

On the other hand, $w = \bar{x}xu\bar{x}_1x_1 \in M'$ and $v\bar{y}_1y_1 \in M'$, hence $y_1v\bar{y}_1 \in M'$, so $y_1v\bar{y}_1w \in M'$ and $v\bar{y}_1wy_1 \in M'$. But this is just the central factor in (4), so $xy \sim x_1y_1$ and we have proved that '$\sim$' is a congruence.

Let $G = M/\sim$ be the quotient semigroup and denote the class corresponding to M' by e. Then in G, $ex = xe = x$, $\bar{x}x = x\bar{x} = e$ for all $x \in G$ (where $\bar{x}$ is the class of all $\bar{a}$, for a in x; this makes sense by (2)). This shows G to be a group, so Lemma 2 is proved.

We can now prove the theorem. Let us call a square matrix over R <u>singular</u> if it lies in Π, and non-singular otherwise. We shall construct a semigroup M whose elements are the first component u_1 of the formal solution of a system

$$Au + a = 0,$$

where A is non-singular. Formally, M is to consist of all $n \times (n+1)$ matrices $(A|a) = (a_1,a_2,\ldots,a_n,a)$ such that $A = (a_1,\ldots,a_n)$ and $A_1 = (a,a_2,\ldots,a_n)$ are non-singular;

here the <u>order</u>, n, of the system can have any value ≥ 1.

We shall use the notation introduced here for elements of M quite generally. Thus if $(B|b)$ is another element, say $B = (b_1,\dots,b_m)$, we write $B_1 = (b,b_2,\dots,b_m)$. Multiplication in M is defined by the formula

$$(5) \qquad (A|a).(B|b) = \left(\begin{array}{c|c|c} A & a\ 0 & 0 \\ 0 & B & b \end{array}\right).$$

If $(A|a)$ and $(B|b)$ have orders n,m respectively, the right-hand side has order $m+n$. To verify that we get an element of M we note that $\begin{pmatrix} A & a\ 0 \\ 0 & B \end{pmatrix}$ is non-singular, and on writing $A' = (a_2,\dots,a_n)$, we have $\left(\begin{array}{cc|c} 0 & A' & a\ 0 \\ b & 0 & B \end{array}\right) \to \left(\begin{array}{c|c} A_1 & 0 \\ b_1 0 & B_1 \end{array}\right)$ by column interchange, and the latter is non-singular.

We shall use notations like A_1, A' for any elements of M, without further explanation. Further, let us put

$$(6) \quad A^+ = (a_1+a,a_2,\dots,a_n), \quad A^* = (-a_1,a_2,\dots,a_n) \quad a^* = a_1+a.$$

Then if $(A|a)$ has the solution u_1, the system $(A_1|a_1)$ has the solution u_1^{-1}, $(A^+|a)$ has solution $u_1(1-u_1)^{-1}$, and $(A^*|a^*)$ has the solution $1-u_1$. These ways of interpreting the solutions will not be used in what follows, but if the reader keeps them in mind we will find the proof easier to follow.

The multiplication introduced in M (by (5)) is associative: we have

$$(A|a)(B|b)(C|c) = \left(\begin{array}{ccc|c} A & a0 & 0 & 0 \\ 0 & B & b0 & 0 \\ 0 & 0 & C & c \end{array}\right)$$

for either bracketing.

Let us define M' as the subset of M consisting of all $(A|a)$ for which A^+ (as defined in (6)) is singular. This is a subsemigroup of M: Given $(A|a)$ and $(B|b)$ in M', if we form the +-operation on their product, we get

$$\left(\begin{array}{c|c} A & a0 \\ b0 & B \end{array}\right) \to \left(\begin{array}{c|c} A+ & a0 \\ b*0 & B \end{array}\right) = \left(\begin{array}{c|c} A+ & 0 \\ b*0 & B \end{array}\right) \nabla \left(\begin{array}{c|c} A^+ & a\ 0 \\ b*0 & 0\ B' \end{array}\right).$$

The arrow indicates a column operation, and on the right we have a determinantal sum. The first term is singular because A^+ is, and the second term, after a column interchange, becomes

$$\left(\begin{array}{c|c} A_1 & a*\ 0 \\ 0 & B^+ \end{array}\right)$$

which is also singular.

We claim that M' satisfies the conditions of Lemma 2. For, given $(A|a)$, $(B|b)$, if their product lies in M', then $\left(\begin{array}{c|c} A & a0 \\ b0 & B \end{array}\right)$ is singular. Hence by column interchanges and row interchanges, $\left(\begin{array}{c|c} A & a0 \\ b0 & B \end{array}\right) \to \left(\begin{array}{c|c} a0 & A \\ B & b0 \end{array}\right) \to \left(\begin{array}{c|c} B & b0 \\ a0 & A \end{array}\right)$, so the matrix on the right is singular, i.e. $xy \in M' \Rightarrow yx \in M'$.

We now define

$$\text{(7)} \qquad \overline{(A|a)} = (A_1|a_1).$$

Clearly $\bar{\bar{x}} = x$ and if $x = (A\ a)$, then $x\bar{x} = \left(\begin{array}{cc|c} A & a0 & 0 \\ 0 & A_1 & a_1 \end{array}\right)$.

By row and column operations we have

$$\left(\begin{array}{c|c} A & a0 \\ a_1 0 & A_1 \end{array}\right) \to \left(\begin{array}{c|c} A+ & a0 \\ a*0 & A_1 \end{array}\right) \to \left(\begin{array}{c|c} 0\ A' & 0\ -A' \\ a*\ 0 & A_1 \end{array}\right) \to \left(\begin{array}{c|c} 0\ A' & 0 \\ a*\ 0 & A_1 \end{array}\right)$$

and the matrix on the right is clearly singular.

Thus all the conditions of Lemma 2 are satisfied, and we obtain a group G. To complete the proof of the theorem we need to define an operation θ as in Lemma 1. Let us write $G_1 = \{x \in G \mid x \neq 1\}$, $M_1 = \{x \in M \mid x \notin M'\}$ and for any $(A|a)$ in M_1, write

$$(A|a)f = (A^*|a^*),$$

where the right-hand side is defined as in (6). The resulting system lies again in M, for $A^* = (-a_1, a_2, \ldots, a_n)$ and $(a^*, a_2, \ldots, a_n)$ are both non-singular, and the system is not in M' because $-a_1 + a^* = a$, and $(a, a_2, \ldots, a_n)$ is non-singular.

Let $x \sim y$; we must show that $xf \sim yf$. write $x = (A|a)$, $y = (B|b)$; then we are given that for some $(C|c)$, C^+ and $\begin{pmatrix} A & a0 & 0 \\ 0 & C & c0 \\ b_10 & 0 & B_1 \end{pmatrix}$ are singular, and we shall show that $\begin{pmatrix} A^* & a^*0 & 0 \\ 0 & C & c0 \\ -b_10 & 0 & B^+ \end{pmatrix}$ is singular, which will prove that $xf \sim yf$. We have

$$\begin{pmatrix} A^* & a^*0 & 0 \\ 0 & C & c0 \\ -b_10 & 0 & B^+ \end{pmatrix} \to \begin{pmatrix} A & a^*0 & 0 \\ 0 & C & c0 \\ b_10 & 0 & B^+ \end{pmatrix} \to \begin{pmatrix} A & a^*0 & 0 \\ 0 & C^+ & c0 \\ b_10 & b^*0 & B^+ \end{pmatrix} \to \begin{pmatrix} A & a0 & 0 \\ 0 & C^+ & c0 \\ b_10 & b0 & B^+ \end{pmatrix}$$

$$= \begin{pmatrix} A & a0 & 0 \\ 0 & C^+ & 0 \\ b_10 & b0 & B \end{pmatrix} \nabla \begin{pmatrix} A & a0 & 0 \\ 0 & C^+ & c0 \\ b_10 & b0 & B_1 \end{pmatrix}.$$

Here the first term is singular because C^+ is, and the second is
$$\begin{pmatrix} A & a0 & 0 \\ 0 & C & c0 \\ b_1 0 & 0 & B_1 \end{pmatrix} \bar{\nabla} \begin{pmatrix} A & 0 & 0 \\ 0 & C_1 & c0 \\ b_1 0 & b0 & B_1 \end{pmatrix} .$$
Here the first term is singular by hypothesis, while the second has two columns equal.

This shows that $x \sim y \Rightarrow xf \sim yf$, and it follows that f induces a mapping θ of G_1 into itself. Clearly $f^2 = 1$, so $\theta^2 = 1$ and it remains to verify the other conditions of Lemma 1. In order to do this we note that we can perform the following operations on a system $(A|a)$ in M without affecting its congruence class mod M': (a) any row operation (b) any column operation of the form $c \to c + d\lambda$ where c is any column of $(A|a)$ and d is any column other than c and the first and last columns in $(A|a)$, (c) $c \to -c$, where c is not the first or last column of $(A|a)$, or (i) $(yxy^{-1})\theta = y(x\theta)y^{-1}$. Let $x = (A|a)$, $y = (B|b)$, then $x\theta = (A^*|a^*)$,
$$yx\bar{y} = \left(\begin{array}{ccc|c} B & b0 & 0 & 0 \\ 0 & A & a0 & 0 \\ 0 & 0 & B_1 & b_1 \end{array}\right), \quad (yx\bar{y})\theta = \left(\begin{array}{ccc|c} B^* & b0 & 0 & b_1 \\ 0 & A & a0 & 0 \\ 0 & 0 & B_1 & b_1 \end{array}\right).$$
$$y(x\theta)\bar{y} = \left(\begin{array}{ccc|c} B & b0 & 0 & 0 \\ 0 & A^* & a^*0 & 0 \\ 0 & 0 & B_1 & b_1 \end{array}\right).$$
Using operations (a), (b), (c), we have

$$(y x \bar{y})\theta = \left(\begin{array}{ccc|c} B^* & b0 & 0 & b_1 \\ 0 & A & a0 & 0 \\ 0 & 0 & B_1 & b_1 \end{array}\right) \to \left(\begin{array}{ccc|c} B^* & b0 & -B_1 & 0 \\ 0 & A & a0 & 0 \\ 0 & 0 & B_1 & b_1 \end{array}\right) \to \left(\begin{array}{ccc|c} -B & -b0 & -B_1 & 0 \\ 0 & A^* & a0 & 0 \\ 0 & 0 & B_1 & b_1 \end{array}\right)$$

$$\to \left(\begin{array}{ccc|c} B & b0 & B_1 & 0 \\ 0 & A^* & a0 & 0 \\ 0 & 0 & B_1 & b_1 \end{array}\right) \to \left(\begin{array}{ccc|c} B & b0 & b0 & 0 \\ 0 & A^* & a0 & 0 \\ 0 & 0 & B_1 & b_1 \end{array}\right) \to \left(\begin{array}{ccc|c} B & b0 & 0 & 0 \\ 0 & A^* & a^*0 & 0 \\ 0 & 0 & B_1 & b_1 \end{array}\right)$$

$$= y(x\theta)\bar{y}$$

We have already proved (ii), to establish (iii) we must show

(8) $\quad (xy^{-1})\theta = [x\theta(y\theta)^{-1}]\theta.(y^{-1})\theta \quad$ for $x,y \neq 1,\ x \neq y$.

Let $x = (A|a)$, $y = (B|b)$ then

$$(x\bar{y})\theta = \left(\begin{array}{cc|c} A^* & a0 & a_1 \\ 0 & B_1 & b_1 \end{array}\right) \qquad (x\theta)\overline{(y\theta)} = \left(\begin{array}{cc|c} A^* & a^*0 & 0 \\ 0 & B^+ & -b_1 \end{array}\right)$$

$\bar{y}\theta = (B_1^*|b^*)$

where $B_1^* = (-b, b_2, \ldots, b_m)$. Now $[(x\theta)\overline{(y\theta)}]\theta = \left(\begin{array}{cc|c} A & a^*0 & -a_1 \\ 0 & B^+ & -b_1 \end{array}\right)$,

hence for the right-hand side of (8) we have $[(x\theta)\overline{(y\theta)}]\theta.\bar{y}\theta =$

$$\left(\begin{array}{ccc|c} A & a^*0 & -a_10 & 0 \\ 0 & B^+ & -b_10 & 0 \\ 0 & 0 & B_1^* & b^* \end{array}\right) \to \left(\begin{array}{ccc|c} A & a^*0 & a_10 & 0 \\ 0 & B^+ & b_10 & 0 \\ 0 & 0 & B_1 & b^* \end{array}\right) \to \left(\begin{array}{ccc|c} A & a^*0 & a_10 & 0 \\ 0 & B^+ & B^+ & b^* \\ 0 & 0 & B_1 & b^* \end{array}\right)$$

$$\to \left(\begin{array}{ccc|c} A & a^*0 & -a0 & 0 \\ 0 & B^+ & 0 & b^* \\ 0 & 0 & B_1 & b^* \end{array}\right) \to \left(\begin{array}{ccc|c} A & a^*0 & -a0 & -a_1 \\ 0 & B^+ & 0 & 0 \\ 0 & 0 & B_1 & b_1 \end{array}\right) \to \left(\begin{array}{ccc|c} A & -a0 & a^*0 & -a_1 \\ 0 & 0 & B^+ & 0 \\ 0 & B_1 & 0 & b_1 \end{array}\right)$$

$$\rightarrow \left(\begin{array}{ccc|c} A & -a0 & a^*0 & -a_1 \\ 0 & B_1 & 0 & b_1 \\ 0 & 0 & B^+ & 0 \end{array}\right) \rightarrow \left(\begin{array}{ccc|c} A^* & -a0 & a^*0 & a_1 \\ 0 & B_1 & 0 & -b_1 \\ 0 & 0 & B^+ & 0 \end{array}\right) \rightarrow$$

$$\left(\begin{array}{ccc|c} A^* & -a0 & a^*0 & a_1 \\ 0 & -B_1 & 0 & b_1 \\ 0 & 0 & B^+ & 0 \end{array}\right) \rightarrow \left(\begin{array}{ccc|c} A^* & a0 & a^*0 & a_1 \\ 0 & B_1 & 0 & b_1 \\ 0 & 0 & B^+ & 0 \end{array}\right) .$$

To show that this is in the same class (mod M') as $(x\bar{y})\theta$, we consider the matrix for $(x\bar{y})\theta.\bar{z}$, where z is the system just found; it is

$$\begin{pmatrix} A^* & a0 & a_1 0 & 0 & 0 \\ 0 & B_1 & b_1 0 & 0 & 0 \\ -a_1 0 & 0 & A & a0 & a^*0 \\ 0 & 0 & b_1 0 & B_1 & 0 \\ 0 & 0 & 0 & 0 & B^+ \end{pmatrix}$$

We must show that this matrix is singular. Since $y \neq 1$, B^+ is non-singular, and we may omit the last row and column block. This leaves us with

$$\begin{pmatrix} A^* & a0 & a_1 0 & 0 \\ 0 & B_1 & b_1 0 & 0 \\ -a_1 0 & 0 & A & a0 \\ 0 & 0 & b_1 0 & B_1 \end{pmatrix} \rightarrow \begin{pmatrix} 0A' & a0 & a_1 0 & 0 \\ b_1 0 & B_1 & b_1 0 & 0 \\ 0 & 0 & A & a0 \\ b_1 0 & 0 & b_1 0 & B_1 \end{pmatrix} \rightarrow$$

$$\begin{pmatrix} 0A' & a0 & a_1 0 & 0 \\ b_1 0 & B_1 & b_1 0 & 0 \\ 0 & 0 & A & a0 \\ 0 & -B_1 & 0 & B_1 \end{pmatrix} \rightarrow \begin{pmatrix} 0A' & a0 & a_1 0 & 0 \\ b_1 0 & B_1 & b_1 0 & 0 \\ 0 & a0 & A & a0 \\ 0 & 0 & 0 & B_1 \end{pmatrix} .$$

Again B_1 is non-singular, so there remains only

$$\begin{pmatrix} 0A' & a0 & a_1 0 \\ b_1 0 & B_1 & b_1 0 \\ 0 & a0 & A \end{pmatrix} \to \begin{pmatrix} 0A' & a0 & a_1 0 \\ b_1 0 & B_1 & b_1 0 \\ 0-A' & 0 & 0A' \end{pmatrix} \to \begin{pmatrix} A^* & a0 & a_1 0 \\ 0 & B_1 & b_1 0 \\ 0 & 0 & 0A' \end{pmatrix}$$

and this is clearly singular.

To prove (iv) we show that $e = x(\bar{x}\theta)\overline{(x\theta)}$ is represented by the system $(1|1)$ of order 1. Let $x = (A|a)$, then $\bar{x}\theta = (A_1^*|a^*)$, $\overline{x\theta} = (A^+|-a_1)$ and the required result will follow if we prove that

$$\begin{pmatrix} A & a0 & 0 & 0 \\ 0 & A_1^* & a^*0 & 0 \\ 0 & 0 & A^+ & -a_1 \\ 0 & 0 & 0 & 1 \end{pmatrix}$$

is singular.

By subtracting the last column from the first, we reduce the last row to zero except in the last place, and so obtain

$$\begin{pmatrix} A & a0 & 0 \\ 0 & A_1^* & a^*0 \\ a_1 0 & 0 & A^+ \end{pmatrix} \to \begin{pmatrix} A & 0A' & a^*0 \\ 0 & A_1^* & a^*0 \\ a_1 0 & 0 & A^+ \end{pmatrix} \to \begin{pmatrix} 0A' & 0A' & 0-A' \\ 0 & A_1^* & a^*0 \\ a_1 0 & 0 & A^+ \end{pmatrix} \to$$

$$\begin{pmatrix} 0A' & 0 & 0 \\ 0 & A_1^* & a^*0 \\ a_1 0 & 0 & A^+ \end{pmatrix}$$

and the latter matrix is clearly singular.

Thus the conditions of Lemma 1 are satisfied and we obtain a field, K say. We define a mapping $\lambda : R \to K$ as follows. If $\alpha \in R$ is singular (as $|x|$ matrix), $\alpha\lambda = 0$, otherwise let

$$\alpha \longmapsto (1|-\alpha),$$

and define $\alpha\lambda$ to be the congruence class (mod M') of $(1|-\alpha)$. To show that this is a homomorphism, let us check that it preserves multiplication: $(xy)\lambda = x\lambda.y\lambda$. This is clear if x or y is singular. Otherwise x and y are both non-singular and we must show that $\left(\begin{smallmatrix}1 & -x\\0 & 1\end{smallmatrix}\middle|\begin{smallmatrix}0\\-y\end{smallmatrix}\right) \sim (1|-xy)$,

i.e. $\begin{pmatrix}1 & -x & 0\\0 & 1 & -y\\1 & 0 & -xy\end{pmatrix}$ is singular. But

$$\begin{pmatrix}1 & -x & 0\\0 & 1 & -y\\1 & 0 & -xy\end{pmatrix} \to \begin{pmatrix}1 & 0 & -xy\\0 & 1 & -y\\1 & 0 & -xy\end{pmatrix} \to \begin{pmatrix}1 & 0 & -xy\\0 & 1 & -y\\0 & 0 & 0\end{pmatrix}$$

and the latter is singular, as claimed. Moreover, if $x = (1|-a)$ and $a, 1-a$ are non-singular, then $x\theta = (-1|1-a)$ and this is $(1-a)\lambda$. Thus $a\lambda\theta = (1-a)\lambda$, i.e. θ corresponds to the operation $a \longmapsto 1-a$, hence λ is a homomorphism of R into K.

Finally we must show that a matrix A over R maps to a singular matrix over K precisely when $A \in \Pi$, i.e. when A is 'singular'. Let Π' be the set of all square matrices that map to singular matrices over K. If $A \notin \Pi$, then the system $Au + a = 0$ has a solution in K for all a, by the construction of K. Taking $a = -e_i$ $(i = 1,2,\ldots,n)$ in turn, we see that A has a right inverse over K, and so is invertible. Thus $A \notin \Pi'$ and we have shown that $\Pi' \subseteq \Pi$. To establish equality we use induction on n, the order of A.

Let $A \in \Pi$ be of order n, say $A = (a_1,\ldots,a_n)$, but suppose that there exists a column a such that $A_1 = (a,a_2,\ldots,a_n) \notin \Pi$. Then the equation

$$A_1 u + a_1 = 0 \tag{9}$$

has a solution in K, and since $A \in \Pi$, it follows that $u_1 = 0$, hence by (9) the columns of A are linearly dependent over K: $a_2u_2 + \ldots + a_nu_n + a_1 = 0$; hence $A \in \Pi'$. On the other hand, if $(a,a_2,\ldots,a_n) \in \Pi$ for all choices of a, let us take $a = e_i$ $(i = 1,\ldots,n)$. Each of the resulting n matrices is in Π, hence each $(i,1)$-minor of A is in Π, and, by induction, in Π'. By expanding A by its first column, we find that $A \in \Pi'$, and this shows $\Pi' = \Pi$. For $n = 1$ there is no problem, as the $|x|$ elements of Π and Π' agree by definition of λ. This completes the proof of the theorem.

Before listing some consequences of this result we note that the definition of pre-ideal set on p.32 can be simplified: condition 2) is a consequence of 1) and 3). For if $A \in \mathcal{P}$, say $A = (a_1,\ldots,a_n)$, then for any $c \in R$,

$$(a_1+a_2c,a_2,\ldots,a_n) = A \nabla (a_2,\ldots,a_n)\begin{pmatrix} c & \\ 0 & I_{n-1}\end{pmatrix}$$

and on the right we have the determinantal sum of A and a non-full matrix, so the result lies in $\mathcal{P}$. Similarly for row operations.

In any ring R, denote by z the set of all determinantal sums of non-full matrices. Then z satisfies 1) 3) and 4), and hence 2) p.33, and so it is a pre-ideal set, clearly the least pre-ideal set in R. Let $\bar{z}$ be the ideal set generated, and put $\eta = \sqrt{\bar{z}}$ then η, the radical of the least ideal set, is the intersection of all prime ideal sets. So η is proper iff R has prime ideal sets. By the theorem we therefore obtain the

Corollary 1. A ring R has an R-field (i.e. a homomorphism into some field) iff the unit matrix I (of any size) cannot be written as a determinantal sum of non-full matrices.

Let A be any square matrix; then η, the radical of the least ideal set contains A iff a diagonal sum of I and copies of A lies in z. Hence we have

Corollary 2. Given any ring R and any square matrix A over R, there is a homomorphism of R into a field K which maps A to a non-singular matrix iff no diagonal sum of I and copies of A can be written as a determinantal sum of non-full matrices.

Using this condition it is easy to find a criterion for the embeddability of rings in fields. We recall that an integral domain is embeddable in a field iff it can be embedded in a product of fields. To embed R in a product of fields we must for each $a \neq 0$ in R find a homomorphism into a field inverting a. Thus by Corollary 2 we find

Corollary 3. A ring R is embeddable in a field iff it is an integral domain and for each $a \neq 0$, no diagonal sum of copies of 1 and a can be written as a determinantal sum of non-full matrices.

An alternative form of this condition is

Corollary 4. A ring R is embeddable in a field iff no diagonal matrix with non-zero elements on the main diagonal can be written as a determinantal sum of non-full matrices.

For if $ab = 0$, then $\begin{pmatrix} a & 0 \\ 0 & b \end{pmatrix} = \begin{pmatrix} a & 0 \\ 1 & b \end{pmatrix} \nabla \begin{pmatrix} 0 & 0 \\ -1 & b \end{pmatrix} =$ $\begin{pmatrix} a \\ 1 \end{pmatrix}(1\ b) \nabla \begin{pmatrix} 0 \\ 1 \end{pmatrix}(-1\ b)$ and here both matrices on the right are non-full. Thus the condition of Corollary 4 is enough to exclude zero-divisors, and now the result follows by Corollary 3. It may be of interest to note that the condition in Corollary 3, apart from the requirement about zero-divisors, is in the form of quasi-identities, as required by general theory (cf. e.g. [4] p.235).

Frequently we are interested in the existence of a universal field of fractions in which all full matrices can be inverted:

Corollary 5. Let R be any ring. Then R has a field of fractions in which every full matrix can be inverted (in which is therefore universal) iff (i) the determinantal sum of non-full matrices is non-full and (ii) the diagonal sum of full matrices is full.

Again the necessity is clear while (i)-(ii) insure that the set of non-full matrices is a prime ideal set.

We end with an application of Corollary 5. Let R be a semifir, (i.e. a ring in which each finitely generated right ideal is free, of unique rank) then full matrices correspond to torsion modules. This is proved for firs in [6], but the proof is easily seen to hold also for semifirs. Since the class of torsion modules is closed under extensions, the diagonal sum of full matrices over R is full. Let A be an $n\times n$ matrix; it defines a mapping

$$R^n \to R^n,$$

whose image is an n-generator submodule H of R^n, and A is full precisely if H is not contained in any (n-1)-generator submodule of R^n. Suppose A,B are two non-full matrices whose determinantal sum with respect to the first column exists, and write

$$C = A \nabla B.$$

Then A,B correspond to submodules H,K respectively of R^n; the columns of A,B represent the coordinates of the vectors spanning H,K respectively. By hypothesis these vectors may be taken to be $a_1,\ldots,a_n$ and $a_1',a_2,\ldots,a_n$. Moreover, since A,B are not full, we can find submodules H',K' of R^n each generated by $n-1$ elements such that $H \subseteq H'$, $K \subseteq K'$. Clearly $H' + K'$ is finitely generated and hence free.

Any free module has a well-defined rank, denoted by rk. If $rk(H'+K') < n$, then the submodule corresponding to the matrix C, which is spanned by $a_1+a_1', a_2, \ldots, a_n$, is contained in $H'+K'$ of rank $< n$, and it follows that C is not full. So we may assume

$$rk(H'+K') \geq n. \tag{10}$$

If $rk(H' \cap K') < n-1$, observe that $a_2, \ldots, a_n \in H' \cap K'$ hence $a_2, \ldots, a_n$ lie in a module of rank $< n-1$, and therefore $a_1+a_1', a_2, \ldots, a_n$ lie in a module of rank $< n$, so again C is not full. Thus we may also assume

$$rk(H' \cap K') \geq n-1. \tag{11}$$

But we clearly have

$$rk(H' + K') + rk(H' \cap K') = rk\ H' + rk\ K'.$$

Here the left-hand side is at least $2n-1$, by (10) and (11) while the right-hand side is at most $2n-2$ because each of H', K' can be generated by $n-1$ elements. This is a contradiction, and it follows that C is not full. We have thus verified the conditions of Corollary 5 and so obtain the following generalization of the result at the foot of p. 25.

Theorem. Every semifir has a universal field of fractions in which each full matrix over R is invertible.

References

1. G. M. Bergman, Skew fields of noncommutative rational functions, after Amitsur, to appear.
2. P. M. Cohn, Rings with a weak algorithm, Trans. Amer. Math. Soc. 109 (1963), 332-356.
3. P. M. Cohn, Free ideal rings, J. Algebra 1 (1964), 47-69.
4. P. M. Cohn, Universal Algebra (New York, London, Tokyo 1965).
5. P. M. Cohn, Some remarks on the invariant basis property Topology 5 (1966), 215-228.
6. P. M. Cohn, Torsion modules over free ideal rings, Proc. London Math. Soc. (3) 17 (1967), 577-599.
7. P. M. Cohn, The embedding of firs in skew fields, Proc. London Math. Soc., to appear.
8. P. M. Cohn, Universal skew fields of fractions, Sympos. Math. to appear.
9. J. Dieudonné, Les determinants sur un corps non-commutatif, Bull. Soc. Math. France 71 (1943), 27-45.
10. J. L. Fisher, Title uncertain (Different embeddings of a free algebra in skew fields), Proc. Amer. Math. Soc., to appear.
11. A. A. Klein, A remark on the embedding of rings in fields, J. Algebra, to appear.
12. A. I. Malcev, On the immersion of an algebraic ring into a field, Math. Ann. 113 (1937), 686-691.
13. A. I. Malcev, Über die Einbettung von assoziativen Systemen in Gruppen I. Mat. Sbornik 6 (48) (1939) 331-336, II ibid. 8 (50) (1950) 251-264.
14. R. Moufang, Einige Untersuchungen über geordnete Schiefkorper J. reine angew. Math. 176 (1937) 203-223.
15. M. Nivat, Séries rationnelles et algébriques en variables non-commutatives, Cours DEA (unpublished).
16. C. Procesi, Rendiconti Circ. Mat. Palermo (2) 17 (1968).
17. P. M. Cohn, On the embedding of rings in skew fields, Proc. London Math. Soc. (3) 11 (1961) 511-30.
18. P. M. Cohn, Un critere d'immersibilite d'un anneau dans un corps gauche, C. R. (Parks) Acad. Sci. (1971).

BALANCED RINGS

by

Vlastimil Dlab and Claus Michael Ringel

Department of Mathematics

Carleton University

Ottawa, Canada

The aim of these notes is to report on recent investigations in the structure of balanced rings. Here, a ring R is called left balanced, if every left R-module is balanced, i.e. has the double centralizer property *). The study of balanced modules can be traced back to C. J. NESBITT and R. M. THRALL: in [23] they showed that uniserial rings in the sense of T. NAKAYAMA [22] are left and right balanced. Later, R. M. THRALL [27] introduced the class of QF-1 rings generalizing quasi-Frobenius rings as those rings R over which all finitely generated faithful R-modules are balanced. Some progress in the study of balanced and QF-1 rings was made in the papers [20], [21] and [24] of K. MORITA and H. TACHIKAWA. In particular, they established in [21] that the property to be balanced is Morita equivalent.

A further progress in the theory of balanced rings has been done quite recently. Considering the question whether every balanced ring is uniserial,

---

*) The term "balanced ring" has been introduced by V. P. CAMILLO in [3].

D. R. FLOYD [14] proved this statement for finite dimensional commutative algebras. Later, S. E. DICKSON and K. R. FULLER [8] extended the result to artinian commutative rings and J. P. JANS [18] proved it for finite dimensional algebras over algebraically closed fields *). In [18], J. P. JANS formally conjectured that the result holds for every artinian ring. In his paper [3], V. P. CAMILLO showed that under certain conditions (if R is commutative or left noetherian) a left balanced ring is left artinian, and that, in general, a left balanced ring is always semiperfect and its radical is nil. K. R. FULLER proved in [16] that a semiprimary ring is left balanced if and only if it is a finite direct sum of full matrix rings over local left balanced rings.

These results were extended in [9]. In addition to the fact that every left balanced ring is left artinian, some necessary conditions on the structure of local left balanced rings were also established. Making use of these structural conditions, every left balanced ring which is finitely generated over its centre was shown to be uniserial. Independently, V. P. CAMILLO and K. R. FULLER [4] obtained this result for algebras. In [10], on the other hand, JANS' conjecture was shown to be false: Local rings R with the radical W such that $Q = R/W$ is commutative, $W^2 = 0$ and $\dim {}_Q W \times \dim W_Q = 2$ are balanced **), and essentially these are the only non-uniserial balanced rings with a commutative radical quotient [11]. This is a special case of the complete characterization of left balanced rings in terms of exceptional rings given in [12]. One of the immediate

---

*) The crucial result that a local algebra over an algebraically closed field with zero radical square is uniserial was previously proved by D. R. FLOYD [14].

**) In a recent letter, Professor H. Tachikawa informs us that he has known of counterexamples to the conjecture, as well.

consequences is the fact that a ring is left balanced if and only if it is right balanced. Moreover, a ring is balanced if and only if it is left artinian and every finitely generated module is balanced. Thus, an artinian ring R is balanced if and only if every factor ring of R is a QF-1 ring. The paper [12] and [13] include also a characterization of balanced rings in terms of their module categories.

These notes are virtually self-contained. Besides the basic properties of Morita equivalent rings, only a theorem of N. JACOBSON concerning division rings which are finitely generated over their centres will be used at one instance (Theorem III.4.4). Section I.3 provides, in fact, a proof that quasi-Frobenius rings are QF-1 rings. However, since there was no need to introduce these classes of rings (see e.g. [7]), we restrict ourselves to the class of uniserial rings. But we note that certain results of Chapters II and III can be generalized to QF-1 rings (cf. [4] and [9]). Let us also note that the duality theory of H. TACHIKAWA [25] (closely related with Lemma I.2.5) could be used to prove Proposition III.6.1. This has been done in [13]; here, we prefer to give an elementary proof.

Some of the proofs given in the present notes have not appeared in the literature. Let us mention that our proof of Morita equivalence of the property to be balanced, which follows the idea of the original proof in [21] yields, in fact, a more general statement (Proposition III.1.3). Also, the concept of "bi-T-nilpotency" of V. P. CAMILLO [3] used previously in [9] to prove the fact that a balanced ring is artinian has been avoided; as a result, the proof has become simpler.

I. *PRELIMINARIES*

I.1. NOTATION AND TERMINOLOGY

Throughout these notes, R denotes an (associative) ring with unity, R* its opposite. By an R-module we always understand a unital R-module; the symbols ${}_RM$ or M_R will be used to underline the fact that M is a left or a right R-module, respectively. We usually consider left R-modules, and, in this case, speak simply about an R-module. It should be noted that homomorphisms always act from the opposite side as the operators; in particular, every (left) R-module M defines a right $\mathcal{C}$-module, where $\mathcal{C}$ is the endomorphism ring of the R-module M. The ring $\mathcal{C}$ or, more precisely $\mathcal{C}(M)$ is called the *centralizer* of M. The *double centralizer* $\mathcal{D}(M)$ (or simply $\mathcal{D}$) is the endomorphism ring of $M_{\mathcal{C}}$. Again, $\mathcal{D}$ operates from the opposite side as $\mathcal{C}$, that is from the left. There is a canonic ring homomorphism from R into $\mathcal{D}$; if this homomorphism is surjective, then M is called *balanced*. The ring R is called *left finitely balanced*, if every finitely generated left R-module is balanced. If every left R-module is balanced, then R is called *left balanced*.

For an element m of an R-module M, the *annihilator* $\{r \in R \mid rm = 0\}$ of m will be denoted by Ann(m). If the R-module M has a composition series, denote by ∂M its *length*. The *radical* Rad M of M is the intersection of all maximal submodules of M; it is the set of all non-generators of M. Here, an element $m \in M$ is called a non-generator, if it can be ommitted from any generating set of M. The radical of the ring R is by definition $\mathrm{Rad}\,{}_RR$; it will be denoted consistently by W. For the R-module M, we always have the inclusion $WM \subseteq \mathrm{Rad}\,M$; moreover, if R/W is artinian, then $WM = \mathrm{Rad}\,M$. If $\mathrm{Rad}\,M$ is the only (proper) maximal submodule of M, then M will be called *local*. Thus, all local modules are monogenic. And, if ${}_RR$ (and, for the matter R_R) is local, then R is said to be a local ring. Note that the ring R is local if and only if the non-units of R form an ideal. If the R-module M has minimal submodules, the *socle* $\mathrm{Soc}\,M$ is defined as their union. We always have $\mathrm{Soc}\,M \subseteq \{m \in M \mid Wm = 0\}$; moreover, if R/W is artinian, then $\mathrm{Soc}\,M = \{m \in M \mid Wm = 0\}$. Considering R as a left R-module, we get the concept of the *left socle* $\mathrm{Soc}\,{}_RR$ of R. A module is said to be *uniserial*, if all its submodules form a chain with respect to inclusion. Hence, a uniserial module of finite length is local. A local ring R is called uniserial, if both ${}_RR$ and R_R are uniserial modules of finite length. An arbitrary ring is called uniserial [22], if it is a finite direct sum of full matrix rings over local uniserial rings. It is not difficult to see that a left artinian ring R is uniserial if and only if R/W^2 is uniserial.

The R-module M is called *indecomposable*, if M cannot be written as the direct sum of two proper submodules. If M is indecomposable and of finite length, then the centralizer $\mathcal{C}$ is a local ring. Moreover, there exists a composition series

$$0 = M_o \subset M_1 \subset \dots \subset M_n = M$$

such that $M_i\mathcal{W} \subseteq M_{i-1}$ for all i, where $\mathcal{W}$ is the radical of $\mathcal{C}$ (see [2], Ex. 3, pp. 26-27). Thus, $\mathcal{W}^n = 0$ and, furthermore,

$$M_1 \subseteq \mathrm{Soc}\,M_{\mathcal{C}} \quad \text{and} \quad \mathrm{Rad}\,M_{\mathcal{C}} \subseteq M_{n-1}\,.$$

Let M_1 and M_2 be two R-modules. If $\varphi : M_1 \to M_2$ is a homomorphism, we denote by $\operatorname{Ker}\varphi$ and $\operatorname{Im}\varphi$ the kernel and the image of φ, respectively. The module M_1 is called a *generator for* M_2, if the images of all homomorphisms $M_1 \to M_2$ generate M_2. Dually, M_1 is called a *cogenerator for* M_2, if the intersection of kernels of all homomorphisms $M_2 \to M_1$ is zero. Finally, M is a generator or a cogenerator for a class of modules, if it is a generator or a cogenerator for every module of this class.

The category of all (left) R-modules will be denoted by $\operatorname{Mod} R$. Two rings R and R' are called Morita equivalent, if $\operatorname{Mod} R$ and $\operatorname{Mod} R'$ are equivalent categories. The rings R and R' are Morita equivalent if and only if there exists a right R-module P_R which is finitely generated, projective and a generator for all right R-modules such that R' is isomorphic to the centralizer of P_R ([19] or [1]). In particular, a ring R is Morita equivalent to every full matrix ring over R.

I.2. GENERATORS AND COGENERATORS

Results of this section belong in part to the folklore of the subject. They are presented here in such a way as to be easily applicable in the next chapters.

LEMMA I.2.1. *Let* M_1 *and* M_2 *be* R-*modules such that* M_1 *is either a generator or a cogenerator for* M_2. *If* M_1 *is balanced, then* $M_1 \oplus M_2$ *is balanced, as well.*

Proof. The elements of the centralizer of $M_1 \oplus M_2$ can be written as matrices $\begin{pmatrix}\varphi_{11} & \varphi_{12}\\ \varphi_{21} & \varphi_{22}\end{pmatrix}$, where $\varphi_{ij} : M_i \to M_j$ with $1 \leq i, j \leq 2$ are R-homomorphisms. Let ψ be in the double centralizer of $M_1 \oplus M_2$. If $x \in M_1$ and $\psi(x, 0) = (x', y')$, then

$$\psi(x, 0) = \psi[(x, 0)\begin{pmatrix}1 & 0\\ 0 & 0\end{pmatrix}] = [\psi(x, y)]\begin{pmatrix}1 & 0\\ 0 & 0\end{pmatrix} = (x', y')\begin{pmatrix}1 & 0\\ 0 & 0\end{pmatrix} = (x', 0) ,$$

and thus $y' = 0$.

Similarly, if $y \in M_2$ and $\psi(0, y) = (x'', y'')$, then $x'' = 0$. It is easy to see that $\psi_1 : M_1 \to M_1$ defined by $(\psi_1 x, 0) = \psi(x, 0)$ belongs to the double centralizer of M_1. Therefore, there is $\rho \in R$ with

$$\psi(x, 0) = \rho(x, 0) \quad \text{for all} \quad x \in M_1 .$$

Assume now that M_1 generates M_2. Then, taking an element of the form $(0, x\varphi_{12})$, where $x \in M_1$ and $\varphi_{12} \in \mathrm{Hom}(M_1, M_2)$, we get the equality

$$\psi(0, x\varphi_{12}) = \psi[(x, 0)\begin{pmatrix}0 & \varphi_{12}\\ 0 & 0\end{pmatrix}] = [\psi(x, 0)]\begin{pmatrix}0 & \varphi_{12}\\ 0 & 0\end{pmatrix} = \rho(0, x\varphi_{12}) .$$

Since the elements of the form $(x, 0)$ and of the form $(0, x\varphi_{12})$ with $x \in M_1$ and $\varphi_{12} \in \mathrm{Hom}(M_1, M_2)$ generate $M_1 \oplus M_2$ additively, it follows that ψ is induced by multiplication by ρ.

If we assume that M_1 cogenerates M_2 consider $\varphi_{21} \in \mathrm{Hom}(M_2, M_1)$ and apply $\begin{pmatrix}0 & 0\\ \varphi_{21} & 0\end{pmatrix}$ to $\psi(0, y) - \rho(0, y)$ with $y \in M_2$; we get

$$[\psi(0, y)-\rho(0, y)]\begin{pmatrix}0 & 0\\ \varphi_{21} & 0\end{pmatrix} = \psi[(0, y)\begin{pmatrix}0 & 0\\ \varphi_{21} & 0\end{pmatrix}]-\rho(0, y)\begin{pmatrix}0 & 0\\ \varphi_{21} & 0\end{pmatrix} =$$

$$\psi(y\varphi_{21}, 0)-\rho(y\varphi_{21}, 0) = 0 .$$

Therefore, $\psi(0, y) = \rho(0, y)$. Again, since the elements $(x, 0)$ and $(0, y)$ with $x \in M_1$ and $y \in M_2$ generate $M_1 \oplus M_2$ additively, we conclude that ψ is induced by multiplication by ρ . The proof is completed.

LEMMA I.2.2. *Let* M_1 *and* M_2 *be* R*-modules such that* M_1 *is both a generator and a cogenerator for* M_2 *. If* $M_1 \oplus M_2$ *is balanced, then* M_1 *is balanced, as well.*

Proof. Let ψ_1 be an element of the double centralizer of M_1 . We are going to construct ψ_2 in the double centralizer of M_2 , such that $\psi_1 \oplus \psi_2 = \begin{matrix} \psi_1 & 0 \\ 0 & \psi_2 \end{matrix}$ belongs to the double centralizer of $M_1 \oplus M_2$.

Since M_1 is a generator for M_2 , every element of M_2 has the form $\Sigma x_i \gamma_i$ with $x_i \in M_1$ and $\gamma_i \in \mathrm{Hom}(M_1, M_2)$. Define ψ_2 as follows

$$\psi_2(\Sigma x_i \gamma_i) = \Sigma(\psi_1 x_i)\gamma_i .$$

Here, ψ_2 is well defined. For, if $\Sigma x_i \gamma_i = 0$ and $\varphi_{21} : M_2 \to M_1$, then the fact that $\gamma_i \varphi_{21} \in C(M_1)$ implies

$$[\psi_2(\Sigma x_i \gamma_i)]\varphi_{21} = \Sigma(\psi_1 x_i)(\gamma_i \varphi_{21}) = \Sigma \psi_1 (x_i \gamma_i \varphi_{21}) = 0 .$$

And, since M_1 cogenerates M_2 , it follows that $\psi_2(\Sigma x_i \gamma_i) = 0$.

Furthermore, ψ_2 belongs to the double centralizer of M_2 . This is an immediate consequence of the relation

$$[\psi_2(\Sigma x_i \gamma_i)]\varphi_{22} = \Sigma(\psi_2 x_i)\gamma_i \varphi_{22} = \Sigma \psi_2(x_i \gamma_i \varphi_{22}) = \psi_2[\Sigma x_i \gamma_i)\varphi_{22}] ,$$

where $\varphi_{22} \in \mathrm{Hom}(M_2, M_2)$. Also, for $x \in M_1$ and for $\Sigma x_i \gamma_i \in M_2$ with $x_i \in M_1$ and $\gamma_i \in \mathrm{Hom}(M_1, M_2)$, we have

$$(\psi_1 x)\varphi_{12} = \psi_2(x\varphi_{12}) \quad \text{and} \quad (\psi_2 \Sigma x_i \gamma_i)\varphi_{21} = \psi_1(\Sigma x_i \gamma_i \varphi_{21}) \ ,$$

with arbitrary $\varphi_{12} \in \mathrm{Hom}(M_1, M_2)$ and $\varphi_{21} \in \mathrm{Hom}(M_2, M_1)$. This shows that $\psi_1 \oplus \psi_2$ belongs to the double centralizer of $M_1 \oplus M_2$. Since $M_1 \oplus M_2$ is balanced, $\psi_1 \oplus \psi_2$ is induced by multiplication by an element $\rho \in R$. Therefore, ψ_1 is induced by multiplication by ρ , and M_1 is balanced, as required.

LEMMA I.2.3. *Every generator of* ${}_R\mathfrak{M}$ *is balanced.*

Proof. If M is a generator of ${}_R\mathfrak{M}$, then ${}_RR$ is an epimorphic image of the direct sum $M^{(n)}$ of n copies of M . Since ${}_RR$ is projective, we can find a complement K such that ${}_RR \oplus K \cong M^{(n)}$. Applying Lemma I.2.1, we get that $M^{(n)}$ is balanced; for, ${}_RR$ is balanced and ${}_RR$ is a generator for K . Now, $M^{(n)}$ is a direct sum of copies of M , and we can therefore apply Lemma I.2.2 to conclude that M is balanced, as well.

Under certain assumptions, a similar statement can be proved for cogenerators. In what follows, we shall need the following result:

LEMMA I.2.4. *Every cogenerator of* ${}_R\mathfrak{M}$ *which is finitely generated over its centralizer is balanced.*

Proof. Let M be a cogenerator of ${}_R\mathfrak{M}$. Then, for any element ψ of the double centralizer $\mathcal{D}$ of M and any $x \in M$, we have $\psi x \in Rx$. For, assuming $\psi x \notin Rx$, we can find an R-homomorphism $\varphi : M/Rx \to M$ with $(\psi x + Rx)\varphi \neq 0$. And, denoting by ϵ the canonic epimorphism $\epsilon : M \to M/Rx$, and observing that $\epsilon\varphi$ belongs to the centralizer $\mathcal{C}$ of M , we get

$$0 = \psi[x(\varepsilon\varphi)] = (\psi x)(\varepsilon\varphi) \neq 0 ,$$

a contradiction.

Now let $x_1, \ldots, x_n$ be elements of M which generate M as a C-Module. If we form the direct sum $M^{(n)}$ of n copies of M, then it is easy to see that every element $\psi \in D$ defines an element $\psi^{(n)}$ of the double centralizer of $M^{(n)}$

$$\psi^{(n)}(m_1, \ldots, m_n) = (\psi m_1, \ldots, \psi m_n) \quad \text{for} \quad m_i \in M .$$

But $M^{(n)}$ is again a cogenerator of ${}_R M$, and we can apply the first result of our proof for $M^{(n)}$ and the element $(x_1, \ldots, x_n) \in M^{(n)}$. In this way, we get

$$(\psi x_1, \ldots, \psi x_n) = \psi^{(n)}(x_1, \ldots, x_n) = \rho(x_1, \ldots, x_n) = (\rho x_1, \ldots, \rho x_n) ,$$

for some $\rho \in R$. But, since the element x_i generate the C-module M_C, the relations $\psi x_i = \rho x_i$ yield that ψ is induced by multiplication.

The above result can be applied, in particular, for an injective cogenerator of ${}_R M$, when the ring R is left artinian.

LEMMA I.2.5. *Let* R *be a left artinian ring. Then any injective cogenerator of* ${}_R M$ *is balanced.*

Proof. Let M be an injective cogenerator of ${}_R M$ and C its centralizer. For any left ideal L of R, denote by $r(L)$ the following C-submodule of M

$$r(L) = \{m \in M \mid Lm = 0\} .$$

If $L' \subseteq L$, then we have a C-homomorphism

$$\alpha : r(L') \to \mathrm{Hom}_R(L/L' , M_C) ,$$

where, for $m \in r(L')$, the R-homomorphism $m\alpha$ maps $\lambda + L' \in L/L'$ into λm . The kernel of this morphism is just $r(L)$, and therefore $r(L')/r(L)$ can be considered as a C-submodule of $\mathrm{Hom}_R(L/L', M_C)$.

If S is a simple R-module, then $\mathrm{Hom}_R(S , M_C)$ is a simple C-module. For, $\mathrm{Hom}_R(S , M_C) \neq 0$ in view of the fact that M is a cogenerator and, given any two elements γ and γ' in $\mathrm{Hom}_R(S , M_C)$ with $\gamma \neq 0$, we see easily that γ is a monomorphism and that there is an R-endomorphism φ of M with $\gamma\varphi = \gamma'$; the latter conclusion follows from the injectivity of M .

Since ${}_RR$ is artinian, we have a composition series

$$0 = L_o \subset L_1 \subset \ldots \subset L_n = {}_RR$$

of left ideals L_i of ${}_RR$. Since L_i/L_{i-1} is a simple R-module, $\mathrm{Hom}_R(L_i/L_{i-1} , M_C)$ is a simple C-module, and thus the C-submodule $r(L_{i-1})/r(L_i)$ of $\mathrm{Hom}_R(L_i/L_{i-1} , M_C)$ is either trivial or simple. Hence

$$M_C = r(L_o) \supseteq r(L_1) \supseteq \ldots \supseteq r(L_n) = 0$$

defines a composition series of the C-module M . In particular, M_C is finitely generated and Lemma I.2.4 yields the result.

I.3. UNISERIAL RINGS ARE BALANCED

The purpose of this section is to offer a brief proof of the statement in the title.

LEMMA I.3.1. *If* R *is a local uniserial ring, then* $_R R$ *is injective.*

Proof. Let L be a left ideal of R . Consider an R-homomorphism $\varphi : L \to {}_R R$. If $L = Rx$, then $\partial(Rx\varphi) \leq \partial(Rx)$, and thus $x\varphi$ belongs to $Rx = xR$ consequently, we can find $r \in R$ such that $x\varphi = xr$. Then, the right multiplication by r gives a homomorphism $_R R \to {}_R R$ which extends φ .

LEMMA I.3.2. *Any direct sum of left balanced rings is left balanced.*

Proof. Assume $R = \bigoplus_{i=1}^{n} R_i$, where all R_i's are left balanced rings. Every R-module M has a unique decomposition as a direct sum of submodules M_i , where M_i can be considered as an R_i-module. The centralizer $\mathcal{C}(M)$ of M is the direct sum of the rings $\mathcal{C}(M_i)$, and because R_i map surjectively onto the double centralizers $\mathcal{D}(M_i)$, R maps surjectively onto $\mathcal{D}(M)$ (which is the direct sum of the $\mathcal{D}(M_i)$'s), as well.

PROPOSITION I.3.3. (C. J. NESBITT AND R. M. THRALL [2]) *Every uniserial ring is both left and right balanced.*

Proof. Because of Lemma I.3.2, it is sufficient to consider the case, where $\underline{\underline{R}}$ is a full matrix ring over a local uniserial ring R . If L is a principal indecomposable left ideal of $\underline{\underline{R}}$, then L is the image of $_R R$ under category isomorphism from Mod R to Mod $\underline{\underline{R}}$. Therefore, Lemma I.3.1 implies that L is an injective $\underline{\underline{R}}$-module.

Assume that M is a faithful $\underline{\underline{R}}$-module. Since L is a generator and a uniserial module, there exists a monomorphism $L \to M$; furthermore, since L is injective, we have $M \cong L \oplus K$ for some $\underline{\underline{R}}$-module K . Now, $L \oplus K$ is also a generator, and thus Lemma I.2.3 yields that M is balanced.

Finally, an arbitrary **R**-module can be considered as a module over some factor ring of **R** , which again is a full matrix ring over a local uniserial ring. Hence, **R** is left balanced. And, similarly **R** is right balanced. The proof is completed.

II. LOCAL RINGS

II.1. A NECESSARY LENGTH CONDITION

In this section, we shall get the first information on the structure of the local left balanced rings.

PROPOSITION II.1.1. *Let* R *be a local left finitely balanced ring with the radical* W . *Then, for each natural* n ,

$$\partial_R(W^n/W^{n+1}) \leq 2 .$$

Proof. Obviously, without loss of generality, we can suppose that $W^{n+1} = 0$ and $W^n \neq 0$. Observe that ${}_R(W^n)$ is then completely reducible.

First, assume that W^n contains a minimal left ideal U which is not a two-sided ideal. Then

$$T = \{\tau \mid \tau \in R \text{ and } U\tau \subseteq U\}$$

is a proper subring of R . It is easy to see that T is again a local ring with the radical W . Therefore, Q = R/W can be viewed as a right vector space over T/W and we have

$$\dim Q_{(T/W)} \geq 2 \ ;$$

let $1 + W$ and $r + W$ be linearly independent elements of $Q_{(T/W)}$.

Now, consider the monogenic left R-module $X = R/U$; X is obviously faithful. For every $x \in R$, write $\bar{x} = x + U \in X$. The elements of the centralizer C of X can be lifted to endomorphisms of ${}_R R$ (that is to say, to right multiplications by elements of R) and, in this way, we get just those elements $\tau \in R$ which satisfy $U\tau \subseteq U$. Thus, the ring C is isomorphic to T/U and its radical W corresponds to W/U in this isomorphism. Consequently the C-module $Q = R/W \cong X/(W/U)$ has the same structure as $Q_{(T/W)}$; let $\epsilon : X_C \to Q_C$ be the canonic epimorphism. Also, observe that, given an arbitrary element $z \in W^n$, $\bar{z}$ belongs to $\mathrm{Soc}(X_C)$; denote by ι the inclusion of $\mathrm{Soc}(X_C)$ into X_C . Now, consider a C-endomorphism

$$\psi : X_C \xrightarrow{\epsilon} Q_C \xrightarrow{\psi'} \mathrm{Soc}(X_C) \xrightarrow{\iota} X_C$$

of X_C such that

$$(1 + W)\psi' = \bar{0} \quad \text{and} \quad (r + W)\psi' = \bar{z} \ .$$

Since X is balanced, ψ is induced by the ring multiplication, say by an element $\rho \in R$:

$$\psi(\bar{x}) = \rho\bar{x} \ \text{ for all } \ \bar{x} \in X \ .$$

Then, $\rho \cdot \bar{1} = \bar{0}$ implies $\rho \in U$ and $\rho\bar{r} = \bar{z}$ implies $z \in \rho r + U \subseteq$ $\subseteq U + Ur$. Hence,

$$W^n \subseteq U + Ur \ ,$$

as required.

In order to complete the proof of Proposition II.1.1, we are going to show that if all left ideals of R contained in W^n are two-sided, then $\partial_R(W^n) = 1$. For, assume the contrary and let Ru and Rv be non-zero two-sided ideals of R such that

$$Ru \subseteq W^n, \ Rv \subseteq W^n \ \text{ and } \ Ru \cap Rv = 0 .$$

Consider the finitely generated faithful left R-module

$$Y = {}_R(R \oplus R)/D \quad \text{with} \quad D = R(u, v) .$$

Every endomorphism of Y can be lifted to an endomorphism of ${}_R(R \oplus R)$ and, in this way, we get just those endomorphisms of the left R-module ${}_R(R \oplus R)$ which map D into D. Let $\begin{matrix} \alpha_{11} & \alpha_{12} \\ \alpha_{21} & \alpha_{22} \end{matrix}$ be the matrix representation of such an endomorphism of ${}_R(R \oplus R)$; here α_{ij} denote endomorphisms of ${}_RR$, that is to say, right multiplications by elements of R. For $(u, v) \in D$, we get

$$(u, v) \begin{matrix} \alpha_{11} & \alpha_{12} \\ \alpha_{21} & \alpha_{22} \end{matrix} = (u\alpha_{11} + v\alpha_{21}, u\alpha_{12} + v\alpha_{22}) = (\lambda u, \lambda v)$$

for a suitable $\lambda \in R$. This yields that both α_{12} and α_{21} belong to W. For, if $\alpha_{12} \notin W$, then

$$u = (\lambda v - v\alpha_{22})\alpha_{12}^{-1} \in Rv$$

and, similarly, if $\alpha_{21} \notin W$, then

$$v = (\lambda u - u\alpha_{11})\alpha_{21}^{-1} \in Ru ,$$

a contradiction in either case.

Now, take a non-zero element $z \in W^n$ and define an additive homomorphism $\phi : R \oplus R \to R \oplus R$ by

$$\psi(r_1, r_2) = (zr_1, 0) \quad \text{for all} \quad (r_1, r_2) \in R \oplus R .$$

Evidently, ψ is a non-zero morphism, $\psi(D) = 0$ and

$$[\psi(r_1, r_2)] \begin{matrix} \alpha_{11} & \alpha_{12} \\ \alpha_{21} & \alpha_{22} \end{matrix} = (zr_1\alpha_{11}, zr_1\alpha_{12}) = (zr_1\alpha_{11}, 0) =$$

$$= (zr_1\alpha_{11} + zr_2\alpha_{21}, 0) = \psi\left[(r_1, r_2)\begin{pmatrix} \alpha_{11} & \alpha_{12} \\ \alpha_{21} & \alpha_{22} \end{pmatrix}\right],$$

because both $zr_1\alpha_{12}$ and $zr_2\alpha_{21}$ belong to $W^n W = 0$. Thus, ψ induces an element ψ^* of the double centralizer of Y. Since Y is balanced, ψ^* is induced by an element $\sigma \in R$; moreover, since ψ^* is non-zero, $\sigma \neq 0$. However, this results in an immediate contradiction; for,

$$\psi^*[(0, 1) + D] = D = \sigma[(0, 1) + D]$$

implies that $(0, \sigma) \in D$, and thus $\sigma = 0$.

The proof of Proposition II.1.1 is completed.

II.2. RINGS WITH $W^2 = 0$

Let R be a local ring with the radical W such that $W^2 = 0$. Denote the skew field R/W by Q, and considering W as a left or right vector space over Q write ${}_QW$ or W_Q, respectively. In this section, we are going to show that every such left balanced ring is either uniserial or satisfies the dimension relation

$$\dim {}_QW \times \dim W_Q = 2$$

together with some conditions binding the elements of W ; these are formulated in Propositions II.2.3 and II.2.4. In the next section, the latter rings will be termed exceptional.

LEMMA II.2.1. *Let* M *be a balanced indecomposable* R-*module of finite length and* m *an element of* M *such that* $\mathrm{Ann}(m) = 0$. *Then, denoting by* C *the centralizer of* M , $mC = M$.

Proof. Let $\mathcal{W}$ be the radical of C . Since M is of finite length, $\mathcal{W}$ is nilpotent and M_C has a non-trivial socle $\mathrm{Soc}\, M_C$ and a non-trivial radical $M\mathcal{W}$. Now, $M/(mC + M\mathcal{W})$ is a completely reducible right C-module; therefore, if we show that any C-homomorphism ψ of the form

$$M_C \xrightarrow{\epsilon} M/(mC + M\mathcal{W}) \rightarrow \mathrm{Soc}\, M_C \xrightarrow{\iota} M_C$$

(where ϵ is the canonical epimorphism and ι the embedding) is trivial, then we have

$$M = mC + M\mathcal{W} \quad .$$

But M is balanced, so $\psi x = \rho x$ for some $\rho \in R$ and any $x \in M$. It follows from $\rho m = \psi m = 0$ and $\mathrm{Ann}\,(m) = 0$, that $\rho = 0$; therefore ψ is trivial. The equality $M = mC + M\mathcal{W}$ yields $M = mC + (mC + M\mathcal{W})\mathcal{W} = mC + M\mathcal{W}^2$, and by induction,

$$M = mC + M\mathcal{W}^n \quad .$$

Since $\mathcal{W}$ is nilpotent, $M = mC$, as required.

LEMMA II.2.2. *Let* R *be a local left artinian ring,* W *its radical and* $Q = R/W$ *; let* $W^2 = 0$ *. Let* $\{w_i\}_{i=1}^{n}$ *be linearly independent elements of the vector space* W_Q *. Then*

$$I_n = R^{(n)}/D \quad with \quad D = R(w_1, \ldots, w_n) ,$$

where $R^{(n)}$ *is a direct sum of* n *copies of* ${}_RR$ *, is an indecomposable* R*-module of finite length. If, moreover,* I_n *is balanced, then*

$$W^{(n)}/D \subseteq R[1, 0, \ldots, 0) + D] .$$

Proof. First, there is no homomorphism of I_n onto ${}_RR$. For assuming the converse, we get a homomorphism

$$R^{(n)} \xrightarrow{\begin{pmatrix} r_1 \\ r_2 \\ \vdots \\ r_n \end{pmatrix}} {}_RR$$

such that D is mapped into 0 . Thus $\sum_{i=1}^{n} w_i r_i = 0$ and, in view of our hypothesis, all $r_i \in W$ and therefore the homomorphism cannot be surjective. In order to show that I_n is indecomposable, assume that $I_n = A \oplus B$. Then, both A and B are finitely generated and $A/\mathrm{Rad}\,A \oplus B/\mathrm{Rad}\,B$ is an n-dimensional vector space over Q . The well-known fact that elements of $\mathrm{Rad}\,A$ and $\mathrm{Rad}\,B$ are non-generators implies that

$$A = \sum_{i=1}^{p} Ra_i \quad \text{and} \quad B = \sum_{j=1}^{q} Rb_j$$

for some elements a_i and b_j, where $p + q = n$. Therefore, applying a length argument, either A or B is a direct sum of copies of ${}_RR$, and consequently either $A = 0$ or $B = 0$, because there is no homomorphism of I_n onto ${}_RR$.

To prove the second assertion we may assume $n \geq 2$. We lift every element $\varphi \in C$ to an endomorphism of R^n and write it as a matrix (α_{ij}) where the α_{ij} are endomorphisms of ${}_RR$. First we show, that for an element φ of the radical $\mathcal{W}$ of C , all α_{ij} belong to W . Denote by e_i the element $(0, \ldots, 0, 1, 0, \ldots, 0) + D$ with 1 at the i-th position. Then $e_i\varphi = (\alpha_{i1}, \ldots, \alpha_{in}) + D$. If $\alpha_{ij} \notin W$ for some (i, j) , then $e_i\varphi$ together with all the elements e_k, $k \neq j$, generates I_n . If $\mathrm{Ann}(e_i\varphi) \neq 0$, then a length argument shows that $R(e_i\varphi)$ is a direct summand, contradicting the fact that I_n is indecomposable. So $\mathrm{Ann}(e_i\varphi) = 0$. An application of Lemma II.2.1 leads to the equality $e_i\varphi C = I_n$, and, a fortiori, $I_n\mathcal{W} = I_n$. Since this is impossible, we conclude that for $\varphi \in \mathcal{W}$, all elements $\alpha_{ij} \in W$. From this it follows that $(W^{(n)}/D)\varphi = 0$ for all $\varphi \in \mathcal{W}$; thus W^n/D is contained in the socle of I_C , where we abbreviate I_n by I . Also, according to Lemma II.2.1, $(1, 0, \ldots, 0) + D$ does not belong to $I\mathcal{W}$. So for any $x \in W^n/D$, we can find a C-homomorphism ψ of the form

$$I_C \xrightarrow{\epsilon} I/I\mathcal{W} \rightarrow \mathrm{Soc}\, I_C \xrightarrow{\iota} I_C$$

(ϵ the canonic epimorphism, ι the inclusion), mapping $(1, 0, \ldots, 0) + D$ onto x . But ψ is induced by left multiplication and thus there is $\rho \in R$ with $x = \rho[(1, 0, \ldots, 0) + D]$. This proves the second part of Lemma II.2.2.

Now, in order to facilitate formulations of the following Propositions II.2.3 and II.2.4, let us define, for a given ring R and an element $v \in R$, the following two subrings of R :

$$T_v = \{\tau \mid \tau \in R \text{ and } v\tau \in Rv\} \text{ and } S_v = \{\sigma \mid \sigma \in R \text{ and } \sigma v \in vR\} .$$

PROPOSITION II.2.3. *Let* R *be a local ring and* W *its radical such that*

$$W^2 = 0 \quad \text{and} \quad \dim_Q W = 2 .$$

If R *is left finitely balanced, then* $\dim W_Q = 1$ *and, for any two linearly independent elements* u *and* v *of* ${}_Q W$,

$$W = Rv + uT_v .$$

Proof. First, we shall prove that $\dim W_Q = 1$. Assume the contrary and choose $0 \neq w_1 \in W$. Since the set-theoretical union of Rw_1 and $w_1 R$ is a proper subset of W , there is an element $w_2 \in W$ which is neither in Rw_1 nor in $w_1 R$. Consider the indecomposable R-module I_2 of Lemma II.2.2. Since I_2 is balanced ${}_R(W \oplus W)/D \subseteq R[(1, 0) + D]$. Therefore, taking $(0, w_1) + D$, there is $r_o \in R$ such that $(-r_o, w_1) \in D$. Thus, in particular, $w_1 = \lambda w_2$ for some $\lambda \in R$. But λ is necessarily a unit and thus $w_2 = \lambda^{-1} w_1$, in contradiction to $w_2 \notin Rw_1$.

Now, to complete the proof, take two linearly independent elements u and v of ${}_Q W$ and verify that $W = Rv + uT_v$. To this end, consider the R-module $N = R/Rv$; let C be its centralizer. Obviously, the rings C and T_v/Rv are isomorphic, and thus $(Rv + uT_v)/Rv$ is a non-zero C-submodule of N . Therefore, there is a non-zero C-homomorphism $\psi : N \to N$ mapping N into $(Rv + uT_v)/Rv$. Since N is a balanced R-module, ψ is induced by the ring multiplication:

$$\psi n = \rho n \quad \text{for all} \quad n \in N \quad \text{with a suitable non-zero} \quad \rho \in R .$$

Consequently, $\rho R \subseteq Rv + uT_v$ and, since $\dim W_Q = 1$, $W = Rv + uT_v$, as required.

PROPOSITION II.2.4. *Let* R *be a local ring and* W *its radical such that*

$$W^2 = 0 \quad and \quad \dim_Q W = 1 .$$

If R *is left finitely balanced, then either* $\dim W_Q = 1$ *or* $\dim W_Q = 2$ *and, for any two linearly independent elements* u *and* v *of* W_Q ,

$$W = vR + S_v u .$$

Proof. First, we are going to show that $\dim W_Q \leq 2$. Assuming the contrary, choose three linearly independent elements w_1, w_2, w_3 in W_Q , and consider the indecomposable R-module I_3 of Lemma II.2.2. Since I_3 is balanced, ${}_R(W \oplus W \oplus W)/D \subseteq R[(1, 0, 0) + D]$. Taking $(0, w_1, 0) + D$, we get $(r_o, 0, 0) + D = (0, w_1, 0) + D$ for a suitable $r_o \in R$, and thus $(-r_o, w_1, 0) \in D$. This is impossible, and therefore $\dim W_Q \leq 2$.

Now, in order to complete the proof, assume that $\dim W_Q = 2$ and take two linearly independent elements u and v of W_Q . Writing $u = w_1$ and $v = w_2$ and denoting by C the centralizer of the indecomposable R-module I_2 of Lemma II.2.2, we have $I_2 = [(1, 0) + D]C$; this follows from Lemma II.2.1, because $\mathrm{Ann}\,[(1, 0) + D] = 0$. Therefore, taking an arbitrary $r \in R$, there is $\varphi \in C$ such that

$$[(1, 0) + D]\varphi = (r, 0) + D .$$

Lifting φ to

$${}_R(R \oplus R) \xrightarrow{\begin{pmatrix}\alpha_{11} & \alpha_{12}\\ \alpha_{21} & \alpha_{22}\end{pmatrix}} {}_R(R \oplus R) ,$$

we get that $(\alpha_{11}-r, \alpha_{12}) \in D$ and thus, in particular, $\alpha_{11}-r \in W$ and $\alpha_{12} \in W$. Also, applying this homomorphism to $(u, v) \in D$, we obtain

$$(u\alpha_{11} + v\alpha_{21}, u\alpha_{12} + v\alpha_{22}) = (ur + v\alpha_{21}, v\alpha_{22}) \in D .$$

Hence,

$$ur + v\alpha_{21} = \lambda u \quad \text{and} \quad v\alpha_{22} = \lambda v \quad \text{for some} \quad \lambda \in R .$$

Therefore $\lambda \in S_v$ and $wr \in vR + S_v u$. Consequently

$$W = uR + vR \subseteq vR + S_v u ,$$

as required.

II.3. EXCEPTIONAL RINGS

Let R be a local ring with the radical W such that $W^2 = 0$. If v is a non-zero element of W, then the subrings

$$T_v = \{\tau | \tau \in R \text{ and } v\tau \in Rv\} \quad \text{and} \quad S_v = \{\sigma | \sigma \in R \text{ and } \sigma v \in vR\}$$

contained obviously W. Moreover, if τ is a unit belonging to T_v, then

$$v\tau = rv \quad \text{for a suitable} \quad r \in R$$

implies $r^{-1}v = v\tau^{-1}$ and thus $\tau^{-1} \in T_v$, as well. Consequently, T_v/W is a division subring of $Q = R/W$. If ρ is a unit of R, then

$$T_{v\rho} = \{\tau | \tau \in R \text{ and } v\rho\tau \in Rv\rho\} = \{\tau | \tau \in R \text{ and } v\rho\tau\rho^{-1} \in Rv\} = \rho^{-1}T_v\rho$$

and thus, in particular,

$$\dim Q_{(T_v/W)} = \dim Q_{(T_{v\rho}/W)} .$$

In a similar fashion, S_v/W is a division subring of Q and, for an arbitrary unit λ of R,

$$\dim_{(S_v/W)} Q = \dim_{(S_{\lambda v}/W)} Q .$$

DEFINITION II.3.1. *A local ring* R *with the radical* W *is said to be exceptional if*

$$W^2 = 0, \dim_Q W \times \dim W_Q = 2$$

and, if $\dim_Q W = 2$, *then*

$$\dim Q_{(T_v/W)} = 2$$

whereas, if $\dim_Q W = 1$, *then*

$$\dim_{(S_v/W)} Q = 2$$

for a non-zero element $v \in W$.

Let us point out the fact that the notion of an exceptional ring is self-dual : *A ring* R *is exceptional if and only if the opposite ring* R* *is exceptional.*

PROPOSITION II.3.2. *Let* R *be a local ring, and* W *its radical with*

$$W^2 = 0, \dim_Q W = 2 \text{ and } \dim W_Q = 1 .$$

Then the following statements are equivalent:

(i) R *is exceptional;*

(ii) *there exist two linearly independent elements* u, v *of* ${}_QW$ *such that*

$$W = Rv + uT_v \ ;$$

(iii) *the indecomposable injective left* R*-module is of length* 2.

Proof. In order to prove the implication (i) → (ii) , let v be a non-zero element of W , $T = T_v$ and let $\dim Q_{(T/W)} = 2$. Thus, there exists $r \in R \backslash S$ such that $R = T + rT + W$. Taking $u = vr$, one gets

$$W = vR = v(T + rT + W) = vT + uT = Rv + uT \ ,$$

because $vT = Rv$ in view of $\dim W_Q = 1$.

Conversely, if

$$vR = W = Rv + uT = vT + uT \ ,$$

then $vR = v(T + rT)$ for a suitable $r \in R$ and hence

$$R = T + rT + W \ .$$

Now, $r \notin T$; for, otherwise $W = vR = vT = Rv$ in contradiction to $\dim {}_QW = 2$. As a consequence, $\dim Q_{(T/W)} = 2$, and we get the equivalence of (i) and (ii) .

In order to show that (ii) implies (iii) , let u, v be the elements given in (ii) . We are going to prove that R/Rv is an injective R-module. To this end, let $\varphi : {}_RW \to R/Rv$ be a non-zero homomorphism. Since right multiplication by elements of R is transitive on W , we can evidently assume that $\operatorname{Ker}\varphi = Rv$. Thus, φ is determined by the conditions $v\varphi = 0$ and $u\varphi = w + Rv$ for a suitable $w \in W$. In view of the relation $W = Rv + uT_v$, we have

$$w = rv + u\tau \quad \text{for some} \quad r \in R \quad \text{and} \quad \tau \in T_v .$$

Consequently, the homomorphism

$$_RR \xrightarrow{\tau} {_RR} \xrightarrow{\epsilon} R/Rv$$

maps Rv into 0 , u into $w - rv + Rv = w + Rv$, and is thus a required extension of φ to $_RR$.

To complete the proof, let us verify the implication (iii) → (ii) . Let M be an indecomposable injective left R-module of length 2; hence $M \cong R/Rv$ for some non-zero element $v \in W$. Let $u \in {_QW}$ so that u and v are linearly independent. Take an arbitrary element $w \in W$ and consider the homomorphism $\varphi : {_RW} \to M$ such that

$$u\varphi = w + Rv \quad \text{and} \quad v\varphi = 0 .$$

Since M is injective, φ can be extended to a homomorphism from $_RR$ to M , and therefore lifted to $_RR \xrightarrow{\tau} {_RR}$. From here, it follows that $v\tau \in Rv$, and thus $\tau \in T_v$; moreover, $w - u\tau \in Rv$. Consequently,

$$W = Rv + uT_v ,$$

completing the proof of Proposition II.3.2.

In a similar way, we can formulate

PROPOSITION II.3.3. *Let* R *be a local ring and* W *its radical with*

$$W^2 = 0, \ \dim {}_Q W = 1 \ \text{and} \ \dim W_Q = 2 \ .$$

Then the following statements are equivalent:

(i) R *is exceptional;*

(ii) *there exist two linearly independent elements* u, v *of* W_Q *such that*

$$W = vR + S_v u \ ;$$

(iii) *the indecomposable injective left* R*-module is of length* 3.

Proof. Both statements (i) and (ii) are dual to those of Proposition II.3.2. and thus they are equivalent. In order to establish that (ii) implies (iii), we are going to show that the indecomposable R-module (cf. Lemma II.2.2)

$$I = {}_R(R \oplus R)/D, \text{ where } D = R(u, v)$$

is injective. Thus, assume that a homomorphism $\varphi : {}_R W \to I$ is given and we are required to extend it to a homomorphism from ${}_R R$ to I. Obviously, φ is determined by the image of v :

$$v\varphi = (v_1, v_2) + D \ \text{ for some } \ v_1, v_2 \ \text{ of } \ W \ .$$

But $v_2 = \lambda v$ and thus, for some $w \in W$,

$$(v_1, v_2) + D = (v_1 - \lambda u, 0) + (\lambda u, \lambda v) + D = (w, 0) + D \ .$$

Now $w \in vR + S_v u$, and therefore there are elements $r_1 \in R$, $\sigma \in S_w$ and $r_2 \in R$ such that

$$w = vr_1 + \sigma u \quad \text{and} \quad \sigma v = -vr_2 .$$

We claim that the homomorphism

$$_R R \xrightarrow{(r_1,\, r_2)} {}_R(R \oplus R) \xrightarrow{\epsilon} I ,$$

where ϵ is the canonical epimorphism, is an extension of φ. Indeed, the element v is mapped into

$$v(r_1, r_2) + D = (w - \sigma u, -\sigma v) + D = (w, 0) + D = v\varphi ,$$

as required. Consequently, I is injective and, being of length 3, necessarily indecomposable.

To complete the proof, let us verify the implications (iii) $\rightarrow$ (ii). An indecomposable injective left R-module I of length 3 is necessarily an amalgam of two copies of $_R R$ over its socle. Thus,

$$I = {}_R(R \oplus R)/D \quad \text{with} \quad D = R(u, v)$$

for suitable u and v of W. Now, take an arbitrary $w \in W$ and consider the homomorphism $\varphi : {}_R W \rightarrow I$ mapping v into $(w, 0) + D$. Extend φ to a homomorphism from $_R R$ to I and lift the latter to

$$_R R \xrightarrow{(r_1,\, r_2)} {}_R(R \oplus R) .$$

Hence,

$$(vr_1, vr_2) - (w, 0) \in D ,$$

and thus

$$(vr_1 - w, vr_2) = (\sigma u, \sigma v) \quad \text{for some} \quad \sigma \in R .$$

Therefore,

$$\sigma \in S_v \quad \text{and} \quad w = vr_1 - \sigma u \in vR + S_v u ,$$

as required.

II.4. EXCEPTIONAL RINGS ARE BALANCED

First, in order to prove that every R-module over an exceptional ring R with $\dim_Q W = 2$ is a direct sum of indecomposable R-modules, we formulate the following two technical lemmas.

LEMMA II.4.1. *Let* R *be a local ring with the radical* W *such that* $W^2 = 0$ *and* $\dim W_Q = 1$. *Let* F *be a free left* R*-module and* $s \neq 0$ *an element of the socle of* F . *Then* s *belongs to a monogenic submodule which is isomorphic to* ${}_R R$.

Proof. The elements of F can be represented by indexed families (r_i) with $r_i \in R$ and the restriction that all but a finite number of the r_i's to be zero. An element (r_i) belongs to the socle $\operatorname{Soc} F$ of F is and only if $r_i \in W$ for all i . Let

$$s = (w_i) \in \operatorname{Soc} F .$$

Let $u \neq 0$ be a fixed element of W . Since $uR = W$, there exists $\rho_i \in R$ such that $w_i = u\rho_i$; here, we take $\rho_i = 0$ if $w_i = 0$. Now, right multiplication by ρ_i yields a homomorphism $\rho_i : {}_R R \to {}_R R$, and thus the family (ρ_i)

defines a homomorphism

$$\varphi : {}_R R \to F .$$

Clearly, $u\varphi = s$, and hence $s \in \operatorname{Im}\varphi$. Furthermore, since $s \neq 0$, there is a unit ρ_{i_o} such that $w_{i_o} = u\rho_{i_o}$; as a consequence, $\operatorname{Im}\varphi \cong {}_R R$.

LEMMA II.4.2. *Let* R *be a local ring with the radical* W *such that* $W^2 = 0$, $\dim_Q W = 2$ *and* $\dim W_Q = 1$. *Let* M *be an* R*-module with submodules* X *and* Y *isomorphic to* ${}_R R$ *such that* $X + Y = M$ *and* $X \cap Y$ *is a minimal submodule. Then* M *contains an indecomposable submodule of length* 2.

Proof. M is obviously isomorphic to the pushout P of the following diagram

$$\begin{array}{ccc} {}_R L & \xrightarrow{\ \mu\ } & {}_R R \\ \iota \big\downarrow & & \big\downarrow \iota' \\ {}_R R & \longrightarrow & P \end{array} ,$$

where L is a minimal left ideal of R , ι the inclusion mapping and μ a monomorphism. If $x \neq 0$ is an element of L , then

$$x\mu = (x\iota)\rho \quad \text{for some} \quad \rho \in R ,$$

because $xR = W$. Thus right multiplication by ρ is a mapping from ${}_R R$ into ${}_R R$ satisfying $\iota\rho = \mu$. But this implies, in view of the properties of a pushout, that ι' splits and that the complement is just the cokernel R/L of ι which is obviously of length 2.

Now, we are ready to prove the following

PROPOSITION II.4.3. *Let* R *be a local ring with* $W^2 = 0$, $\dim_Q W = 2$ *and* $\dim W_Q = 1$. *If the indecomposable injective* R*-module is of length* 2, *then every indecomposable* R*-module is either simple, injective or isomorphic to* ${}_R R$, *and any* R*-module is a direct sum of these indecomposable modules.*

Proof. To prove our proposition, we shall show that every R-module can be expressed as a direct sum of modules of the isomorphism types A_1, A_2, A_3 represented by the R-modules ${}_R(R/W)$, ${}_R(R/Ru)$ with a non-zero $u \in W$ and ${}_R R$, respectively. Here, A_2 is the injective indecomposable type.

Let M be a left R-module. Take a submodule X of M which is maximal with respect to the property of being a direct sum of modules of type A_2. Since X is injective, $M = X \oplus M'$, where M' is a submodule of M which contains no submodules of type A_2.

Now, let Y be a submodule of M' which is maximal with respect to the property of being a direct sum of modules of type A_3. Let Z be a complement of $\operatorname{Soc} Y$ in $\operatorname{Soc} M'$. Then, Z is a direct sum of modules of type A_1 and, evidently, $Y \cap Z = 0$. We want to show that

$$Y \oplus Z = M' .$$

To this end, assume that there is an element $m \in M' \setminus (Y \oplus Z)$. Then Rm must be of type A_3, because $m \notin \operatorname{Soc} M'$ and M' contains no submodule of type A_2. The submodule $Y \cap Rm$ is non-zero; for, otherwise $Y + Rm$ would be a direct sum of modules of type A_3, contradicting the maximality of Y. Take $s \neq 0$ of $Y \cap Rm$. Since $s \in \operatorname{Soc} Y$, Lemma II.4.1 implies that there is a submodule $N \subseteq Y$ of type A_3 containing s. In view of Lemma II.4.2,

$N \cap Rm$ cannot be simple and therefore the length of $N \cap Rm$ is 2 .

If we now assume that Soc $(N + Rm)$ is of length 2, then $N + Rm$ is isomorphic to the injective hull of Soc $(N + Rm)$ (because both modules are of length 4). However, since M' has no submodules of type A_2 , this is impossible. Thus, Soc $(N + Rm)$ has to be of length 3, and therefore

$$N + Rm = N + \mathrm{Soc}\,(N + Rm) \quad .$$

But this means that

$$Rm \subseteq Y + \mathrm{Soc}\,M' \subseteq Y \oplus Z \ ,$$

and we get a contradiction to our hypothesis. The proof is completed.

In analogy to the preceeding result, we shall prove also that every R-module over an exceptional ring R with $\dim_Q W = 1$ is a direct sum of indecomposable R-modules of the types B_1, B_2 and B_3 represented by the R-modules ${}_R(R/W)$, ${}_RR$ and the injective module I_2 of Lemma II.2.2. Here again, the index of B_i refers to the length of the respective module. Note however that, contrary to the previous situation, B_3 is not a monogenic module.

First, let us prove by induction the following

LEMMA II.4.4. *Let* R *be a local ring with the radical* W *such that* $W^2 = 0$, $\dim_Q W = 1$ *and* $\dim W_Q = 2$.

(a) *Let* M *be an* R*-module of length* $2n + 1$ *generated by* $n + 1$ *monogenic submodules. Let* N *be a submodule of* M *which is a direct sum of* n *copies of modules of type* B_2 . *If, furthermore,* M *does not contain a submodule of type* B_3 *, then*

$$M = N + \mathrm{Soc}\,M \ .$$

(b) *The only indecomposable* R*-modules of length* $\leq 2n + 1$ *are modules of type* B_1, B_2 *and* B_3 .

Proof. If the length of M is 3 , and if M contains a monogenic submodule N of length 2 , then either $\operatorname{Soc} M$ is simple in which case the injectivity of B_3 yields that M is of type B_3 , or $\operatorname{Soc} M$ is of length ≥ 2 ; in the latter case, evidently

$$M = N + \operatorname{Soc} M \ .$$

This establishes the validity of both (a) and (b) for $n = 1$.

Now, assume that both assertions hold for all $m \leq n - 1$.

(a) Without loss of generality, we may assume that the $n + 1$ monogenic submodules which generate M are all of length 2. We can consider M as the amalgamation of N with a monogenic module of length 2 with simple submodules identified. Thus, M is isomorphic to the pushout P of the following diagram

$$\begin{array}{ccc} {}_RW & \xrightarrow{\ \ \eta\ \ } & {}_RR \oplus {}_RR \oplus \cdots \oplus {}_RR \\ \iota\downarrow & & \downarrow\iota' \\ {}_RR & \xrightarrow{\ \ \eta'\ \ } & P \end{array}$$

where ι is the inclusion of W in R , η is a monomorphism and ι' corresponds to the inclusion $N \subseteq M$. Let us take a non-zero element $w \in W$; hence, $w\eta$ is of the form $(x_1, x_2, \ldots, x_n)$ with at least one non-zero x_i . Assume that $x_1 \neq 0$ and distinguish three cases:

(i) Let $x_i \in wR$ for all $1 \leq i \leq n$. Then we can find elements σ_i such that $x_i = w\sigma_i$, and thus the morphism

$$(\sigma_1, \sigma_2, \ldots, \sigma_n) : {}_RR \to {}_RR \oplus {}_RR \oplus \ldots \oplus {}_RR ,$$

representing right multiplication, maps w into $(x_1, x_2, \ldots, x_n) = w\eta$. But this means that ${}_RR \oplus {}_RR \oplus \ldots \oplus {}_RR$ is a direct summand of P. Consequently, the complement is simple and therefore $M = N + \mathrm{Soc}\, M$.

(ii) Let $x_1 \notin wR$ and $x_i \in x_1R$ for all $1 \leq i \leq n$. Then, we can find elements σ_i with $x_i = x_1\sigma_i$; observe that σ_1 is a unit. Now, both $1\eta'$ and $(\sigma_1, \sigma_2, \ldots, \sigma_n)\iota'$ generate submodules of length 2 and the equality

$$w(1\eta') = w\eta' = w\iota\eta' = w\eta\iota' = (x_1, x_2, \ldots, x_n)\iota' =$$

$$= (x_1\sigma_1, x_1\sigma_2, \ldots, x_1\sigma_n)\iota' = x_1(\sigma_1, \sigma_2, \ldots, \sigma_n)\iota'$$

shows that

$$w\eta' \in R(1\eta') \cap R(\sigma_1, \sigma_2, \ldots, \sigma_n)\iota' .$$

Let $X = R(1\eta') + R(\sigma_1, \sigma_2, \ldots, \sigma_n)\iota'$. Assuming that $R(\sigma_1, \sigma_2, \ldots, \sigma_n)\iota'$ is a direct summand of X, we deduce that there is a morphism $R(1\eta) \to R(\sigma_1, \sigma_2, \ldots, \sigma_n)\iota'$ mapping $w\eta'$ into $(x_1\sigma_1, x_1\sigma_2, \ldots, x_1\sigma_n)\iota'$, and thus a morphism ${}_RR \to {}_RR \oplus {}_RR \oplus \ldots \oplus {}_RR$ mapping w into $(x_1\sigma_1, x_1\sigma_2, \ldots, x_1\sigma_n)$. In particular, there is a morphism ${}_RR \to {}_RR$ mapping w into $x_1\sigma_1 = x_1$ and since such a morphism must be induced by right multiplication we get that $x_1 \in wR$, contradicting our hypothesis. Thus, X has to be an indecomposable R-module of length 3 and therefore of type B_3. Since M has no submodule of type B_3, we conclude that the case (ii) cannot happen.

(iii) Let $x_1 \notin wR$ and there is x_i such that $x_i \notin x_1R$. We may assume that $x_2 \notin x_1R$. Thus, $W = x_1R + x_2R$ and therefore there are elements σ_1, σ_2 such that

$$w = x_1\sigma_1 + x_2\sigma_2 .$$

In this case, the pushout P can be considered as the quotient module of $n+1$ copies of ${}_RR$ by the submodule generated by $(w, -x_1, -x_2, \ldots, -x_n)$. Under the morphism

$$(1, \sigma_1, \sigma_2, 0, \ldots, 0) : {}_RR \oplus {}_RR \oplus \ldots \oplus {}_RR \to {}_RR$$

representing right multiplication, the element $(w, -x_1, -x_2, \ldots, -x_n)$ is mapped into $w - w_1\sigma_1 - x_2\sigma_2 = 0$ and thus the morphism factors through P. As a consequence P has a homomorphic image of type B_2. The latter splits off and we deduce that M is a direct sum of a module of type B_2 and a module M' of length $2n - 1$.

Now, using the induction argument, M' is a direct sum of modules of types B_1, B_2 and B_3. However, since M has no submodules of type B_3, M' is a direct sum of monogenic modules of length 1 and 2. In particular, $\mathrm{Soc}\, M'$ has to be of length at least n and therefore $\mathrm{Soc}\, M$ has to be of length at least $n + 1$. Consequently, $M = N + \mathrm{Soc}\, M$, as required.

The statement (a) is established.

(b) Given an indecomposable R-module M of length $\leq 2n + 1$, we deduce immediately that M has no proper submodule of type B_3; this follows from the fact that B_3 is injective. Now, take a submodule N which is maximal with respect to the property of being a direct sum of copies of modules of type B_2, and let K be a complement of $\mathrm{Soc}\, N$ in $\mathrm{Soc}\, M$. In order to verify (b), it is sufficient to show that $M = N \oplus K$, i.e. to show that every element $x \in M$ generating a submodule of length 2 belongs to $N \oplus K$. Let $M' = N + Rx$. If $x \notin N$, then the length of M' is $2m + 1$, where m is the number of direct summands of type B_2 in N. Since $m \leq n$, we get by induction

$$M' = N + \operatorname{Soc} M' .$$

But this means that $x \in N + K$.

The proof of Lemma II.4.4 is completed.

PROPOSITION II.4.5. *Let* R *be a local ring with* $W^2 = 0$, $\dim_Q W = 1$ *and* $\dim W_Q = 2$. *If the indecomposable injective* R*-module is of length* 3, *then every indecomposable* R*-module is either simple, injective or isomorphic to* ${}_R R$, *and any* R*-module is a direct sum of these indecomposable modules.*

Proof. It is sufficient to show that every R-module M can be expressed as a direct sum of modules of types B_1, B_2 and B_3.

Following the method of proving Proposition II.4.3, we denote by X a submodule of M which is maximal with respect to the property of being a direct sum of modules of type B_3 and observe that $M = X \oplus M'$. In M', take a submodule Y which is a maximal direct sum of modules of type B_2, and denote by Z a complement of $\operatorname{Soc} Y$ in $\operatorname{Soc} M'$. We intend to show that

$$M = X \oplus Y \oplus Z .$$

Assume the contrary, i.e. that there is an element $m \in M' \setminus (Y \oplus Z)$ which generates a submodule of length 2. Clearly, because of maximality of Y, $Y \cap Rm \neq 0$. Thus, there is a direct sum Y' of a finite number of copies of B_2 contained in Y such that

$$Y' \cap Rm \neq 0 .$$

Now, applying Lemma II.4.4 (a) to the module $Y' + Rm$ and the submodule Y' we get readily that

$$Y' + Rm = Y' + \mathrm{Soc}(Y' + Rm) .$$

Consequently, $m \in Y' + \mathrm{Soc}(Y' + Rm) \subseteq Y + \mathrm{Soc}\, M' = Y \oplus A$, a contradiction. Proposition II.4.3 follows.

PROPOSITION II.4.6. *Every exceptional ring is both left and right balanced.*

Proof. Because the opposite ring of an exceptional ring is again exceptional, it is sufficient to prove that exceptional rings are left balanced.

Both in the case when $\dim_Q W = 2$, as well as when $\dim_Q W = 1$, it is easy to verify that all indecomposable modules are balanced. This is trivial for modules of the types A_3 and B_2 and also for modules of the simple types A_1 and B_1; and, it follows for modules of the injective types A_2 and B_3 immediately from Lemma I.2.5.

Now, in view of Propositions II.4.3 and II.4.5, we can apply Lemma I.2.1 and complete the proof.

II.5. STRUCTURE OF LOCAL BALANCED RINGS

LEMMA II.5.1. *Let* R *be a local left finitely balanced ring with the radical* W *such that* $W^3 = 0$. *Let* R *be right uniserial and not left uniserial. Then* R *is exceptional.*

Proof. First, observe that, according to Proposition II.1.1, $\partial_R(W/W^2) = 2$. Thus, in order to prove our Lemma, it is sufficient, in view of Proposition II.2.3 and Definition II.3.1, to show that $W^2 = 0$.

Assume that $W^2 \neq 0$. First, we can see that ${}_RW$ is the direct sum of two monogenic submodules Ru and Rv , where u and v belong to $W \setminus W^2$. This follows from the fact that ${}_RW$ can be considered as a left R/W^2-module and, according to Lemma II.4.3, it is a direct sum of monogenic modules. And, since radical of ${}_RW$ is ${}_R(W^2)$ and since ${}_R(W/W^2)$ is of length 2 , it is a direct sum of two monogenic modules.

Now, since u and v belong to $W \setminus W^2$ and R is right uniserial, there is $\rho \in R$ such that $u\rho = v$. Thus, in particular, $Ru \cong Rv$. Therefore, in view of Proposition II.1.1, $\partial_R(W^2) \leq 2$ and hence, $\partial(Ru) = \partial(Rv) = \partial_R(W^2) = 2$. Write $L = \mathrm{Ann}(u)$. Then ${}_R(R/L) \cong Ru$ and thus $\partial_R L = 3$. But R is right uniserial and therefore

$$LW = L(uR) = 0 ;$$

consequently, L is contained in the right socle W^2 of R . We arrive at a contradiction and conclude that $W^2 = 0$.

LEMMA II.5.2. *Let* R *be a local left finitely balanced ring with the radical* W *such that* $W^3 = 0$. *Let* R *be left uniserial and not right uniserial. Then* R *is exceptional.*

Proof. In view of Proposition II.2.4 and Definition II.3.1, R/W^2 is an exceptional ring and thus it is sufficient to prove that $W^2 = 0$.

Let us give again an indirect proof. Assume that $W^2 \neq 0$ and consider the *right* R/W^2-module W_{R/W^2} . Applying the dual statements of Proposition II.3.3 and Proposition II.4.3 to this *right* R/W^2-module and taking into account the fact that W_{R/W^2} possesses a completely reducible quotient $(W/W^2)_{R/W^2}$ of length 2 , we can easily conclude that there are elements u and v in W with $uR \cap vR = 0$ such that $u + W^2$ and $v + W^2$ are linearly independent in $(W/W^2)_{R/W} = (W/W^2)_Q$.

Now, let us construct two non-isomorphic R-modules M_1 and M_2 of length 4 such that $M_1 \oplus M_2$ is not balanced.

First, consider

$$M_1 = {}_R(R \oplus R)/D_1, \quad \text{where} \quad D_1 = R(u, v) .$$

The R-module M_1 has no monogenic quotient of length 2 . For, given a homomorphism $\varphi : M_1 \to {}_R(R/W^2)$, we can lift it to a homomorphism

$$_R(R \oplus R) \xrightarrow{\binom{r_1}{r_2}} {}_RR .$$

Since D_1 is mapped into W^2 , $ur_1 + vr_2$ belongs necessarily to W^2 . But $u + W^2$ and $v + W^2$ are linearly independent in $(W/W^2)_{R/W}$ and therefore both ur_1 and vr_2 belong to W^2 . Consequently, both r_1 and r_2 lie in W and hence φ cannot be surjective. From here it follows easily that $\operatorname{Soc} M_1$ is simple; for, otherwise $_R(R \oplus W)/D_1 \cong {}_RR$ would be a direct summand.

Secondly, take a non-zero element $w \in vR \cap W^2$ and define the R-module

$$M_2 = {}_R(R \oplus R)/D_2, \quad \text{where} \quad D_2 = R(u, w) .$$

Again, $\operatorname{Soc} M_2$ is simple. For, if $\operatorname{Soc} M_2$ is not simple, then ${}_R(W \oplus R)/D_2 \cong {}_RR$ is a direct summand of M_2 and M_2 possesses an epimorphic image which is a monogenic R-module of length 3. But a homomorphism $\varphi : M_2 \to {}_RR$ can be lifted to a homomorphism

$$ {}_R(R \oplus R) \xrightarrow{\begin{pmatrix} r_1 \\ r_2 \end{pmatrix}} {}_RR $$

mapping D_2 into 0. Therefore, $ur_1 + wr_2 = 0$. This relation shows that both r_1 and r_2 belong to W and thus the homomorphism φ cannot be surjective.

Now, M_1 and M_2 are two non-isomorphic R-modules of length 4. This follows from the fact that M_2 has a monogenic quotient ${}_R(R \oplus R)/(R \oplus W^2)$ of length 2. Consequently, any homomorphism between M_1 and M_2 must have a non-trivial kernel. Since both $\operatorname{Soc} M_1$ and $\operatorname{Soc} M_2$ are simple, such a homomorphism $\varphi_{ij} : M_i \to M_j$ (with $i \neq j$) satisfies $(\operatorname{Soc} M_i)\varphi = 0$. But then the R-module $M = M_1 \oplus M_2$ is not balanced. For, represent the elements of the centralizer of M by the matrices

$$ \begin{pmatrix} \varphi_{11} & \varphi_{12} \\ \varphi_{21} & \varphi_{22} \end{pmatrix}, \quad \text{where} \quad \varphi_{ij} : M_i \to M_j . $$

Take a non-zero element z of $\operatorname{Soc} {}_RR$, and define an additive homomorphism $\psi : M \to M$ by

$$ \psi(m_1, m_2) = (zm_1, 0) \quad \text{for} \quad (m_1, m_2) \quad \text{in} \quad M_1 \oplus M_2 . $$

Now, zm_i belongs to $\operatorname{Soc} M_i$; so, for $i \neq j$, we have

$$ z(m_i\varphi_{ij}) = (zm_i)\varphi_{ij} = 0 . $$

This implies, that ψ belongs to the double centralizer of M , because of

$$[\psi(m_1, m_2)]\begin{pmatrix}\varphi_{11} & \varphi_{12}\\ \varphi_{21} & \varphi_{22}\end{pmatrix} = (zm_1\varphi_{11}, zm_1\varphi_{12}) = (zm_1\varphi_{11} + zm_2\varphi_{21}, 0) =$$

$$= \psi[(m_1, m_2)\begin{pmatrix}\varphi_{11} & \varphi_{12}\\ \varphi_{21} & \varphi_{22}\end{pmatrix}] .$$

Assuming that ψ is induced by left multiplication by $\rho \in R$, the equation $(zm_1, 0) = (\rho m_1, \rho m_2)$ for all $m_i \in M_i$ implies $\rho = 0$, because M_2 is faithful. But $zM_1 \neq 0$, because M_1 is faithful. Hence M is not balanced.

We conclude that $W^2 = 0$ completing the proof of Lemma II.5.2.

Now, we are ready to formulate the following

PROPOSITION II.5.3. *Let* R *be a local ring with the radical* W *such that* $W^n = 0$ *for some natural* n . *Then* R *is left finitely balanced if and only if it is either uniserial or exceptional.*

Proof. Let R be left finitely balanced. Obviously, without loss of generality, we may assume that $W^3 = 0$. Then the conclusion follows immediately from Proposition II.1.1 and Lemmas II.5.1 and II.5.2.

The opposite implication follows from Proposition I.3.3 and Proposition II.4.6.

III. GENERAL THEOREMS

III.1. MORITA EQUIVALENCE

In this section, we shall give a short proof of the fact that the ring property of being balanced is Morita equivalent. We start with two technical lemmas.

LEMMA III.1.1. *Let* P_R *be a finitely generated projective right* R-*module. Let* ${}_RM_C$ *be an* R-C-*bimodule. Then the homomorphism*

$$\alpha : P_R \otimes \mathrm{Hom}_C({}_RM_C, {}_RM_C) \to \mathrm{Hom}_C({}_RM_C, P_R \otimes {}_RM_C)$$

given by

$$[\alpha(p \otimes \varphi)]m = p \otimes (\varphi m) \quad \textit{for} \quad p \in P_R, \ \varphi \in \mathrm{Hom}_C({}_RM_C, {}_RM_C) \quad \textit{and} \quad m \in M,$$

is an isomorphism of right R-*modules.*

Proof. It is easy to see that $\alpha(p \otimes \varphi)$ is a C-homomorphism of M_C into $P_R \otimes {}_RM_C$, and that α is an R-homomorphism. For $P_R = R_R$, the homomorphism α is trivially an isomorphism. It turns out that α is also an isomorphism for a direct summand P_R of a finite direct sum of copies of R_R.

LEMMA III.1.2. *Let* P_R *be a finitely generated projective right* R*-module. Let* ${}_RM$ *be a balanced* R*-module with the centralizer* C. *Then the morphism*

$$\beta : P_R \to \mathrm{Hom}_C({}_RM_C, P_R \otimes {}_RM_C)$$

defined by

$$(\beta p)m = p \otimes m \quad \textit{for} \quad p \in P \quad \textit{and} \quad m \in M ,$$

is an epimorphism of right R*-modules.*

Proof. The morphism β is the composition of

$$P_R \xrightarrow{\beta_1} P_R \otimes {}_RR_R \xrightarrow{\beta_2} P_R \otimes \mathrm{Hom}_C({}_RM_C, {}_RM_C) \xrightarrow{\alpha} \mathrm{Hom}_C({}_RM_C, P_R \otimes {}_RM_C) ,$$

where $\beta_1 p = p \otimes 1$ for $p \in P$, β_2 is induced by the canonic homomorphism ${}_RR_R \to \mathrm{Hom}_C({}_RM_C, {}_RM_C)$ mapping $r \in R$ onto the left multiplication by r, and α is the morphism defined in Lemma III.1.1. In fact, $\beta_2\beta_1 p = p \otimes \iota$, where ι is the identity automorphism of M_C, and thus $(\alpha\beta_2\beta_1 p)m = [\alpha(p \otimes \iota)]m = p \otimes \iota(m) = p \otimes m$. It is well-known that β_1 is an isomorphism, and the fact that ${}_RM$ is balanced means that β_2 is an epimorphism. Consequently, we deduce that β is an epimorphism, as required.

PROPOSITION III.1.3. *Let* P_R *be a finitely generated projective right* R*-module with the centralizer* A. *If an* R*-module* ${}_RM$ *is balanced, then also the* B*-module* ${}_AP_R \otimes {}_RM$ *is balanced.*

Proof. Let C be the centralizer of ${}_RM$. Then C induces endomorphisms of ${}_AP_R \otimes {}_RM$ and, in this way, we get a right C-module $P_R \otimes {}_RM_C$. We shall show that all endomorphisms of $P_R \otimes {}_RM_C$ are induced by left multiplication by the elements of A ; as an immediate consequence, we obtain that the canonic mapping of A into the double centralizer of ${}_AP_R \otimes {}_RM$ is surjective.

Since P_R is projective, the epimorphism β of Lemma III.1.2 induces an epimorphism

$$\beta' : \mathrm{Hom}_R(P_R , P_R) \to \mathrm{Hom}_R(P_R , \mathrm{Hom}_C ({}_RM_C , P_R \otimes {}_RM_C) ,$$

where β' is given by $[(\beta' \lambda)p]m = [\beta(\lambda p)]m = (\lambda p) \otimes m$ for $\lambda \in \mathrm{Hom}_R(P_R , P_R)$, $p \in P$ and $m \in M$. Also, we have the canonic isomorphism

$$\gamma : \mathrm{Hom}_R(P_R , \mathrm{Hom}_C ({}_RM_C, P_R \otimes {}_RM_C)) \to \mathrm{Hom}_C (P_R \otimes {}_RM_C, P_R \otimes {}_RM_C) ,$$

defined by $(\gamma\varphi)(p \otimes m) = (\varphi p)m$ for $\varphi \in \mathrm{Hom}_R(P_R , \mathrm{Hom}_C ({}_RM_C, P_R \otimes {}_RM_C))$, $p \in P$ and $m \in M$. Therefore, under β' and γ, an element $\lambda \in \mathrm{Hom}_R(P_R , P_R)$ is mapped onto the C-endomorphism $\gamma\beta'\lambda$ of $P_R \otimes {}_RM_C$, which, by definition, equals

$$(\gamma\beta'\lambda)(p \otimes m) = [(\beta'\lambda)p]m = (\lambda p) \otimes m .$$

But λ is an element of the centralizer $A = \mathrm{Hom}_R(P_R , P_R)$, and hence $\gamma\beta'\lambda$ is just left multiplication by the element $\lambda \in A$ on $P_R \otimes {}_RM_C$. The fact that $\beta'\lambda$ is surjective shows that every C-endomorphism of $P_R \otimes {}_RM_C$ is induced by an element of A. Consequently, ${}_AP_R \otimes {}_RM$ is balanced.

PROPOSITION III.1.4. (K. MORITA & H. TACHIKAWA [21]). *Let* S *be a category isomorphism from* Mod R *onto* Mod A. *If the* R*-module* M *is balanced, then the* A*-module* $S(M)$ *is balanced.*

In particular, let $\underline{\underline{R}}$ *be a full matrix ring over a ring* R. *Then* $\underline{\underline{R}}$ *is left balanced, or left finitely balanced, if and only if* R *is left balanced, of left finitely balanced, respectively.*

Proof. Indeed, there is a bimodule ${}_AP_R$ such that $S(M) \cong {}_AP_R \otimes {}_RM$. But P_R is finitely generated and projective, A is its endomorphism ring, and hence $S(M)$ is balanced by Proposition III.1.3.

As a consequence, the existence of a category isomorphism between Mod R and Mod A implies that the ring R is left balanced, or left finitely balanced, if and only if A is left balanced, or left finitely balanced, respectively.

III.2. LEFT BALANCED RINGS ARE LEFT ARTINIAN

The main purpose of this section is to prove the statement in the title. In order to facilitate the proof we are going to prove first several auxiliary results.

LEMMA III.2.1. *Let* R *be a left balanced ring. Let* M *be a faithful and* S *a simple* R*-module. Then there exists either an injection* $\iota : S \to M$ *or a surjection* $\varepsilon : M \to S$. *In particular, every module over a left balanced ring possesses either a minimal or a maximal submodule.*

Proof. If there is no non-zero homomorphism between M and S, then the elements of the centralizer of $M \oplus S$ have the form $\begin{pmatrix} \alpha & 0 \\ 0 & \beta \end{pmatrix}$, where $\alpha \in C(M)$ and $\beta \in C(S)$. Hence, the morphism

$$\psi : M \oplus S \to M \oplus S$$

defined by $\psi(m, s) = (0, s)$ is an element of the double centralizer of $M \oplus S$ which is not induced by the ring multiplication. The lemma follows.

PROPOSITION III.2.2. (V. P. CAMILLO [3]). *Let* R *be a left balanced ring with the radical* W . *Then* R/W *is artinian.*

Proof. Without loss of generality, assume that $W = 0$. Consider the direct sum $N = \bigoplus_{\omega \in \Omega} V_\omega$ of all non-isomorphic simple R-modules V_ω . The module N is obviously faithful and thus R is isomorphic to the double centralizer $\mathcal{D}$ of N . It is easy to see that $\mathcal{D}$ is a cartesian product of full endomorphism rings of vector spaces over the centralizers of the V_ω's . In particular, $\mathrm{Soc}\,{}_R R$ is essential in ${}_R R$. To complete the proof it suffices to show that $R = \mathrm{Soc}\,{}_R R$. Assume the contrary. Then there is a maximal left ideal L of R containing $\mathrm{Soc}\,{}_R R$. Since $\mathrm{Soc}\,{}_R R$ is essential in R , the annihilator of the simple R-module $S = R/L$ is essential and therefore S cannot be isomorphic to a submodule of $M = \mathrm{Soc}\,{}_R R$. Since R is left balanced and M is faithful we arrive at a contradiction of Lemma III.2.1.

PROPOSITION III.2.3. (K. R. FULLER [16]). *Let* R *be a left finitely balanced ring with the radical* W . *Then, for every natural* n , *the quotient ring* R/W^n *is a finite direct sum of full matrix rings over local left finitely balanced rings. Moreover, if* R *is left balanced, the local rings are also left balanced.*

Proof Without loss of generality, assume that $W^n = 0$. In view of Lemma III.2.2, we may choose a direct decomposition ${}_R R = \bigoplus_{i=1}^{t} L_i$ of ${}_R R$ into indecomposable left ideals. In order to establish the lemma, we want to show that the simple composition factors of every L_i are isomorphic.

For $n = 2$, this follows from the fact that an extension of a simple R-module S_1 by a non-isomorphic simple R-module S_2 always splits. Indeed, assuming that such an extension M does not split, we can see easily that the centralizer of M is a skew field. However, then M_C is a vector space and C- homomorphisms act on it transitively. Therefore, since M is balanced and S_1 is a proper R-submodule of M, we arrive at a contradiction. If $n > 2$, we proceed by induction. If we assume that every composition factor of L_i/W^nL_i is isomorphic to L_i/WL_i, we may define for any $x \in W^{n-1}L_i \setminus W^nL_i$ a homomorphism

$$\eta_x : L_i/W^2L_i \to L_i/W^{n+1}L_i$$

with $x \in (L_i/W^2L_i)\eta_x$. The fact that the images of all these homomorphisms cover $W^nL_i/W^{n+1}L_i$ implies that also the composition factors of $W^nL_i/W^{n+1}L_i$ are isomorphic to L_i/WL_i. This proves that the composition factors of L_i are isomorphic.

Therefore, R is a finite direct sum of rings $\underline{\underline{R}}_i$ which are full matrix rings over local rings R_i. If R is left balanced, or left finitely balanced, also the rings $\underline{\underline{R}}_i$ are left balanced or left finitely balanced, respectively. For, any $\underline{\underline{R}}_i$-module M_i can be considered as an R-module and is balanced as an R-module if and only if it is balanced as an $\underline{\underline{R}}_i$-module. If $\underline{\underline{R}}_i$ is left balanced or left finitely balanced, then, by Proposition III.1.4, R_i has the respective property, too. This completes the proof.

LEMMA III.2.4. *Let* M *be a module with an endomorphism* φ *such that* $M\varphi \subseteq \operatorname{Rad} M$. *Then the direct limit* X *of the diagram*

$$M \xrightarrow{\varphi} M \xrightarrow{\varphi} M \xrightarrow{\varphi} \cdots$$

has no maximal submodules.

Proof. Let $K = \bigcup_{n \geq 1} \operatorname{Ker} \varphi^n$. Then $K\varphi \subseteq K$, and φ induces an endomorphism $\varphi' : M/K \to M/K$. It is easy to see that φ' is a monomorphism. Moreover, $(M/K)\varphi' \subseteq \operatorname{Rad} M/K$ and X can be considered as the direct limit of the diagram

$$M/K \xrightarrow{\varphi'} M/K \xrightarrow{\varphi'} M/K \xrightarrow{\varphi'} \cdots .$$

This shows that we may assume that φ is a monomorphism. We denote by $\iota_n : M \to X$ the canonic homomorphisms; for these homomorphisms we have commutative diagrams

$$\begin{array}{ccc} M & \xrightarrow{\varphi} & M \\ & {\scriptstyle \iota_n} \searrow \quad \swarrow {\scriptstyle \iota_{n+1}} & \\ & X & \end{array}$$

Let $M_n = M\iota_n$, and $\varphi_n : M_n \to M_{n+1}$ be induced by φ . Then M_n is a submodule of X and X is the union of the M_n 's .

Assume that Y is a maximal submodule of X . Because $Y \neq X$, we find n such that $M_n \nsubseteq Y$. Take $x \in M_n \setminus Y$. Then we have

$$(Y \cap M_{n+1}) + Rx = (Y + Rx) \cap M_{n+1} = X \cap M_{n+1} = M_{n+1} .$$

but $x = (x\varphi_n)$ belongs to $\operatorname{Rad} M_{n+1}$ and therefore $Y \cap M_{n+1} = M_{n+1}$. This implies that $M_n \subseteq M_{n+1} \subseteq Y$, a contradiction.

PROPOSITION III.2.5. (V. P. CAMILLO [3]). *The radical of a left balanced ring is a nil ideal.*

Proof. Let R be a left balanced ring, and W its radical. For $w \in W$, consider the direct limit X of the diagram

$${}_R R \xrightarrow{w} {}_R R \xrightarrow{w} {}_R R \xrightarrow{w} \cdots ,$$

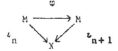

where w denotes the right multiplication. According to Lemma III.2.4, X has no maximal submodule. If we assume that $w^n \neq 0$ for all natural n, then $X \neq 0$, and every non-zero factor module of X has minimal submodules. This follows from Lemma III.2.1. Thus, if we define by transfinite induction a sequence of submodules X_α, with $X_o = 0$, $X_\alpha / X_{\alpha-1} = \mathrm{Soc}\,(X/X_{\alpha-1})$ for non-limit ordinal α and $X_\alpha = \bigcup_{\beta<\alpha} X_\beta$ for a limit ordinal α, then $X = \bigcup X_\alpha$. For each $x \in X_\alpha$, let $h(x)$ be the least ordinal α such that $x \in X_\alpha$. Note, that for $x \neq 0$ and $r \in W$, we have $h(rx) < h(x)$. Let x_n be the image of $1 \in {}_RR$ under the canonic homomorphism $\iota_n : {}_RR \to X$. Then $wx_n = x_n w = x_{n+1}$ and therefore we have, for some n, $x_n = 0$. Indeed, otherwise we would get a strictly decreasing sequence

$$h(x_1) > h(x_2) > \dots > h(x_n) > \dots .$$

But $x_n = 0$ implies that $1 \in {}_RR$ is mapped under some morphism $w^m : {}_RR \to {}_RR$ into 0, that is $w^m = 0$. This contradiction shows that W is, in fact, a nil ideal.

LEMMA III.2.6. *Let* R *be a ring with the radical* W. *Assume that, for some natural* n, $W^n = W^{n+1} \neq 0$. *Then there exists a non-zero module which has neither a minimal nor a maximal submodule.*

Proof. Define $X = \{r \in W^n | W^n r = 0\}$. Then X is a left R-module and has the property $W^n X = 0$. The module $M = {}_R(W^n/X)$ is non-zero, because otherwise $W^n = X$ would imply $W^n = W^{2n} = W^n X = 0$. Also, M has no maximal submodule. For, assuming the contrary, we get that ${}_R(W^n)$ has a maximal submodule. But $W \cdot W^n$ is contained in every maximal submodule of ${}_R(W^n)$, and therefore $W^{n+1} = W^n$ implies that ${}_R(W^n)$ has no maximal submodule. Let $s + X$

be in the socle of M. Then $W(s + X) \subseteq X$, and thus $Ws \subseteq X$. But $W^n = W^{n+1}$ implies

$$W^n s = W^n \cdot Ws \subseteq W^n X = 0 ;$$

consequently, $s \in X$, and M has no minimal submodule. This shows that M has the required properties.

PROPOSITION III.2.7. *A left balanced ring is left artinian.*

Proof. Let W be the radical of the left balanced ring R. Our first aim is to show the existence of a natural n with $W^n = W^{n+1}$. Let w_i be finitely many elements in W such that the set of the elements $w_i + W^2$ generates ${}_R(W/W^2)$. Such a set exists because R/W^2 is a finite direct sum of full matrix rings over local left balanced rings R_i (Proposition III.2.3), and if W_i is the radical of R_i, then ${}_R(W_i)$ is finitely generated (Proposition II.5.3). Observe also that the same references show that the set $\{w_i + W^n\}$ generates ${}_R(W/W^n)$, for any n. According to Lemma III.2.5, the elements w_i are nilpotent; therefore, there is a natural n with $w_i{}^n = 0$ for all i. It remains to show that $W^n/W^{n+1} = 0$. But this follows from the fact that R/W^{n+1} is a direct sum of full matrix rings over local rings and that its radical W/W^{n+1} is generated by the elements $w_i + W^{n+1}$ which satisfy $(w_i + W^{n+1})^n = 0$.

Now, $W^n = 0$. Otherwise, according to Lemma III.2.6, there exists a non-zero module which has neither minimal nor maximal submodules. But this is impossible, because of Lemma III.2.1.

Finally, the fact that R is artinian follows from another application of Proposition III.2.3 and Proposition II.5.3.

III.3. THE STRUCTURE OF BALANCED RINGS

The preceeding investigations allow to give a complete description of left balanced rings. The following theorem summarizes these results.

THEOREM III.3.1. *The following properties of a ring* R *are equivalent:*

(i) R *is left balanced.*

(ii) R *is left artinian and left finitely balanced.*

(iii) R *is a direct sum of a uniserial ring and finitely many full matrix rings over exceptional rings.*

Proof. The fact, that (i) implies (ii) is shown in Proposition II.2.7. The implication (ii) $\rightarrow$ (iii) follows from the Propositions III.2.3 and II.5.3. Finally, in order to prove (iii) $\rightarrow$ (i) let R be a direct sum of a uniserial ring R_o and finitely many full matrix rings $\underline{\underline{R}}_i$ over exceptional rings R_i (with $1 \leq i \leq n$) . According to Proposition I.3.3, R_o is left balanced. Each of the rings R_i is left balanced because of Proposition II.4.6, and therefore Proposition III.1.4 implies that the rings $\underline{\underline{R}}_i$ are left balanced. Since R is the direct sum of R_o and the rings $\underline{\underline{R}}_i$, also R is left balanced. This establishes the theorem.

An immediate consequence of the structure theorem above is the fact that the opposite ring of a left balanced ring is again left balanced.

THEOREM III.3.2. *A ring is left balanced if and only if it is right balanced.*

Proof. The opposite ring of a uniserial or an exceptional ring is again uniserial or exceptional, respectively. Therefore, if a ring R satisfies the condition (iii) of Theorem III.3.1, also the opposite ring of R satisfies this condition.

Thus, we may simply speak of a balanced ring and drop the adjectives "left" and "right" in the notion of a balanced ring.

Another corrollary is the following statement.

THEOREM III.3.3. *A ring* R *with the radical* W *is balanced if and only if* R *is left artinian and* R/W^3 *is balanced.*

Proof. According to Theorem III.3.1, we have to show that a left artinian ring R satisfies the condition (iii), if R/W^3 satisfies the condition (iii). But if R is left artinian and R/W^3, or even R/W^2, is a direct sum of full matrix rings over local rings, then also R itself is a direct sum of full matrix rings $\underline{\underline{R}}_i$ over local rings R_i. If W_i is the radical of R_i, then the condition (iii) for R/W^3 implies, that for all i, R_i/W_i^3 is either uniserial or exceptional. In the first case, R_i itself has to be uniserial. But if R_i/W_i^3 is exceptional, then $W_i^2/W_i^3 = 0$, so $W_i^2 = W_i^3$. Because R_i is left artinian, we may conclude $W_i^2 = 0$ and R_i itself is exceptional. This proves the theorem.

It should be noted that contrary to the case where R is finitely generated over its centre (see below), it is not enough in general to assume here that R/W^2 is balanced. In fact, it is easy to construct local left artinian rings R with radical W such that R/W^2 is exceptional, whereas $W^2 \neq 0$; this will be done in Section III.7. Such a ring is, of course, not balanced.

REMARK III.3.4. The assumption in Theorem III.3.1 (ii) on the ring R to be left artinian is essential. It is well-known (see [2]) that every principal ideal domain is finitely balanced. In fact, it is not difficult to show that if R is a noetherian integral domain, then R is finitely balanced if and only if R is a Dedekind domain. Or, more generally, *a noetherian commutative ring is finitely balanced if and only if it is a direct sum of a uniserial ring and a finite number of Dedekind domains.* The sufficiency of the condition follows, in view of Lemma I.2.3, from the fact that every faithful finitely generated R-module over a Dedekind domain R possesses a direct summand which is isomorphic to a fractional ideal (see [7]) and a fractional ideal is always a generator. And, an R-module over a Dedekind domain R which is not faithful can be considered as a module over a uniserial ring and is therefore balanced, as well. Conversely, every finitely balanced ring is arithmetical, i.e. has a distributive ideal lattice. *) For, if I is a maximal ideal of such a ring R , then R/I^2 is, by Proposition II.1.1, uniserial. Thus, the ideals of the localization R_I are linearly ordered by inclusion and therefore R is arithmetical. But, a noetherian arithmetical ring is a direct sum of a uniserial ring and a finite number of Dedekind domains.

---

*) The authors are indebted to Professor CH. U. JENSEN for bringing to their attention his paper on Arithmetical rings, *Acta Math. Acad. Sci. Hung.* 17 (1966), 115-123 and suggesting the proof of the necessity of the result.

III.4. RINGS FINITELY GENERATED OVER THEIR CENTRES

In the case, where the ring R or at least the factor ring R/W, where W is the radical of R, is finitely generated over its centre, the description of balanced ring becomes simpler.

THEOREM III.4.1. *Let* R *be a ring finitely generated over its centre. Then* R *is balanced if and only if* R *is uniserial.*

Proof. We first show, that an exceptional ring cannot be finitely generated over its centre. For, assume R is exceptional with the radical W and the centre Z. Clearly, $(Z + W)/W$ is contained in the centre of $Q = R/W$ Let F be the quotient field of $(Z + W)/W$, considered as a subring of Q.

Again, we consider W as a Q-Q-bimodule, so we get the equation

$$fw = wf \quad \text{for} \quad w \in W ,$$

first for the elements $f \in (Z + W)/W$, and therefore also for all $f \in F$. If we assume that R is finitely generated as a Z-module, then R/W is a finite dimensional vector space over F. Let n be the dimension $\dim_F Q$. If $\dim_Q W = m$, then $\dim_F W = mn$. Since the dimension of W over F does not depend on whether we consider the left action or the right action of F on W, we conclude

$$\dim_Q W = \dim W_Q \quad ,$$

contrary to the definition of an exceptional ring.

Let us now assume that R is an arbitrary balanced ring. According to Theorem III.3.1, R is a direct sum of a uniserial ring and a finite number of full matrix rings over exceptional rings R_i. If R is finitely generated over its centre, then any one of the rings R_i is finitely generated over its centre, but as we have seen above, this is impossible for an exceptional ring. So we conclude that R is uniserial.

Let us mention that Theorem III.3.4 applies immediately to the case of a finite dimensional algebra and to the case of a finite ring.

REMARK III.4.2. *Let* R *be a ring finitely generated over its centre; let* W *be its radical. Then* R *is balanced if and only if* R *is left artinian and* R/W^2 *is balanced.*

Proof. A left artinian ring R is uniserial if and only if R/W^2 is uniserial. Therefore, the theorem follows from Theorem III.4.1.

Our next aim is to consider the case, where R/W is finitely generated over its centre. We will need the following lemma which establishes some properties of the subring T_v of a local ring R (for the definition, see II.3).

LEMMA III.4.3. *Let* R *be a local ring with the radical* W *such that* $W^2 = 0$ *and* $\dim W_Q = 1$. *Let* v *be a non-zero element of* W. *Then*

$$T_v/W \cong R/W = Q \quad \textit{and} \quad \dim_{(T_v/W)} Q = \dim_Q W .$$

Proof. Write $T = T_v$ and define a mapping $\alpha : Q \to W$ by

$$\alpha(r + W) = vr \quad \text{for} \quad r \in R .$$

Furthermore, define a mapping $\beta : T/W \to Q$ by

$$\beta(\tau + W) = s + W \quad \text{for} \quad \tau \in T, \text{ where } s \in R \text{ satisfies } v\tau = sv .$$

Obviously, both α and β are well-defined bijections because of $\mathrm{Ann}(v) = W$ and $vR = W$. And, they are additive. Moreover, β is multiplicative; for, if $\beta(\tau_i + W) = s_i + W$ for τ_1 and τ_2 from T, then

$$v\tau_1\tau_2 = s_1 v\tau_2 = s_1 s_2 v ,$$

and therefore $\beta[(\tau_1 + W) \cdot (\tau_2 + W)] = \beta(\tau_1 + W) \cdot \beta(\tau_2 + W)$. This shows that β defines a ring isomorphism of T/W and Q. Now, the pair

$$(\alpha, \beta) : {}_{(T/W)}Q \to {}_Q W$$

satisfies, for any $r \in R$, $\tau \in T$ and $s \in R$ with $v\tau = sv$,

$$\alpha[(\tau + W)(r + W)] = \alpha(\tau r + W) = v\tau r =$$

$$= svr = (s + W) \cdot vr = \beta(r + W)\alpha(r + W) ,$$

which implies the required equality of dimensions.

If R is a local ring with

$$W^2 = 0 , \dim W_Q = 2 \quad \text{and} \quad \dim {}_Q W = 1 ,$$

then the Lemma III.4.3 shows that $\dim {}_{(T_v/W)}Q = 2$. But R is exceptional if and only if $\dim Q_{(T_v/W)} = 2$. Therefore, if Q has the property that any division subring of left index 2, which is isomorphic to Q, has also right index 2, then the above conditions imply that R is exceptional. In particular, this leads to the following theorem.

THEOREM III.4.4. *Let* R *be a ring with radical* W. *Assume that* R/W *is finitely generated over its centre. Then* R *is the direct sum of a uniserial ring and finitely many full matrix rings over local rings* R_i *with*

$$W_i^2 = 0 \text{ and } \dim_{Q_i} W_i \times \dim W_{i\,Q_i} = 2 ,$$

where W_i *is the radical of* R_i *and* $Q_i = R_i/W_i$.

Proof. It is sufficient to prove the statement for local rings because, if R is a direct sum of finitely many full matrix rings over local rings R_i and if R/W is finitely generated over its centre, then R_i/W_i is finitely generated over its centre for all i.

But if R is a local ring with radical W and R/W is finite dimensional over its centre, then for any division subring of $Q = R/W$, its right index is 2 if and only if its left index is 2 ([17], p. 158). So R is exceptional if and only if $W^2 = 0$ and $\dim_Q W \times \dim W_Q = 2$.

The last section will deal with the question whether there are rings with $W^2 = 0$ and $\dim_Q W \times \dim W_Q = 2$, which are not exceptional.

III.5. THE MODULE CATEGORY OF A BALANCED RING

It is shown in Proposition III.1.4 that the property of being left balanced is Morita equivalent. This means that, if the category Mod R of all left R-modules is equivalent to the category Mod R' of all left R'-modules, then R is balanced if and only if R' is balanced. Here, we characterize explicitely the balanced rings R in terms of the module categories Mod R .

LEMMA III.5.1. *Let* R *be a local left artinian ring. Then* R *is balanced if and only if any two indecomposable left* R*-modules of length* 3 *are isomorphic and all other indecomposable left* R*-modules are uniserial. In this case, any two indecomposables of the same length are isomorphic.*

Proof. If R is balanced, then according to Proposition II.5.3, R is either uniserial or exceptional. For a local uniserial ring R , every indecomposable module is uniserial and any two indecomposables of the same length are isomorphic. For an exceptional ring, we may apply Propositions II.3.2 and II.4.3, or Propositions II.3.3 and II.4.5 to show that every indecomposable module of length $\neq 3$ is uniserial and that all indecomposables of the same length are isomorphic. This proves the necessity of the conditions, as well as the last statement.

In order to prove the sufficiency, let us first assume that all indecomposable left R-modules are uniserial. Then R is trivially left uniserial. And, R is right uniserial, too. For, otherwise $({}_RR \oplus {}_RR)/D$, where $D = R(u, v) + (W^2 \oplus W^2)$ with linearly independent elements u, v in $(W/W^2)_{R/W^2}$, is a non-uniserial indecomposable left R-module, according to Lemma II.2.2.

If there is a non-uniserial indecomposable R-module X of length 3 with a simple socle, then X is necessarily injective and R (being a monogenic indecomposable R-module) is left uniserial. Consequently, ${}_RR$ can be embedded in X and therefore $W^2 = 0$ and $\dim_Q W = 1$, where $Q = R/W$. If $\dim W_Q \geq 3$, then Lemma II.2.2 would give us an indecomposable module I_3 of length 5 which is not uniserial. Hence $\dim W_Q = 2$. Proposition II.3.3 shows that R has to be exceptional and therefore balanced.

If there is a non-uniserial indecomposable R-module Y of length 3 with a non-simple socle, then Y/Rad Y is simple and thus, necessarily, $Y \cong {}_R R$. Consequently, $W^2 = 0$, $\dim {}_Q W = 2$ and the indecomposable injective is uniserial (and of length 2). If $\dim W_Q \geq 2$, then Lemma II.2.2 would give us an indecomposable module I_2 of length 5 which is not uniserial. Hence we also have $\dim W_Q = 2$. Proposition II.3.2 shows that R has to be exceptional and therefore balanced.

As a consequence, we can describe the module category Mod R of a balanced ring. Here, we restrict to left artinian rings, because the property of a ring R to be left artinian can easily be described in terms of the module category Mod R .

THEOREM III.5.2. *Let* R *be a left artinian ring. Then* R *is balanced if and only if the category* Mod R *of all left* R*-modules is equivalent to a category* K *with the following properties*

(i) *the composition factors of any indecomposable object of* K *are isomorphic,*

(ii) *every indecomposable object of* K *with length* > 3 *is uniserial, and*

(iii) *any two indecomposable objects of* K *with length* 3 *and isomorphic composition factors are isomorphic.*

Proof. First, note that R is the direct sum of full matrix rings over local rings R_i if and only if condition (i) is satisfied in Mod R . And then, Mod R satisfies (ii) and (iii) if and only if, for all i , the categories Mod R_i satisfy these conditions. Therefore, Theorem III.5.2 is an immediate consequence of Lemma III.5.1.

THEOREM III.5.3. *A balanced ring has only finitely many isomorphism types of indecomposable modules.*

Proof. A balanced ring is left artinian, therefore the length of the uniserial left R-modules is bounded. For any simple module S , all indecomposable left R-modules with composition factors isomorphic to S are uniserial or of length 3, and any two of them are isomorphic, if there length is equal. This follows from Lemma III.5.1. Since there is only a finite number of non-isomorphic simple R-modules, the number of isomorphism types of indecomposable left modules is finite. By considering the opposite ring, also the number of isomorphism types of indecomposable right modules is finite.

III.6. CENTRALIZERS OF INDECOMPOSABLE MODULES

The main result of this section asserts that, if R is a balanced ring, then the centralizer of every indecomposable R-module is balanced, as well. This is obvious when R is a local uniserial ring; for, in this case an indecomposable R-module is isomorphic to $_R(R/I)$ for a certain ideal I of R and the centralizer C(R/I) is isomorphic to the uniserial (and hence balanced) ring R/I .

PROPOSITION III.6.1. *Let* R *be a local balanced ring. Then the centralizer of every indecomposable* R*-module* M *is a balanced ring.*

Proof. In view of the remark preceeding the proposition, we can assume that R is exceptional.

First, let W be the radical of R, $Q = R/W$, $W^2 = 0$, $\dim_Q W = 2$, $\dim W_Q = 1$ and $W = Rv + uT_v$ for two linearly independent elements u, v of $_Q W$. According to Proposition II.4.3, there are three types of indecomposable R-modules, viz. A_1, A_2 and A_3 and, obviously, only the type A_2 needs a consideration. Thus, let $M = R/Rv$ be the injective R-module. Clearly, its centralizer C equals to T_v/Rv; the latter is a local ring with the radical $\mathcal{W} = W/Rv$ and thus $\mathcal{W}^2 = 0$. Denote the elements $x + Rv$ of C simply by $\bar{x}$.

Now, in view of Lemma III.5.1, $\dim_{(T_v/W)} Q = 2$; thus, let $1 + W$ and $r + W$ be a basis of $_{(T_v/W)}Q$. Obviously, it is also a basis of $Q_{(T_v/W)}$. Therefore,

$$R = T_v + T_v r + W = T_v + rT_v + W .$$

From here, $Ru = T_v u + T_v ru$ and thus,

$$\mathcal{W} = (Ru \oplus Rv)/Rv = C\,\bar{u} \oplus C\,\overline{ru} .$$

Also, $W = Rv + uT_v$ implies readily that

$$\mathcal{W} = \bar{u}\,C .$$

Consequently, writing $\mathcal{Q} = C/\mathcal{W}$,

$$\dim_{\mathcal{Q}} \mathcal{W} = 2 \quad \text{and} \quad \dim \mathcal{W}_{\mathcal{Q}} = 1 .$$

Now, for $\bar{u} \in C$, define

$$T_{\bar{u}} = \{\bar{\tau} \mid \bar{\tau} \in C \text{ and } \overline{u\tau} \in C\bar{u}\}$$

and deduce from $Ru = T_v u + rT_v u$ that

$$\mathcal{W} = W/Rv = C\bar{u} + \bar{r}C\bar{u} = C\bar{u} + \overline{ru}\,T_{\bar{u}} ,$$

as required. We conclude that C is a balanced (exceptional) ring.

Secondly, in a similar manner, let the exceptional ring R satisfy $\dim_Q W = 1$, $\dim W_Q = 2$ and $W = vR + S_v u$ for two linearly independent elements u, v of W_Q. Again, this time in view of Proposition II.4.5, only R-modules of type B_3 require attention. Thus, let $M = {}_R(R \oplus R)/R(u, v)$ be the injective R-module. Lifting the endomorphisms of M to endomorphisms of ${}_R(R \oplus R)$, we deduce immediately that the elements of the centralizer C of M are induced by

$$(\varkappa_{ij}) : {}_R(R \oplus R) \to {}_R(R \oplus R)$$

with $\varkappa_{ij} \in R$, $1 \leq i, j \leq 2$ such that

$$u\varkappa_{11} + v\varkappa_{21} = \lambda u \quad\text{and}\quad u\varkappa_{12} + v\varkappa_{22} = \lambda v \quad\text{for some } \lambda \in R .$$

It is easy to see that the radical $\mathcal{W}$ of the local ring C consists of all endomorphisms induced by such matrices with $\varkappa_{ij} \in W$. In fact, since in this case there are $\rho_i \in R$, $i = 1, 2$, such that $\varkappa_{12} = \rho_i v$, one can see easily that

$$\begin{pmatrix} \varkappa_{11} & \varkappa_{12} \\ \varkappa_{21} & \varkappa_{22} \end{pmatrix} \quad\text{and}\quad \begin{pmatrix} \varkappa_{11} - \rho_1\varkappa_{12} & 0 \\ \varkappa_{21} - \rho_2\varkappa_{22} & 0 \end{pmatrix}$$

induce the same endomorphisms. Consequently, there is a one-to-one correspondence between the elements of $\mathcal{W}$ and the matrices $\begin{pmatrix} w_1 & 0 \\ w_2 & 0 \end{pmatrix}$ with $w_i \in W$, and thus, we can identify them.

Now, put

$$\varphi_1 = \begin{pmatrix} v & 0 \\ 0 & 0 \end{pmatrix} \in \mathcal{W} \quad \text{and} \quad \varphi_2 = \begin{pmatrix} 0 & 0 \\ v & 0 \end{pmatrix} \in \mathcal{W} .$$

Obviously, given $w_i \in W$, there are $\varkappa_{i1} \in R$ satisfying $\varkappa_{i1}v = w_i$, $i = 1, 2$, and there are $\varkappa_{i2} \in R$ such that

$$u\varkappa_{12} + v\varkappa_{22} = \lambda v ,$$

where λ is determined by $u\varkappa_{11} + v\varkappa_{21} = \lambda u$. Hence,

$$\begin{matrix} \varkappa_{11} & \varkappa_{12} \\ \varkappa_{21} & \varkappa_{22} \end{matrix} \; \varphi_1 = \begin{matrix} w_1 & 0 \\ w_2 & 0 \end{matrix} ,$$

and writing $\mathcal{Q} = C/\mathcal{W}$, we have

$$\dim_{\mathcal{Q}} \mathcal{W} = 1 .$$

Also, given $w_1 \in W$, we can find $\mu_{11} \in R$ and $\sigma \in S_v$ such that $w_1 = v\mu_{11} + \sigma u$, and $\mu_{12} \in R$ such that $v\mu_{12} = -\sigma v$. Moreover, it is easy to determine μ_{12} and μ_{22} to satisfy the respective equation and thus

$$\varphi_2 \begin{matrix} \mu_{11} & \mu_{12} \\ \mu_{21} & \mu_{22} \end{matrix} = \begin{matrix} w_1 & 0 \\ 0 & 0 \end{matrix} .$$

In a similar manner, given $w_2 \in W$, one can show that

$$\varphi_2 \begin{pmatrix} \nu_{11} & \nu_{12} \\ \nu_{21} & \nu_{22} \end{pmatrix} = \begin{pmatrix} 0 & 0 \\ w_2 & 0 \end{pmatrix} \quad \text{for suitable } \nu_{ij}\text{'s} .$$

Hence,

$$\dim \mathcal{W}_{\mathcal{Q}} = 2 .$$

Finally, define

$$S_{\varphi_1} = \{\varphi | \varphi \in C \text{ and } \varphi\begin{pmatrix} v & 0 \\ 0 & 0 \end{pmatrix} \in \begin{pmatrix} v & 0 \\ 0 & 0 \end{pmatrix} C\} .$$

Thus, if $\sigma_{21} \in W$, $\sigma_{22} \in R$ arbitrary, $\sigma_{12} \in R$ and $\lambda_o \in S_n$ satisfying $u\sigma_{12} + v\sigma_{22} = \lambda_o v$, and σ_{11} such that $\lambda_o u = u\sigma_{11}$, then obviously (σ_{ij}) induces an element of S_{φ_1}.

Now, given $w_1, w_2 \in W$, determine σ_{22} by the relation $\sigma_{22}u = w_2$. Then

$$\begin{pmatrix} \sigma_{11} & \sigma_{12} \\ \sigma_{21} & \sigma_{22} \end{pmatrix} \varphi_2 = \begin{pmatrix} \sigma_{12}v & 0 \\ w_2 & 0 \end{pmatrix} .$$

Therefore, since we can choose $\varphi \in C$ such that

$$\varphi_1 C = \begin{pmatrix} w_1 - \sigma_{12}v & 0 \\ 0 & 0 \end{pmatrix} ,$$

we conclude that

$$\varphi_1 C + S_{\varphi_1} \varphi_2 = W ,$$

as required. The proof of Proposition III.6.1 is completed.

THEOREM III.6.2. *Let* R *be a balanced ring and* M *an indecomposable* R*-module. Then the centraliser of* M *is a local balanced ring.*

Proof. In view of Theorem III.3.1 we may assume that R is the full matrix ring over a local ring R', where R' is either uniserial or exceptional. The rings R and R' are Morita equivalent, so M and the image M' of M under a categorical isomorphism of Mod R onto Mod R' have isomorphic endomorphism rings. Applying Proposition III.6.1 for M' we see that the centralizer of M' is a local balanced ring.

THEOREM III.6.3. *Let* R *be a balanced ring and* M *a finitely generated injective* R-*module. Then the centralizer of* M *is a balanced ring.*

Proof. A finitely generated injective R-module M is a finite direct sum of indecomposable injective R-modules. It follows from Theorem III.3.1 that any two indecomposable direct summands are either isomorphic or have no non-trivial homomorphism from one to the other. Consequently, the endomorphism ring of M is a direct sum of full matrix rings over endomorphism rings of indecomposable injective R-modules. The latter are local balanced rings, and thus, again by Theorem III.3.1, the centralizer of M is balanced.

III.7. EXISTENCE OF EXCEPTIONAL RINGS

In this last section, exceptional rings are constructed and it is shown the relation to a problem in the theory of division rings.

LEMMA III.7.1. *Let* D *be a division ring with an isomorphic subring* D' *such that* $\dim_{D'} D = 2$. *Let* $\gamma : D \to D'$ *be an isomorphism. Denote by* R *the ring of all pairs* (a, b) *of elements of* D *with component-wise addition and the following multiplication*

$$(a_1, b_1)(a_2, b_2) = (a_1 a_2, \gamma(a_1) b_2 + b_1 a_2) .$$

Then R *is a local ring, and denoting its radical by* W *and* $Q = R/W$ *, we have*

$$W^2 = 0,\ \dim_Q W = 2 \quad \textit{and} \quad \dim W_Q = 1 \ .$$

Moreover, R *is exceptional if and only if* $\dim D_{D'} = 2$.

Proof. It is easy to see that (a, b) is a unit if and only if $a \neq 0$. Therefore, the radical W is given by $W = \{(0, b) \mid b \in D\}$ and R is a local ring. Obviously, $W^2 = 0$. If $(0, b) \in W$, then

$$R(0, b) = \{(0, d'b) \mid d' \in D'\} \quad \text{and} \quad (0, b)R = W \ .$$

The first equation shows that $\dim_Q W = \dim_{D'} D = 2$, the second that $\dim W_Q = 1$.

Now let $v = (0, 1) \in W$; then $T_v = \{\tau \in R \mid v\tau \in Rv\}$ is given by

$$T_v = \{(a, b) \mid a \in D' \ \text{and} \ b \in B\} \ ,$$

and therefore, $\dim Q_{(T_v/W)} = \dim D_{D'}$. This implies that R is exceptional if and only if $\dim D_{D'} = 2$.

A division ring D with an isomorphic subring D' such that $\dim_{D'} D = \dim D_{D'} = 2$ can easily be constructed: Let D" be an arbitrary division ring and denote by D the ring of quotients of the polynomial ring D"[x] in one (commuting) indeterminate. The ring endomorphism $D''[x] \to D''[x]$ which fixes D" and maps x onto x^2 can be extended to an endomorphism $D \to D$, and we denote its image by D' . Then, obviously, D and D' are isomorphic and $\dim_{D'} D = \dim D_{D'} = 2$. This yields the existence of exceptional rings.

As a consequence of the remark above we get the following theorem.

THEOREM III.7.2. *The following assertions are equivalent:*

(i) *There exists a local ring* R *with the radical* W *and* $Q = R/W$ *such that* $W^2 = 0$ *and* $\dim_Q W \times \dim W_Q = 2$ *, which is not exceptional.*

(ii) *There exists a division ring* D *with an isomorphic subring* D' *such that* $\dim_{D'} D = 2$ *and* $\dim D_{D'} \neq 2$.

Proof. If (ii) is satisfied, then the ring constructed in LEMMA III.7.1 has all the properties mentioned in (i). Conversely, assume there exists a local ring R with the properties described in (i) . We may assume $\dim_Q W = 2$ and $\dim W_Q = 1$; for, otherwise we may consider the ring opposite to R . If w is a non-zero element of W and $T_v = \{\tau \in R \mid v\tau \in Rv\}$, then it follows from Lemma III.4.3 that Q and T_v/W are isomorphic and that $\dim_{(T_v/W)} Q = 2$. Since R is not exceptional, we have $\dim Q_{(T_v/W)} \neq 2$. Therefore, $D = Q$ and $D' = T_v/W$ satisfy the conditions (ii) .

In [5], P. M. Cohn has constructed an example of a division ring D with a division subring D' such that $\dim_{D'} D = 2$ and $\dim D_{D'} \neq 2$. Thus, the question is whether such a subring D' exists which is, in addition, isomorphic to D .

The last remark shows that the condition in Theorem III.3.3 that R/W^3 is balanced cannot be replaced by the condition that R/W^2 is balanced.

REMARK III.7.3. *There exists a local ring* R *with the radical* W *such that* $W^3 = 0$ *, which is not balanced, although* R/W^2 *is balanced.*

Proof. We start with a division ring D with an isomorphic subring D' such that $\dim_{D'} D = 2 = \dim D_{D'}$, and $\gamma : D \to D'$ is an isomorphism. Denote by R the ring of all triples (a, b, c) of elements of D with component-wise addition and the following multiplication

$$(a_1, b_1, c_1)(a_2, b_2, c_2) = (a_1a_2, \gamma(a_1)b_2 + b_1a_2, \gamma[\gamma(a_1)]c_2 + \gamma(b_1)b_2 + c_1a_2) .$$

Then the radical W is given by $W = \{(0, b, c) \mid b, c \in D\}$. It is easy to see that $W^2 = \{(0, 0, c) \mid c \in D\}$ and $W^3 = 0$. Since R/W^2 is isomorphic to the ring constructed in LEMMA III.7.1, R/W^2 is exceptional. But $W^2 \neq 0$, and thus Proposition II.5.3 shows that R itself is not balanced.

IV. REFERENCES

[1] BASS, H. *Algebraic K-theory.* Benjamin, New York, 1968.

[2] BOURBAKI, N. *Algèbre*, Ch. VIII. Hermann, Paris, 1958.

[3] CAMILLO, V. P. Balanced rings and a problem of Thrall. *Trans. Amer. Soc.* 149 (1970), 143-153.

[4] CAMILLO, V. P. AND FULLER, K. R. Balanced and QF-1 algebras. *To appear.*

[5] COHN, P. M. Quadratic extensions of skew fields. *Proc. London Math. Soc.* 11 (1961), 531-556.

[6] CUNNINGHAM, R. S. AND RUTTER, E. A., JR. The double centralizer property is categorical. *To appear.*

[7] CURTIS, C. W. AND REINER, I. *Representation theory of finite groups and associative algebras*, Interscience Publ., New York, 1962.

[8] DICKSON, S. E. AND FULLER, K. R. Commutative QF-1 rings are QF. *Proc. Amer. Math. Soc.* 24 (1970), 667-670.

[9] DLAB, V. AND RINGEL, C. M. Rings with the double centralizer property. *To appear.*

[10] DLAB, V. AND RINGEL, C. M. On a class of balanced non-uniserial rings. *Math. Ann.* To appear.

[11] DLAB, V. AND RINGEL, C. M. Balanced local rings with commutative residue fields. *To appear.*

[12] DLAB, V. AND RINGEL, C. M. The structure of balanced rings. *To appear.*

[13] DLAB, V. AND RINGEL, C. M. Exceptional rings. *To appear.*

[14] FLOYD, D. R. On QF-1 algebras. *Pacific J.* 27 (1968), 81-94.

[15] FULLER, K. R. Double centralizers of injectives and projectives over artinian rings. *Illinois J.* 14 (1970), 658-664.

[16] FULLER, K. R. Primary rings and double centralizers. *Pacific J.* 34 (1970), 379-383.

[17] JACOBSON, N. *Structure of rings.* Coll. Publ. Amer. Math. Soc., Vol. 37, 1956.

[18] JANS, J. P. On the double centralizer condition. *Math. Ann.* 188 (1970), 85-89.

[19] MORITA, K. Duality for modules and its application to the theory of rings with minimum condition. *Sci. Rep. Tokyo Kyoiku Daigaku,* Sec. A 6 (1958), 83-142.

[20] MORITA, K. On algebras for which every faithful representation is its own second commutator. *Math. Z.* 69 (1958), 429-434.

[21] MORITA, K. AND TACHIKAWA, H. On QF-3 rings. *To appear.*

[22] NAKAYAMA, T. On Frobeniusean algebras. II. *Ann. of Math.* 42 (1941), 1-21.

[23] NESBITT, C. J. AND THRALL, R. M. Some rings theorems with applications to modular representations. *Ann. of Math.* 47 (1946), 551-567.

[24] TACHIKAWA, H. Duality theorem of character modules for rings with minimum condition. *Math. Z.* 68 (1958), 479-487.

[25] TACHIKAWA, H. On rings for which every indecomposable right module has a unique maximal submodule. *Math. Z.* 71 (1959), 200-222.

[26] TACHIKAWA, H. Double centralizers and dominant dimensions. *Math. Z.* 116 (1970), 79-88.

[27] THRALL, R. M. Some generalizations of quasi-Frobenius algebras. *Trans. Amer. Math. Soc.* 64 (1948), 173-183.

MODULES FINITE OVER ENDOMORPHISM RING[1]

by

Carl Faith[2]

Department of Mathematics
Rutgers, The State University
New Brunswick, New Jersey
and
The Institute for Advanced Study
Princeton, New Jersey

<u>ABSTRACT</u>

A right R-module M is said to be *finendo* provided that as a left module over its endomorphism ring it is finitely generated. Let $B = \text{End}\, M_R$. Then $B^n \to M \to 0$ is exact for some finite integer n. This implies (and when M is quasi-injective is implied by) the exactitude of $0 \to R/\text{ann}_R M \to M^n$. This implies that *a finendo module is quasi iff injective modulo annihilator* (1). Similarly for any finitely generated module over a commutative ring. *Any finendo quasi-injective module is injective over its biendomorphism ring* (17B). *A ring* R *is right Artinian iff every quasi-injective right module is finendo* (17). This supplies the converse of a theorem of K. Fuller [69]. Moreover, *a ring* R *is right self-injective iff every finendo faithful (injective) right R-module generates* mod-R (18B). Azumaya defined a right PF ring as one such that every faithful right R-module generates mod-R. *We characterize a right* PF *ring as a right self-injective ring such that every faithful* (*quasi*) *injective right* R-*module is finendo* (38). Then, *if every factor ring is* PF, *the ring must be right Artinian, hence uniserial* (40). In general, *a factor ring* R/A *of a right* PF *ring is right* PF *iff the left annihilator of* A *is* $Rz = zR$ *for some* $z \in R$ (39).

We study Σ-injective modules over a regular ring R, and show *that* M *has* Σ-*injective hull in mod*-R *iff* M *is semisimple, injective, and finendo* (16). This generalizes a result of J. Levine [71] assuming that M is simple.

We characterize a prime right Goldie ring as *the class of rings over which there exists an indecomposable injective finendo faithful right module* E, *with no fully invariant nontrivial submodule, such that* End E_R *a field* (34). In this case, the biendomorphism ring Q of E is the right quotient ring of R, and E is isomorphic to the unique minimal right ideal of Q (35A). Semiprime Goldie rings can be similarly characterized. Employing a recent characterization of quasi-injective abelian groups by Fuchs, we can describe all the finendo ones: thus, M *must be a divisible group containing* $\mathbb{Q}$, *or else* M *is a torsion group of bounded order the p-components of which are direct sums of isomorphic cyclic groups* (16).

A ring is a right *V-ring* if every simple right R-module is injective. We abbreviate quasi-injective by QI. A ring R is right QI (quasi => injective) provided that every right QI module is injective. Thus, every right QI ring is a V ring. Moreover, every right QI ring is a right Noetherian (Koehler [70]). Cozzen's example [70] of a right V-domain R is actually a right QI ring, as A. Boyle [71] observed. *Any right Goldie right* V-*ring is a finite product of simple rings* (31). This reduces the structure theory for a right V, or QI, ring to simple rings. The latter are characterized as those right Noetherian rings for which the endomorphism ring of any indecomposable injective right module is a field (see the paper of the author [72].

ACKNOWLEDGEMENTS

About half of this was presented in lectures at the Tulane Symposium in Ring Theory during my month-long tenure, Fall, 1970, as Visiting Professor under the Ford Program for Tulane. I am grateful to the faculty for the opportunity to participate not only in the Symposium but also in the delights and pleasures its location at Audubon Park offers. The present form, and length of the manuscript, an expanded version of my lectures, owes in large measure to the stimulation which my questioning auditors provoked. For this, I have to thank A. H. Clifford (and his good-natured forbearance of my paedogogic deficiencies!), J. Dauns, L. Fuchs, K. Koh, W. Nico, H. Storrer, and others. (The reader may be indebted to Laszlo Fuchs for editing my jokes out of the manuscript, including a Hungarian one: (input-coffee, output-mathematics)!)

I read (without proofs) most of this paper in an hour long lecture to the Symposium in Ring Theory at Arkansas State University, Jonesboro, Spring, 1971, and I have to thank that Faculty, especially the Symposium's director, R. L. Tangeman, for this opportunity.

I am grateful to the Faculty of the Institute for Advanced Study for a sanctuary in Summer 1971, which enabled me to complete this long postponed paper. I have also the pleasure to thank a fellow Kentuckian at the Institute, Miss Evelyn Laurent, for many typing pains taken with the manuscript.

INTRODUCTION

A right A-module M is injective in case the exactitude of

(1) $$0 \to X \to Y \to Z \to 0$$

in mod-A implies that of

(2) $$0 \to \mathrm{Hom}_A(Z, M) \to \mathrm{Hom}_A(Y, M) \to \mathrm{Hom}_A(X, M) \to 0.$$

Furthermore, M is quasi-injective, or QI, when (1) => (2) whenever Y = M. If M is a module which is injective modulo annihilator, that is, if M is injective as an $A/\mathrm{ann}_A M$ module, then M is a QI canonical right A-module, a fact which doubtlessly contributes to the importance of QI-modules. One objective of this article is to investigate QI-modules of this kind.

1. Theorem and Definition. *Any right A-module M which is finitely generated over End M_A is said to be finendo. A finendo right A-module M is QI if and only if M is an injective canonical $A/\mathrm{ann}_A M$ module.*

This has been observed by Fuchs [70] under a hypothesis (called K-boundedness) equivalent to the requirement that M is cyclic over End M_A.

The class of right Artinian rings satisfies the condition of Theorem 1, as observed by Fuller [69].

2. Theorem. *Let A be a right Artinian ring. Then every QI right A-module M is finitely generated over End M_A.*

Thus, the only quasis are those injective modulo annihilator. Every faithful QI right A-module is injective.

3. Theorem. Over a commutative ring R, any finitely generated QI-module M is injective modulo annihilator. Thus, if M is faithful, then M is injective.[3]

A ring R is said to be right QI provided that:

(I) Every QI right A-module is injective. ("Q implies I rings".)
A right QI ring A has the property for any ideal B.

(II) Every injective right A/B module M is injective as a canonical right A-module

This is equivalent to the requirement on ideals B:

(III) The inclusion functor mod-A/B~> mod-A preserves injectives.

(III) has been discussed in the setting of categories, namely:

4. Proposition. If T:C~> D is a functor of abelian categories with left adjoint S:D ~>C, and if C has enough injectives, then T preserves injectives if and only if S is exact.

See Bucur and Deleanu [68] for a proof.

5A. Corollary. (II) and (III) are each equivalent to the requirement that A/B is a flat left A-module. A *necessary and sufficient condition for this is that the inclusion functor* mod-A/B ~> mod-A *preserves injective hulls of simple right* A/B-*modules*.

Proof. The first stated equivalence follows from 4, as does the necessity of the second statement. Now suppose that the inclusion functor mod-A/B preserves injective hulls of simple right A/B-modules. If M is an injective right A/B-module, then M embeds in some product E^I of any injective cogenerator E. By injectivity of M, $E^I = M \oplus X$, for some right A/B-module. Choose E to be the product $\Pi_{j\in J}\hat{V}_j$, where $\{V_j\}_{j\in J}$ is a representative class of simple right A/B-modules. Now E is a product in mod-A of injective right A-modules, so E is injective in mod-A, and hence so is E^I. Then, M, a summand of an injective, is injective, and therefore the inclusion functor preserves injectives. □

5B. *Corollary*. *If* A *is a right* V-*ring*, *then*, *for any ideal* B, A/B *is a flat left* A-*module*. □

5A also yields a proof of Kaplansky's theorem (see 24A).

ANNIHILATORS

If M is a right R-module with endomorphism ring S, then an additive subgroup X of M such that $sx \in X\ \forall\ s \in S$ is an S-module, and a submodule of the canonical left S-module M. Thus, we may refer to S-submodules of M which are not necessarily R-submodules. An R-submodule which is also an S-submodule is called a *fully invariant* submodule.

For any subset X of M, the set

$$\text{ann}_R M = \{r \in R \mid Xr = 0\} = X^{\perp}$$

is a right ideal of R, called the annihilator of X. Dually, for any subset A of R, the set

$$\text{ann}_M A = \{m \in M \mid mA = 0\} = {}^{\perp}A$$

an S-submodule of M, called the annihilator of A in M. Then, $\text{ann}_M \text{ann}_R X$ is called the double annihilator of X, and $\text{ann}_R \text{ann}_M A$ is called the double annihilator of A. Moreover, A is an annihilator of a subset of M if and only if $A = \text{ann}_R \text{ann}_M A$, in which case A is said to satisfy the double annihilator condition with respect to M. This is abbreviated by M-d.a.c., or simply by d.a.c., when M is a fixed module in a discussion. Dually, for annihilators in M of subsets of R. A collection of subsets (of M or R) is said to satisfy the d.a.c. provided that every set in the collection satisfies the d.a.c., that is, is an annihilator.

In reference to the lemma below, a summand of R in mod-R is just a right ideal generated by an idempotent.

6. Lemma. *If M is any faithful right R-module, then any summand of R in mod-R is the annihilator of a subset of M. Moreover, if $S = \text{End}\, M_R$, and if $R \approx \text{End}_S M$ canonically, then any S-summand of M satisfies the d.a.c.*

Proof. Let $A = eR$, where $e = e^2 \in R$. Clearly $\text{ann}_M A \supseteq M(1-e)$. But if $m \in M$, and $mA = 0$, then $me = 0$, and so, $m(1-e) = m \in M(1-e)$. This proves the first statement. If X is an S-summand of M, say $M = X \oplus Y$, for some S-submodule Y, then there is an S-endomorphism p such that p induces the projection $M \to X$ parallel to Y. Since $R \approx \text{End}_S M$ canonically, then there is an idempotent $e \in R$ such that e induces p. Thus, $X = \text{ann}_M A$, where $A = (1-e)R$. ⌈

Next there are two theorems needed; the first is by Johnson and Wong [61], and the second is by Jacobson [56, 64; p. 27, Lemma], and independently, Johnson and Wong (*loc. cit.*).

7A. *Theorem. A right R-module M is QI if and only if M is fully invariant in its injective hull* $\hat{M}$.

7B. Theorem. If M is a right QI-module, then every finitely generated S-submodule of M satisfies the d.a.c., where $S = \text{End } M_R$.

The next proposition is known.

8. *Proposition. If E is any injective right A-module, then for any ideal B of A,* $\text{ann}_E B$ *is an injective A/B-module.*

Proof. Let F be the injective hull of $X = \text{ann}_E B$ taken in mod-A/B. Then F is a canonical A-module, and is essential over X. Therefore, F can be embedded in the injective hull $\hat{X}$ of X, hence $X \subseteq F \subseteq E$. But $FB = 0$, so $F \subseteq X = \text{ann}_E B$, and $F = X$ as asserted. □

9A. *Proposition and Definition* (Faith [66]): *An injective right R-module M is Σ-injective provided that the equivalent conditions hold:*

(1) *Any direct sum* $M^{(I)}$ *is injective in* mod-R.

(2) *Any countable direct sum* $M^{(\omega)}$ *is injective in* mod-R.

(3) M *satisfies the d.c.c. on* End M_R*-submodules annihilated by subsets of* R.

(4) R *satisfies the a.c.c. on right ideals annihilated by subsets of* M.

If the injective hull of M is Σ-injective, we say that M has Σ-injective hull.

9B. *Corollary. A ring* R *is right* Noetherian *iff* mod-R *possesses a* Σ-*injective cogenerator.*

Proof. If R is Noetherian, every injective is Σ-injective. Conversely, if E is a cogenerator, then every right ideal of R is the annihilator of a subset of E (cf. Rosenberg and Zelinsky [61].) Then, by 9A, E is Σ-injective only if R satisfies the a.c.c. on right ideals. □

9C. Corollary (Kurshan [70]). *A ring* R *is right Noetherian iff the direct sum of the injective hulls of any collection of simple modules is injective.*

Proof. For then mod-R has a Σ-injective cogenerator. □

9D. *Corollary* (Kurshan). *If every semisimple right* R-*module is injective, then* R *is right Noetherian.*

The rings of 9D are, then, precisely the right Noetherian right V rings.

Note: Kurshan's result is stronger than stated in 9D in that he requires the condition only for countable collections of simple modules.

10. *Lemma*. *If* M *has* Σ-*injective hull* $\hat{M}$ *in* mod-R, *then* M *has* Σ-*injective hull in* mod-$R/\mathrm{ann}_R M$.

Proof. Let $A = \mathrm{ann}_R M$. Then, $E = \mathrm{ann}_{\hat{M}} A$ is an injective R/A-module which is essential over M, and so E is the injective hull of M in mod-R/A. Similarly, the annihilator of A in $\hat{M}^{(\omega)}$ is the injective hull of $M^{(\omega)}$, but this is simply $E^{(\omega)}$. □

11. *Corollary*. *If* M *has* Σ-*injective hull in* mod-R, *then* $R/\mathrm{ann}_R M$ *satisfies the a.c.c. on right ideals annihilated by subsets of* M, *and on right ideals generated by idempotents*.

Proof. Immediate from 9, and the lemma □

MODULES FINITE OVER ENDOMORPHISM RING

Adopt the convention of writing maps opposite scalars. Thus, a right module is canonically a left module over its endomorphism ring.

12. *Proposition*. *For any right* R-*module* M, *set* $B = \mathrm{End}\, M_R$, *and*

$$A = \mathrm{End}_B M = \mathrm{Biend}\, M_R.$$

(1) (M *finitely generated in* mod-R)

$$R^n \to M \to 0 \text{ exact in mod-}R \Rightarrow 0 \to B \to M^n \text{ exact in } B\text{-mod}$$

(2) (M *is finendo*).

$B^n \to M \to 0$ *exact in* B-mod $\Rightarrow$ $0 \to A \to M^n$ *exact in* mod-A

and

$0 \to A \to M^n$ *exact in* mod-A $\Rightarrow$ $0 \to R/\mathrm{ann}_R M \to M^n$ *exact in* mod-R

(3) *A faithful right* R-*module* M *is finendo only if* $0 \to R \to M^n$ *is exact in* mod-R *for some integer* $n > 0$.

Proof. (1) Apply the left exact functor $(\ ,M) = \mathrm{Hom}_R(\ ,M)$ to the exact sequence on the left, and use the natural isomorphisms

$$(B^n, M) \approx (B, M)^n \approx M^n \qquad \text{(in B-mod)}.$$

(2) then follows from (1), and the fact that $R/\mathrm{ann}_R M$ embeds in A canonically. □

13. *Proposition. Let* M *be a* QI *faithful right* R-*module. The following conditions are equivalent:*

(1) M *is finendo.*

(2) $0 \to R \to M^n$ *is exact in* mod-R *for some integer* $n > 0$.

When this is so, then M *is injective in* mod-R.

Proof. (1) $\Rightarrow$ (2) by (3) of the last proposition. Moreover, applying $(\ ,M)$ to the exact sequence in (2), and using the fact that M is QI yields $B^n \to M \to 0$ exact in B-mod. Thus, (2) $\Rightarrow$ (3).

Baer's criterion for injectivity requires only that every map $I \to M$ of a right ideal I of R can be extended to a map $R \to M$. This is indeed the case for any QI module containing a copy of R. Thus, (2) implies that M^n, whence M, is injective. □

PROOF OF THEOREM 1. If M is any QI right R-module, then M is a QI right $R/\mathrm{ann}_M R$ module, and End M_R coincides with the endoring of M over $R/\mathrm{ann}_M R$. Thus, the proposition implies that M is injective over $R/\mathrm{ann}_M R$. □

PROOF OF THEOREM 2. If $x_1,\ldots,x_n$ are elements of M chosen such that the annihilator in A of M is minimal among the set of all right ideals of A annihilated by finite subsets of M, then $\cap_{i=1}^n x_i^{\perp} = \mathrm{ann}_A M$. Therefore, there is an embedding $0 \to A/\mathrm{ann}_A M \to M^n$ sending $[a + \mathrm{ann}_A M] \longrightarrow (x_1 a,\ldots,x_n a)$ for any $a \in A$. Thus, by 12, M is finendo, so 1 applies. □

PROOF OF THEOREM 3. If A is commutative, then the set $x_1,\ldots,x_n$ can be any finite set of generators of M. □

SECOND PROOF OF THEOREM 1. Let M be a QI module with injective hull E, let $S = \mathrm{End}\, E_A$, and $R = \mathrm{End}\, M_A$. By 7, M is an S-submodule of E. Since E is injective, then there is a ring surjection $S \to R$. Since M is finitely generated over R, this shows that the canonical left S-module M is finitely generated, and hence, by 6, M satisfies the d.a.c. Thus, if $I = \mathrm{ann}_A M$, then I is an ideal such that $M = \mathrm{ann}_E I$. Then M is an injective A/I-module by 8. □

15A. Proposition. Let M be a right R-module, and M $\overline{R} = R/\mathrm{ann}_R M$.

(1) If $\overline{R}$ is semisimple, and a flat left R-module, then M

is Σ*-injective and finendo.*

(2) *If* M *has* Σ*-injective hull* $\hat{M}$, *and if* $\bar{R}$ *is a regular ring, then* $\bar{R}$ *is semisimple.*

(3) *If* M *has* Σ*-injective hull* $\hat{M}$, *and if* $R/\mathrm{ann}_R\hat{M}$ *is a regular ring, then* $M = \hat{M}$ *is finendo, and* $\bar{R}$ *is semisimple.*

(4) *If* M *is any semisimple finendo right* R*-module, then* $\bar{R} = R/\mathrm{ann}_R M$ *is semisimple. Furthermore,* M *injective in* mod-R *iff* $\bar{R}$ *is a flat left* R*-module.*

Proof. (1) Any module over a semisimple ring is injective. By 4, flatness of R/A implies that the inclusion mod-$R/A \subseteq$ mod-R preserves injectives, hence preserves Σ-injectivity. Moreover, M is finendo by 1.

(2) By 10, and the regularity of $\bar{R}$, the ring $\bar{R}$ is Noetherian, hence Artinian semisimple. Furthermore, 11 similarly suffices for semisimplicity $R/\mathrm{ann}\ M$ in (3).

(3) Since $R/\mathrm{ann}_R\hat{M}$ is semisimple, then $\hat{M}$ is semisimple, so that M, being a summand of $\hat{M}$, equals $\hat{M}$. Then M is finendo by (1).

(4) Since M is quasi, and finendo, then $\bar{R}$ embeds in M^n by 13, for some integer $n > 0$, hence $\bar{R}$ is also semisimple. Then flatness of $\bar{R}$ preserves injectivity of M in mod-R. Conversely, since every simple $\bar{R}$ module is a summand of M, then injectivity of M in mod-R implies that every simple $\bar{R}$-module is an injective R-module, and therefore $\bar{R}$ is left flat by 5A. □

15B. Corollary. *For a right module* M *over a regular ring* R, *the following conditions are equivalent:*

(1) M *has* Σ-*injective hull in* mod-R.

(2) M *has* Σ-*injective hull in* mod-R/ann_MR.

(3) R/ann_RM *is semisimple*.

(4) M *is a semisimple, injective right* R-*module which is finite over endomorphism ring*.

(5) M *is a semisimple finendo right* R-*module*.

Proof. Over a regular ring R, any left R-module is flat, and every factor ring is regular, so the proposition implies the equivalences (1) - (5). (cf. 24B.) □

This corollary contains the result of J. Levine [71], namely, for the case M is a simple module, the equivalence of (1) and (4).

Let R be a local commutative Noetherian ring with radical J ≠ 0. Then, the simple module M = R/J has Σ-injective hull in mod-R, but R/J is not flat. (cf. (3) and (4) of 15A.)

Fuchs [69] proved that an abelian group M is a QI module over $\mathbb{Z}$ if and only if M is either divisible, or a torsion group each p-component of which is a direct sum of isomorphic cyclic or quasi cyclic groups $\mathbb{Z}_{p^n}$ $(n \leq \infty)$.

16. Proposition. *An abelian group* M *is a* QI *group finite over endomorphism ring if and only if* M *is a divisible group containing a summand isomorphic to* $\mathbb{Q}$, *or* M *is a torsion group of bounded order the* p-*components of which are direct sums of isomorphic cyclic groups*.

Proof. Let M be a QI abelian group. If M is not a torsion group, then M contains a copy of $\mathbb{Z}$, and hence, by 9, M is finite over endomorphism ring and is injective (= divisible). Conversely, by 12, any QI abelian group M finite over endoring must be such that M^n, contains $\mathbb{Z}/\text{ann}\ M$ for some integer $n > 0$. If ann $M = m\mathbb{Z}$, and if $m = 0$, then M must be a divisible group containing $\mathbb{Q}$ as a summand. Otherwise, M must have bounded order m. Since M^n contains $\mathbb{Z}/m\mathbb{Z}$, then 12 implies that M is finendo. □

17A. *Theorem. The following conditions on a ring* R *are equivalent:*

(1) R *is right Artinian.*

(2) *For each right* R-*module* M, *there corresponds an integer* $n = n(M)$, *and an exact sequence*

$$0 \to R/\text{ann}_R M \to M^n.$$

(3) *Every* QI *right* R-*module is finendo.*

Proof. (1) => (2) and (2) => (3) by 13; (3) => (1). Let M denote the direct sum of a complete set of representatives of simple right R-modules. Since (3) => (2) for QI modules by 13, it follows that R/rad R embeds in the semisimple module M^n, that is, R/rad R is semisimple.

Next let E denote the injective hull in mod-R of M. Then, E is faithful, so that 13 implies that $0 \to R \to E^m$ is exact for a suitable integer $m > 0$. Since E, and E^m, have essential socles, then R has essential socle. Moreover, since R/J is semisimple, where J = rad R, then M is a module of finite length, so that E, E^m, and R all have finite socles. We propose to show that any right R-module $N \neq 0$ has socle $\neq 0$. If $\text{ann}_R N = I \neq 0$, then R is faithful over R/I, and since R/I inherits the hypotheses, it suffices to prove this for the case N is faithful. Then, 13 implies an embedding of R into some finite power F^n of the injective hull F of N. If V is a simple module contained in the socle of R, then the injective hull $\hat{V}$ of V is a summand of F^n. Since the endomorphism ring of V is a local ring, then by the Krull-Schmidt theorem, $\hat{V}$ is a summand of F. This proves that the socle of $F \neq 0$, whence socle of $N \neq 0$. Then a theorem of Bass [60] implies that J = rad R is right T-nilpotent (= right vanishing). Since R/J is semisimple, then Bass's theorem [60] implies that R is left perfect, that is, has d.c.c. on principal right ideals.

The fact that every factor ring of R inherits the hypotheses implies that every factor ring has finite right socle. Thus, J/J^2 is finite, which in a left perfect ring R implies that R is right Artinian (Osofsky [66], Ornstein [67]). □

17B. Corollary. *Any finendo* QI *right* R-*module is injective over* Biend M_R.

Proof. Any QI module in mod-R is QI in mod-A, for A = Biend M_R. Moreover, End M_R = End M_A, and M finendo implies that M is injective in mod-A by 13.

Definition. A right R-module M is said to be projectivendo, provided that M is projective as a left module over End M_R.

18A. *Theorem* (Morita [58]). *An object M of* mod-R *generates* mod-R *iff M is finendo, projectivendo, and* $R \approx$ Biend M_R *canonically.*

For a very short proof of this, see a paper of the author [67f].

18B. *Proposition. A ring R is right self-injective if and only if every finendo faithful (injective) module E is a generator of* mod-R. *(In this case, then E is projectivendo.)*

Proof. A ring R is right self-injective if and only if mod-R has an injective generator. For any ring R, the injective hull of R in mod-R is cyclic over endomorphism ring (as indeed is any QI-module containing R by the proof of 13). Thus, the condition of the proposition implies R is right self-injective.

Conversely, assume that R is injective in mod-R, and let $M \in$ mod-R be a faithful module finite over $S = \text{End } M_R$, say

$$S^n \to M \to 0$$

is exact in S-mod. Then Proposition 12 implies that $0 \to R \to M^n$ is exact in mod-R, and injectivity of R implies an isomorphism $M^n \approx R \oplus X$. Therefore, M is a generator.

In this case, 18A implies that M is projective over endomorphism ring. □

REGULAR RINGS

A ring R is *Dedekind finite* provided that the implication

$$xy = 1 \Rightarrow yx = 1$$

holds in R. Any ring with no infinite set of orthogonal idempotents is Dedekind finite by a theorem of Jacobson [50]. The next proposition has been proved by Utumi [66].

19. *Proposition. If* R *is a right and left self-injective ring, then* R *is Dedekind finite.*

20. *Proposition. Let* R *be a ring satisfying the two equivalent conditions:*

(1) $R \approx \text{End}_D V$, *where* V *is a left vector space over a field* D.

(2) R *is a prime left self-injective ring with a simple right ideal* $\approx V$, *and* $D = \text{End}\ V_R$.

Then, V *is injective in* mod-R *if and only if* V *is finite dimensional over the field* $D = \text{End}\ V_R$.

Proof. Assuming (1), then the canonical right R-module V is simple by the density theorem and R is left self-injective by a theorem of Utumi [56]. Since $D = \text{End}\ V_R$ is a field, then D embeds in V in D-mod, and so V embeds canonically in R. Thus, (1) $\Rightarrow$ (2).

Conversely, assuming (2), then R embeds canonically in $L = \text{Biend } V_R$. If $a \in L$, and $a \neq 0$, then $Va \neq 0$, and so $Ra \cap R \neq 0$. This implies that L is contained in the maximal left quotient ring $\hat{R}$ of R. Since R is left self-injective, and regular, then $R = \hat{R}$, which proves (1) that $L = R = \text{End}_D V$.

If V is finite dimensional, then R is semisimple Artinian, and every R-module is injective. Conversely, assume (1) and V are injective. If $\hat{R}$ is the right quotient ring of R, then the right ideal $V\hat{R}$ of $\hat{R}$ generated by V is the injective hull of V. Then, injectivity of V implies $V = V\hat{R}$, which shows that $\hat{R}$ embeds canonically in $R = \text{End}_D V$, and hence that $R = \hat{R}$ is right self-injective. Since R is also left self-injective, then Utumi's theorem 13 implies that R is Dedekind finite, which is possible if and only if V is finite dimensional. □

The equivalence (1) ⇔ (2) is a theorem of Utumi [56].

21. Example (a). Let $R = k[y, D]$ denote the ring of differential polynomials over a universal field k, and let V be the unique simple ring R-module. Then, by Cozzen's theorem [70], V is injective in mod-R. Nevertheless, V is not injective in $L = \text{Biend } V_R$, since V is not finite dimensional over $D = \text{End } V_R$.

(b) R <u>is a right</u> QI-<u>ring</u>. Let Q be any QI right R-module, and $E = \Sigma_{i \in I} \oplus E_i$ its injective hull, where E_i is indecomposable for every $i \in I$. Since $Q \cap E_i \neq 0$, for every $i \in I$, it suffices to assume that E is indecomposable. But by Cozzen's results, either $E = Q$ is simple, or else E is the right quotient field of R. Since Q is by 7 a fully invariant R-submodule of E, then Q is a left ideal of E. Since E is a field, then $Q = E$ in this case too.

(c) *Every* QI *module over* R *is* Π-QI. Thus, R affords an example of a Noetherian ring not Artinian having the stated property. This answers a question raised by Fuller [69] (following his Corollary 1.3).

22. *Corollary*. *A right neat ring* R *is canonically isomorphic to the biendomorphism ring of an injective right ideal* $V \neq 0$ *if and only if* R *is right self-injective and regular*.

Proof. Any regular ring R is right neat (Johnson [51]) and $R \approx \text{End}_R R = \text{Biend } R_R$. The converse is a corollary of the proof of the last part of 20 which shows that $R = \hat{R}$ is right self-injective. □

In 14, any simple left ideal of $R = \text{End}_D V$ is generated by an idempotent, hence a summand of R in R-mod, and consequently is injective. The question of whether or not there are any other simple left modules of left full linear rings has been discussed by Osofsky [66].

23. *Proposition*. *If* R *is a regular ring, then every injective right* R/A-*module is canonically an injective right*

R-module, for any ideal A. If R is a commutative ring, then conversely.

Proof. If R is regular, then every R-module is flat, so Corollary 5 applies. Conversely, if R is commutative, and if R/A is flat, for every ideal A, that is, if every cyclic module is flat, then R is regular (Auslander [57]).

24A. Corollary (Kaplansky). A commutative ring A is regular iff A is a V-ring.

Proof. Assume A is regular, and let V = A/M be a simple A-module. Then, V is injective over A/M, hence by 23, V is injective over A. Thus, A is a V-ring. Conversely, assume that A is a V-ring, and let B be any ideal. If V is any simple A/B-module, then V is injective over A, and hence over A/B. This proves that the inclusion functor mod-A/B ⇝ mod-A preserves the injective hulls of simple A/B-modules, so A/B is flat over A by 5A, and hence A is regular by 23. □

24B. Corollary. Over a regular ring A, a right A-module E is injective in mod-A if and only if E is injective modulo annihilator. □

25. Example. A regular right V-ring may have infinite dimensional simple modules.

The example, surprisingly perhaps, is one which the author cited in his Lectures [67] of a regular ring not a left V-ring! Let $L = \text{End}\ V_F$ be the full right linear ring over an infinite dimensional right vector space V over a field F, let S be the ideal consisting of l.t.'s of finite dimensions, and let $R = S + F$ be the subring generated by S and the subring F consisting of scalar transformations (sending every $v \longrightarrow va$ for some $a \in F$). Clearly, R is a regular ring. Moreover, V is canonically isomorphic to a simple left ideal of L, say $V = Le$, for $e = e^2 \in L$. Now L is injective in mod-L hence $W = eL$ (the L-dual of V) is a simple injective in mod-L, but, by 15A, V is not injective in L-mod, and hence neither is L. (Thus, L is not a left V-ring.[4])

<u>First we show that</u> R <u>is a right</u> V <u>ring</u>. Let W be a simple right (or left ideal) or R, and let $f : I \to W$ be a mapping of a right (or left) ideal of R. In order to determine if W is injective, it suffices to assume that I is an essential one-sided ideal, and thus, that I contains the socle S of R. Since R/S is simple, either $I = S$, or $I = R$. If $I = S$, and f is a morphism of mod-R, then f is a morphism of mod-L, since

$$[f(xa) - f(x)a]s = 0 \qquad \forall s,\ x \in S,\ a \in L.$$

Since W is a summand of L in mod-L, then W is injective along with L, and so f has an extension to a mapping $L \to W$ which induces a mapping $R \to W$ extending f. This proves that W is injective in mod-R. If W is a simple right R-module not contained in S, then $W \approx R/S$ is injective over R/S, whence by 23, W is injective over R. This proves that R is a right V-ring.[5]

In order to prove that R is not a left V-ring, assume for the moment that V is injective in R-mod, and that $f : I \to V$ is a morphism in L-mod, where I is a left ideal of L. In order to extend f, we may assume that I is an essential left ideal, that is, that $I \supseteq S$. Then, f has an extension f' in R-mod, and f' is a morphism of L-mod, since

$$s[(ax)f' - a(x)f'] = 0 \quad \forall s \in S,\ x \in I,\ a \in L.$$

But this implies that V is injective in L-mod, a contradiction. Thus, R is not a left V-ring. (This fact is also a consequence of the author's [67a, p. 103, Theorem 3.1] which implies that the maximal left quotient ring of R is also a right quotient ring. This involves a contradiction.)

Two additional remarks on this example:

(a) Let Q denote the maximal left quotient ring L. Then the injective hull in L-mod of V is QV = Qe, and

$$\mathrm{End}_L \hat{V} = \mathrm{End}_L Qe = \mathrm{End}_Q Qe = eQe = eLe = \mathrm{End}_L V$$

(the last equality since eQe is the quotient ring of eLe).

(b) V is a fully invariant proper L-submodule of its injective hull. (cf. (3) of the following Theorem 34.)□

QI RINGS

The next obvious remark is:

26. *Obvious Remark. If R is a ring, and M in mod-R is such that $M \oplus \hat{R}$ is QI, then M is injective.*

Proof. For then $M \oplus \hat{R}$ is a QI module containing R, hence is injective. □

This proves the equivalence of (1) and (2) in the next proposition.

27. *Proposition*. (Koehler [70]). *The two conditions on a ring* A *are equivalent:*

(1) A *is a right* QI *ring*.

(2) *The direct sum of any two* QI *right* A-*modules is again a* QI *module*.

When this is so, then A *is right Noetherian*.

Proof. Since any semisimple module is injective, then A is right Noetherian by 9C. □

29. Proposition. *If A is a right V-ring, and if B is a proper ideal* $\neq 0$, *then* $xA \not\approx A$ (*xA is not isomorphic to A*) *for any* $x \in B$.

Proof. Let M be a maximal right ideal containing B. Then $V = A/M$ is simple, hence injective. The condition $xA \approx A$ is equivalent to the existence of elements $b, c \in A$ such that $y = xa$ has nonzero right annihilator $y^{\perp}$ in A, and $x = yb$. (Then $ba = 1$ and $yA = xA$.) In this case, for any $v \in V$ there is a map $xA \to V$ sending yr onto vr for any $r \in A$. Since V is injective, there exists $m \in V$ such that $v = my \in Vy$. Thus, $V = Vy$, which shows that $y \notin B$. Then $x \notin B$. □

30. **Proposition**. *A ring A is a finite product of simple rings if and only if the three conditions hold*:

FPS_1: *A has finite ideal lattice*.

FPS_2: *A has no proper ideals which are essential right ideals*.

FPS_3: *A is semiprime*.

Proof. The necessity is obvious. Conversely, assume FPS, and let B be a maximal ideal. Since B is not essential, then $B \cap I = 0$ for some nonzero right ideal I. Thus, $IB = 0$ (since $IB \subseteq I \cap B$) and then $(BI)^2 = 0$, so that semiprimeness of A implies $BI = 0$. Thus $A_1 = \text{ann}_A B$ is a nonzero ideal. Semiprimeness of A implies $A_1 \cap B = 0$, and maximality of B yields a ring product $A = B \times A_1$. Since B inherits FPS, by induction on the number of ideals of A, the proof is done. □

The next theorem was first proved by Faith in 1964 (published in [67a]) assuming A is prime, and Ornstein [66] assuming A is right Noetherian.

31. Theorem. *Any right Goldie right V-ring is a finite product of simple right V-rings.*

Proof. Any right V-ring A has rad A = 0, hence is semiprime (FPS_3). Since A is right Goldie, A has a semisimple right full quotient ring Q. If Q is a direct sum of n simple right modules, then every set of independent right ideals of A has cardinality $\leq n$. Furthermore, every essential right ideal contains a regular element of A (Goldie [58,60]). Thus, 29 shows that A has FPS_2. Then the bound n on the summands of A, and the proof of 30, show that A has FPS_1. Thus, A is a finite product of simple rings by 30.□

32. Proposition. *If A is right QI, then End E_A is a field for any indecomposable right A-module E.*

Proof. Let S = End E_A. By 6, E has the d.a.c. for finitely generated S-modules. Since A is right Noetherian then E satisfies the d.c.c. on finitely generated S-submodules, and thus, the socle W of E in S-mod is nonzero. Since W is an (S,R)-submodule of E, then W is a QI, hence injective, right A-module. Since E is indecomposable, this implies that W = E. Since the radical J of S annihilates the S-socle of E, then $JW = JE = 0$, and $J = 0$. Since S is a local ring, then S is a field.

The proof is more general than the proposition. See Statement 1 of the following paper [72].

33. Proposition. *Let R be a simple right QI ring. Let E_i be non-isomorphic indecomposable injective right R-modules, $i = 1,2$, such that $\text{Hom}_R(E_1, E_2) \neq 0$, and let $F = E_1 \oplus E_2$. Then:*

(1) $\text{Hom}_R(E_2, E_1) = 0$.

(2) *$B = \text{End } F_R$ has "square zero" radical J such that $E_2 = JF = \text{ann}_F J = \text{socle}_B F$.*

(3) *B is a semiperfect ring, and if $e_i : F \to E_i$ is the projection endomorphism, then $D_i = e_i B e_i$ a field, $i = 1,2$, and $J = e_2 B e_1$.*

(4) *$A = \text{End}_B F$ embeds canonically in $\text{End}_{D_i} E_i$, $i = 1,2$.*

(5) *In the case E_1 is isomorphic to the minimal right ideal of $Q = Q(R)$, then $A = \text{End}_B F$ is a simple right Goldie ring with $Q(A) = Q$.*

Proof. An endomorphism ring $B = \text{End } F_R$ has a decomposition into a direct sum of indecomposable right ideals $e_i B$ such that $e_i B e_i$ is a local ring, $i = 1,\ldots,n$, iff F has a decomposition into a finite direct sum of indecomposable modules having local endomorphism ring. Thus, for the module $F = E_1 \oplus E_2$ described, certainly $B = \text{End } F_R$ is semilocal, and idempotents of B/J lift. Thus, B is semiperfect in the sense of Bass [60].

The trace $T(E_1, E_2)$ of E_1 in E_2 is a nonzero fully invariant submodule of E_2, and hence equals E_2. This proves that $BE_1 = F$, since any $f : E_1 \to E_2$ extends to an element of B. Since the endomorphism ring D_i of E_i is a field isomorphic to e_iBe_i, and $\text{rad}(e_iBe_i) = e_iJe_i$, $i = 1,2$, then $e_iJe_i = 0$, $i = 1,2$. Now any nonzero $f : E_1 \to E_2$ is induced by an element (also denoted f) of B with essential kernel. Then $f \in J$, and hence $J \neq 0$. Since JF is fully invariant, then $JF = F$, or E_1, or E_2. But since $F = BE_1$, then $JF = F$, or $JF = E_1$, would imply $E_1 = e_1Je_1E_1 = 0$, a contradiction. Then $JF = E_2$. Similarly, $\text{ann}_F J = E_2$. (Also, for the latter, the intersection of the kernels of all $b \in B$ such that $bE_2 = 0$ and $bE_1 \neq 0$ is a fully invariant submodule $\neq F$, hence $= E_2$. But these b all belong to J.) Direct computation shows that $J = e_2Be_1$. This proves (1) - (3).

Any summand S of a module M is canonically a module over the biendomorphism ring A of M, that is, $SA \subseteq S$. Moreover, in the case $F = E_1 \oplus E_2$ being considered, $E_1a = 0$ for some $a \in A$ implies that $Fa = BE_1a = 0$. Thus, since R is simple, then F is faithful, so $a = 0$, and A embeds canonically in $\text{Biend } E_1$. Similarly, if $E_2a = 0$, then

$$JE_1a = JBE_1a = JFa = E_2a = 0,$$

so that

$$E_1a \subseteq E_1 \cap \text{ann}_F J = E_1 \cap E_2 = 0,$$

and hence $a = 0$. This proves that a embeds canonically in Biend E_2.

For any ring R, $\mathrm{Hom}_R(R, E_2) \approx E_2 \neq 0$. Since R is prime, then the right quotient ring Q is an essential extension of R, so any nonzero $f:R \to E_2$ extends to an element (also denoted f) of Q. Thus, if E_1 is the minimal right ideal of Q, then E_1 is indecomposable injective in mod-R, and f induces a nonzero map $E_1 \to E_2$. Then, by the following proposition 34, $Q = \mathrm{End}_{D_1} E_1$, and hence there are inclusions $R \subseteq A \subseteq Q$. Thus, simplicity of R implies that of A. Moreover, A is a right Goldie ring with $Q(A) = Q$. □

(5) of the proposition is interesting because A is a "good" ring which is the endomorphism ring of a "bad" module over B. Also, it is easy to see that E_1 and E_2 remain injective and indecomposable over A, that is, the hypotheses are satisfied when R is replaced by A. (Actually, is $R = A$?)

34. Theorem. *The following conditions on a ring R are equivalent:*

(1) *R is a prime right Goldie ring.*

(2) *R has simple Artinian right quotient ring $Q(R)$.*

(3) *R has a faithful finendo indecomposable injective right R-module E, having no nontrivial fully invariant submodules, and $D = \mathrm{End}\, E_R$ is a field.*

When this is so, then $Q(R) \approx \mathrm{End}_D E$, and E is isomorphic in mod-Q(R) to a minimal right ideal of $Q(R)$.

Proof. (1) <=> (2) is a theorem of Goldie [58, 60], and

Lesieur-Croisot [59]. However, we shall prove (1) <=> (2) and (2) <=> (3) without recourse to these theorems.

(3) ⟹ (2). Let $n = \dim_D E$. Then $A = \mathrm{End}_D E$ is a full matrix ring D_n over the field D. Moreover, there is a canonical embedding of the right A-module E in A. Since E is a simple right A-module, then E is a minimal right ideal of A. Since E is faithful in mod-R, then R embeds in A canonically. Now $A \approx E^n$ in mod-A. Write $E^n = E_1 \oplus \ldots \oplus E_n$, a direct sum of n isomorphic A-submodules of A. If, for example, $R \cap E_1 = 0$, then the projection $E^n \to E_2 \oplus \ldots \oplus E_n$ maps R isomorphically in mod-R into E^{n-1}, which, by the proof of 13, implies that $D^{n-1} \to E \twoheadrightarrow 0$ is exact in D-mod, contrary to the assumption that $\dim_D E = n$. Thus, if $U_i = R \cap E_i$, $i = 1,\ldots,n$, then R contains an essential R-submodule $U_1 \oplus \ldots \oplus U_n$ of $A = E^n$. This proves that the right R-module A is the injective hull of R in mod-R. By the Krull-Schmidt theorem applied to E^n, this proves that R satisfies the a.c.c. on direct sums of right ideals contained in R. Moreover, any subring of a right Noetherian ring satisfies the a.c.c. on right annulets, so we have proved that R is a right Goldie ring, which together with primeness of R (see below), implies (1).

Write $E = eA$, for some idempotent $e \in A$, and let $D = eAe$. We invoke for the first (and last) time the hypothesis on fully invariant submodules: If I is any nonzero right ideal of R, then DI is a fully invariant submodule of E, so that $DI = E$, and there exists a basis of E over D consisting of n elements $x_1,\ldots,x_n$ of I. Then $E = \Sigma_{i=1}^n Dx_i$, and faithfulness of E implies that $\cap_{i=1}^n x_i^{\perp} = 0$. Thus, (R is prime and) R embeds as a right R-module in a direct sum I^n (under a map $r \longrightarrow (x_1r,\ldots,x_nr)$) of n copies of any nonzero right ideal. Now, if H is any essential right ideal of R, then $V_i = E_i \cap H \neq 0$, $i = 1,\ldots,n$, so that H contains the direct sum $V_1 \oplus \ldots \oplus V_n$ of uniform right ideals. Moreover, since R is prime, $V_iV_1 \neq 0$, for any i. If $y_i \in V_i$ is such that $y_iV_1 \neq 0$, then V_1 embeds in V_i under the map $v \longmapsto y_iv$, since $y_iv = 0$ for some nonzero $v \in V_1$ would imply that $y_iE_i = 0$, whence $y_iV_1 = 0$, by virtue of the fact that E_i is a minimal right ideal of A. This proves that H contains a direct sum V^n of right ideals isomorphic to $V = V_1$, and hence that R embeds in H as a right R-module. Therefore, any essential right ideal H of R contains an element x such that x has zero right annihilator in R, and, by the fact that R is essential in A, has zero right annihilator in A. But in an Artinian ring A, this implies that x is a unit in A (and hence that x is a regular element of R). Similarly any regular element of R is a unit of A.

The proof that A is the right quotient ring of R is is immediate from this. If q is any element of A, then, since R is an essential R-submodule of A, the right ideal $(q:R)$ is essential in R, and therefore contains a regular element x. Thus, $q = ax^{-1}$, where $a = qx \in R$, proving that A is the right quotient ring $Q(R)$ of R.

(1) $\Rightarrow$ (3). By the a.c.c. on complement right ideals (or equivalently, the a.c.c. on direct sums contained in R), R contains a uniform right ideal, which can be chosen to be a right complement U of any maximal right complement right ideal. (Then, U is a minimal complement right ideal.)

Assertion: The injective hull $E = \hat{U}$, of any uniform right ideal U has the property stated in (3).

Proof of the assertion. We first show that the right singular ideal Z of R is nil. If $x \in Z$, then there is a finite integer n such that $y = x^n, x^{n+1},\dots,x^{2n}$, all have the same right annihilator. It follows that $y^{\perp} \cap yR = 0$, whence $y = x^n = 0$. Next, if A is a nil left ideal contained in Z, then one can show (as Utumi [63]) that the element $a \in A$ which is maximal in $\{x^{\perp} \mid 0 \neq x \in A\}$ generates a nilpotent ideal of index ≤ 3. Then (semi)primeness of R forces $A = 0$, and $Z = 0$.

Next, there is a finite integer n such that R contains an essential right ideal $W = U_1 \oplus \dots \oplus U_n$, with $U_i \approx U$, $i = 1,\dots,n$. Thus, the injective E^n hull of U^n is the injective $\hat{R}$ of R. Write $\text{End } E^n_R = B_n$, where $B = \text{End } E_R$. Since E^n is injective, the radical J of B_n consists of those b with essential kernels (theorem of Utumi [56]). Since $R \subseteq E^n$, and $b \in J$, then $x = b(1) \in E^n$ belongs to the singular submodule S of E^n. But $S = 0$, since R is an essential submodule with 0 singular submodule. Then, $\ker b \supseteq R$, and if $y \in E^n$, there is an essential right ideal I with $yI \subseteq R$, and $b(y)I = 0$, so that $b(y) \in S = 0$. Thus, $b = 0$, and therefore, B_n, hence B, has zero radical. Since B is a local ring, then B must be a field. Then, the fact that $R \subseteq E^n$ implies by 13 that $n = \dim_B E$, and hence, that $A = \mathrm{End}_B E \approx B_n$ canonically. Moreover, $E^n \approx A$ in mod-A is an essential extension of R in mod-R, and then it follows from the proof of the next proposition that E has no nontrivial fully invariant submodules. This completes the proof of the assertion, and of (1) $\Rightarrow$ (3).

Now (3) $\Rightarrow$ (1) (parenthetically in the proof of (3) $\Rightarrow$ (2)) and (2) $\Rightarrow$ (3) by the next proposition, so the equivalence of (1), (2) and (3) is completed by the next proposition. □

35A. <u>Proposition</u>. <u>Let R be a ring with simple Artinian right quotient ring Q. Let E denote a minimal right ideal of Q. Then, E is the (up to isomorphism) unique finendo indecomposable injective faithful right R-module. Moreover, E has no nontrivial fully invariant R-submodules.</u>

<u>Proof</u>. By the Wedderburn theorem, $Q \approx \mathrm{Biend}\, E_Q$. Let $D = \mathrm{End}\, E_Q$, and let $n = \dim_D E$. Since Q is Artinian, then n is finite. Since $D = \mathrm{End}\, E_R$ (in fact, $Q = \mathrm{End}\, Q_R$), it follows that the right R-module E is finendo. Since Q is the injective hull of R, and E is a summand of Q, then E is an injective, faithful, right R-module. Moreover, E is indecomposable in mod-R, since any R-summand of E is a Q-summand.

Let H be a nonzero fully invariant R-submodule of E, and write $E = H \oplus G$, for a D-submodule G of E. Let $g \in Q$ be the idempotent which induces the projection $E \to H$ parallel to G in D-mod. Then, $Eg = H$, and $H(1-g) = 0$. Since H is an R-submodule, then $gr = grg$, for any $r \in R$, that is, $gR(1-g) = 0$. But, then $(gR \cap R)((1-g)R \cap R) = 0$. Since $H \neq 0$, then $g \neq 0$, and $gR \cap R \neq 0$. Similarly $g \neq 1$ implies $(1-g)R \cap R \neq 0$, so that primeness of R implies that $g = 1$. Thus, $H = E$, so E has no fully invariant non-trivial submodules.

That any finendo faithful indecomposable injective right R-module is isomorphic to E follows from the proof of 34.□

The next theorem may be proved similarly as 34.

35B. Theorem. *A ring* R *has a semisimple right quotient ring if and only if* R *is a semiprime ring containing finite faithful set* $\{E_1, \ldots, E_n\}$ *of indecomposable finendo injective right* R-*modules, none of which have nontrivial fully invariant submodules, and the endomorphism ring of each is a field.*

The condition of the theorem is equivalent to the existence of a *faithful finendo injective module* F *such that* End F *is a finite product of* n *fields, and* F *has precisely* n *fully invariant indecomposable submodules* $\neq 0$.

A prime ring may have nonfaithful finendo indecomposable injective modules, yet not be right Goldie, as the example, Example 25, of a regular V-ring R shows, since the simple module $W = R/S$ is injective, and 1-dimensional over its endomorphism ring R/S.

36A. Corollary. *If R is a right QI-ring, and if I is any co-irreducible right ideal such that R/I is finendo, then I is a maximal complement right ideal.*

Proof. Any right QI ring is right Noetherian, and right V, hence semiprime, with semisimple classical right quotient ring Q. Moreover, by 35, every finendo indecomposable injective right R-module E embeds in Q. Thus, $E = \widehat{R/I}$ embeds in Q under a map f such that I is the right annihilator in R of $x = f[1+I]$. Since Q is semisimple, the right annihilator of x in Q is a principal right ideal eQ, that is, a complement right ideal of Q, and hence $I = eQ \cap R$ is a complement right ideal of R. Clearly, I is a maximal right complement. □

36B. Remarks. (1) More generally, if R is any right neat ring with quotient ring $\hat{R}$, then a cyclic module R/I embeds in $\hat{R}$ if and only if I is a complement right ideal.

(2) If R is semiprime right Goldie, then any torsion free module M is right QI if and only if it is injective. This follows since then M is canonically a right $Q = Q(R)$ module, and hence $\hat{M}$ is a direct sum of finendo indecomposable injective right R-modules, each of which is canonically isomorphic to a principal indecomposable right ideal (right prindec) of Q. But a right prindec E of Q has no fully invariant submodules, as the proof of 35 shows. □

PF-RINGS

For the background on the next proposition, consult Azumaya [66], Kato [68], Onodera [68], Osofsky [66], and Utumi [66].

37. Definition and Proposition. A ring R is said to be right PF provided that the equivalent conditions hold:

(1) R is injective cogenerator in mod-R.

(2) Every faithful right R-module is a generator.

(3) R is injective in mod-R, with finitely generated essential socle.

(4) R is a semiperfect right self-injective ring with essential socle.

38. Theorem. Let R be a right self-injective ring. Then the conditions which follow are equivalent:

(1) Every faithful (injective) right R-module is finendo.

(2) R is right PF.

(3) Every faithful (QI) right R-module is finendo.

Proof. The equivalence (1) <==> (3) with parentheses follows from Theorem 1. Also (2) ==> (1) by 18A, in view of (2) of 37. Hence, it suffices to prove (1) ==> (2). Let E denote the least injective cogenerator of mod-R, that is, let E be the direct sum of the injective hulls of a full set of representatives of simple right R-modules. Then, E is finendo by the hypothesis (1), and hence 18B implies that E generates mod-R. Therefore, $E^m \approx R \oplus X$, for some finite integer n, and $X \in$ mod-R. Since R is a cyclic module, then R is contained in a finite direct sum of injective hulls of simple modules, and so R has finite essential socle. Then R is right PF by (3) of 37.

The next proposition is a generalization of the well-known result for quasi-Frobenius rings of Nakayama [39, 40, 41].

39. _Proposition. Let_ R _be a basic ring with radical_ J.

(1) _Then_ R _is right_ PF _if and only if_ R _is isomorphic to the injective hull of its top,_ $R \approx \widehat{R/J}$ _in_ mod-R.

(2) _Let_ A _be an ideal of a self-basic right_ PF _ring_ R. _Then the following conditions are equivalent:_

(A) R/A _is right_ PF.

(B) _There exists_ $z \in R$ _such that_ ${}^{\perp}A = zR = Rz$.

(C) _There exists_ $z \in R$ _with_ $z^{\perp} = A$, _and_ $zR = Rz$.

(D) _There is an isomorphism_ ${}^{\perp}A \approx R/A$ _in_ mod-R.

Proof. (1) If $R \approx \widehat{R/J}$, then R is right PF by (3) of the proposition. Conversely, when R is basic and right PF, the proof of (3) ⟹ (1) of the proposition implies that $R \approx \widehat{R/J}$.

(2) The top of any module M is isomorphic to a summand of the top of M/MA, for any ideal A, since top M = M/MJ is semisimple, and there is an exact sequence

$$0 \to (MJ + MA)/MJ \to M/MJ \to \text{top } M/MA \to 0.$$

(A) $\Longrightarrow$ (B). Since $R = R/J$ by (1), and since the top of R/A embeds in R/J, then there is an embedding of the top of R/A in R. Again by (1), and the assumption (A), R/A is an essential extension of its top, and hence R/A also embeds in R. Then A is the right annihilator of the image z of $[1+A]$ under this embedding. Since R is injective in mod-R, then Rz is a left annulet, and so $Rz = {}^{\perp}A$. Since $zR \approx R/A$ is an injective cogenerator in mod-R/A, then every simple submodule W of ${}^{\perp}A$ in mod-R/A is isomorphic to a submodule V of zR. However, since R is basic, and right self-injective, necessarily $V = W$, so that socle ${}^{\perp}A$ = socle zR. Since R has essential right socle, so does ${}^{\perp}A$. Since the right socles of zR and ${}^{\perp}A$ are equal, this proves that ${}^{\perp}A$ is the injective hull of the socle of zR. But $zR \approx R/A$ is the injective hull of the socle of zR. Therefore, ${}^{\perp}A = Rz = zR$.

(B) $\Longleftrightarrow$ (C) is obvious, since every right and left ideal is an annulet. Thus, $z^{\perp} = A$ implies ${}^{\perp}A = Rz$, and conversely.

(D) $\Longleftrightarrow$ (A). Since R has finitely generated essential socle, then (D) implies that R/A has finitely generated essential socle. Since R/A is right self-injective by 8, then (3) of the last proposition implies that R/A is right PF. Conversely, (A) $\Rightarrow$ (D) by the proof of (A) $\Rightarrow$ (B).

40. _Proposition_. _The following conditions on a ring_ R _are equivalent_:

(1) R _is uniserial_.

(2) R _is a finite product of matrix rings over right and left principal ideal local rings_.

(3) Every factor ring of R is QF.

(4) Every factor ring of R is right PF.

Proof. The equivalence of (1) <=> (3) is a theorem of Nakayama [40]. Since a right PF is QF if and only if it is Artinian, for (3) <=> (4) it suffices to prove that R is Artinian. But, by (3) of 38, every right QI module is finendo, and then R is right Artinian by 17A.

The proof of (2) => (3) is easy; the proof of (3) => (2) has some interest (e.g. Fuller [69, p. 496].) It is possible to give a short proof of (3) => (2) employing some foregoing ideas. It suffices to assume that R is self-basic. Let e be a right prindec of R, and let A be the annihilator ideal of eR on the right. Then eR is canonically isomorphic to the right ideal of R/A generated by [e+A]. Since eR is faithful over a QF ring, then eR is a generator of mod-R/A, and therefore by 18A, finendo and projectivendo. But $B = \text{End } eR_R = eRe$ is a local ring, and over local rings projectives are free by Kaplansky's theorem. Say $eR \approx B^n$ in B-mod. This proves that eR is a finitely generated projective generator both in B-mod, and in mod-R/A, and by Morita's theorem 18A,

$$R/A \approx \text{End}_B eR \approx \text{End}_B B^n \approx B_n.$$

Since R is self-basic, so is R/A, and therefore $n = 1$. Then, $B = eRe \approx eR$ in B-mod, which suffices to show that e is central, and that A is the ideal generated by the central idempotent 1-e. By induction, A is a finite product of local rings, and hence so is R. Each of the local rings inherits the hypothesis on R, and trivially are left and right principal ideal rings. □

FOOTNOTES

1. The bibliographic references such as Bass [60] refer to a 1960 paper of Bass listed in the bibliography. References in parenthesis e.g. (1), (17A), etc. refer to numbered propositions, corollaries, or lemma in the text.
2. The research on this article was supported in part by a grant from the National Science Foundation.
3. This was independently observed by David A. Hill.
4. Cf. Kato [71, p. 141, Korollar 2], where the result is ascribed to Villamayor.
5. This was drawn to my attention by G. Wagner.

REFERENCES

[57] Auslander, M., "On regular group rings," Proc. Amer. Math. Soc. 8 (1957), 658-664.

[66] Azumaya, G., "Completely faithful modules and self-injective rings," Nagoya Math. J. 27 (1966), 697-708.

[60] Bass, H., "Finitistic dimension, and a homological generalization of semiprimary rings," Trans. Amer. Math. Soc. 95 (1960), 466-488.

[71] Boyle, A. K., Ph.D. Thesis, Rutgers, The State University, New Brunswick, N.J., 1971.

[70] Cozzens, J. H., "Homological properties of the ring of differential polynomials," Bull. Amer. Math. Soc. 76 (1970), 75-79.

[66] Faith, C., "Rings with ascending condition on annihilators,' Nagoya Math. J. 27 (1966), 179-191.

[67a] ________, "Lectures on Injective Modules and Quotient Rings," Lecture Notes in Mathematics, Springer, New York-Berlin-Heidelberg, 1967.

[67b] ________, "A general Wedderburn theorem," Bull. Amer. Math. Soc. 73 (1967), 65-67.

[72] ________, "On the structure of indecomposable injective modules over simple rings,

[64] ________ and Utumi, Y., "Quasi-injective modules and their endomorphism rings," Arch. Math. 15 (1964), 166-174.

[70] Fuchs, L., "On quasi-injective modules," Sc. Norm. Sup. Pisa, 23 (1969), 541-546.

[68] Fuller, K. R., "Generalized uniserial rings and their Kuppisch series," Math. Zeit. 106 (1968), 248-260.

[69] ____________, "On direct representations of quasi-injectives and quasi-projectives," Arch. Math. 20 (1969), 495-502.

[58] Goldie, A. W., "The structure of prime rings under ascending chain conditions," Proc. London Math. Soc. VIII (1958), 589-608.

[60] ____________, "Semiprime rings with maximum condition," Proc. London Math. Soc. X (1960), 201-220.

[55,64] Jacobson, Structure of Rings, Colloquium Publication vol. 37, revised, Providence R.I., 1964.

[59] Johnson, R. E., and Wong, E. T., "Self-injective rings," Can. Math. Bull. 2 (1959), 167-173.

[61] ____________, "Quasi-injective modules and irreducible rings," J. Lond. Math. Soc. 36 (1961), 260-268.

[68] Kato, T., "Some generalizations of QF-rings," Proc. Japan Acad. 44 (1968), 114-119.

[70] Koehler, A., "Quasi-projective covers and direct sums," Proc. Amer. Math. Soc. (1970), 655-658.

[70] Kurshan, R. P., "Rings whose cyclic modules have finitely generated socles," J. Algebra 15 (1970), 376-386.

[70] Levine, J., Ph.D. Thesis, Rutgers, The State University, New Brunswick, N. J., 1970.

[59] Lesieur, L., and Croisot, R., Sur les anneaux premiers Noetheriens à gauche, ann. Sci. École Norm. Sup. 76 (1959), 161-183.

[58] Morita, K., "Duality theory for modules and its applications to the theory of rings with minimum condition," Sci. Reps. Tokyo Kyoiku Daigaku, 6 (1958), 83-142.

[39,41] Nakayama, T., "On Frobeniusean Algebras, I, II," Ann. of Math. 40 (1939), 611-633; 42 (1941), 1-21.

[40] ____________, "Note on uniserial and generalized uniserial rings, Proc. Imp. Acad. Tokyo 16 (1940), 285-289.

[68] Onodera, T., "Über Kogeneratoren," Arch. Math. 19 (1968), 402-410.

[71] __________, "Eine Bemerkung über Kogeneratoren," Proc. Japan Acad. 47 (1971), 140-141.

[66] Osofsky, B., "Cyclic injective moduels of full linear rings," Proc. Amer. Math. Soc. 17 (1966), 247-253.

[67] Ornstein, A. J., Rings with restricted minimum condition, Ph.D. Thesis, Rutgers, The State University, New Brunswick, N. J., 1967.

[68] ____________, "Rings with restricted minimum condition," Proc. Amer. Math. Soc. 19 (1968), 1145-50.

[61] Rosenberg, A., and Zelinsky, D., Annihilators, Portugaliae Math. 20 (1961), 53-65.

[56] Utumi, Y., "On quotient rings," Osaka Math. J., 8 (1956), 1-18.

[63] ________, A theorem of Levitzki, Amer. Math. Monthly, 70 (1963), 286.

[67] ________, "Self-injective rings," J. Algebra 6 (1967), 56-64.

THE CANCELLATION PROPERTY FOR MODULES

by

L. Fuchs

Department of Mathematics
Tulane University
New Orleans, Louisiana

By a ring will be meant throughout an associative ring R with identity and by a module a unital left R-module.

A module A is said to have the <u>cancellation property</u> if, for modules H and K,

$$A \oplus H \cong A \oplus K$$

implies

$$H \cong K .$$

Equivalently, if $A \oplus H = B \oplus K$ with $A \cong B$ implies $H \cong K$.

The cancellation property was introduced by Jónsson and Tarski in their fundamental memoir [8]. They proved that various finite algebraic structures, including groups, have the cancellation property. Answering a question posed by I. Kaplansky, Cohn [1] and Walker [14] proved the cancellation property for finitely generated abelian groups. To certain countable abelian groups, the result has been extended by Rotman and Yen [12] and Crawley [2]. The cancellation property for finitely generated modules over Dedekind domains was established by Hsü [7] and slightly generalized by Vasconcelos [13] and Kaplansky [9]. The case of torsion-free modules of rank 1 over Dedekind domains has been discussed in a paper by Fuchs and Loonstra [5], and an almost complete solution has been given for their cancellation property.

Our primary purpose here is to exploit the idea based on an observation made in [5], namely: the cancellation property is closely related to subdirect sums of modules, i.e., to certain pullbacks. For arbitrary modules, we can use pushouts and pullbacks in order to derive necessary conditions for A to share the cancellation property. These conditions turn out to be sufficient in a few important special cases, for instance, when A is quasi-injective or when all submodules of A are projective. More explicit sufficient conditions can be established if A is such that all of its endomorphisms $\neq 0$ are monic, in particular, if A is torsion-free of rank 1 over a commutative domain.

§1. Quasi-injective modules

The following simple-minded lemma is fundamental. It is an immediate consequence of Mac Lane [11, p.72], but for completeness' sake we prove it here.

Lemma 1. *Let* γ *and* δ *be monic in the pushout square*

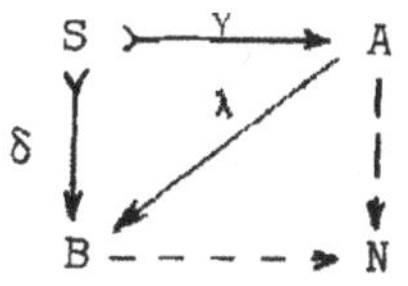

If there is a homomorphism $\lambda : A \longrightarrow B$ *making the upper triangle commute, then the pushout* N *satisfies*

(1) $N \cong \operatorname{Coker} \gamma \oplus B.$

Recall that $N \cong (A \oplus B)/H$ where $H = \{(\gamma s, -\delta s) \in A \oplus B \mid s \in S\}$. If $A' = \{(a, -\lambda a) \mid a \in A\}$, then $A \oplus B = A' \oplus B$ and $H \subseteq A'$, thus $N \cong A'/H \oplus B$. But $\operatorname{Coker} \gamma = A/\gamma S \cong A'/H$ under the map $a + \gamma S \longmapsto (a, -\lambda a) + H$, proving (1).

We shall say that two submodules, S and T, of a module A are endo-equivalent if there exist endomorphisms λ, μ of A such that the restrictions $\lambda|S$ and $\mu|T$ are inverse isomorphisms between S and T. Manifestly, isomorphic direct summands of A are endo-equivalent. In the special case when A is quasi-injective, any two isomorphic submodules of A are endo-equivalent.

Theorem 1. A module A with the cancellation property satisfies:

(a) if S and T are endo-equivalent submodules of A, then $A/S \cong A/T$.

Let λ, μ be endomorphisms of A such that $\lambda|S$ and $\mu|T$ are inverse to each other. Applying Lemma 1 to the case where $A = B$, γ is the injection of S into A and $\delta = \lambda|S$, we obtain: $N \cong A/S \oplus A$. In a similar fashion, we conclude that $N \cong A/T \oplus A$, and the cancellation property of A implies the stated isomorphism.

For direct summands S and T of A, condition (a) amounts to saying that isomorphic summands have isomorphic complements. We intend to show that for quasi-injective modules A, this already guarantees that A has the cancellation property.

Theorem 2. A quasi-injective module A (and more generally, a module A with the exchange property) has the cancellation property exactly if isomorphic summands of A have isomorphic complements.

In order to establish sufficiency, suppose A is a quasi-injective module satisfying the stated condition. If $A \oplus H = B \oplus K$ with $A \cong B$, then in view of the exchange property of quasi-injective modules (see [3]) we can write $A \oplus H = A \oplus B' \oplus K'$, $B = B' \oplus B''$ and

$K = K' \oplus K''$ for suitable submodules B', B'' of B and submodules K', K'' of K. By hypothesis, $A \cong B'' \oplus K''$ and $A \cong B = B' \oplus B''$ imply $B' \cong K''$. Hence $H \cong B' \oplus K' \cong K'' \oplus K' = K$, and A has the cancellation property, in fact.

§2. Projective modules

Now our discussion begins with the dual of Lemma 1 (cf. Mac Lane[11, p.74]). (The reader will easily recognize that from Lemma 2, Schanuel's Lemma can easily be derived.)

Lemma 2. *Let α and β be epic in the pullback square*

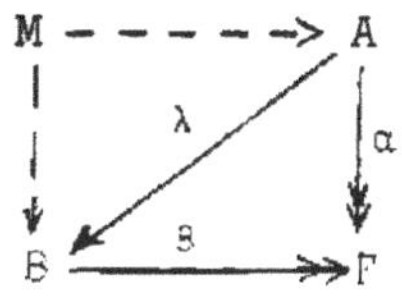

If some homomorphism $\lambda: A \longrightarrow B$ makes the lower triangle commutative, then the pullback M satisfies

(2) $\qquad M \cong A \oplus \operatorname{Ker} \beta.$

We know that M can be thought of as a submodule of $A \oplus B$ such that $M = \{(a, b) \in A \oplus B \mid \alpha a = \beta b\}$. Now $A' = \{(a, \lambda a) \mid a \in A\}$ satisfies $A \oplus B = A' \oplus B$ and $A' \subseteq M$ whence $M = A' \oplus (M \cap B)$. Since $M \cap B = \operatorname{Ker} \beta$, the proof is completed.

Let U and V be submodules of A. They will be called *quotient-equivalent* in A if there is an isomorphism $\phi : A/U \longrightarrow A/V$ such that, for suitable endomorphisms λ, μ of A, the two squares in the diagram

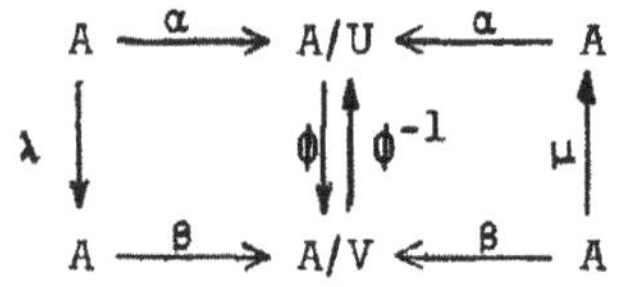

commute where α and β are the canonical maps. It is clear that direct summands with isomorphic complements are quotient-equivalent. Furthermore, it is readily seen that, in a quasi-projective module A, submodules U, V with isomorphic quotients $A/U \cong A/V$ are necessarily quotient-equivalent.

The following is dual to Theorem 1.

Theorem 3. *If a module A has the cancellation property, then*

(b) *quotient-equivalent submodules of A are isomorphic.*

Let U and V be submodules of A, and suppose the last diagram has commutative squares. Then in the diagram

(3)

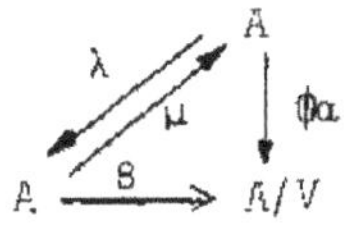

both triangles commute, therefore Lemma 2 implies that the pull-back M of $\{\phi\alpha, \beta\}$ will satisfy both $M \cong A \oplus \operatorname{Ker} \beta$ and $M \cong A \oplus \operatorname{Ker} \phi\alpha$. By the cancellation property of A, we obtain $V = \operatorname{Ker} \beta \cong \operatorname{Ker} \phi\alpha = U$.

Corollary 1. *If the quasi-projective module A has the cancellation property, then*

(b*) *for submodules U and V of A, $A/U \cong A/V$ implies $U \cong V$.*

We shall call an R-module A hereditarily projective if all submodules of A are projective. Evidently, projective modules over a hereditary ring are hereditarily projective.

Theorem 4. A hereditarily projective module A has the cancellation property if and only if it satisfies (b*).

To establish sufficiency, suppose A satisfies (b*), and let $M = A \oplus H = B \oplus K$ with $A \cong B$. Since $H/(H \cap K)$ is isomorphic to the submodule $(H + K)/K$ of $M/K \cong B$, it is projective. Hence $H = (H \cap K) \oplus U$, and similarly, $K = (H \cap K) \oplus V$ for suitable submodules U and V, each being isomorphic to some submodule of A. We obtain $\bar{A} \oplus \bar{U} = \bar{B} \oplus \bar{V}$ where bars indicate cosets mod $H \cap K$. Manifestly, the bars can be suppressed and we can start with $N = A \oplus U = B \oplus V$ where, in addition to $A \cong B$, we also have $U \cap V = 0$. Let π and σ denote the coordinate projections of N onto A and B, respectively; then $\pi|V$ and $\sigma|U$ are monomorphisms. From $\pi V \oplus U = U + V = \sigma U \oplus V$ we deduce the (natural) isomorphisms

$$A/\pi V \cong N/(U + V) \cong B/\sigma U .$$

Condition (b*) implies $\pi V \cong \sigma U$, that is, $V \cong \pi V$ and $U \cong \sigma U$ are isomorphic. Thus $H \cong K$, and A has the cancellation property.

Corollary 2. Let R be a left hereditary ring. The module $_RR$ has the cancellation property exactly if, for left ideals L_1 and L_2 of R, $R/L_1 \cong R/L_2$ implies $L_1 \cong L_2$.

Notice that if R is commutative, then for any two left ideals L_1, L_2, the isomorphism $R/L_1 \cong R/L_2$ moreover implies $L_1 = L_2$. Consequently, Corollary 2 shows that a commutative hereditary ring (in particular, a Dedekind domain) R enjoys the cancellation property;

cf. Kaplansky [9, p.75].

The second part of the proof of Theorem 4 works for arbitrary modules, hence we have:

Corollary 3. For an arbitrary module A, $A \oplus H = B \oplus K$ with $A \cong B$ and $H \cap K = 0$ always implies $H \cong K$ if and only if (b*) holds.

§3. Further necessary conditions

By making use of our Lemmas 1 and 2 in the cases when Coker $\gamma \cong$ Coker δ and Ker $\alpha \cong$ Ker β are supposed to have the cancellation property, it is easy to derive more necessary conditions for a module to have the cancellation property. In contrast to the intrinsic character of criteria (a) and (b), these are far less explicit and thus less satisfactory; as a matter of fact, they are formulated in terms of 'all' extensions of [by] 'all' modules by [of] the module under consideration. Nevertheless, they can be used under certain circumstances in order to derive criteria which are more intrinsic in nature, and thus more useful in applications.

The following two lemmas are dual to each other; it will suffice to give a proof for one of them only.

Lemma 3. Let γ, γ' and δ be monomorphisms and let the two squares be pushouts in the diagram

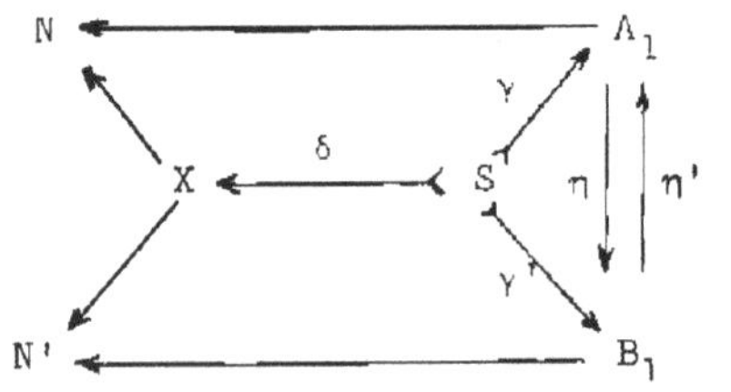

<u>If</u> $\eta: A_1 \longrightarrow B_1$ <u>and</u> $\eta': B_1 \longrightarrow A_1$ <u>make the two right triangles commute, then</u>

$$\text{Coker } \gamma \oplus N' \cong \text{Coker } \gamma' \oplus N.$$

Lemma 4. <u>Assume α, α' and β are epimorphisms and the two squares are pullbacks in the diagram</u>

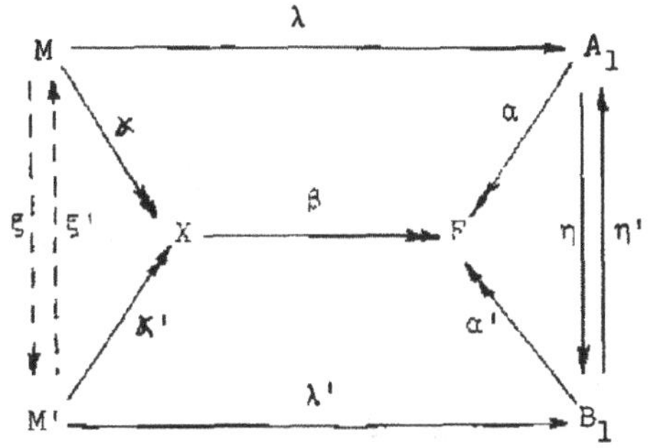

<u>If there are maps</u> $\eta: A_1 \longrightarrow B_1$ <u>and</u> $\eta': B_1 \longrightarrow A_1$ <u>making the two right triangles commutative, then</u>

$$M \oplus \text{Ker } \alpha' \cong M' \oplus \text{Ker } \alpha .$$

To prove Lemma 4, we note that $\beta\varkappa = \alpha\lambda = \alpha'\eta\lambda$ implies, in view of the pullback character of M', that there exists a $\xi : M \longrightarrow M'$ satisfying $\varkappa'\xi = \varkappa$ [and $\lambda'\xi = \eta\lambda$]. In a similar fashion, the existence of a $\xi': M' \longrightarrow M$ follows such that $\varkappa\xi' = \varkappa'$. Since both $\varkappa$ and $\varkappa'$ must be epic, Lemma 2 implies $M \oplus \text{Ker } \varkappa' \cong M' \oplus \text{Ker } \varkappa$. A reference to the pullback character of M and M' shows that $\text{Ker } \varkappa' \cong \text{Ker } \alpha'$ and $\text{Ker } \varkappa \cong \text{Ker } \alpha$, completing the proof.

Now if in the situation of Lemma 4, $\text{Ker } \alpha \cong A \cong \text{Ker } \alpha'$ and if A has the cancellation property, then $M \cong M'$ must hold. This shows the following condition is necessary for the cancellation property of A(the extra hypothesis $A_1 \cong B_1$ makes it milder, but still sufficiently general to work in the proof of sufficiency):

    (c) <u>if the diagram</u>

$$
\begin{array}{ccccccccc}
0 & \longrightarrow & A & \longrightarrow & A_1 & \overset{\alpha}{\longrightarrow} & F & \longrightarrow & 0 \\
& & & & \big\updownarrow{\scriptstyle\eta'} & & \big\| & & \\
0 & \longrightarrow & B & \longrightarrow & B_1 & \overset{\alpha'}{\longrightarrow} & F & \longrightarrow & 0
\end{array}
$$

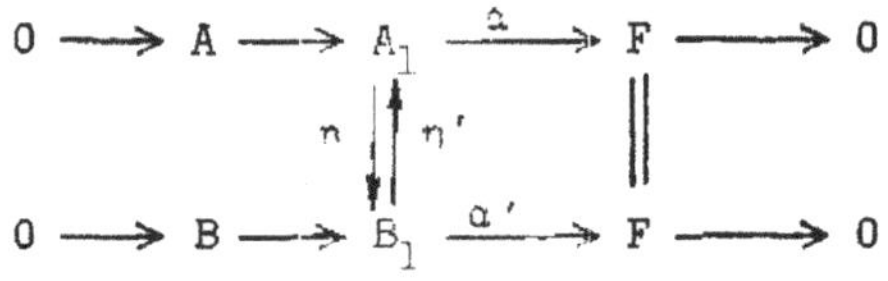

<u>with $A \cong B$, $A_1 \cong B_1$ has exact rows, and the two squares commute, then for any epimorphism</u> $\beta : X \longrightarrow F$, <u>the pullbacks of</u> $\{\alpha,\ \beta\}$ <u>and</u> $\{\alpha',\ \beta\}$ <u>are isomorphic.</u>

§4. Equipollence of pullbacks

    We shall need the following equivalence relation between certain pullbacks.

    Let the epimorphisms α, α', β and β' have the same codomain F, and suppose that the two squares in the diagram

(4)

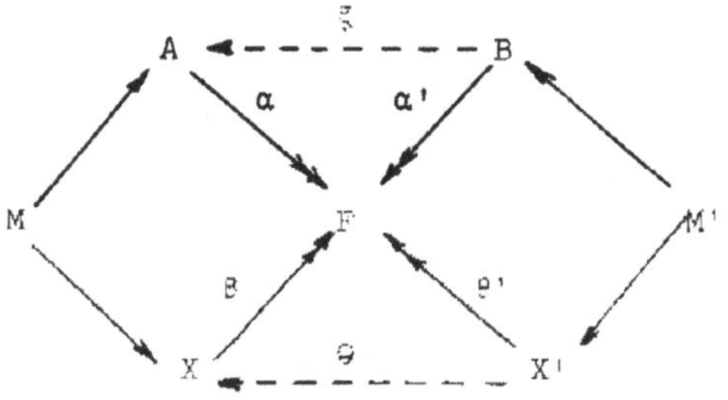

are pullbacks. They will be called <u>equipollent pullbacks</u> if there exist isomorphisms $\xi : B \longrightarrow A$ and $\theta : X' \longrightarrow X$ making two triangles commute, i.e., $\alpha\xi = \alpha'$ and $\beta\theta = \beta'$. This equipollence is a symmetric relation, because ξ^{-1} and θ^{-1} also make the arising triangles commutative.

From the definition of pullbacks, it is easy to see that equipollent pullbacks M and M' ought to be isomorphic. This follows more directly if M and M' are defined as $M = \{(a,x) \in$ $\in A \oplus X \mid \alpha a = \beta x\}$ and $M' = \{(b, x')' \in B \oplus X' \mid \alpha' b = \beta' x'\}$, for then

$$(b, x')' \longmapsto (\xi b, \theta x')$$

yields an isomorphism between M' and M.

A glance at the kernels $\operatorname{Ker} \alpha = A_0$, $\operatorname{Ker} \alpha' = B_0$, $\operatorname{Ker} \beta = X_0$ and $\operatorname{Ker} \beta' = X_0'$ reveals that ξ and θ carry B_0 and X_0' onto A_0 and X_0, respectively. Therefore, they induce automorphisms $\bar{\xi}$ and $\bar{\theta}$ of F such that $\bar{\xi}\alpha' = \alpha\xi$, and $\bar{\theta}\beta' = \beta\theta$. In the special case when $A = B$, $A_0 = B_0$, $X' = X$ and $X_0' = X_0$, somewhat more can be said.

Let $\Gamma = \operatorname{Aut} F$ denote the automorphism group of F, and let Γ_A, Γ_X consist of automorphisms in Γ which are induced by automorphisms of A and X, respectively [i.e., $\bar{\xi} \in \Gamma_A$ means that $\bar{\xi}\alpha = \alpha\xi$ for some $\xi \in \Gamma$].

Lemma 5. _The pullbacks of_ $\{\alpha, \beta\}$ _and_ $\{\varphi\alpha, \psi\beta\}$ _where_ $\varphi, \psi \in \Gamma$ _are equipollent if and only if_ $\varphi \in \Gamma_A$ _and_ $\psi \in \Gamma_X$.

In fact, setting $\alpha' = \varphi\alpha$ and $\beta' = \psi\beta$ in (4), the membership relations $\varphi \in \Gamma_A$ and $\psi \in \Gamma_X$ are equivalent to the existence of $\xi \in \operatorname{Aut} A$ and $\theta \in \operatorname{Aut} X$, rendering the triangles in (4) commutative.

We investigate cases when the mere isomorphy of the fourth corners implies the equipollence of the pullbacks.

Pairs (A, A_1) of modules will be considered such that A is always a submodule of A_1. Suppose we are given three pairs

(A, A_1), (B, B_1) and (X, X_1) such that

(i) $A_1/A \cong B_1/B \cong X_1/X$ [for the sake of convenience, these will be identified, under some fixed isomorphisms, with a module F];

(ii) $\mathrm{Hom}_R(A, X_1) = 0 = \mathrm{Hom}_R(X, A_1)$ and $\mathrm{Hom}_R(B, X_1) = 0 = \mathrm{Hom}_R(X, B_1)$.

We shall need the special case $A = B$, $A_1 = B_1$ only, but in this way our result is more general.

We have the following theorem:

Theorem 5. Let the modules A, A_1, B, B_1, X, X_1 satisfy (i) - (ii), and let M and M' be the pullbacks of $\{\alpha, \beta\}$ and $\{\alpha', \beta\}$, respectively, where $\alpha: A_1 \longrightarrow A_1/A = F$, $\alpha': B_1 \longrightarrow B_1/B = F$ and $\beta: X_1 \longrightarrow X_1/X = F$ are the canonical maps. If $M \cong M'$, then the two pullbacks are equipollent.

In the proof, it is convenient to regard M and M' as submodules of $A_1 \oplus X_1$ and $B_1 \oplus X_1$, respectively; then $A \oplus X$ and $B \oplus X$ will be contained in M and M', respectively. Assume $\mu: M' \longrightarrow M$ is an isomorphism. By (ii), the composite map $X \longrightarrow M' \xrightarrow{\mu} M \xrightarrow{\nu} A_1$ is 0 where the first map is the injection and ν is the natural map. Thus μ sends $X(\subseteq M')$ into $\mathrm{Ker}\, \nu = X(\subseteq M)$. Arguing with μ^{-1}, it follows at once that $\mu|X$ is an automorphism of X. Consequently, μ induces an isomorphism $\xi: M'/X \longrightarrow M/X$ which can be regarded as an isomorphism of B_1 with A_1, because e.g. $M/X \cong A_1$ holds canonically [under $(a, x) + X \longleftrightarrow a$]. By the same token, μ gives rise to an isomorphism $\theta: M'/B \longrightarrow M/A$ which can be viewed as an automorphism of X_1. Inspection of the maps shows that μ acts as follows:

$(b, x)' \longmapsto (\xi b, \theta x)$. Here ξ, θ are induced by μ which induces an automorphism of X and an isomorphism of B with A, thus ξ and θ induce some $\bar{\xi} \in \Gamma$ and $\bar{\theta} \in \Gamma$, i.e., $\bar{\xi}\alpha' = \alpha\xi$ and $\bar{\theta}\beta = \beta\theta$.

From $\alpha' b = \beta x$ and $\alpha\xi b = \beta\theta x$ we obtain $\alpha' b = \bar{\theta}^{-1}\beta\theta x = \bar{\theta}^{-1}\bar{\xi}\alpha' b$ whence $\bar{\xi} = \bar{\theta}$. Now M' is the pullback of $\{\alpha', \beta\}$ which is the same as the pullback of $\bar{\xi}\alpha' = \alpha\xi$ and $\bar{\theta}\beta = \beta\theta$. Hence M' and M are equipollent.

§5. The lifting property

The discussion of the cancellation property in [5] led to the notion of what was called lifting property. Here we modify the definition in order to suit better to our present general setting (but in the special case considered in [5], the two definitions coincide; cf. §8).

We say that the pair (A, A_1) has the lifting property if the following holds: if (B, B_1) is another pair with $B \cong A$, $B_1 \cong A_1$ and $B_1/B = A_1/A$ (identified under some isomorphism), and if $\lambda : A_1 \longrightarrow B_1$, $\mu : B_1 \longrightarrow A_1$ make the squares

$$\begin{array}{ccc} A_1 & \xrightarrow{\ \alpha\ } & A_1/A \\ \lambda \big\downarrow \big\uparrow \mu & & \big\| \\ B_1 & \xrightarrow{\ \beta\ } & B_1/B \end{array}$$

commutative (where α, β are the canonical maps), then there exists an isomorphism $\xi : B_1 \longrightarrow A_1$ such that $\alpha\xi = \beta$.

With the aid of the lifting property, we can show that condition (c) is satisfied. More precisely, we have:

Lemma 6. <u>If the pair</u> (A, A_1) <u>has the lifting property, then for this pair, condition</u> (c) <u>is satisfied</u>.

Let (A, A_1) have the lifting property, and let M and M' be the pullbacks of $\{\alpha, \beta\}$ and $\{\alpha', \beta\}$, as in the formulation of (c). By hypothesis, there is a $\xi: B_1 \longrightarrow A_1$ such that $\alpha\xi = \alpha'$. We then have a diagram like (4) with $X = X'$, $\beta = \beta'$ and $\theta = 1_X$; therefore M and M' are equipollent, and so isomorphic. This establishes (c) for the pair (A, A_1).

The converse of the last lemma is not true in general; in fact, the lifting property is not a consequence of the cancellation property. This can be illustrated by pairs (nZ, Z) with $n \geqq 5$ which do not share the lifting property. In order to derive the lifting property from the cancellation property, an additional condition is needed.

We call the pair (A, A_1) <u>isolated</u> if there exists an other pair (X, X_1) with the following properties:

1) $A_1/A = X_1/X = F$(identified under a fixed isomorphism);

2) $\mathrm{Hom}_R(A, X_1) = 0 = \mathrm{Hom}_R(X, A_1)$.

We can now prove:

Theorem 6. <u>Let A have the cancellation property. Then every isolated pair</u> (A, A_1) <u>enjoys the lifting property</u>.

Assume (A, A_1) is an isolated pair and A has the cancellation property. Let $\varphi \in \mathrm{Aut}\, F$ and let the endomorphisms λ, μ of A_1 satisfy $\varphi\alpha = \alpha\lambda$ and $\varphi^{-1}\alpha = \alpha\mu$, where $\alpha: A_1 \longrightarrow A_1/A = F$ is the canonical map. If (X, X_1) is a pair satisfying 1) - 2), and if $\beta: X_1 \longrightarrow X_1/X = F$ is again the canonical map, then we form the pullbacks M and M' of the maps $\{\alpha, \beta\}$ and $\{\varphi\alpha, \beta\}$, respectively.

In view of the existence of λ, μ, Lemma 4 is applicable, and we obtain $M \oplus A \cong M' \oplus A$. Cancelling by A, we have $M \cong M'$. We find ourselves in the situation of Theorem 5, whence the equipollence of the pullbacks follows. Lemma 5 implies $\varphi \in \Gamma_{A_1}$, as we wished to show.

§6. Consequences of the necessary conditions

Our next goal is to find out how much can be said about the cancellation property of a module A if it satisfies conditions (a) and (b).

In this section and in the next one we shall adhere to the following notation. We write

(5) $\quad M = A \oplus H = B \oplus K \qquad \text{with } A \cong B \neq 0,$

and for the corresponding projections we use the notations

$$\pi : M \longrightarrow A, \ \rho : M \longrightarrow H, \ \sigma : M \longrightarrow B, \ \tau : M \longrightarrow K.$$

Evidently, $\pi K \oplus H = H + K = \sigma H \oplus K$ whence

$$A/\pi K \cong M/(H + K) \cong B/\sigma H.$$

Here the isomorphisms, given by $a + \pi K \longleftrightarrow a + (H + K) = b + (H + K) \longleftrightarrow b + \sigma H$ ($a = b + k$ with $a \in A$, $b \in B$, $k \in K$), are natural, and $A/\pi K$ can be identified in the obvious way with $B/\sigma H$. The arising two triangles in

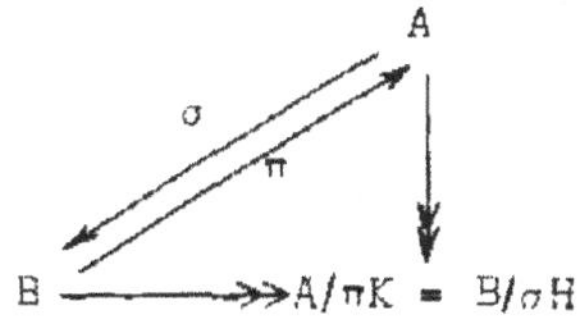

are commutative (the unnamed maps are the natural maps). Condition (b) implies $\pi K \cong \sigma H$. Since obviously $\pi K \cong K/(H \cap K)$ and $\sigma H \cong H/(H \cap K)$, the isomorphism

(6) $\qquad H/(H \cap K) \cong K/(H \cap K)$

results. Thus condition (b) alone guarantees that H and K are extensions of the same module $H \cap K$ by isomorphic modules (6). [Consequently, they are necessarily isomorphic whenever $H \cap K$ is 0 or is a summand of both H and K; cf. Corollary 3.]

Observing the commutativity of the triangles in the diagram

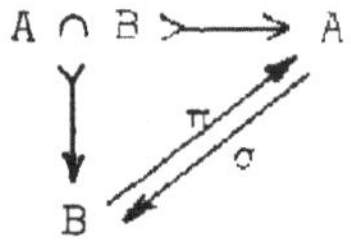

from condition (a) we deduce that $A/(A \cap B)$ and $B/(A \cap B)$ are isomorphic. These quotients are isomorphic to ${}_\tau A$ and ρB, respectively, hence

(7) $\qquad {}_\tau A \cong \rho B.$

From $A \oplus \rho B = A + B = B \oplus {}_\tau A$ we obtain the natural isomorphisms $H/\rho B \cong M/(A + B) \cong K/{}_\tau A$, as given by $h + \rho B \longleftrightarrow h + (A + B) = k + (A + B) \longleftrightarrow k + {}_\tau A$ $(h = b + k$ with $h \in H,\ b \in B,\ k \in K)$. Notice that the elements of $H \cap K$ are left fixed under these isomorphisms. Furthermore, we have

$$\rho B \cap K = (A + B) \cap H \cap K = {}_\tau A \cap H.$$

From this point on, the general case becomes very difficult to handle. A tractable particular case arises when the last intersection is 0. This holds under a stronger hypothesis which will be discussed in the next section.

§7. Modules whose endomorphisms are monic

We restrict our considerations to modules A satisfying the following condition:

(m) <u>every non-zero endomorphism of</u> A <u>is monic</u>.

We continue supposing (a) and (b) for A. Starting again with (5), first assume $\sigma\rho B = 0$. This means $\rho B \subseteq K$ whence $\pi B + K = \pi B + \rho B + K \supseteq B + K = M$. Here $\pi B \cap K = 0$, for otherwise $\sigma\pi|B$ is not monic, thus by (m) necessarily $\sigma\pi B = 0$, $\pi B \subseteq K$, in in contradiction to $\pi B + K = M$. In this case,

$$M = \pi B \oplus K \ .$$

Thus πB is a summand of A, $A = \pi B \oplus A'$ for some $A' \subseteq A$, such that $\pi B \cong M/K \cong B \cong A$. Condition (a) implies $A' = 0$, hence $M = A \oplus K$ and $H \cong M/A \cong K$.

In a similar fashion, $\pi\tau A = 0$ implies the isomorphy of H and K.

Thus we may, and from now on we shall, assume that both $\sigma\rho B \neq 0$ and $\pi\tau A \neq 0$. Then hypothesis (m) implies that $\sigma\rho|B$ and $\pi\tau|A$ are monomorphisms. We deduce that $\rho|B$ and $\tau|A$ are monic and $\rho B \cap K = 0 = \tau A \cap H$.

Notice the inclusions

(8) $\qquad \pi\tau A \subseteq \pi K \subseteq A$ and $\sigma\rho B \subseteq \sigma H \subseteq B$ where $\pi K \cong \sigma H$,

see §6. This shows that $K/(H \cap K) \cong \pi K$ and A are <u>equivalent</u> in the sense that each of them contains a submodule isomorphic to the other. In like manner, $H/(H \cap K) \cong \sigma H$ and B are equivalent.

The following result is a converse to Theorem 6, under the strong hypothesis (m).

Theorem 7. Let A be a module satisfying conditions (m), (a) and (b). If all pairs (A, A_1) with A_1 equivalent to A have the lifting property, then A has the cancellation property.

Let all pairs (A, A_1) with A_1 equivalent to A share the lifting property. Then in the non-trivial case from (5) we derive, on using (a), (b) and (m), that (6), (7) and $\rho B \cap K = 0 = \tau A \cap H$ are satisfied. The triviality of the last intersections implies that we have a diagram

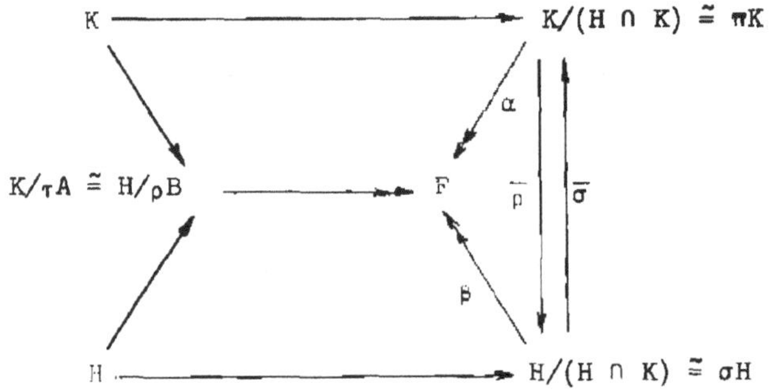

with pullback squares and two commutative triangles where F is the quotient $H/[\rho B \oplus (H \cap K)] \cong K/[\tau A \oplus (H \cap K)]$, $\bar{\rho}$ and $\bar{\sigma}$ are induced by $\rho|K$ and $\sigma|H$, and all the other maps are canonical. Since Ker $\alpha \cong {}_\tau A \cong A$ and Ker $\beta = {}_\rho B \cong B$, hypotheses of (c) hold and we infer from Lemma 6 that $H \cong K$. This completes the proof.

§8. Torsion-free modules of rank 1

We further specialize and suppose throughout this section that R is a commutative domain with K as field of quotients, and A is torsion-free of rank 1, i.e., a submodule of K. Then condition (m) is trivially fulfilled.

Assuming again (5), we can easily deduce--without referring to (a) or (b)--that only the alternatives $\sigma_\rho B \neq 0$ and $\pi_T A \neq 0$ must be considered. In this case, condition (a) can be dropped, because this was only needed to derive the isomorphism $A/(A \cap B) \cong B/(A \cap B)$ which is obviously satisfied, because $A \cap B$ is either 0 or equal to A and B.

Corollary 4. A torsion-free module A of rank 1 over a commutative domain R has the cancellation property exactly if it satisfies (c) with A_1, B_1 equivalent to A, but without assuming $A_1 \cong B_1$ in the hypothesis of (c).

The proof is essentially the same as for Theorem 7. As we noticed, (a) can be dropped, while (b) is needed only to derive the isomorphy of πK and σH. This becomes superfluous if $A_1 \cong B_1$ is dropped from the hypothesis of (c).

The last corollary holds true if R is an arbitrary left Ore domain.

In the special case when R is a (commutative) Dedekind domain, the lifting property takes a simplified form. In this case, A_1/A is a cyclic R-module for equivalent A and A_1 (see Kolettis [10]), thus the every homomorphism $\beta: A_1 \longrightarrow A_1/A$ with kernel A is the composition of the natural homomorphism α with a multiplication $\hat{r}$ by some $r \in R$ (which acts as an automorphism on A_1/A). If multiplication $\hat{s}$ by $s \in R$ is inverse to $\hat{r}$ on A_1/A, then choosing $\mu = \hat{r}$ and $\lambda = \hat{s}$ in the diagram of §5, the squares will be commutative. The existence of an isomorphism ξ simply says that ξ

induces $\hat{\gamma}$, so the lifting property is equivalent to $\Gamma_{A_1} = \Gamma =$ $= \text{Aut } A_1/A$ in the notation of §4.

For Dedekind domains R, condition (b) can be dropped. In the case when (b) is needed, the quotient A/V is cyclic, so the existence of the maps λ, μ in the diagram can be established exactly in the same way as in the preceding paragraph. Now $V = IA$ for some ideal I of R, and if I is the largest ideal with this property, then $A/V \cong R/I$ (see Kolettis [10]). Thus if U is a submodule of A, quotient-equivalent to V, then $U = IA$ holds, thus $U = V$, and condition (b) is fulfilled. Thus we have proved the following result:

Theorem 8 (Fuchs and Loonstra [5]). Let R be a Dedekind domain and A a torsion-free R-module of rank 1. If all pairs (A, A_1) with A_1 equivalent to A have the lifting property, then A has the cancellation property.

For $R = Z$, A and $A_1 (\geq A)$ are equivalent exactly if $A = mA_1$ for some integer $m > 0$ and every pair (mA_1, A_1) is isolated except for the pairs (mZ, Z) and (Q, Q). We conclude that a torsion-free abelian group A of rank 1 has the cancellation property exactly if either $A \cong Z$ or for every integer $m > 0$, the automorphisms of A/mA are induced by automorphisms of A (see Fuchs and Loonstra [5]).

§9. Final comments

1) A property which is somewhat stronger than the cancellation property has been discussed in [4]. This was actually established by Crawley [2] for the groups he considered, and this also holds for semilocal rings as modules over themselves (their

cancellation property is proved in algebraic K-theory).

We say that the module A has the <u>substitution property</u> if $M = A \oplus H = B \oplus K$ with $A \cong B$ implies that there is a submodule C of M such that $M = C \oplus H = C \oplus K$. It is easy to see that Z as an abelian group fails to have this property.

In [4], it is proven that A has the substitution property if and only if, for every R-module H and every splitting epimorphism $\sigma: A \oplus H \longrightarrow A$ there exists a $\chi: A \longrightarrow H$ such that the composite

$$A \xrightarrow{1 \oplus \chi} A \oplus H \xrightarrow{\sigma} A$$

is an automorphism of A. This condition has several applications in special cases, see [4].

2) Corollary 3 and the substitution property suggest that the following property of the module A might be worthwhile to look at: if $A \oplus H = B \oplus K$ with $A \cong B$, then there exists an isomorphism $H \cong K$ fixing the elements of $H \cap K$. It is readily checked that this is weaker than the substitution property.

3) A property which is, for indecomposable modules A, much stronger than the substitution property is the <u>exchange property</u>. Warfield [15] and [16] proved remarkable results for modules with the exchange property.

4) It should be pointed out that the infinite cyclic group fails to have the cancellation property in non-commutative groups, as noticed by Hirshon [6].

REFERENCES

1. P. M. Cohn, The complement of a finitely generated direct summand of an abelian group. Proc. Amer. Math. Soc. 7 (1956), 520-521.

2. P. Crawley, The cancellation of torsion abelian groups in direct sums, J. Algebra 2(1965), 432-442.

3. L. Fuchs, On quasi-injective modules, Annali Scuola Normale Sup. Pisa 23(1969), 541-546.

4. L. Fuchs, On a substitution property of modules, Monatshefte f. Math. 75(1971).

5. L. Fuchs and F. Loonstra, On the cancellation of modules in direct sums over Dedekind domains, Indagationes Math. 33 (1971), 163-169.

6. R. Hirshon, On cancellation in groups, Amer. Math. Monthly 76 (1969), 1037-1039.

7. C. S. Hsü, Theorems on direct sums of modules, Proc. Amer. Math. Soc. 13(1962), 540-542.

8. B. Jónsson and A. Tarski, Direct decompositions of finite algebraic systems, Notre Dame Math. Lectures, No. 5 (1947).

9. I. Kaplansky, Infinite abelian groups (Ann Arbor, 1969).

10. G. Kolettis, Homogeneously decomposable modules, Studies on Abelian Groups, Symposium, Montpellier 1967, 223-238.

11. S. Mac Lane, Homology (New York, 1963).

12. J. Rotman and T. Yen, Modules over a complete valuation ring, Trans. Amer. Math. Soc. 98 (1961), 242-254.

13. W. V. Vanconcelos, Ideals and cancellations, Math. Zeitschr. 102 (1967), 353-355.

14. E. A. Walker, Cancellation in direct sums of groups, Proc. Amer. Math. Soc. 7 (1956), 898-902.

15. R. B. Warfield, Jr., A Krull-Schmidt theorem for infinite sums of modules, Proc. Amer. Math. Soc. 22 (1969), 460-465.

16. R. B. Warfield, Jr., Exchange rings and decompositions of modules, to appear.

THE STRUCTURE OF NOETHERIAN RINGS

by

Alfred W. Goldie

University of Leeds

Leeds, England

Chapter 1. Semi-prime rings

R is a ring which need not have 1, and M a right R-module which is faithful

$$mR = 0 \, , \, m \in M \Longrightarrow m = 0.$$

(1.01). A submodule $N \subset M$ is <u>essential in M</u> if $N \cap X = 0$ implies that $X = 0$; X is a submodule of M.

In other words, M is an <u>essential extension</u> of N. Immediate consequences of the definition are:

(1.02). If N_1, N_2 are essential in M then so is $N_1 \cap N_2$. If N is essential in P and P is essential in M then N is essential in M.

(1.03). <u>Lemma</u>. Let $M = M_1 \oplus \dots \oplus M_k$ be a direct sum and N_i is essential in M_i for $i = 1,\dots,k$. Then $N = N_1 \oplus \dots \oplus N_k$ is essential in M.

<u>Proof</u>. Let $M' = N_1 \oplus M_2 \oplus \dots \oplus M_k$ and $m = m_1 + \dots + m_k \neq 0$, where $m_i \in M_i$. If $m_1 = 0$ then $m \in M'$ and if not, then $m_1R \cap N_1 \neq 0$. Choose $x \in R$ and $0 \neq m_1x \in N_1$. Then $mx \neq 0$ and $mx \in M'$. Hence, in any case, $mR \cap M' \neq 0$ and M' is essential in M.

Since we have transferred one component without loss of the property, this successive transfer shows that N is essential in M.

(1.04). A module U is _uniform_ if every non-zero submodule is essential in U.

A submodule $N \subset M$ is _irreducible_ (or _meet-irreducible_) if M/N is uniform.

When U is a uniform module and $u_1 \neq 0$, $u_2 \neq 0$ are elements then $u_1R \cap u_2R \neq 0$ so elements $a,b \in R$ exist with

$$u_1a = u_2b \neq 0.$$

This is the _common right multiple property_.

(1.05). A module M has finite _dimension_ (or _rank_) if there are no infinite direct sums of non-zero submodules in M.

(1.06). _Lemma_. A non-zero submodule N of a module M of finite rank contains a uniform submodule.

Proof. Either N is uniform or it contains a direct sum $N_1 \oplus N_1'$. Either N_1' is uniform or it contains a direct sum $N_2 \oplus N_2'$ and so on. We have a chain

$$N \supset N_1 \oplus N_1' \; ; \; N_1' \supset N_2 \oplus N_2' \; ; \; N_i' = N_{i+1} \oplus N_{i+1}' \; \ldots \; .$$

However $N_1 + N_2 + \ldots + N_i$ is a direct sum for $i = 2,3,\ldots$ and must stop for some value $i = k$. Then N_k is uniform.

This lemma makes it clear that a module M which has finite rank possesses essential submodules which are direct sums of uniform submodules. For let $S = U_1 \oplus \dots \oplus U_k$, where the U_i are essential in M. Then either S is essential in M or has zero intersection with a submodule, which contains a uniform submodule called U_{k+1}. Then S can be lengthened, a process which is not allowed to go on indefinitely. It stops when the sum is essential.

(1.07). Theorem. Let M be a module of finite rank. There is an integer $n \geq 0$ such that

(1) a direct sum of uniform submodules which is essential in M has n terms;

(2) a direct sum of non-zero submodules in M always has $\leq n$ terms;

(3) a submodule is essential in M if and only if it contains a direct sum of n uniform submodules.

Proof. In order to prove (1), consider

$$S = U_1 \oplus \dots \oplus U_m \ ; \ T = V_1 \oplus \dots \oplus V_p$$

which are essential in M and U_i, V_j are uniform. Suppose that $m \leq p$. Taking $S' = U_2 \oplus \dots \oplus U_m$, if $S' \cap V_j \neq 0$ for all j then S' is essential in T by (1.03) and hence is essential in M, which is false. Thus $S' \cap V_j = 0$ for some j and we can renumber so that $S' \cap V_1 = 0$. Set $S_1 = V_1 \oplus U_2 \oplus \dots \oplus U_m$. If $S_1 \cap U_1 = 0$ then $S_1 \oplus U_1 = V_1 \oplus S$

which is not allowed, because S is essential in M. Thus $S_1 \cap U_1 \neq 0$ and hence S_1 is essential in S and in M.

In passing from S to S_1, we have replaced U_1 by V_1 and retained the property of being essential in M. The process is repeated with S_1 and T, replacing U_2 by V_2 (with renumbering of the V_j as necessary). Eventually we see that $S_m = V_1 \oplus \ldots \oplus V_m$ is essential in M and obtain $m = p$.

The common integer $m = p$ is taken to be n.

In proving (2) let $N = N_1 \oplus \ldots \oplus N_k$ be a direct sum of non-zero terms and let $N_i \supset U_i$ for $i = 1,\ldots,k$, where U_i is uniform. Either $U_1 \oplus \ldots \oplus U_k$ is essential in M, or can be lengthened to be so. Then (1) implies that $k \leq n$.

In order to prove (3), let N be essential in M and let $U_1 \oplus \ldots \oplus U_n$ be also essential, the U_i being uniform. Now $N \supset (N \cap U_1) \oplus \ldots \oplus (N \cap U_n)$ and each $N \cap U_i$ is uniform. The converse is clear from (1).

(1.08). The <u>rank</u> of M is the integer n (defined above) when it exists. Otherwise rank $M = \infty$. Some properties are evident. A submodule N has rank $N \leq$ rank M and equality holds in the case of finite rank if and only if N is essential in M. Also rank $N = 0$ means that $N = 0$ and rank $N = 1$ means that N is uniform. For direct sums $\operatorname{rank}(\sum_{i=1}^{k} N_i) = \sum_{i=1}^{k} \operatorname{rank} N_i$.

However a factor module may have higher rank.

(ex) Take $R = \mathbb{Z}$, $M = R_R$, $N = mR$, $m = p_1 \ldots p_k$ is a product of distinct primes. Then

$$\text{rank } M = 1 \ ; \ \text{rank } (^M/_N) = k.$$

This anomaly brings out the value of a particular type of submodule, the <u>complement</u>.

Let S be a submodule of M and consider the set of submodules N of M such that $N \cap S = 0$. This set has a maximal member by Zorn's lemma. Such a submodule is a complement of S in M.

(1.09). Let S be a submodule of M. A complement K of S in M is defined by the properties:

$$K \cap S = 0 \quad \text{and} \quad L \cap S \neq 0 \quad \text{for any submodule} \quad L \supsetneq K.$$

Note that $K \oplus S$ is essential in M and in fact this is the only mark of identity of S when K alone is given. For suppose that $K \oplus T$ essential in M and let $L \supset K$ with $L \cap T = 0$. Then $L \supset K \oplus (L \cap S)$ and $(L \cap S) \cap (K \oplus T) = 0$. Thus $L \cap S = 0$ and consequently $L = K$, so that K is a complement of T.

We can now rephrase the definition of a complement K in M without specifying the S of (1.09).

(1.10) K is a complement in M if whenever $K \oplus S$ is essential in M and $L \supsetneq K$ then $L \cap S \neq 0$.

(ex) Let M be a semi-simple R-module. A submodule of M is uniform if and only if it is simple. The only essential submodule is M itself. The rank $M = n$ if and only if M is a direct sum of n simple submodules. Every submodule of M is a complement. We deduce that a module is semi-simple if and only if it has no proper essential submodules.

(1.11). Lemma. A module M has rank n if and only if it has a strictly ascending chain of submodules of length n, but none longer.

Proof. Let rank $M = n$ and $U_1 \oplus \ldots \oplus U_n$ be a sum of uniform submodules. Define a chain of complements

$$0 = K_o \subset K_1 \subset \ldots \subset K_n = M$$

by $K_{i+1} \supset K_i \oplus U_i$; $K_{i+1} \cap (U_{i+1} \oplus \ldots \oplus U_n) = 0$; $i = 0,\ldots,n-1$. Clearly $K_{i+1} \supsetneq K_i$ so that the chain has length n. No longer chain is possible, because $K \supsetneq K'$, where K, K' are complements and K' complements N, means that $K \supset (K \cap N) \oplus K'$ hence rank $K >$ rank K'.

Conversely, when we suppose that M has a complement chain of length n, then the previous remark makes it clear that rank $M \geq n$. If rank $M > n$ then M has a sum $V_1 \oplus \ldots \oplus V_{n+1}$ and if we knew that rank $M < \infty$ then one could finish with the help of the first part. In general let $V_1 \oplus \ldots \oplus V_{n+1}$ have a complement K in M and construct the complement chain

$$0 = K_o \subset K_1 \subset \ldots \subset K_i \subset \ldots \subset K_{n+1}$$

$K_{i+1} \supset K_i \oplus V_{i+1}$ and $K_{i+1} \cap (V_{i+1} \oplus \cdots \oplus V_{n+1} \oplus K) = 0$; $i=0,\ldots,n$

$K_{n+1} \supset K_n \oplus V_{n+1}$ and $K_{n+1} \cap K = 0$.

Then $K_i \subsetneqq K_{i+1}$ so that the chain has length n+1, which is not allowed. Hence rank $M = n$.

(1.12). <u>Theorem</u> Let rank $M < \infty$ and K be a complement in M. Then
$$\text{rank} \left(^M/K\right) = \text{rank } M - \text{rank } K.$$

<u>Proof</u>. Set $\overline{M} = {^M/K}$ and $\overline{N}$ as the image of a submodule N where $K \subset N \subset M$, under the natural map $M \rightarrow \overline{M}$. Let L be a complement with $L \supset K$ and V be a complement of L in M. $L \cap V = 0 \rightarrow \overline{L} \cap \overline{V} = \overline{0}$. Let $\overline{L'}$ be a complement of $\overline{V}$ in $\overline{M}$ such that $\overline{L'} \supset \overline{L}$. Let L' be its inverse image in M. Now $L' \cap V = 0$ and $L' \supset L \rightarrow L' = L \rightarrow \overline{L} = \overline{L'}$. Thus $\overline{L}$ is a complement submodule of $\overline{M}$.

On the other hand, let $\overline{P}$ be a complement of a submodule $\overline{Q}$ in $\overline{M}$ and P, Q their inverse images, so that $P \cap Q = K$. Let N be a complement of K in M and P' a complement of $Q \cap N$ in M with $P' \supset P$. Then

$P' \cap Q \cap N = 0 \rightarrow P' \cap Q = K \rightarrow \overline{P'} \cap \overline{Q} = \overline{0} \rightarrow \overline{P'} = \overline{P} \rightarrow P' = P$.

Hence P is a complement in M.

Thus the complements in M which contain K are in (1-1) correspondence with the complements in $\overline{M}$, and this correspondence preserves inclusion. The maximum length of a chain of distinct complements with endpoints K and M is rank M - rank K by (1.11) and, equally well, it is rank M/K.

(1.13). A submodule K of M is a maximal complement of any submodule of M, which properly contains K, is essential in M.

The definition ensures that K is a complement and M is the only larger complement. K is irreducible and rank K = rank $M - 1$.

(1.14). <u>Theorem</u>. A complement H of rank h in a module M of rank n is an intersection of s maximal complements, where $s \geq n - h$. If the intersection is irredundant, then $s = h - h$.

<u>Proof</u>. Because of (1.12), we can take $H = 0$. Let $U_1 \oplus \ldots \oplus U_n$ be essential in M and $N_i = U_1 \oplus \ldots \oplus U_{i-1} \oplus U_{i+1} \oplus \ldots \oplus U_n$ for $i = 1, \ldots, n$. Let N_i' be a complement with

$$N_i' \supset N_i \quad \text{and} \quad N_i' \cap U_i = 0.$$

Each N_i' is a maximal complement in M, because rank N_i' = rank $N_i = n - 1$. Using the lattice modular law repeatedly we obtain

$$\bigcap_{i=1}^{n} (N_i' \oplus U_i) = (\bigcap_{i=1}^{n} N_i') \oplus U_1 \oplus \ldots \oplus U_n.$$

As $U_1 \oplus \ldots \oplus U_n$ is essential, we have

$$0 = N_1' \cap \ldots \cap N_n'.$$

This representation of 0 is irredundant, because each U_i is common to N_j'; $j \neq i$.

Now suppose that $0 = K_1 \cap \ldots \cap K_r$, the K_i being maximal complements and the intersection irredundant. Set $U_i = K_1 \cap \ldots \cap K_{i-1} \cap K_{i+1} \cap \ldots \cap K_r$; $i = 1,\ldots,r$. Since $U_i \oplus K_i$ is essential, we see that U_i is uniform, and the modular law gives

$$\bigcap_{i=1}^{r} (U_i \oplus K_i) = U_1 \oplus \ldots \oplus U_r .$$

The left hand side is essential in M, hence $r = \text{rank } M$.

Next we consider torsion in R-modules, using essential right ideals of R to provide singularities. This gives a particular case of a torsion theory. See Lambek [24].

(1.15). Submodules N, P of M are related provided that

$$N \cap X = 0 \Longleftrightarrow P \cap X = 0 ,$$

where X is an arbitrary submodule of M.

This is clearly an equivalence relation $N \sim P$ and its classes are called <u>intersection classes</u>, written $[\sim N]$. The members of a class have the same rank (it may be infinite). The zero submodule is a class $[\sim 0]$. The essential submodules are a class $[\sim M]$. Each class has a maximal element, these are the complement submodules which belong to the class.

Let N be a submodule of M and $m \in M$. Denote

$$(m \backslash N) = (x \in R \mid mx \in N).$$

$(m\backslash N)$ is a right ideal of R.

(1.16). A right ideal E of R is <u>essential</u> if E is an essential submodule of R_R.

Let $a \in R$ and E, F be essential right ideals, then $E \cap F$ and $a\backslash E$ are essential right ideals. More generally

(1.17). <u>Lemma</u>. Let N, P be related submodules of M and $p \in P$. Then $p\backslash N$ is an essential right ideal of R.

<u>Proof</u>. Let I be a right ideal of R and suppose that $pI \neq 0$. If $pI \cap N \neq 0$ then $I \cap (p\backslash N) \neq 0$. If however $pI = 0$ then $I \subset (p\backslash N)$. Thus in either case $p\backslash N$ meets I nontrivially, so that $p\backslash N$ is an essential right ideal.

We next define a closure operation on submodules of M. Let N be a submodule and set

$$N^* = (m \in M \mid (m\backslash N) \text{ is an essential right ideal})$$

N^* is a submodule of M and we have the following.

(1.18). <u>Lemma</u>. Let N, P be submodules of M, then

(1) $N^* \sim N + 0^*$,

(2) $P \sim N \rightarrow P \subset N^*$,

(3) $N^{***} = N^{**}$ and N^{**} is a complement submodule,

(4) $N^* \cap P^* = (N \cap P)^*$; $N^{**} \cap P^{**} = (N \cap P)^{**}$.

<u>Proof</u>. Let X be a submodule of M such that $(N + 0^*) \cap X = 0$.

Let $m \in N^* \cap X$ and $mE \subset N \cap X = 0$, where E is an essential right ideal. Then $m \in 0^* \cap X$ and so $m = 0$. Since $N^* \supset N + 0^*$, we have proved (1). Now (2) follows at once from (1.17).

Now apply (1) with N^* and N^{**} replacing N. This gives $N^{***} \sim N^{**} \sim N^*$ and by (2) $N^{***} \subset N^{**}$, so that $N^{***} = N^{**}$. Because of (2) we know that N^{**} is the only maximal member of its intersection class, hence N^{**} is a complement. Finally (4) is left to the reader.

These properties enable a closure operation to be defined by

$$\text{cl } N = N^{**} .$$

Clearly $\text{cl cl } N = \text{cl } N$ and $\text{cl}(N \cap P) = \text{cl } N \cap \text{cl } P$. In a later chapter on localization it is shown that this concept fits into the topological ideal system of Gabriel [7].

(1.19). The <u>singular submodule</u> of M is 0^* and the <u>torsion submodule</u> is 0^{**}.

The (right) <u>singular ideal</u> of R is the singular submodule of R_R.

Clearly $^M/_{0^{**}}$ is torsion-free in this sense; it has zero singular (hence torsion) submodule. In general the same cannot be said of $^M/_{0^*}$. However 0^* is easier to evaluate and $0^* = 0 \Rightarrow 0^{**} = 0$.

(1.20). <u>Proposition</u>. Let R have zero singular ideal. Then the torsion and singular submodules of M coincide.

Proof. Let $m \in 0^{**}$ and $mE \subset 0^{*}$, where E is an essential right ideal. Let I be any $\neq 0$ right ideal and let $e \in E \cap I$, $e \neq 0$. There is an essential right ideal F with $meF = 0$. Since $eF \neq 0$, we have $\operatorname{ann} m \cap I \neq 0$. Hence $\operatorname{ann} m$ is an essential right ideal, so that $m \in 0^{*}$.

The singular ideal $Z(R)$ of a ring R plays an important part in the theory of non-commutative rings, so we give some examples.

(ex) Let R be an Artinian ring. The right socle E is the sum of the minimal right ideals, it is the unique minimal essential right ideal and $Z(R) = \ell(E) = (x \in R \mid xE = 0)$. The left singular ideal $Z'(R) = r(E)$. In general $Z(R)$ and $Z'(R)$ differ, but equality can sometimes be proved. See Goldie [11].

(ex) Let R be a commutative Noetherian ring. Then $Z(R)$ is the nil radical $N(R)$. For $a \in Z(R) \rightarrow r(a^n) \cap a^nR = 0$ for some $n > 0$ by max. on right annihilators and $r(a^n) \supset r(a)$ is essential. Thus $a^nR = 0$. Hence $Z(R) \subset N(R)$.

Conversely, let $a \in N(R)$ then $a^n = 0$ for some $n > 0$ and if I is any non-zero ideal, there exists $m \geq 0$ with $a^mI = 0$ and $a^{m-1}I \neq 0$. Hence $r(a)$ meets I non trivially, so $r(a)$ is an essential ideal; $N(R) \subset Z(R)$, $N(R) = Z(R)$.

(ex) Let R be the ring of all $n \times n$ lower triangular matrices with integer coefficients. Then $Z(R) = 0$ and $N(R) \neq 0$. The reader may be interested in finding $Z(R)$ for the subring of matrices with all the main diagonal entries equal.

The notion of torsion introduced above can be used to modify the definition of the rank of a module in order to bring it into line with that defined in abelian groups. The present definition coincides with the classical one only for torsion-free abelian groups.

(1.21). The reduced rank of a module M is rank M - rank 0^{**}. It is denoted by $\rho(M)$ and may be calculated as follows

$$\rho(M) = \text{rank}\left(\frac{M}{0^{**}}\right) = \text{rank } M - \text{rank } 0^{*},$$

since 0^{**} is a complement in M and since rank 0^{*} = rank 0^{**}. See (1.12) and (1.18). The improvement is apparent because instead of theorem (1.12) we now have

(1.22). Theorem. Let N be a submodule of M. Then

$$\rho(M/N) = \rho(M) - \rho(N).$$

Proof. Set $\overline{M} = M/N$ and $P \supset N$ be a submodule such that $\overline{P}$ is the torsion submodule of $\overline{M}$. Now $P = \text{cl } N = N^{**}$, hence P is a complement in M. Hence by (1.12)

$$\rho(M/N) = \text{rank}\,\frac{\overline{M}}{\overline{P}} = \text{rank}\,\frac{M}{P} = \text{rank } M - \text{rank } P.$$

Moreover we have

$$P \sim N + 0^{*} \sim N + 0^{**}$$

so that

$$\rho(^M/N) = \text{rank } M - \text{rank } (N+0^{**}) = \rho(M) - \rho(\frac{N+0^{**}}{0^{**}})$$

$$= \rho(M) - \rho(\frac{N}{N \cap 0^{**}}) = \rho(M) - \rho(N).$$

When M is torsion-free the intersection classes are more readily identified by the following result.

(1.23). Theorem. Let M be torsion-free and N, P be submodules of M. Then we have

    (1) cl N = (m ε M | (m \ N) is an essential right ideal).

    (2) N ~ P if and only if cl N = cl P ,

    (3) the class [~N] has a single maximal element cl N,

    (4) the relation ~ is a congruence relation on the
        lattice of submodules of M.

Proof. We need to prove that N* = N**. Now 0* = 0 and hence N* ~ N by (1.18) so that N** ~ N* ~ N. Again by (1.18) we have N** ⊂ N*, hence N** = N*. This proves (1). It follows that N ~ P if and only if N* ~ P*. These are complements and are related, hence N* = P*. This proves (2) and (3).

    Let N' ~ N and P' ~ P. We need to prove that

$$N' \cap P' \sim N \cap P \quad \text{and} \quad N' + P' \sim N + P.$$

Now cl(N ∩ P) = cl N ∩ cl P = cl N' ∩ cl P' = cl(N' ∩ P') also N + P ⊂ cl N + cl P = cl N' + cl P' ⊂ cl(N'+P'). Hence cl(N+P) ⊂ cl(N'+P'). The opposite inclusion follows by symmetry and completes the proof.

Quotient Rings

(1.24). An element of R is regular if it is neither a left nor right zero divisor. A set S of regular elements of R which is multiplicatively closed is a right Ore set (satisfies the right Ore condition) if $a \in R$, $c \in S$ always implies that $a_1 \in R$, $c_1 \in S$ exist with $ac_1 = ca_1$.

Naturally enough, we would like to generalize this, as in the commutative case.

Let S be as before, but without supposing that the elements are regular, merely that $0 \notin S$; let us say, S is a generalized Ore set. Define the S-singular ideal of R to be

$$(0 : S) = (x \in R \mid xs = 0, \quad \text{some} \quad s \in S).$$

$(0{:}S)$ is an ideal, because $xs_1 = ys_2 = 0$ and $a \in S$ imply that $(x + y)\, s_1\bar{s} = 0$ and $(xa)\hat{s} = 0$, whenever

$$s_1\bar{s} = s_2 h : a\hat{s} = s_1 a'. \quad \bar{s},\ \hat{s} \in S.$$

(1.25). Proposition. The image $\bar{S}$ of S in the ring $R/(0{:}S)$ consists of elements which are not right zero divisors, and $\bar{S}$ is a generalized Ore set.

Proof. Trivial

Unlike the commutative case we are not yet finished because the elements of $\bar{S}$ are only left regular $(xs = 0 \rightarrow x = 0)$ and apparently in general it cannot be finished. However we have

(1.26) Lemma. Let R have maximum condition on right

annihilators and suppose that a generalized right Ore set S consists of left regular elements. Then the elements of S are regular.

<u>Proof</u>. Let $s \in S$ and $r(s) \subset r(s^2) \subset \ldots \subset r(s^k) = r(s^{k+1}) = \ldots$ Supposing that $sy = 0$ and $ys' = s^k y'$, where $s' \in S$, then $0 = s^{k+1}y' = s^k y' = ys'$, and hence $y = 0$.

(1.27). A ring Q is a <u>right quotient ring</u> of R, with respect to a set $\mathcal{C}$ of regular elements of R, if

(1) $Q \supseteq R$;

(2) the elements of $\mathcal{C}$ are units in Q ;

(3) the elements of Q have the form ac^{-1}, where $a \in R$, $c \in \mathcal{C}$.

Notation. $Q = R_{\mathcal{C}}$.

When $\mathcal{C}$ is the set of all regular elements of R, we say that Q is the right quotient ring of R. Some authors add the adjective 'classical' because there are other quotient rings obtained using the injective hull of R_R. See Lambek [23].

The concept of a quotient ring of R presupposes that R has regular elements and, indeed, in some sense that it has 'many' regular elements. These requirements are hard to fulfil in general. It can be shown that $R_{\mathcal{C}}$ exists if and only if $\mathcal{C}$ is a right Ore set in R . The necessity of this condition is at once clear, because $c^{-1}a \in R$ and has to be of the form $a_1c_1^{-1}$. Sufficiency is usually established by an elaborate form of the construction of classes of ordered pairs (a,c) used in commutative algebra. See Jacobson [19]. This programme can also be carried out for a left quotient ring and requires that R satisfy the left Ore condition. If both conditions hold then the left and right quotient rings of R are isomorphic. Examples occur, see later, of rings with a quotient ring on one side but not the other.

We consider the full quotient ring of R, when $\mathcal{C}$ is the set of all regular elements.

(1.28). The (right) rank of R is the rank of the module R_R. A right ideal of R is uniform if it is uniform as submodule of R_R.

(1.29) Theorem. The right rank of an integral domain R is either 1 or ∞. R has a right quotient ring if and only if the rank is 1.

Proof. $R - (0)$ satisfies the right Ore condition if and only if $aR \cap cR \neq 0$, $a \in R$, $c \in R$. This is equivalent to saying that R itself is a uniform right ideal. Thus R has a right quotient ring if and only if its rank is 1.

Now suppose that rank $R = n < \infty$. R has a uniform right ideal U by lemma (1.06). Let $I \cap J = 0$: I, J right ideals. Let $u \in U$, $u \neq 0$ then $uI \cap uJ = 0$ and hence either $I = 0$ or $J = 0$. Thus R is a uniform right ideal and rank $R = 1$.

Note that when R is a right Ore domain, the quotient ring

is a division ring. The finite rank requirement is a weak form of the max-r condition, indeed, is vacuous for commutative domains. There do exist domains of infinite rank, because of the classical example of A. Malcev [28], which is a domain not embeddable in a division ring.

ex) (Well-behaved).
Let L be a Lie algebra of finite dimension over a field K and $U(L)$ the universal enveloping algebra. See Jacobson [21], chapter 5. Choose a basis $u_1,\ldots,u_n$ of L over K and suppose that

$$[u_i u_j] = \sum_k \gamma_{ijk} u_k : \gamma_{ijk} \in K : i,j = 1,\ldots,n .$$

Then $U(L)$ is isomorphic to the ring of polynomials over K in indeterminants $x_1,\ldots,x_n$ subject to the relations

$$x_i x_j - x_j x_i = \Sigma \gamma_{ijk}\, x_k \; ; \; i,j = 1,\ldots,n .$$

$U(L)$ is a domain, has max-r and max-ℓ, and has a left and right quotient division ring.

ex) (Left and right go separate ways).
Let F be a field with an isomorphism $a \to \bar{a}$ which is not an automorphism and $\bar{F}$ be the subfield of images. Denote $F(x,-)$ the ring of polynomials $a_o + xa_1 + \ldots + x^k a_k$, $k \geq o$, $a_i \in F$, subject to $ax = x\bar{a}$. The right ideals of this domain are principal so it has a right quotient ring. It does not have a left quotient ring. The left Ore condition breaks down due to $F \neq \bar{F}$.

ex) (Simple domains, which are not division rings).
$F(\xi)$ is the ring of rational functions over a field F of characteristic zero, $F[x,\xi]$ is the ring of differential polynomials

$$\alpha x = x\alpha + \frac{d\alpha}{d\xi} \quad ; \quad \alpha \in F(\xi).$$

$F[x,\xi]$ is an integral domain, is simple and is a principal right and left ideal ring.

Take F again and let $A_1 = F[x,y]$, $xy - yx = 1$. Set $A_n = A_1 \otimes_F \ldots\ldots \otimes_F A_1$ (n times) (the Jordan-Weyl algebra). It is a factor algebra of $U(L)$, L a nilpotent Lie algebra (certainly when F is the complex field) A_n is a simple domain with max-r and max-ℓ. We need it again in chapter 3.

Semi-prime rings

(1.30). R a ring and S a subset of R

$$r(S) = (x\in R \mid Sx = 0) \ ; \ \ell(S) = (x\in R \mid xS = 0).$$

$r(S)$ is a right ideal, $\ell(S)$ a left ideal, the right and left annihilators of S in R.

(1.31). Lemma. Let R be a ring which has max. for right annihilators. Then a nil non-zero left (right) ideal contains a nilpotent non-zero left (right) ideal.

Proof. $L \neq 0$ is given. If $a \in L$, $a \neq 0$, $aR = 0$ then the principal left ideal generated by a, say R^1a, is nilpotent and in L. So let $r(a) \neq R$, for all $a \in L$, $a \neq 0$. The set of right annihilators of elements ($\neq 0$) in L has a maximal member $r(b)$, $b \in L$. Let $y \in R$ and $(yb)^{k-1} \neq 0$, $(yb)^k = 0$. Now $r(b) = r((yb)^{k-1})$ by maximality, hence $byb = 0$. Thus $bRb = 0$ and $(Rb)^2 = 0$.

For right ideals note that Ra is nil (nilpotent) if and only if aR is nil (nilpotent).

(1.32) Lemma. Let R be a ring which has max. for right annihilators and has finite rank. Let $a \in R$, then there is an $n > 0$ with $a^nR \oplus r(a^n)$ is an essential right ideal.

Proof. The chain $r(a) \subset r(a^2) \subset \dots$ stops at n, say $r(a^n) = r(a^{n+1}) = \dots$. Then $r(a^n) \cap a^nR = 0$; let I be a right ideal with $I \cap (a^nR + r(a^n)) = 0$.

The sum $I + a^nI + a^{2n}I + \dots$ is direct and hence $I = 0$, because R has finite rank.

(1.33). In particular. Let $c \in R$ and $r(c) = 0$ then cR is an essential right ideal.

(1.34). R is semi-prime if it has no non-zero nilpotent ideals (right, left, two-sided are equivalent suppositions). R is prime if $aRb = 0$. $a,b \in R$ implies that either $a = 0$ or $b = 0$.

(1.35). Let R be a semi-prime ring of finite rank and having max. on right annihilators. Then the singular ideal $Z(R)$ is zero.

Proof. Let $z \in Z(R)$ and $z^n R \cap r(z^n) = 0$ for some $n > 0$. Now $z^n \in Z(R)$, hence $r(z^n)$ is essential as a right ideal, so that $z^n R = 0$; $z^{n+1} = 0$. Thus $Z(R)$ is a nil ideal and must be zero by (1.31).

(1.36). A ring R has a right quotient ring Q.

(1) Let $c_1,\ldots,c_k$ be regular elements of R, then $c, d_1,\ldots,d_k$, which are regular in R exist such that

$$c_i^{-1} = d_i c^{-1}, \qquad i = 1,\ldots,k.$$

(2) Let I be a right ideal, then $IQ = (ac^{-1} \mid a \in I,$ c regular in $R)$.

(3) Let J be a right ideal of Q, then $J = (J \cap R)Q$.

(4) A right ideal E of R is essential in R if and only if EQ is essential in Q.

Proof. Part (1) is proved by induction. When $k = 2$ we know that $c_1^{-1}c_2 = bd^{-1}$; $b,d \in R$. Let $c = c_2 d = c_1 b$, then c is also regular and so is b, because $b = c_1^{-1}c$ is a unit in Q. Set $d_1 = b$, $d_2 = d$ and have $c = c_1 d_1 = c_2 d_2$ which does it.

For $k > 2$, suppose that $c', d_1', \ldots, d_{k-1}'$ exists with $c' = c_1 d_1' = \ldots = c_{k-1}d_{k-1}'$, these elements being regular, and let $c = c'd' = c_k d_k$, where d', d_k are regular. Then $c = c_1 d_1 = \ldots = c_k d_k$, where $d_i = d_i' d'$ for $i = 1,\ldots,k-1$. This gives $c_i^{-1} = d_i c^{-1}$ for $i = 1,\ldots,k$.

Part (2) follows from (1), because an element $q \in IQ$ is

of the form

$$q = i_1 a_1 c_1^{-1} + \ldots + i_k a_k c_k^{-1}; \; i_j \in I, \; a_j \in R, \; c_j \text{ regular in } R.$$

Let $c_j^{-1} = d_j c^{-1}$ $(j = 1,\ldots,k)$ then $q = ac^{-1}$, where

$$a = i_1 a_1 d_1 + \ldots + i_k a_k d_k .$$

Part (3) is now obvious.

In (4), let E be essential in R and J be a right ideal of Q such that $EQ \cap J = 0$. Then $EQ \cap R \supseteq E$, hence $J \cap R = 0$. Now $J = (J \cap R)Q = 0$, hence EQ is essential in Q.

Conversely, suppose that EQ is essential in Q and let I be a right ideal of R. Then by (2) we have $EQ \cap IQ = (E \cap I)Q$. Then $I \cap E = 0 \rightarrow IQ = 0 \rightarrow I = 0$, so that E is essential in R.

(1.37) Theorem. A ring R has a semi-simple, Artinian ring Q as its right quotient ring if and only if

(1) R is a semi-prime ring,

(2) R has finite right rank,

(3) R has max. on right annihilators.

Proof. Assume that (1), (2), (3) hold. A nil right ideal of R is zero by (1.31). Let E be an essential right ideal of R and $a \in E$ be a non-nilpotent element. Let $a^n R \cap r(a^n) = 0$ and set $a_1 = a^n$, then $r(a_1) = r(a_1^2)$. Now either $r(a_1) \cap E = 0$ or $r(a_1) \cap E \neq 0$. In the latter case, since $r(a_1) \cap E$ is not nil, we choose $a_2 \in r(a_1) \cap E$ with $a_2 \neq 0$ and

$r(a_2) = r(a_2^2)$. If $r(a_1) \cap r(a_2) \cap E \neq 0$, then continue. At the general stage, there is a direct sum

$$a_1R \oplus \ldots \oplus a_kR \oplus (r(a_1) \cap \ldots \cap r(a_k) \cap E),$$

where $a_i \in r(a_1) \cap \ldots \cap r(a_k) \cap E$ and $a_i \neq 0$, $r(a_i) = r(a_i^2)$; $i = 1,\ldots,k$. The process has to stop because R has finite rank and suppose it does at the k-th stage. Then

$$r(a_1) \cap \ldots \cap r(a_k) \cap E = 0 = r(a_1) \cap \ldots \cap r(a_n),$$

because E is essential. Then

$$r(a_1^2 + \ldots + a_k^2) = r(a_1) \cap \ldots \cap r(a_k) = 0,$$

because Σa_iR is a direct sum. (The sum of squares is used for convenience, because R may not have 1). Set $c = a_1^2 + \ldots + a_k^2$, then $r(c) = 0$. Then cR is an essential right ideal by (1.33) and, consequently, $\ell(c) \subseteq Z(R) = 0$, which means that c is a regular element in the ideal E.

Let $a,d \in R$ and d be a regular element. Taking $E = (x \in R \mid ax \in dR)$ we observe that dR is essential by (1.33) hence so is E by (1.17). We know that E has a regular element c and then $ac = da_1$, say, which gives the right Ore condition in R. Then R has a right quotient ring Q.

Now let F be an essential right ideal of Q, then $F \cap R$ is essential in R. Since $F \cap R$ has a regular element of R and this is a unit of Q, it follows that $F = Q$. Let J be any right ideal of Q and K be a right complement of J in Q, then $J \oplus K$ is essential in Q, hence $J \oplus K = Q$. This implies that the module Q_Q is semi-simple (its lattice of submodules is complemented). Now Q has a unit element, hence it is a semi-simple Artinian ring.

Conversely, let R be a ring which has a semi-simple right quotient ring Q. Let E be an essential right ideal of R, then EQ is essential in Q by (1.36). It follows that $EQ = Q$, because Q is a sum of a finite number of minimal right ideals, and $1 = ec^{-1}$, $e \in E$, c regular in R. Then $e = c$ is itself regular, so E contains a regular element. Let N be a nilpotent ideal of R, then $\ell(N)$ is an essential right ideal, which we have shown must have a regular element, hence $\ell(N) = R$, $N = 0$. Let S be a direct sum of a set of non-zero right ideals $(I_\alpha;\alpha \in \Lambda)$ which is an essential right ideal of R. Any regular element $c \in S$ can be written as a finite sum $c = x_1 + \ldots + x_n$; $x_i \in I_{\alpha_i}$, say.

Now cR is an essential right ideal, because $(cR)Q = Q$ allows us to use (1.36). As $cR \subset I_{\alpha_1} \oplus \ldots \oplus I_{\alpha_n}$, this sum is already essential in R, hence we can deduce that R has finite right rank.

Finally, the maximum condition holds for right annihilators because the relation $r_Q(S) \cap R = r_R(S)$, holds between the annihilators in R and Q of an arbitrary subset S of R. Since Q is semi-simple it has the maximum condition for right annihilators and R inherits this.

(1.38) Corollary. R is a prime ring if and only if Q is a simple ring.

Proof. Let R be a prime, then Q is easily seen to be prime, and it is simple, because it is also an Artinian ring.

Conversely, let $B \lhd R$ and $A = \ell(B)$. Then $AQ = Q$. To see this, suppose that $d^{-1}a = xc^{-1}$ where $a \in A$ and c,d regular. Then $dx = ac$ and $dxB = 0 \to xB = 0 \to x \in A$. Hence $QAQ = AQ$ and $QAQ = Q$, because Q is simple. Thus we have $AB = 0$ for $A,B \lhd R$, if and only if $A = 0$ or $B = 0$; R is prime.

Chapter 2. The Quotient Problem

The extension of theorem (1.37) to rings with nilpotent ideals is not possible for any Noetherian ring, as Small has shown in [33]. The following example taken from [33] illustrates some of the differences from the semi-prime case.

ex) R is the ring of matrices of the form $\begin{pmatrix} a & b \\ 0 & c \end{pmatrix}$ $\begin{matrix} a \in \mathbb{Z}, \\ b,c \in \frac{\mathbb{Z}}{(p)} \end{matrix}$,

where p is a prime number.

(1) R has max-r and max-ℓ and a two-sided quotient ring.

(2) If $x \in R$ and $r(x) = 0$ then $\ell(x) = 0$; x is regular.

(3) $Z(R) = 0$, $Z'(R) = (x \in R \mid \ell(x)$ essential left ideal$) \neq 0$

(4) Let $x = \begin{pmatrix} p & 0 \\ 0 & 1 \end{pmatrix}$, then $\ell(x) = 0$ and $r(x) \neq 0$.

(2.1) Let $A \lhd R$ and denote

$$\mathcal{C}'(A) = (c \in R \mid cx \in A,\ x \in R \rightarrow x \in A),$$
$$'\mathcal{C}(A) = (c \in R \mid xc \in A,\ x \in R \rightarrow x \in A),$$
$$\mathcal{C}(A) = \mathcal{C}'(A) \cap {}'\mathcal{C}(A).$$

(2.2) $A \lhd R$ is a <u>semi-prime ideal</u> if the factor ring R/A is a semi-prime ring.

$A \lhd R$ is a <u>prime ideal</u> if R/A is a prime ring.

(2.3) Let R have max-r and T be a semi-prime ideal of R. Then $\mathcal{C}'(T) = \mathcal{C}(T)$ and is non-empty. Let $x \in R$, and $c \in \mathcal{C}(T)$, then $xc' = cx' + t$ holds for suitable $t \in T$, $x' \in R$, $c' \in \mathcal{C}(T)$.

Proof. Apply theorem (1.37) to the semi-prime ring R/T. The fact that $\mathcal{C}'(0) = \mathcal{C}(0)$ in the ring R/T, is apparent from the proof of that theorem; since the singular ideal of R/T is zero.

(2.4) The nilpotent radical of R is denoted by N.

(2.5) Theorem. Let R have max-r then $\mathcal{C}'(0) \subseteq \mathcal{C}'(N)$. When $a \in R$, $c \in \mathcal{C}'(0)$ there exist $a_1 \in R$, $c_1 \in \mathcal{C}(N)$ with $ac_1 = ca_1$.

Proof. Set $T_k = \ell(N^k)$; $k = 0,1,\ldots,\rho$ where $T_0 = 0$ and $N^{\rho-1} \neq 0$, $N^\rho = 0$.
Since $cy \in \ell(N^k) \to cy\, N^k = 0 \to y \in \ell(N^k)$ we know that $c \in \mathcal{C}'(T_k)$. Applying (1.33) to R/T_k we see that the natural image of cR is an essential right ideal in the ring R/T_k, thus "$cR + T_k$ is essential over T_k". Now suppose that $aN^{k+1} = 0$ and $E = (x \in R \mid ax \in cR + T_k)$. Clearly $E \geq N$. Let $I \triangleleft_r R$ with $E \cap I = N$. If $aI \not\subset T_k$ then

$$aI + T_k \cap cR + T_k \supsetneq T_k,$$

because $cR + T_k$ is essential over T_k. Hence we have $i \in I$ with

$$ai = cy + t_k\ ,\ t_k \in T_k\ ,\ y \in R\ ,\ ai \not\in T_k.$$

Now $i \not\in N$ and $i \in E \cap I = N$, which is a contradiction. It follows that $aI \subset T_k$ and $I = N$. Then E is essential over N, so that E/N is an essential right ideal in the semi-prime ring R/N. It follows that E contains an element of $\mathcal{C}(N)$.

We have an equation

$$ad_k = cy_1 + b_k \; ; \; t_k \in T_k \; , \; d_k \in \mathcal{C}(N), \; y_1 \in R.$$

The method is applied successively to give

$$t_k d_{k-1} = cy_2 + t_{k-1} \; , \; t_{k-1} \in T_{k-1} \; , \; d_{k-1} \in \mathcal{C}(N) \; , \; y_2 \in R \; ,$$

$$\vdots$$

$$t_1 d_o = cy_{k+1} \; , \; t_1 \in T_1 \; , \; d_o \in \mathcal{C}(N) \; , \; y_{k+1} \in R.$$

Then $ac_1 = ca_1$, where $c_1 = d_k d_{k-1} \ldots d_o \in \mathcal{C}(N)$, since this set is multiplicatively closed.

Now choose $a \in \mathcal{C}(N)$, then $ca_1 \in \mathcal{C}(N)$. Let $a, x \in N$ then $ac_1 x \in N$, hence $x \in N$. Thus $a_1 \in \mathcal{C}(N)$. Now let $cy \in N$ and suppose that $yc_2 = a_1 z + n$, $c_2 \in \mathcal{C}(N)$, $n \in N$, $z \in R$. Thus $ca_1 z \in N$, hence $z \in N$ and $yc_2 \in N$, $y \in N$. It follows that $c \in \mathcal{C}(N)$.

(2.6) Corollary. Let R have max-r and be a right order in a ring Q. Then Q has max-r and the nilpotent radicals N, N' of R, Q are related by

$$N' = NQ = QNQ \; ; \; N = N' \cap R \; ;$$
$$N'^k = N^k Q \qquad ; \; k = 1,2,\ldots \; .$$

Proof. Let $J \underset{r}{\triangleleft} Q$ then $J = (J \cap R)Q$ because elements of J have the form jc^{-1}; $j \in J$, c regular in R. Now $J \cap R$ is finitely generated over R and hence so is J over Q. Thus Q has max-r.

Now $QN \subset NQ$ because $c^{-1}n = ad^{-1}$ for $n \in N$, $c \in \mathcal{C}(0)$ and $ca \in N \to a \in N$ because $c \in \mathcal{C}(N)$. We deduce at once that $(NQ)^k = N^kQ$ for $k > 0$, so that NQ is nilpotent in Q, hence $NQ \subset N'$. Now $N' \cap R \subset N$ is also clear and the rest is obvious.

(2.7) Theorem (Small [34]). Let R have max-r, then R is a right order in a right Artinian ring if and only if $\mathcal{C}(0) = \mathcal{C}(N)$.

Proof. Let $\mathcal{C}(0) = \mathcal{C}(N)$, then the right Ore condition holds in R because of theorem (2.5). It follows that R has a right quotient ring Q which has max-r and radical $N' = NQ$. The right quotient ring $\overline{Q}$ of the semi-prime ring R/N is isomorphic to Q/N' under the correspondence

$$(a+N)(d+N)^{-1} \longleftrightarrow (ad^{-1} + N') \quad ; \quad a \in R,\ d \in \mathcal{C}(0).$$

This is verified directly and the argument depends on $\mathcal{C}(0) = \mathcal{C}(N)$. Since Q/N' is semi-simple the Loewy series

$$Q \supset N' \supset N'^2 \supset \dots \supset N'^{\rho} = 0$$

can be refined to a composition series of right Q-modules as the factor module $N'^k - N'^{k+1}$ is a f.g. Q/N'-module and so has a composition series (indeed is semi-simple). Hence Q is a right Artinian ring.

Conversely, suppose that R is a right order in the right Artinian ring Q. Let $c \in \mathcal{C}(N)$ and $cq \in N'$, $q = yd^{-1}(y \in R,$ $d \in \mathcal{C}(0))$. Then $cy \in N' \cap R = N$ and, using $c \in \mathcal{C}(N)$, we have

$y \in N$ and $q \in NQ = N'$. Thus $\mathcal{C}(N) \subset \mathcal{C}(N')$. Now $\mathcal{C}(N')$ consists of units of Q/N' and units can be raised over a nilpotent ideal, so $\mathcal{C}(N')$ is the set of units of Q. Hence $\mathcal{C}(N)$ is a subset and its elements have to be regular in R, so $\mathcal{C}(N) \subset \mathcal{C}(0)$, hence $\mathcal{C}(N) = \mathcal{C}(0)$.

ex) A right hereditary ring R with max-r is a right order in a right Artinian ring.

Let $c \in \mathcal{C}(N)$, cR projective and $0 \to r(c) \to R \to cR \to 0$ splits. Hence $r(c) = eR$; $e = e^2$ and as $r(c) \subset N$ we have $r(c) = 0$. Also $r(a)$ cannot be essential as right ideal if $a \neq 0$, because $r(a)$ is generated by an idempotent. Hence $Z(R) = 0$ and $\ell(c) = 0$, so $\mathcal{C}(N) = \mathcal{C}(0)$.

ex) Let R be a commutative Noetherian ring, let $P_1 \ldots P_n$ be the associated primes of zero. Then $\mathcal{C}(0)$ is $R - (P_1 \cup P_2 \cup \ldots \cup P_n)$ so $\mathcal{C}(0)$ is empty only when $aR = 0$ for some $a \neq 0$. There is always a quotient ring if $r(R) = 0$ and it is a right Artinian ring if and only if $P_1, \ldots, P_n$ are all minimal primes of $\mathcal{C}(0)$, that is, no embedded primes occur.

Now $F[[x,y]]/(x^2,xy)$, F a field, is Noetherian, its own quotient ring, and not Artinian. Also $y \in \mathcal{C}(N)$, $y \notin \mathcal{C}(0)$; there is an embedded prime, of course.

(2.8). A prime ideal P of R is <u>minimal</u> if $P \supset P'$ with P' a prime ideal implies that $P = P'$.

The set of minimal primes of R is denoted by min-$\mathcal{P}$.

(2.9). An ideal P of R is a <u>left maximal annihilator</u> if $r(P) \neq 0$ and $P = \ell(T)$ for any non-zero ideal $T \subset r(P)$. In particular, $P = \ell r(P)$. This property ensures that P is a prime ideal and this set of primes is denoted by lma-$\mathcal{P}$.

(2.10). A prime ideal P of R is a <u>left annihilator prime</u> if $P = \ell r(P)$. The set of these primes is la-$\mathcal{P}$.

There are corresponding classes rma-$\mathcal{P}$ and ra-$\mathcal{P}$.

(2.11). A prime P which is not essential as a right (left) ideal is in min-$\mathcal{P}$.

<u>Proof</u>. Let $I \cap P = 0$, where $I \triangleleft_r R$, $I \neq 0$. If $P \supset P'$ then $IP = 0 \subset P'$, hence either $P = P'$ or $I \subset P' \subset P$, which is a contradiction.

(2.12). In a semi-prime ring R with a right quotient ring, we have

$$\text{rma-}\mathcal{P} = \text{lma-}\mathcal{P} = \text{min-}\mathcal{P} = \text{ra-}\mathcal{P} = \text{la-}\mathcal{P}.$$

<u>Proof</u>. A prime ideal P of R is essential as a right ideal if and only if it has zero left annihilator. Hence rma-$\mathcal{P}$ = min-$\mathcal{P}$. Now $P \cap \ell(P) = 0$ for any prime P, hence $r(P) = \ell(P)$. The rest follows at once.

Note that when R is a prime ring, (0) is the only prime of these particular types.

(2.13). Let R have max-r. A prime $P \in \text{min-}\mathcal{P}$ if and only if $P \cap X = N$ for some $X \lhd R$, $X \neq N$.

Proof. Let $P \supseteq P'$, where P' is a prime, and $P \cap X = N$, then $PX \subseteq N \subseteq P'$. Thus either $P \subseteq P'$ or $X \subseteq P'$. Hence either $P = P'$ or $X = N$.

Conversely, let $P \in \text{min-}\mathcal{P}$. Then P/N is a minimal prime of the semi-prime ring R/N and it must be a maximal annihilator ideal of R/N. Then $X \lhd R$ exists with $P/N \cap X/N = \bar{0}$, $X \neq N$, and the condition follows.

(2.14). A prime $P \in \text{min-}\mathcal{P}$ if and only if P does not meet $\mathcal{C}(N)$.

Proof. Pass to the semi-prime ring R/N.

(2.15). An ideal $U \lhd R$, $U \neq 0$ is uniform if it does not contain a direct sum of non-zero ideals of R.

This condition differs from that of uniform module presented in chapter 1. However, the argument used there will show that any non-zero ideal contains a uniform ideal.

(2.16). Let $T \lhd R$, $T \neq 0$. There exists a prime ideal $P \in \text{la-}\mathcal{P}$ with $P \supseteq \ell(T)$.

Proof. T contains a uniform ideal U. Set $P = \{x \in R \mid xU' = 0, 0 \neq U' \subseteq U, U' \lhd R\}$. Clearly P is a prime ideal. Note that as P is f.g. as right ideal, $P = p_1R^1 + \ldots + p_kR^1$; p_iR^1 is the least right ideal which contains p_i. Let $p_iU_i' = 0$ then

$P(U_1' \cap \dots \cap U_s') = 0$ and $U_1' \cap \dots \cap U_s' \neq 0$. Then $P = \ell(U_1' \cap \dots \cap U_s')$.

(2.17). Min-$\mathcal{P}$ is a finite set $\{P_1,\dots,P_k\}$, say, and

(1) $({}^{P_1}/_N,\dots,{}^{P_k}/_N) = \min_{R/N}\text{-}\mathcal{P}$.

(2) $P_1 \cap \dots \cap P_k = N$.

(3) $\mathcal{C}(N) = \mathcal{C}(P_1) \cap \dots \cap \mathcal{C}(P_k)$.

Proof. The classes min-$\mathcal{P}$ and $\min_{R/N}\text{-}\mathcal{P}$ are in (1-1) correspondence under $P \longleftrightarrow {}^{P}/_N$. This implies that we need only consider the case $N = 0$. Let $U_1 \oplus \dots \oplus U_k = S$ be a direct sum of $\neq 0$ ideals of maximum length k; the ideals U_i are uniform (two-sided). Each U_i defines a prime P_i, where $P_i = \ell(U_i')$; $U_i' \triangleleft R$, $U_i' \subseteq U_i$. Now $P_1,\dots,P_k \in$ min-$\mathcal{P}$ by (2.12). Now $S' = U_1' \oplus \dots \oplus U_k'$ has $\ell(S') \cap S' = 0$, because R is semi-prime. Hence $\ell(S') = 0$, as otherwise S' could be lengthened. Clearly $P_1 \cap \dots \cap P_k = 0$. Let P be a minimal prime, then $P_1 \dots P_k = 0 \subseteq P$ and hence $P_i \subseteq P$ for some i, so $P_i = P$. Thus $P_1,\dots,P_k$ are the only minimal primes.

Let $c \in \mathcal{C}(0)$ and $cx \in P_i$ then

$$cxU_i' = 0 \to xU_i' = 0 \to x \in P_i.$$

Thus $c \in \mathcal{C}'(P_i) = \mathcal{C}(P_i)$. This holds for $i = 1,\dots,k$. Now $P_1 \cap \dots \cap P_k = 0$ which proves that $\mathcal{C}(P_1) \cap \dots \cap \mathcal{C}(P_k) \subseteq \mathcal{C}'(0) = \mathcal{C}(0)$. Equality follows.

(2.18). Let R be a right order in a ring Q and P be a prime of R.

(1) PQ is a prime ideal of Q and $PQ \cap R = P$ if and only if $\mathcal{C}(0) \subseteq \mathcal{C}(P)$ holds in R. In particular, this is valid if $P \in \text{min-}\mathcal{P}$ or $P \in \text{la-}\mathcal{P}$.

(2) Let P' be a prime of Q then $P' \cap R$ is a prime of R and

$$P' = (P' \cap R)Q.$$

Proof. (1) Let $\mathcal{C}(0) \subseteq \mathcal{C}(P)$ then $QP \subseteq PQ$. For suppose that

$$c^{-1}p = ad^{-1}\ ,\ p \in P\ ,\ c,d \in \mathcal{C}(0)\ ,\ a \in R.$$

Then $ca \in P$ and $a \in P$ so that $c^{-1}p \in PQ$, hence $QP \subseteq PQ$.

Let $x \in PQ \cap R$ then $x = pc^{-1}$, where $p \in P$, $c \in \mathcal{C}(0)$, and $xc \in P$ implies that $x \in P$, so that $P = PQ \cap R$.

Let $A \triangleleft Q$, $B \triangleleft Q$, and $AB \subseteq PQ$. Then $(A \cap R)(B \cap R) \subseteq P$; either $A \cap R \subseteq P$ or $B \cap R \subseteq P$, so either $A \subseteq PQ$ or $B \subseteq PQ$, hence PQ is a prime ideal.

Conversely, let $c \in \mathcal{C}(0)$ and $cx \in P$, $x \in R$. Then $x \in QP \subseteq PQ$, hence $x \in PQ \cap R = P$. It follows that $\mathcal{C}(0) \subseteq \mathcal{C}(P)$.

If $P \in \text{min-}\mathcal{P}$ then $\mathcal{C}(0) \subseteq \mathcal{C}(N) \subseteq \mathcal{C}(P)$. If $P \in \text{la-}\mathcal{P}$ then let $c \in \mathcal{C}(0)$ and have

$$cx \in P \rightarrow cxr(P) = 0 \rightarrow xr(P) = 0 \rightarrow x \in P.$$

so $\mathcal{C}(0) \subseteq \mathcal{C}'(P) = \mathcal{C}(P)$.

(2) Let V be the radical of $P' \cap R$ and $V^{\rho} \subseteq P' \cap R$. Then $\mathcal{C}(0) \subseteq \mathcal{C}(P' \cap R) \subseteq \mathcal{C}(V)$ by theorem (2.5). Hence $QV \subseteq VQ$ and $(VQ)^{\rho} \subseteq P'$, $VQ \subseteq P'$, $V = P' \cap R$. Hence $P' \cap R$ is a semi-prime ideal of R and $P' \cap R$ is a finite intersection of primes of R which are the minimal ones containing $P' \cap R$, say,

$$P' \cap R = P_1 \cap \dots \cap P_s$$

and $\mathcal{C}(0) \subseteq \mathcal{C}(P' \cap R) = \mathcal{C}(P_1) \cap \dots \cap \mathcal{C}(P_s)$.

Then $P' = (P' \cap R)Q = P_1Q \cap \dots \cap P_sQ = P_1' \cap \dots \cap P_s'$, where $P_i' = P_iQ$ are primes of Q by part (1). Then $P_1' \dots P_s' \subseteq P'$, hence $P_1' \subseteq P'$, say, and $P_1' = P'$; thus $P' \cap R = P_1' \cap R = P_1$.

ex) Let $R = Q$ be an Artinian ring with zero singular ideal and not semi-simple. The right socle $E \neq R$ and $\ell(E) = 0$. There is a maximal right ideal $M \supseteq E$ with $\ell(M) = 0$. Yet M is a minimal prime.

Non-commutative local rings

Our main aim is to construct a local ring, beginning with an arbitrary prime ideal P in a ring R. We find it necessary for technical reasons to suppose that R has both max-r and max-ℓ. The basic unsettled question is whether the local ring or, indeed, the complete local ring, has maximum condition or not.

Since a local ring can be defined in many ways, we give the following definition.

(2.19). A ring Q is a <u>local ring</u> if

1. Q has a unit element,
2. The Jacobson radical of Q is a maximal ideal M,
3. Q/M is an Artinian simple ring.
4. $\cap M^n;\ n = 1,2,\ldots = 0$.

Our construction makes use of a particular topological family of right ideals. For these ideals in general terms see Bourbaki [3].

(2.20). Let R be an arbitrary ring. A family $\mathcal{F}$ of right ideals of R is <u>topological</u> if

1. $\mathcal{F}$ is a filter ;
2. $a \in R$, $F \in \mathcal{F}$ imply that $a^{-1}F \in \mathcal{F}$, where $a^{-1}F = \{x \in R;\ ax \in F\}$.

R may be regarded as a topological ring, the open right ideals being the members of $\mathcal{F}$.

(2.21). The $\mathcal{F}$-<u>singular</u> submodule of a right R-module M is defined to be

$$Z_{\mathcal{F}}(M) = \{m \in M\ ;\ r(m) \in \mathcal{F}\}.$$

(2.22). The $\mathcal{F}$-<u>closure</u> of a submodule $N \subset M$ is defined to be $cl_{\mathcal{F}}N = \{m \in M\ ;\ mF \subset N,\ F \in \mathcal{F}\}$.

Let $\mathcal{F}, \mathcal{G}$ be topological families of right ideals and define $\mathcal{F}\mathcal{G} = \{I\ ;\ I \underset{r}{\triangleleft} R,\ cl_{\mathcal{G}}I \in \mathcal{F}\}$,

(2.23). $\mathcal{F}\mathcal{G}$ is a topological family of right ideals which contains FG, whenever $F \in \mathcal{F}$, $G \in \mathcal{G}$.

The proof is left to the reader.

(2.24). A topological family $\mathcal{F}$ is <u>idempotent</u> when $\mathcal{F} = \mathcal{F}\mathcal{F}$. This property is equivalent to the additional axiom on $\mathcal{F}$:

3. $I \triangleleft_r R$ and $cl_{\mathcal{F}} I \in \mathcal{F}$ implies that $I \in \mathcal{F}$.

ex) Let R be a semi-prime ring with max-r. The set of essential right ideals of R is an idempotent topological set.

(2.25). Let $\mathfrak{M}$ be a multiplicatively closed subset of elements of R and define $\mathcal{F}(\mathfrak{M})$ to be the family of right ideals $F \triangleleft_r R$ such that $a^{-1}F$ meets $\mathfrak{M}$ for all $a \in R$. $\mathcal{F}(\mathfrak{M})$ is an idempotent topological family.

(2.26). A multiplicatively closed subset $\mathfrak{M}$ of R is said to satisfy the right <u>Ore condition</u> if, given $a \in R$, $m \in \mathfrak{M}$ there exists $a_1 \in R$, $m_1 \in \mathfrak{M}$ such that $am_1 = ma_1$. In (2.26), $\mathcal{F}(\mathfrak{M})$ is the family of all right ideals which meet $\mathfrak{M}$.

When $\mathfrak{M}$ consists of regular elements alone, there exists a partial right quotient ring $R_{\mathfrak{M}}$, the elements of this ring have the form am^{-1}, $a \in R$, $m \in \mathfrak{M}$. We shall concentrate upon the case when R has max-r and max-ℓ.

(2.27). Suppose that P is given prime ideal of R and define

$$\mathfrak{C} = \mathfrak{C}(P) = \{c \in R \;;\; cx \in P \quad x \in P\}$$
$$= \{c \in R \;;\; xc \in P \quad x \in P\}.$$

Define $\mathcal{F}(P)$ to be the family of all those right ideals F for which $a^{-1}F$ meets $\mathcal{C}$ for each $a \in R$.

Define $\mathcal{G}(P)$ to be the family of all those left ideals G for which Ga^{-1} meets $\mathcal{C}$ for each $a \in R$.

(2.28). $\mathcal{F}(P)$ is an idempotent topological family of right ideals as is $\mathcal{G}(P)$ of left ideals.

(2.29). $cR + P \in \mathcal{F}(P)$ when $c \in \mathcal{C}$.

Proof. The prime ring R/P has a right and left quotient ring and its set of regular elements is $\{[c + P] ; c \in \mathcal{C}\}$. Thus for each pair $a \in R$, $c \in \mathcal{C}$ there exist $a_1 \in R$, $c_1 \in \mathcal{C}$ such that $ac_1 - ca_1 \in P$. Then $c_1 \in a^{-1}(cR + P)$ and $cR + P \in \mathcal{F}(P)$.

(2.30). Let $X \triangleleft_r R$, $X \supset P$ and such that

$$X \cap I = P, \quad I \triangleleft_r R \text{ implies that } I = P.$$

(X is essential over P). Then $X \in \mathcal{F}(P)$.

Proof. Pass to the prime ring R/P and observe that X/P is an essential right ideal there; it has a regular element $[c + P]$, $c \in \mathcal{C}$. Then $X \supset cR + P$ and $X \in \mathcal{F}(P)$.

There are corresponding results for left ideals.

For convenience we now denote $\mathcal{F} = \mathcal{F}(P)$, $\mathcal{G} = \mathcal{G}(P)$.

(2.31). Let $I \triangleleft_r R$ and define $\rho(P) = \mathrm{cl}_{\mathcal{F}} I$;

Let $L \triangleleft_\ell R$ and define $\lambda(L) = \mathrm{cl}_{\mathcal{G}} L$;

Let $A \triangleleft R$ and define $\kappa(A) = \{x \in R;\ GxF \subset A$, some $G \in \mathcal{G}$, $F \in \mathcal{F}\}$.

(2.32). The closure operations ρ, λ, are idempotent.

Proof. Let $I \underset{r}{\triangleleft} R$, we need to prove that $\rho(\rho(I)) = \rho(I)$. Let $x \in \rho^2(I)$ and $xF \subset \rho(I)$, $F \in \mathcal{F}$. Choose $f \in F$ and then $F' \in \mathcal{F}$ such that

$$xfF' \subset I .$$

Let $H \underset{r}{\triangleleft} R$ be $\{h \in R; xh \in I\}$. Then for every $f \in F$, there exists $F' \in \mathcal{F}$ such that $fF' \subset H$. In other words, $f \in \rho(H)$ and it follows that $F \subset \rho(H)$, which means that $\rho(H) \in \mathcal{F}$ and $H \in \mathcal{F}$, since the latter is an idempotent family. Thus $x \in \rho(I)$, which proves it. The other result is obtained similarly.

(2.33). Let $I,J \underset{r}{\triangleleft} R$ then $\rho(I \cap J) = \rho(I) \cap \rho(J)$. Similar results hold for λ and κ. The proofs are obvious.

(2.34). Let $A \triangleleft R$. Then $F \in \mathcal{F}$, $G \in \mathcal{G}$ exists such that

$$\rho(A)F \subset A,$$
$$G\lambda(A) \subset A,$$
$$G\kappa(A)F \subset A.$$

Proof. Let $a_1,\ldots,a_n$ be a set of generators of $\rho(A)$ considered as a left ideal and suppose that $a_iF_i \subset A$; $i = 1,\ldots,n$. Set $F = F_1 \cap \ldots \cap F_n$ and the first result is obtained at once. The others may be obtained in a similar way and the F,G used in the statements can be taken to be common, since $\mathcal{F}$ and $\mathcal{G}$ are closed under finite intersection.

(2.35). The closure operation κ is idempotent and $\kappa = \lambda\rho = \rho\lambda$.

<u>Proof</u>. Let $A \triangleleft R$ then $G \in \mathcal{G}$, $F \in \mathcal{F}$ exists such that $G\,\kappa(A)F \subset A$. Then $\kappa(A)F \subset \lambda(A)$, $\kappa(A) \subset \rho\lambda(A)$. Now $\rho\lambda(A) \subset \kappa(A)$ is clear and we have

$$\kappa(A) = \rho\lambda(A).$$

The other result is obtained similarly.

(2.36). <u>The left symbolic powers</u> of P are the members of the sequence $\{H_n\}$, where

$$H_1 = \kappa(P) = P; \quad H_{n+1} = \kappa(PH_n).$$

The <u>right symbolic powers</u> of P are the members of the sequence $\{K_n\}$, where

$$K_1 = \kappa(P) = P; \quad K_{n+1} = \kappa(K_nP).$$

ex). Let R be a commutative Noetherian ring, P a prime of R. Then $H_n = K_n = P^{(n)}$ is the nth symbolic power of P.

$$= \{x \in R \mid xc \in P^n;\ c \in \mathcal{C}(P)\}.$$

ex). Let R be non-commutative, let P be a maximal ideal. Observe that R/P need not be Artinian simple. Then $H_n = K_n = P^n$. The development of the argument is centered on the H_n and their factor rings.

(2.37). $\mathcal{C} = \mathcal{C}(P) \subset \mathcal{C}(H_n)$.

<u>Proof</u>. Let $c \in \mathcal{C}$ and $cx \in H_n$, as $H_n \subset P$ we obtain $x \in P = H_1$. Suppose that we have proved that $x \in H_m$ $(m < n)$, so that

$$(P + Rc)x \subset PH_m + H_n \subset H_{m+1}.$$

Then $x \in \lambda(H_{m+1}) = H_{m+1}$, using (2.29) and (2.35).

(2.38). The set of cosets $\{[c + H_n]; c \in \mathcal{C}\}$ is the nth <u>associated divisor set of $\mathcal{C}$</u>. Call it $\mathcal{C}_n$.

(2.39) <u>Theorem</u>. Let R have max-r and max-ℓ, P be a prime ideal of R and H_n; $n = 1,2,\ldots$ the left symbolic powers of P. Then

1. $H_rH_s \subset H_n$ for all $r + s = n$,
2. R/H_n satisfies the right Ore condition with respect to the associated divisor set $\mathcal{C}_n$ and the latter is a set of regular elements.

<u>Proof</u>. The proof proceeds by induction on n and by induction on r for given n. Part (1) obviously holds for any $n > 0$ when $r = 1$.

Now examine H_{n+1}, and $H_{r+1}\,H_{n+1-r-1}$ assuming that the theorem holds as a whole for $H_m (m \le n)$ and that $H_{r'}H_{n+1-r'}$ $\subset H_{n+1}$ whenever $0 < r' \le r$.

Consider $H_{r+1}H_{n-r}$, choosing $h \in H_{r+1}$, $h' \in H_{n-r}$. Now

$$GH_{r+1}F \subset PH_r \text{ where } G \in \mathcal{G},\ F \in \mathcal{F}.$$

Let $c \in F \cap \mathcal{C}$, then there exists a relation

$$h'c' = cx + h_n, \quad \text{where } h_n \in H_n, \ c' \in \mathcal{C}, \ x \in R.$$

This follows from our assumption that $\mathcal{C}_n$ satisfies the right Ore condition. Then we have

$$cx \in H_{n-r} + H_n = H_{n-r} \quad \text{and hence} \quad x \in H_{n-r}.$$

Also

$$Ghh'c' = Gh(cx + h_n) \subset PH_rH_{n-r} + H_{r+1}H_n \subset PH_n,$$

since we have assumed that $H_rH_{n-r} \subset H_n$. It follows that

$$Ghh'(c'R+P) \subset PH_n + H_{r+1}H_{n-r}P \subset PH_n + H_{r+1}H_{n-r+1} \subset PH_n + H_nH_{n-r+1} \subset H_{n+1},$$

still employing the inductive hypothesis. Then

$$hh' \in \kappa(H_{n+1}) = H_{n+1}$$

and this means that

$$H_{r+1}H_{n-r} \subset H_{n+1}.$$

It should be remembered that the proof of (1) is incomplete until the induction has been carried through for (2). We first show that $\mathcal{C}_{n+1}$ contains only regular elements of R/H_{n+1}. Let $c \in \mathcal{C}$ and $xc \in H_{n+1}$. Then $x \in H_n$, since the elements of $\mathcal{C}_n$ are taken to be regular. Hence

$$x(cR + P) \subset H_{n+1} + H_nP \subset H_{n+1},$$

so that $x \in \kappa(H_{n+1}) = H_{n+1}$. We know already that $cx \in H_{n+1}$ implies that $x \in H_{n+1}$. Thus $\mathcal{C}_{n+1}$ consists of regular elements.

Now let $a \in R$, $c \in \mathcal{C}$. As R/H_n has the right Ore condition there exist $a_1 \in R$, $c_1 \in \mathcal{C}$ with

$$ac_1 - ca_1 = h_n \in H_n.$$

Pass to the ring $\bar{R} = R/H_{n+1}$, denoting images by the use of bars. Set

$$\bar{X} = \{\bar{x} \in \bar{R};\ \bar{h}_n\bar{x} \in \bar{c}\,\bar{R}\}.$$

As $H_nP \subset H_{n+1}$ we have $\bar{H}_n\bar{P} = \bar{0}$ and hence $\bar{X} \supset \bar{P}$. Now $\bar{c}$ is regular in R/H_{n+1} and it follows that $\bar{c}\,\bar{R}$ is an essential right ideal of $\bar{R}$. Let I be a right ideal of R with $I \supset P$ and suppose that

$$\bar{X} \cap \bar{I} = \bar{P}.$$

If $\bar{h}_n\bar{I} = \bar{0}$ then $\bar{I} \subset \bar{X}$ and $\bar{I} = \bar{P}$. If $\bar{h}_n\bar{I} \neq \bar{0}$ then $\bar{h}_n\bar{I} \cap \bar{c}\,\bar{R} \neq \bar{0}$, since $\bar{c}\,\bar{R}$ is essential. In the latter case there exists $\bar{b} \in \bar{I}$ with say

$$\bar{h}_n\bar{b} = \bar{c}\,\bar{r} \neq \bar{0}.$$

Then $\bar{b} \in \bar{X} \cap \bar{I} = \bar{P}$, so that $\bar{b} \in \bar{P}$. Now $\bar{h}_n\bar{P} = \bar{0}$ and $\bar{h}_n\bar{b} \neq \bar{0}$, which is a contradiction. It follows that $\bar{I} = \bar{P}$.

Apply (2.30) to the ring $\bar{R}$ and deduce that $\bar{X}$ meets $\mathcal{C}_{n+1}$.

Hence there exist $\bar{c}_2 \in \bar{X}$, $c_2 \in \mathcal{C}$ and $\bar{h}_n\bar{c}_2 = \bar{c}\,\bar{r}$, say. Then

$$\bar{a}\,\bar{c}_1\bar{c}_2 = \bar{c}(\bar{a}_1\bar{c}_2 + \bar{r}).$$

It follows that R/H_{n+1} has right Ore condition with respect to $\mathcal{C}_{n+1}$ which completes the proof.

(2.40) Theorem. Under the conditions of theorem (2.39) and denoting by K_n; $n = 1,2,\ldots,$ the nth symbolic power of P, the following are valid.

1. $H_n = K_n$; $n = 1,2,\ldots$;
2. R/H_n satisfies the left Ore condition with respect to $\mathcal{C}_n$;
3. $\mathcal{C}_n$ is the full set of regular elements of R/H_n ;
4. H_n, $n = 1,2,\ldots$ is a P-primary ideal of R.

Proof. Certainly $H_1 = K_1$. Suppose that $H_n = K_n$, then

$$K_{n+1} = \kappa(K_nP) = \kappa(H_nH_1) \subset H_{n+1}$$

by theorem (2.39). Similarly we can say that $H_{n+1} \subset K_{n+1}$ and hence $H_{n+1} = K_{n+1}$. Since $H_1 = K_1 = P$, part (1) has been proved by induction.

Part (2) follows from (2.39) and the symmetry of the conditions. In order to prove (3), we let $a \in R$ and $[a + H_n]$ be a regular element of R/H_n. Then $ax \in H_n$ implies that $x \in H_n$. We need to prove that $a \in \mathcal{C}(P) = \mathcal{C}$, so let $ay \in P$ and consider

$$Z = \{z \in R;\ zH_{n-1} \subset H_n\}.$$

Since $PH_{n-1} \subset H_n$, we have $Z \supset P$. It is also an ideal of R. Moreover,

$$ay \in P \Longrightarrow ayH_{n-1} \subset PH_{n-1} \subset H_n \Longrightarrow yH_{n-1} \subset H_n \Longrightarrow y \in Z.$$

In the case when $Z = P$ we can now assert that $a \in \mathcal{C}$. If $Z \supsetneq P$ is the non-zero ideal in the prime ring R/P and contains a regular element of that ring. Hence Z meets $\mathcal{C}$ and $c \in \mathcal{C}$ exists with $cH_{n-1} \subset H_n$. It follows from (2.37) that $H_{n-1} = H_n$. This means that, either $a \in \mathcal{C}$ or $H_{n-1} = H_n$. In the second case,

$$ax \in H_{n-1} \Longrightarrow x \in H_{n-1}$$

and our argument can be repeated with n replaced by n-1. It follows that either $a \in P$, or

$$H_n = \dots H_1 = P,$$

and in this case we are given that $a \in \mathcal{C}$ anyway.

Before proving part (4) we give the definition of primary ideal.

(2.41). An ideal T R is <u>P-primary</u>, where P is a prime ideal if $AB \subset T$, $A \triangleleft R$, $B \triangleleft R$ implies that if $A \not\subset T$ then $B \subset P$ and if $B \not\subset T$ then $A \subset P$, where $P = \sqrt{T}$.

<u>Proof of (4)</u>. Let $A \triangleleft R$, $B \triangleleft R$ and $AB \subset H_n$. If $A \not\subset P$ then $A + P$ meets $\mathcal{C}$ and hence A meets $\mathcal{C}$. Let $c \in \mathcal{C} \cap A$. Then $cB \subset H_n$ implies that $B \subset H_n$, because of (2.37). Also $P^n \subset H_n$,

so that $P \subset \sqrt{H_n}$. Now $\sqrt{H_n} \supsetneq P$ would mean that $\sqrt{H_n}$ meets $\mathcal{C}$ which is absurd. Hence $P = \sqrt{H_n}$.
The other condition holds by symmetry and this completes the proof of the theorem.

Another definition of the term primary can be given which seems likely to be more useful than the normal one given in (2.41).

(2.42). An ideal $T \triangleleft R$ such that $\sqrt{T} = P$ is a prime ideal is said to be <u>strongly P-primary</u> if $\mathcal{C}(T) = \mathcal{C}(P)$.

(2.43). H_n is strongly P-primary. This is obvious from the argument used in part (4) of (2.40). We also have

(2.44). A strongly primary ideal is primary.

<u>Proof</u>. Let T be strongly P-primary, suppose that

$$AB \subset T \quad A \triangleleft R, \quad B \triangleleft R.$$

Suppose that $B \not\subset P$ then B meets $\mathcal{C}(P)$, as R/P is a prime ring, and this says that B meets $\mathcal{C}(T)$. Then $A \subset T$. The other condition is now evident.

It is not known whether a primary ideal is strongly primary, although it seems to be unlikely. The following result is easy to prove by the methods used here and we need a particular case. We shall prove this and leave the reader to prove (2.45).

(2.45). $T \triangleleft R$ is strongly primary if and only if R/T has a quotient ring (on right and left) which is a primary artinian ring.

(2.46). Theorem. Under the conditions of theorem (2.39),

(1) the quotient ring Q_n of R/H_n is an Artinian primary ring;

(2) There is a canonical epimorphism $Q_{n+1} \to Q_n$; $n = 1,2,\ldots$

Proof. It is easier to prove the second part at the outset. Take $a \in R$, $c \in \mathcal{C}$ and consider the correspondence between Q_{n+1} and Q_n given by

$$[a + H_{n+1}]\,[c + H_{n+1}]^{-1} \longleftrightarrow [a + H_n]\,[c + H_n]^{-1}.$$

We show that this is a mapping. Suppose that

$$[a + H_{n+1}][c + H_{n+1}]^{-1} = [b + H_{n+1}][d + H_{n+1}]^{-1};\quad b \in R,\ d \in \mathcal{C}.$$

There exist $e,f \in \mathcal{C}$ such that

$$[c + H_{n+1}][e + H_{n+1}] = [d + H_{n+1}][f + H_{n+1}]$$
$$[a + H_{n+1}][e + H_{n+1}] = [b + H_{n+1}][f + H_{n+1}],$$

because R/H_n satisfies the Ore condition (on right is used here for its set $\mathcal{C}_n$ of regular elements. Since $ce - df \in H_{n+1}$ and $ae - bf \in H_{n+1}$, these equations may be replaced by corresponding ones having cosets of H_n instead of cosets of H_{n+1}. It follows that

$$[a + H_n][c + H_n]^{-1} = [b + H_n][d + H_n]^{-1},$$

so that the correspondence is a mapping. Similar methods verify that it is a homomorphism from Q_{n+1} onto Q_n.

The product of these maps for the cases $1,2,\ldots,n-1$ shows that

$$[a + H_n][c + H_n]^{-1} \to [a + P][c + P]^{-1}$$

is an epimorphism from Q_n to Q_1. Since Q_1 is an Artinian simple ring, the kernel of this map is a maximal ideal M_n of Q_n. Now

$$M_n = \{[p + H_n][c + H_n]^{-1};\ p \in P,\ c \in \mathcal{C}\}$$

so that M_n^n is generated by elements of the form

$$[x] = [p_1][c_1]^{-1} \ldots [p_n][c_n]^{-1};\ p_i \in P,\ c_i \in \mathcal{C}$$

where $[a] = [a + H_n]$ for each $a \in R$.

Also we have $[c]^{-1}[p] = [b][d]^{-1}$ where $p \in P$, $c \in \mathcal{C}$ are given elements and $b \in R$, $d \in \mathcal{C}$ are then determined.

Since $pd - cb \in H_n$, we see that $b \in P$. Thus $[x]$ can be written as

$$[x] = [p_1'][p_2'] \ldots [p_n'][d]^{-1};\ p_i' \in P,\ d \in \mathcal{C}.$$

This means that $[x] = [0]$ and $M_n^n = (0)$. Thus Q_n is an Artinian primary ring.

(2.47). A local ring Q with maximal ideal M is <u>complete</u> if it is complete in the usual topology defined by M.

This corresponds to the commutative case. We are now in a position to obtain the complete local ring appropriate to our construction. The conditions and notation of the next theorem are those of earlier theorems in this chapter.

(2.48) Theorem. The inverse limit $\hat{Q}$ of the rings Q_n; $n = 1,2,\ldots$ under the canonical epimorphisms $Q_{n+1} \to Q_n$ is a complete local ring. Q is a full $k \times k$ matrix ring over a ring $\hat{L}$. $\hat{L}$ is a complete local ring with maximal ideal $\hat{N}$ such that $\hat{L}/\hat{N}$ is a division ring, isomorphic to D, where D_k is the quotient ring of R/P.

Proof. Let $q \varepsilon \hat{Q}$ with $q = (q_1, q_2, \ldots)$ where $q_n \varepsilon Q_n$. Let $\hat{M}$ be the kernel of the homomorphism $\hat{Q} \to Q_1$ which is defined by $q \to q_1$. Since $\hat{Q}/\hat{M} \cong Q_1$ and Q_1 is an Artinian simple ring, $\hat{M}$ is a maximal ideal of $\hat{Q}$. Now the maximal ideal M_n of Q_n satisfies $M_n^n = (0_n)$; here we denote the zero of Q_n by the symbol 0_n. Also

$$\hat{M}^n = (M_1^n, M_2^n, \ldots) = (0_1, 0_2, \ldots, 0_n, M_{n+1}^n, \ldots)$$

for $n = 1,2,\ldots$. It follows that

$$\{\cap \hat{M}^n; \ n = 1,2,\ldots\} = (0).$$

Let 1_n be the unit element of Q_n and 1 be that of $\hat{Q}$. Let $m \varepsilon M$

$$1 = (1_1, 1_2, \ldots), \quad m = (m_1, m_2, \ldots); \quad m_n \varepsilon M_n.$$

The element $q' = (1_1, 1_2 + m_2, 1_3 + m_3 + m_3^2, \ldots)$ belongs to

Q (check homomorphisms), and

$$(1 - m)q' = q'(1 - m) = 1.$$

It follows that $\hat{M}$ is the Jacobson radical of $\hat{Q}$ and hence is the only maximal ideal of $\hat{Q}$. Thus $\hat{Q}$ is a local ring and its completeness in the $\hat{M}$-topology is apparent from its form. Let $\hat{Q}/\hat{M} = D_k$, D a division ring.

Let $(e_{ij}^{n};\ i,j = 1,\ldots,k)$ be a set of matrix units for the Artinian primary ring Q_n. Under the canonical map of Q_{n+1} to Q_n defined earlier we can find a set $(e_{ij}^{n+1};\ i,j = 1,\ldots,k)$ of matrix units in Q_{n+1} such that

$$e_{ij}^{n+1} \to e_{ij}^{n};\quad i,j = 1,\ldots,k.$$

This can be effected by using the standard techniques for lifting a set of matrix units (see [20], pp. 54-55; the theorem there has to be modified a little in order to apply here). It is clear that the set of elements

$$e_{ij} = (e_{ij}^{1}, e_{ij}^{2},\ldots);\quad i,j = 1,\ldots,k$$

is a full set of matrix units in $\hat{Q}$. Now for $n = 1,2,\ldots$ we have

$$Q_n = \sum_{i,j=1}^{n} e_{ij}^{n}\ell_{ij}^{n};\quad \ell_{ij}^{n} \in L_n,$$

where L_n is a completely primary subring of Q_n. The canonical map $Q_{n+1} \to Q_n$ induces an epimorphism $L_{n+1} \to L_n$ which can be used to form the inverse limit

$$\hat{L} = (L_1, L_2, \ldots).$$

Evidently we obtain

$$\hat{Q} = \sum_{i,j=1}^{n} e_{ij}\ell_{ij}; \quad \ell_{ij} \in \hat{L} .$$

Thus $\hat{Q}$ is a full $k \times k$ matrix ring over the ring L. Now the argument which we have used to obtain the local properties of $\hat{Q}$ can be used to demonstrate the same for $\hat{L}$, so that $\hat{L}$ is a complete local ring. The rest is easy.

Our next aim is to obtain the 'ordinary' local ring R_P and this is done by constructing a suitable subring of Q. The implications and advantages of this step have to be considered. Certainly the elements of $\mathcal{C} = \mathcal{C}(P)$ are units of Q and to obtain this essential feature we have replaced R be its factor ring R/H, where $H = (\cap H_n;\ n = 1,2,\ldots)$. In the commutative case the move is appropriate, since

$$\begin{aligned}(\cap H_n;\ n = 1,2,\ldots) &= (\cap P^{(n)};\ n = 1,2,\ldots)\\ &= \{x \in R;\ xc = 0,\ \text{some}\ c \in \mathcal{C}(P)\}.\end{aligned}$$

In the commutative case the left hand side of this relation contains both $\{x \in R;\ xc = 0,\ \text{some}\ c \in \mathcal{C}(P)\}$ and the corresponding $\{x \in R;\ cx = 0,\ \text{some}\ c \in \mathcal{C}(P)\}$. These are sets made up of left and right ideals respectively and it is not known that H is the smallest ideal which contains them.

Now regarding R as a subring of $\hat{Q}$ its elements are represented in the form

$$([a + H_1], [a + H_2], \ldots); \quad a \in R.$$

$\mathcal{C}$ is a set of units of $\hat{Q}$, the set of inverses is taken to be $\mathcal{C}^{-1}$. Take $\hat{M}$ to be the maximal ideal of $\hat{Q}$. Denote

$$Q = \{R, \mathcal{C}^{-1}\}$$

which is the subring of $\hat{Q}$, generated by R and $\mathcal{C}^{-1}$. This subring has the following properties:

(2.49). $Q = \{R, \mathcal{C}^{-1}\}$ has a maximal ideal $M = QPQ = \hat{M} \cap Q$ such that

(1) $\hat{M}^n \cap R = M^n \cap R = H_n$

(2) $\hat{Q}/\hat{M}^n \cong Q/M^n \cong Q_n$; $n = 1,2,\ldots$.

Proof. An element a R has the form

$$([a + H_1], [a + H_2], \ldots)$$

and belongs to $\hat{M}^n$ if and only if $[a + H_i] = 0$ for $i = 1,\ldots,n$. Thus $\hat{M}^n \cap R = H_n$.

Now $H_n = \kappa(PH_{n-1})$, so that $H_n \cap QPH_{n-1}Q$. Supposing that $H_{n-1} \subset M^{n-1}$, we obtain $H_n \subset M^n$. Now $H_1 = P \subset M$ and hence $H_n \subset M^n$ follows by induction. However

$$M^n \cap R \subset \hat{M}^n \cap R = H_n$$

and these, taken together, give

$$M^n \cap R = \hat{M}^n \cap R = H_n .$$

It is clear from the construction of $\hat{Q}$ as an inverse limit that $\hat{Q}/\hat{M} \cong Q_n$. Let $a \in R$, $c \in \mathcal{C}$; then there exist $a_1 \in R$, $c_1 \in \mathcal{C}$ with $ac_1 - ca_1 \in H_n$ and hence

$$c^{-1}a - a_1c_1^{-1} \in QH_nQ \subset M^n .$$

The elements of Q are generated by those of the form ac^{-1} and this relation makes it clear that they can always be expressed in the form ac^{-1} modulo M^n. Thus an arbitrary element of A/M^n is expressible as

$$[a + M^n][c + M^n]^{-1}; \quad a \in R, \quad c \in \mathcal{C}.$$

Suppose that

$$[a + \hat{M}^n][c + \hat{M}^n]^{-1} = [b + \hat{M}^n][d + \hat{M}^n]^{-1}; \quad b \in R, \quad d \in \mathcal{C}.$$

There exist $c_1, d_1 \in \mathcal{C}$ with $cd_1 - dc_1 \in H_n$ and we have

$$ac^{-1} - bd^{-1} \in \hat{M}^n .$$

It follows that $ad_1 - bc_1 \in H_n$ and hence that

$$[a + M^n][c + M^n]^{-1} = [b + M^n][d + M^n]^{-1} .$$

The converse is evident since $M^n \subset \hat{M}^n$. This means that the correspondence

$$[a + \hat{M}^n][c + \hat{M}^n]^{-1} \longleftrightarrow [a + M^n][c + M^n]^{-1}$$

is (1-1). It can be verified to be an isomorphism and hence

$$\hat{Q}/\hat{M}^n \cong Q/M^n.$$

At once we see that M is a maximal ideal in Q and that Q/M is an Artinian ring.

Now $\hat{M} \supset P$ and $\hat{M} \cap Q \supset QPQ = M$. However, $\hat{M} \cap Q \subset M$ since $\hat{M} \cap Q = Q$ would imply that $1 \in \hat{M}$ which is not so. Thus

$$\hat{M} \cap Q = M.$$

We have shown that $Q = \{R, \mathcal{C}^{-1}\}$ has all the properties of a local ring except that which states that the maximal ideal coincides with the Jacobson radical. This omission is vital and we have to deal with it by enlarging Q, as will be seen by examining (2.53).

(2.50). The elements of Q have the form $ac^{-1} \pmod{M^n}$ for each n where

$$a \in R, \quad c \in \mathcal{C}.$$

This is clear from (2.49). Notice that the elements a,c vary with n.

(2.51). A ring S for which $R \subset S \subset \hat{Q}$ is <u>closed under localization</u> with respect to a set of units Φ of units of Q if $\Phi^{-1} \subset S$.

(2.52). $Q = \{R, \mathcal{C}^{-1}\}$ is closed under localization with respect to $\mathcal{C}(M)$ where $M = \hat{M} \cap Q$, if and only if Q is a local ring.

Proof. First we prove that $\mathcal{C}(M)$ is a set of units in $\hat{Q}$. For the elements of $\hat{Q}$ have the form

$$ac^{-1} + \hat{m}, a \in R, \quad c \in \mathcal{C}, \quad \hat{m} \in \hat{M}.$$

Hence $\mathcal{C}(M) = \mathcal{C}(\hat{M}) \cap Q$. A coset $[\hat{d} + \hat{M}]$, where $\hat{d} \in \mathcal{C}(\hat{M})$, is a regular element in the Artinian ring $\hat{Q}/\hat{M}$ and so is a unit there. Now units can be raised over the Jacobson radical $\hat{M}$ of $\hat{Q}$ and hence $\mathcal{C}(\hat{M})$ is a set of units in $\hat{Q}$. This means that $\mathcal{C}(M)$ has the same property.

Now suppose that $\mathcal{C}(M)^{-1} \subset Q$, then $1 + m$ is in $\mathcal{C}(M)$ for each $m \in M$ and so must be a unit in Q. It follows that M is the Jacobson radical of Q and Q is local. On the other hand, if M is the radical of Q then the argument used for $\hat{Q}$ can be applied to Q and shows that $\mathcal{C}(M)^{-1} \subset Q$.

(2.53) Theorem. Let $\{Q_\alpha ; \alpha \in \Lambda\}$ be the family of local rings with corresponding maximal ideals $M_\alpha ; \alpha \in \Lambda\}$ such that

$$R \subset Q_\alpha \subset \hat{Q}; \quad P \subset M_\alpha \subset \hat{M}; \quad \alpha \in \Lambda.$$

Then for all $\alpha \in \Lambda$

(1) $M_\alpha = \hat{M} \cap Q_\alpha$,

(2) $\{R, \mathcal{C}^{-1}\} \subset Q_\alpha$,

(3) $H_n = M_\alpha^n \cap R$, $n = 1,2,\ldots$,

(4) $Q_\alpha/M_\alpha \cong \hat{Q}/\hat{M}$.

Let $Q = (\cap Q_\alpha ; \alpha \in \Lambda)$ and $M = (\cap M_\alpha ; \alpha \in \Lambda)$. Q is a local ring with maximal ideal M.

(2.54). The <u>local envelope</u> of R in $\hat{Q}$ is the local ring Q defined by this theorem.

<u>Proof of (2.53)</u>. (1) is trivial because M_α and $\hat{M}$ are maximal ideals. To prove (2), let $c \in \mathcal{C}$ then

$$cx \in M_\alpha;\ x \in Q_\alpha \Longrightarrow cx \in \hat{M} \Longrightarrow x \in \hat{M} \Longrightarrow x \in M_\alpha.$$

Then $\mathcal{C} \subset \mathcal{C}(M_\alpha)$ and the argument of (2.52) shows that c is a unit of Q_α and hence that $ac^{-1} \in Q_\alpha$, from which we deduce that $\{R, \mathcal{C}^{-1}\} \subset Q_\alpha$.

To prove (3), observe that $H_n = \kappa(PH_{n-1})$ means that

$$H_n \subset Q_\alpha PH_{n-1}Q_\alpha.$$

Now $H_1 = P \subset M_\alpha$ and then $H_n \subset M_\alpha^n$ follows by induction. Then

$$H_n = \hat{M}^n \cap R \supset M_\alpha^n \cap R$$

and (3) is proved.

To prove (4), first observe that an element of $\hat{Q}$ has the form

$$ac^{-1} + \hat{m},\quad a \in R,\quad c \in \mathcal{C},\quad \hat{m} \in \hat{M},$$

so that those of Q_α have this form with $\hat{m} \in M_\alpha$. The correspondence

$$[ac^{-1} + M_\alpha] \longleftrightarrow [ac^{-1} + \hat{M}],\quad a \in R,\quad c \in \mathcal{C}$$

is easily shown to be an isomorphism of Q_α/M_α onto $\hat{Q}/\hat{M}$.

The same argument applies to Q. Thus $M = \hat{M} \cap Q$ and hence elements of Q have the form

$$ac^{-1} + m; \quad a \in R, \quad c \in \mathcal{C}, \quad m \in M.$$

The correspondence

$$[ac^{-1} + M] \longleftrightarrow [ac^{-1} + \hat{M}]$$

is an isomorphism of Q/M onto $\hat{Q}/\hat{M}$. As $\hat{Q}/\hat{M}$ is an Artinian simple ring, M is a maximal ideal of Q. An element of $1 + m$, $m \in M$ is in $1 + M_\alpha$ for each α and hence $(1 + m)^{-1} \in Q_\alpha$ for all $\alpha \in \Lambda$. Thus $(1 + m)^{-1} \in Q$. It follows that M is the radical of Q and Q is a local ring.

(2.55) <u>Theorem</u>. Define the sequence of rings $Q^{(n)}$, $n = 1,2,\ldots$, where

$$R \subset Q' \subset Q'' \subset \ldots \subset Q^{(n)} \subset \ldots \subset \hat{Q}$$

with $Q' = \{R, \mathcal{C}^{-1}\}$ and

$$Q^{(n+1)} = \{Q^{(n)}, \mathcal{C}^{(n)-1}\}; \quad M^{(n)} = \hat{M} \cap Q^{(n)},$$

$\mathcal{C}^{(n)} = \mathcal{C}(M^{(n)})$ in the ring $Q^{(n)}$.

Set

$$Q^* = \cup\, Q^{(n)}; \quad n = 1,2,\ldots ;$$
$$M^* = \cup\, M^{(n)}; \quad n = 1,2,\ldots .$$

Then Q^* is the local envelope of R in $\hat{Q}$ and M^* is its maximal ideal.

<u>Proof</u>. At each stage the set $\mathcal{C}^{(n)}$ consists of units of $\hat{Q}$,

so that $Q^{(n+1)}$ can be formed.

Evidently $M^* = Q^* \cap \hat{M}$ and hence the elements of Q^* have the form

$$ac^{-1} + m^*, \quad a \in R, \quad c \in \mathfrak{C}, \quad m^* \in M^*.$$

The correspondence

$$[ac^{-1} + M^*] \longleftrightarrow [ac^{-1} + \hat{M}] \quad a \in R, \quad c \in \mathfrak{C}$$

is readily seen to be an isomorphism of Q^*/M^* onto $\hat{Q}/\hat{M}$. Thus M^* is a maximal ideal of Q^*.

Let $m \in M^*$, then $m \in M^{(k)}$ for some k, hence $1 + m \in \mathfrak{C}^{(k)}$ and $(1 + m)^{-1} \in Q^{(k+1)}$. Thus $(1 + m)^{-1} \in Q^*$ and we have M^* is the Jacobson radical of Q^*. It follows that Q^* is a local ring with maximal ideal M^*. Since its construction proceeds by enlargement of $\{R, \mathfrak{C}^{-1}\}$ it is clearly the local envelope of R in $\hat{Q}$.

(2.56). The elements of Q^* have the form $ac^{-1} (\mathrm{mod}\ M^{*n})$ for each n, where $a \in R$, $c \in \mathfrak{C}$.

Proof. Left to the reader.

Next we consider conditions to be imposed on R so that the construction of the local envelope shall follow that for the commutative case; this is the so-called classical construction. Our conditions are sufficient, but are unlikely to be necessary. We suppose at the outset that

$$0 = (\cap H_n;\ n = 1,2,\ldots).$$

The additional property to be assumed is

$$(\cap \rho(E + H_n);\ n = 1,2,\ldots) = \rho(E) \qquad (*)$$

for all $E \underset{r}{\lhd} R$. This is denoted by (*).

(2.57). Assuming condition (*) on the ring R, then for any $E \underset{r}{\lhd} R$ there exists an integer $m > 0$ such that

$$E \cap H_m \subset \rho(EP) \qquad (**)$$

(This is condition (**)).

Proof. Consider the finitely generated module $\bar{E} = E - EP$ over the prime ring R/P. For any submodule $\bar{F} \subset \bar{E}$ define a closure operation cl_p by

$$cl_p\bar{F} = \{\bar{e} \in \bar{E};\ \bar{e}\bar{X} \subset \bar{F},\ \bar{X} \text{ an essential right ideal of } R/P\}.$$

Set

$$\bar{V}_n = cl_p\{[(E \cap H_n) + EP] - EP\};\quad n = 1,2,\ldots$$

Now $\{\bar{V}_n\}$, $n = 1,2,\ldots$ is a descending chain of closed submodules of E and this satisfies the descending chain condition. (See Goldie [10], p. 203). Suppose that $\bar{V}_m = \bar{V}_{m+1} \ldots$. Let $x \in E \cap H_m$, then $c \in \mathcal{C}$ exists with

$$[x + EP][c + P] \in [(E \cap H_{m+s}) + EP] - EP;\quad s = 1,2,\ldots,$$

so that

$$xc \in (E \cap H_{m+s}) + EP.$$

Hence

$$x(cR + P) \subset (E \cap H_{m+s}) + EP$$

and

$$x \in \rho[(E \cap H_{m+s}) + EP] = \rho(E) \cap \rho(EP + H_{m+s}); \quad s = 1,2,\ldots .$$

Thus

$$x \in \rho(E) \cap \{\cap \rho(EP + H_{m+s}); s = 1,2,\ldots\} = \rho(EP) .$$

It follows that

$$E \cap H_m \subset \rho(EP).$$

(2.58) Theorem. Let R have max-r and max-ℓ and P be a prime ideal which satisfies condition (*) and is such that the symbolic powers of P meet in the zero ideal. Then R has the right Ore condition with respect to $\mathcal{C}(P)$ and its right quotient ring with respect to this set is a local ring with max-r.

Proof. Let $a \in R$, $c \in \mathcal{C} = \mathcal{C}(P)$. There exist $a_n \in R$, $c_n \in \mathcal{C}$ with

$$ac_n = ca_n + h_n; \quad h_n \in H_n, \quad n = 1,2,\ldots .$$

Let $E = h_1R + h_2R + \ldots$ and suppose that $h_1,\ldots,h_{m_1}$ generate E. There exists m_0 such that

$$E \cap H_m \subset \rho(EP) \quad \text{for all} \quad m > m_0.$$

Let $n \geq \max(m_1,m_0)$. Then $h_n \in \rho(EP)$, so that we have

$$h_nc' = h_1p_1 + \ldots + h_{m_1}p_{m_1}$$

for some $c' \varepsilon \mathcal{C}$ and $p_i \varepsilon P$. Then

$$a(c_n c' - c_1 p_1 - \dots - c_{m_1} p_{m_1}) = c(a_n c' - a_1 p_1 - \dots \quad a_{m_1} p_{m_1}).$$

However $c_n c' - c_1 p_1 - \dots - c_{m_1} p_{m_1} \varepsilon \mathcal{C}$ and the right Ore condition is satisfied.

Let Q be the right quotient ring of R with respect to $\mathcal{C}$. The elements of Q have the form ac^{-1}; $a \varepsilon R$, $c \varepsilon \mathcal{C}$. That Q has max-r follows as in the commutative case.

(2.59). (1) The right ideals of Q are in (1-1) correspondence with the right closed right ideals of R, where

$$I \underset{r}{\triangleleft} Q \to I \cap R,$$

$$J \underset{r}{\triangleleft} R \to JQ.$$

(2) Any right ideal I of Q is closed in the M-topology

$$(\cap\ I + M^n;\ \ n = 1,2,\dots) = I.$$

These are proved as in commutative algebra, except that (2) also depends on condition (*).

Next we study the important property expressed in (2).

(2.60) <u>Theorem</u>. Let Q be a local ring with max-r and having M as its maximal ideal. The right ideals of Q are closed in the M-topology if and only if, for each $I \underset{r}{\triangleleft} Q$, there exists $s \geqq 1$ such that

$$E \cap M^s \subset EM.$$

Proof. Assume the stated condition. Let $I \underset{r}{\triangleleft} Q$ and denote

$$I' = (\cap(I + M^n);\quad n = 1,2,\ldots).$$

Let $E \underset{r}{\triangleleft} Q$ and be maximal in the class of those right ideals I for which $I \neq I'$. If $K \underset{r}{\triangleleft} Q$ and $K \supsetneq E$ then $K = K' \supset E'$. It follows that the right Q-module $Q - E$ has a unique simple submodule $E' - E$. Now $E'M \subset E$, since the alternative would be that $E'M + E = E'$ which would imply that $E = E'$, because M is the Jacobson radical and E' is finitely generated. Now for some $s \geq 1$

$$E' \cap M^s \subset E'M \subset E$$

and hence

$$E = E + (E' \cap M^s) = E' \cap (E + M^s) = E'.$$

This contradiction means that the right ideals of Q are closed in the M-topology.

The converse is proved by an easy modification of the method of (2.57).

(2.61). Conditions (*) and (**) on the ring R are equivalent.

Proof. We need to prove that (**) implies (*). However (**) was used to prove theorem (2.58) and the first part of (2.59). Thus the classical local ring Q can be formed and we can expect to use theorem (2.60), so we verify its condition. Let

$E \triangleleft_r Q$ then

$$(E \cap R) \cap H_s \subset \rho[(E \cap R)P] \text{ for some } s \geq 1, \text{ using } (**)$$

and

$$E \cap M^s = (E \cap R \cap H_s)Q \subset [\rho(E \cap R)P]Q \subset (E \cap R)PQ \subset EM.$$

Now let $F \triangleleft_r R$ then by (2.60)

$$(\cap FQ + M^n;\ n = 1,2,\ldots) = FQ.$$

Also $FQ \cap R = \rho(F)$ and $(FQ + M^n) \cap R = \rho(F + H_n)$. Thus

$$(\cap \rho(F + H_n);\ n = 1,2,\ldots) = \rho(F)$$

which is condition (*).

The purpose and meaning of condition (*), or (**) can now be settled.

(2.62) Theorem. Let R have max-r and max-ℓ and P be a prime ideal of which the symbolic powers meet in the zero ideal. Then R satisfies condition (**) if and only if R has the right Ore condition with respect to $\mathcal{C}(P)$ and the right ideals of the resulting local ring R_P are closed in the topology of its maximal ideal.

Proof. Given all the results we need to prove that (*) holds in R. Let $E \triangleleft_r R$ and consider

$$E \cap H_n;\ n = 1,2,\ldots .$$

Now $(E \cap H_n)Q = EQ \cap H_nQ = EQ \cap M^n$ and for $n \geqq s$, say,

$$EQ \cap M^n \subset EQM = EM.$$

Then

$$E \cap H_n \subset (E \cap H_n)Q \cap R \subset EM \cap E \subset \rho(EP).$$

For an element $x \in EM \cap R$ has the form

$$x = e_1m_1 + \ldots + e_km_k \quad e_i \in E, \; m_i \in M.$$

Now $e_im_i = p_ic^{-1}$ $i = 1,\ldots,k$ where $p_i \in P$ and $c \in \mathcal{C}$. Thus $xc \in EP$ and $x \in \rho(EP)$. Hence R satisfies condition (**).

It has not so far been possible to show that the general construction of a local ring or of its completion are Noetherian, or indeed to find a counter example. All that we are able to do is to indicate an approach and some consequences of it.

(2.63). The graded ring Q^* of a local ring Q having maximal ideal M is

$$Q^* = Q/M \oplus M/M^2 \oplus \ldots \oplus M^n/M^{n+1} \oplus \ldots$$

Its multiplication is defined by setting

$$[a + M^{r+1}][b + M^{s+1}] = [ab + M^{r+s+1}] \quad \begin{array}{l} a \in M^r, \; a \notin M^{r+1} \\ b \in M^s, \; b \notin M^{s+1}. \end{array}$$

and extending the definition in the obvious way to general elements of Q^*.

(2.64). Q^* is isomorphic to $(L^*)_k$, where $\hat{Q} = \hat{I}_k$ and L^* is the graded ring of $\hat{L}$.

Proof. The reader should examine the proof of theorem (2.48).

In commutative algebra the following sequence of implications is proved.

$$Q \text{ Noetherian} \Rightarrow Q^* \text{ Noetherian} \Rightarrow \hat{Q} \text{ Noetherian}.$$

In our situation we still have

$$Q^* \text{ has max-r} \Rightarrow \hat{Q} \text{ has max-r},$$

and this requires an easy modification of the commutative method (see Zariski and Samuel [37], II, chapter VIII). This requirement of Q^* is very strong, as indicated below.

(2.65) Let Q be a local ring of which every right ideal is closed in the M-topology such that Q^* has max-r. Then the Artin-Rees property holds for M.

Proof. Let $E \underset{r}{\triangleleft} Q$. Its leading ideal is the right ideal $E^* \underset{r}{\triangleleft} Q^*$ given by

$$E^* = \sum_{n=0}^{\infty} \frac{M^n \cap (M^{n+1} + E)}{M^{n+1}} .$$

As E^* is finitely generated, there exists $s > 0$ such that

$$\frac{M^n \cap (M^{n+1} + E)}{M^{n+1}} = \frac{M^s \cap (M^{s+1} + E)}{M^{s+1}} \cdot \left(\frac{M}{M^2}\right)^{n-s} ,$$

for all $n \geq s$.

Hence

$$M^{n+1} + [M^n \cap (M^{n+1} + E)] = [M^s \cap (M^{s+1} + E)]M^{n-s} + M^{n+1}$$

and this gives

$$M^n \cap E \subset (M^{n+1} \cap E) + (M^s \cap E)M^{n-s}; \quad n \geqq s.$$

Fix an $n \geqq s$ and apply this inequality, replacing n by $n+1, n+2, \ldots n+p-1$ successively. Then

$$M^n \cap E \subset M^{n+p} + (M^s \cap E)M^{n-s}.$$

Let $p \to \infty$ and use the fact that $(M^s \cap E)M^{n-s}$ is closed in the M-topology to obtain

$$M^n \cap E = (M^s \cap E)M^{n-s}, \quad n \geqq s$$

which is the Artin-Rees property for M.

(2.66). Let Q be a complete local ring with M as its maximal ideal and suppose that Q* has max-r. Then Q has the Artin-Rees property for M.

Proof. Q has max-r and, as we shall see from (2.67), every right ideal is closed in the M-topology. The result follows.

The following result is a particular case of a theorem due to Y. Hinohari (see [18]).

(2.67) Theorem. Let Q be a complete local ring which has a finitely generated maximal ideal M, and E be a finitely generated right ideal of Q. Then E is closed in the M-topology.

Proof. We begin by proving that

$$E' = E + (EM)'$$

where $'$ denotes closure in the M-topology. Now $E'M \subset (EM)'$ and hence $E'/(EM)'$ is a right Q/M-module. Let $e' \in E'$ and consider the image in this module of the right ideal $e'Q + E$. This image is finitely generated and is a module over an Artinian ring, namely Q/M, hence it has the minimum condition for submodules. Its submodules include the intersections of $\frac{e'Q+E}{(EM)'}$ with $\frac{(EM)'+M^n}{(EM)'}$ for $n = 1,2,\ldots$ and these become stationary at $n = k$, say. Now the intersection of the ideals $(EM)' + M^n$ for $n = k,k+1,\ldots$ is $(EM)'$ and accordingly we have

$$(EM)' = (e'Q + E + (EM)') \cap ((EM)' + M^n); \quad n \geqq k$$

Now $E' \subset E + M^k$ and hence

$$e' = e + m_k; \quad e \in E, \quad m_k \in m^k.$$

Then $m_k \in e'Q + E \cap M^k$ and hence $m_k \in (EM)'$. It follows that $e' \in E + (EM)'$ and hence $E' \subset E + (EM)'$ from which equality is at once obtained.

Before leaving the subject of localization it is appropriate to consider an important particular case. We suppose that

R is the universal enveloping algebra of a finite dimensional Lie algebra L over al algebraically closed field of characteristic zero and that P is a prime ideal of R. It can be shown that when L is solvable each ideal (two-sided) of R has a 'normalizing' set of generators and that, when L is nilpotent, this set can be taken to be a 'centralizing' set of generators. The definitions of these terms are given below but the proofs of the results stated above use a substantial analysis of the enveloping algebra and for this we refer the reader to a paper by McConnell [24].

(2.68). Let R be a ring with 1 and $x \in R$. Then x is <u>normal</u> if $xR = Rx$. The set of normal elements of R is N(R).

(2.69). An ideal $I \triangleleft R$ has a <u>normalizing set of generators</u> if there exist $x_1,\dots,x_n$ such that $I = x_1R + \dots + x_nR$ and

(i) $x_1 \in N(R)$

(ii) $[x_i + x_1R + \dots + x_{i-1}R] \in N(\frac{R}{x_1R + \dots + x_{i-1}R})$

for $i = 2,\dots,n$.

(2.70). An ideal $I \triangleleft R$ has a <u>centralizing set of generators</u> if there exist $x_1,\dots,x_n$ such that $I = x_1R + \dots + x_nR$ and

(i) $x_1 \in C(R)$

(ii) $[x_i + x_1R + \dots + x_{i-1}R] \in C(\frac{R}{x_1R + \dots + x_{i-1}R})$

for $i = 2,\dots,n$; where C(R) is the center of R.

(2.71). Let R be a ring with max-r and max-ℓ and 1 and P be a prime ideal of R. In the following two cases

(i) each ideal of R has a centralizing set of generators

(ii) R/P is a domain, $0 = \cap(H_n;\ n = 1,2,\ldots)$, and each ideal of R has a normalizing set of generators

we have

$$H_n = \{x \in R;\ \ xc \in P^n,\ \ c \in \mathcal{C}(P)\}$$
$$= \{x \in R;\ \ cx \in P^n,\ \ c \in \mathcal{C}(P)\}.$$

(2.72). Let R and P be as in (2.71) and suppose that either (i) or (ii) holds for ideals of R. Then the condition (**) on P is a consequence of the following condition:

For any right ideal E of R, there exists $n > 0$ such that

$$E \cap P^n \subset EP \qquad (***)$$

and R_P is classical.

ex) Let $R = F[x,y]$ be the universal enveloping algebra over an algebraically closed field F of characteristic zero, the Lie algebra being solvable of dimension two. Briefly

$$xy - yx = x.$$

Let $P_1 = xR + yR$, $P_2 = xR$; these are primes.

Localization with respect to P_1

Here $P_2 \subset P_1^n$, $n \geqq 2$, and $H_n = P_1^n$ since P_1 is a maximal ideal. Then $H = xR$ and $R/H = F[y]$ is commutative so that localization is classical. The condition (***) is not satisfied, nor is (2.71). The completion of R_P is Noetherian.

<u>Localization with respect to P_2</u>

Here $H_n = P_2^n$, $H = 0$ and R_{P_2} is classical, but the construction is non-commutative. Thus (2.71) (ii) holds as does condition (***). The complete local ring is Noetherian.

For details of the proofs of (2.71) and (2.72) the reader is referred to [27].

Chapter 3. The structure of orders

We consider a ring R with right quotient ring Q, often the terminology, R is a <u>right order</u> in Q, is employed. For the most part, R needs to have 1, but not at the outset. The right orders in Q are organized by an equivalence relation.

(3.01). Right orders R,S in Q are equivalent, $R \sim S$, if units $a, b, c, d \in Q$ exist with $aRb \subseteq S$, $cSd \subseteq R$. The definition can be used to obtain equivalent orders from a given one, because a subring $S \subseteq Q$, which satisfies these relations, is a right order in Q.

<u>Left equivalence</u> $(R \overset{\ell}{\sim} S)$ is defined in the same way but uses

$$Ra \subseteq S, \quad Sb \subseteq R.$$

<u>Right equivalence</u> $(R \overset{r}{\sim} S)$ uses $aR \subseteq S$, $bS \subseteq R$. These are stronger relationships than equivalence itself.

(3.02). Let R,S be right orders with $R \subseteq S$, $R \sim S$. There exist right orders T, T' in Q such that

$$R \subseteq T \subseteq S, \quad R \overset{\ell}{\sim} T \overset{r}{\sim} S,$$
$$R \subseteq T' \subseteq S, \quad R \overset{r}{\sim} T' \overset{\ell}{\sim} S.$$

<u>Proof</u>. Let $ab^{-1}S\,cd^{-1} \subseteq R$ where a,b,c,d are regular and in R. Set $T = R + aS + RaS$, then $aS \subseteq T$, $Tc \subseteq R$, $R \subseteq T \subseteq S$

and the rest follows. Define T' similarly.

(3.03). Let R be a right order in Q such that no right order, which contains R properly, is equivalent to R. Then R is called a maximal (equivalent) order.

Maximal orders in the classical theory of simple algebras are maximal in an absolute sense, not merely in this relative sense. It is not true that a class of equivalent orders always has maximal members.

(3.04). Let R be a right order in Q. An R-submodule I of Q_R is a fractional right R-ideal if $aR \supseteq I \supseteq bR$ for units a, b of Q.

Notice that $a^{-1}I \subseteq R$. A fractional ideal which lies in R is called integral; it is an 'ordinary' right ideal which has a regular element.

ex) When R is a prime ring with right quotient ring Q, an integral right R-ideal is an essential right ideal and conversely.

(3.05). Let I be a right R-ideal and define

$O_r(I) = (q \in Q \mid Iq \subseteq I)$; the right order of I,

$O_\ell(I) = (q \in Q \mid qI \subseteq I)$; the left order of I.

Corresponding definitions etc. hold for left and two-sided R-ideals.

(3.06). $O_r(I)$ and $O_\ell(I)$ are right orders equivalent to R;

they have 1.

Proof. $0_r(I) \supseteq R$. Let $a \in I$, $bI \subseteq R$, with a,b units; then $ba\, 0_r(I) \subseteq R$. Hence $0_r(I) \sim R$. Also $b0_\ell(I)a \subseteq R$ and $aR = aR1 \subseteq 0_\ell(I)$, so $0_\ell(I) \sim R$.

(3.07). A maximal ($\sim$) right order has 1.

Proof. $R \subseteq 0_r(R)$ and $R \sim 0_r(R)$, so $R = 0_r(R) \ni 1$.

(3.08). The inverse I^{-1} of a fractional right R-ideal I is given by

$$I^{-1} = (q \in Q \mid IqI \subseteq I) = (q \in Q \mid Iq \subseteq 0_\ell(I)) = (q \in Q \mid qI \subseteq 0_r(I)).$$

The following result is well-known (dual bases lemma). [4].

(3.09). Let R be a ring with 1 and A a right R-module. Then A is projective if and only if there exist families

$$\{a_\alpha\}\ ,\ \{\phi_\alpha\}\ ,\ a_\alpha \in A,\ \phi_\alpha \in \mathrm{Hom}(A,R)$$

such that for $a \in A$,

$$a = \sum_\alpha a_\alpha \phi_\alpha(a) \text{ and } \phi_\alpha(a) = 0, \text{ except for finite number of } \alpha.$$

(3.10) Lemma. Let I be a right R-ideal. Then $II^{-1} = 0_\ell(I)$ if and only if I is a projective right $0_r(I)$-ideal and in this case I is f.g. (finitely generated) over $0_r(I)$.

Proof. Let I be projective and the families be as given:

$$a_\alpha \in I, \quad \phi_\alpha \in \mathrm{Hom}(I, O_r(I)).$$

As $O_r(I)$ is a right order in Q, the elements of Q have the form ab^{-1}, $a,b \in O_r(I)$. Now $IQ = Q$, hence they have the form iq $(i \in I, q \in Q)$ and $iq = iab^{-1} = i'b^{-1}$ $(i' \in I)$.

An $O_r(I)$ - hom. $\theta : I \to O_r(I)$ can be extended (uniquely) to $\theta^* \in \mathrm{End}_Q Q$ given by

$$\theta^*(ic^{-1}) = \theta(i)c^{-1} : i \in I, \quad c \in O_r(I).$$

For, let $ic^{-1} = jd^{-1}$, $c^{-1}d = d_1c_1^{-1}$, then $ic_1 = jd_1$ implies that $\theta(i)c_1 = \theta(j)d_1$, hence that $\theta(i)c^{-1} = \theta(j)d^{-1}$. This shows that θ^* is well-defined. Its mapping properties are obtained by similar arguments. However $\theta^* : q \to q_o q$ $(q \in Q)$, where $q_o = \theta^* 1$, and the correspondence $\theta^* \longleftrightarrow q_o$ gives

$$\mathrm{Hom}(I, O_r(I)) \cong (q \in Q \mid qI \subseteq O_r(I)) = I^{-1}.$$

Let $\phi_\alpha \longleftrightarrow q_\alpha$ under this $\cong$, let $a \in I$ and

$$\phi_\alpha a = q_\alpha a = 0$$

Then $a = \sum_\alpha a_\alpha(\phi_\alpha a) = (\sum_\alpha a_\alpha q_\alpha)a$.

Taking a to be a regular element in I, so $\sum_\alpha a_\alpha q_\alpha = 1 \in II^{-1}$, then $II^{-1} = O_\ell(I)$.

Moreover $I = \sum_\alpha a_\alpha(q_\alpha I) = \sum_\alpha a_\alpha O_r(I)$ is f.g. over $O_r(I)$

Conversely, let $II^{-1} = O_\ell(I)$. Finite families $\{q_\alpha\}$, $\{a_\alpha\}$ $q_\alpha \in I^{-1}$, $a_\alpha \in I$ exist with $\sum_\alpha a_\alpha q_\alpha = 1$, then defining

ϕ_α by $\phi_\alpha a = q_\alpha a$ for $a \in I$ the rest follows by (3.09).

(3.11). Let R be a right order and I be a projective right R-ideal. Then I is f.g. over R and $II^{-1} = O_\ell(I)$.

Proof. In the proof of (3.10) replace $O_r(I)$ by R and I^{-1} by $I^* = (q \in Q \mid qI \subseteq R)$; then $II^* = O_\ell(I)$ because $I^* \approx \mathrm{Hom}_R(I,R)$. Consequently, I is f.g. over R and, since $I^{-1} \supseteq I^*$, we have $II^{-1} = O_\ell(I)$.

(3.12). Let R be a maximal right order in Q.

(1) Let I be a right R-ideal, then I is projective over R if and only if $II^{-1} = O_\ell(I)$, and then I is f.g. over R.

(2) Let T be a two-sided R-ideal then T is projective over R as a right ideal if and only if $TT^{-1} = R$ and T is f.g. over R.

Proof. $R = O_r(I) = O_r(T) = O_\ell(T)$.

(3.13). Let R be a simple ring with 1 and a right order in Q. Then R is a maximal order.

Proof. Let $S \supseteq R$, $S \overset{r}{\sim} R$, $cS \subseteq R$. As $RcS \triangleleft R$ we have $RcS = R$, and hence $R = RcS^2 = RS = S$. Similarly for $S \overset{\ell}{\sim} R$ and, putting the two together, for $S \sim R$.

(3.14). A right order R with 1 is an Asano right order if the R-ideals in Q form a group under multiplication (as ideals).

The following theorem originated by Asano, see Jacobson [19], in present form due to Robson [31].

(3.15) <u>Theorem</u>. Let R be a right order with 1. The following are equivalent.

(1) R is Asano,

(2) R is a maximal order and integral R-ideals are right R-projective,

(3) Every integral R-ideal is invertible.

<u>Proof</u>. (1) → (2). Let $S \supseteq R$, $S \overset{r}{\sim} R$, $bS \subseteq R$. Then RbS has an inverse $T(RbS) = (RbS)T = R$. Then

$$S = RS = TRbSS = TRbS = R .$$

Similarly, $S \supseteq R$, $S \overset{\ell}{\sim} R \to S = R$. Hence R is a maximal order. For any R-ideal T we have $O_r(T) = O_\ell(T) = R$ and $TT^{-1} \subseteq R$, $T^{-1}T \subseteq R$. Let T' be the group inverse, $TT' = T'T = R$, then $T' \subseteq T^{-1}$, hence $TT^{-1} = R = T^{-1}T$ which means that T is projective by (3.12).

(2) → (3). As integral R-ideals are f.g. by (3.12), R satisfies acc for these ideals. Let M be a maximal integral ideal. Then $MM^{-1} = R$, hence $R \subsetneq M^{-1}$ and $M^{-1}M \neq M$. As $R \underset{o}{\supseteq} M^{-1}M \supsetneq M$ and M is maximal, we deduce that $M^{-1}M = R$.

Let T be any integral R-ideal and $T \subseteq M_1$, where M_1 is maximal. Then $T \subsetneq M_1^{-1}T \subseteq R$ because $R = TT^{-1}$. If $M_1^{-1}T \subsetneq R$ then $M_1^{-1}T \subseteq M_2$ for a maximal ideal M_2 and $M_1^{-1}T \subsetneq M_2^{-1}M_1^{-1}T$.

Eventually, for an $n > 0$ we have

$$M_n^{-1}M_{n-1}^{-1} \ldots M_2^{-1}M_1^{-1}T = R \qquad M_i \text{ all maximal.}$$

Then $T = M_1 \ldots M_n$ and, with $T^* = M_n^{-1} \ldots . M_1^{-1}$, we have

$$TT^* = T^*T = R.$$

(3) → (1). Let T be an R-ideal and define $S = (x \in R | xT \subseteq R)$. Let q be a unit of Q with $qT \subseteq R$, $q = ab^{-1}$; $a,b \in R$. Then $aT \subseteq R$ so $a \in S$ and S,ST are integral R-ideals. They are given to be invertible, hence so is T.

(3.16) Corollary. Let R be an Asano order. Then

(1) the group of R-ideals is abelian,

(2) prime integral R-ideals are maximal,

(3) an integral R-ideal is a unique product of primes,

(4) R has d.c.c. for integral R-ideals which contain a fixed one,

(5) $TT^{-1} = T^{-1}T = R$ for every R-ideal T,

(6) an R-ideal is f.g. projective as left and right R-module.

Proof. The results summarize the proof of the theorem and need little more discussion. However, (1) needs some. Let M_1, M_2 be maximal, then $M_1 \cap M_2 = M_1A$, A some integral ideal. As $M_1A \subseteq M_2$ which is prime, we have $A \subseteq M_2$, $A = M_2$, hence $M_1 \cap M_2 = M_1M_2$ and equals M_2M_1 by symmetry. The maximal ideals generate the group and, consequently, it is abelian.

(3.17). A right order R is <u>bounded</u> if every integral right R-ideal I contains an integral R-ideal; the largest such is called the bound of I.

ex) Any right order in a finite dimensional central simple algebra is bounded. See [1].

The following theorem, basically due to Asano, modernized by Michler [29], but the present proof is due to T. Lenagan [25].

(3.18) <u>Theorem</u>. Let R be a bounded prime ring with max-r and is an Asano right order. Then R is a left order and is right and left hereditary.

(3.19) <u>Lemma</u>. A prime proper factor ring R/P is a simple Artinian ring.

<u>Proof</u>. We need to prove that the elements of $\mathcal{C}(P)$ are units modulo P. Let $c \in \mathcal{C}(P)$ and $cy = 0$. Then $y \in P$, $yP^{-1} \subseteq R$, $cyP^{-1} = 0$, so $yP^{-1} \subseteq P$, so $y \in P^2$ and so on. Eventually, obtain $y \in Y = \bigcap_1^\infty P^n$. If $Y \neq 0$ then (3.16) means that $Y = P^n = P^{n+1} \ldots$, say, and this is not allowed in a group. Hence $Y = 0$ and $\mathcal{C}(P) \subseteq \mathcal{C}'(0) = \mathcal{C}(0)$. Clearly cR is an essential right ideal and when $B = \text{bound}(cR)$ we can see that $B \not\subseteq P$. For $B \subseteq P$ means that $B \subseteq cP$, $BP^{-1} \subseteq cR$, $BP^{-1} = B$ which is not allowed in a group. Since $B \not\subseteq P$ then $cR + P \supseteq B + P = R$ and c is a unit modulo P.

(3.20) Lemma. R/E has a composition series for any essential right ideal E of R.

Proof. B = bound E is a finite product of maximal ideals of R, whose factors are Artinian, hence R/B has max-r and min-r, so R/E has a composition series.

(3.21) Lemma. If the maximal right ideals of R are projective, then R is right hereditary.

Proof. A right ideal of R has a right complement, their sum is direct and essential as right ideal. So we only need to prove that an essential right ideal E is projective. Since $|R/E| = k < \infty$, (length as R-module), it can be proved by induction on k. Let $R \supset F \supset E$, where F/E is simple, hence $F/E \approx R/M$ where M is a maximal right ideal. Then by Schanuel's lemma

$$R \oplus E \approx F \oplus M.$$

Now M is given to be projective, so that E is also projective.

Proof of theorem. Let M be a maximal right ideal and T = bound M. Now T is a primitive ideal and hence is maximal by (3.19). From $0 \to T \to R \to R/T \to 0$, both natural and exact, with T projective, we deduce that $pd_R(R/T) \leq 1$. Since R/T is a simple Artinian ring, $pd_R(R/M) \leq 1$ (direct summand of R/T), hence using the natural exact sequence $0 \to M \to R \to R/M \to 0$ we see that M is projective and (3.21) shows that R is right hereditary.

It is known that R is semi-hereditary (L. Small [35]) and that R is a bounded left order in Q (N. Jacobson [19]). So left hereditary and max-ℓ go hand-in-hand. Let $L_1 \subseteq L_2 \subseteq L_3 \subseteq \ldots$ be a chain of f.g. left ideals. As R has finite left rank we can find a f.g. left ideal K such that $\overset{\infty}{\underset{1}{\cup}} L_n \oplus K$ is left essential; hence $L_n \oplus K$ is left essential for large N. Let $J_i = L_{n+i} \oplus K$ and $T = \text{bound } J_1$. Then

$$T^{-1} \supseteq J_1{}^* \supseteq J_2{}^* \supseteq \ldots \supseteq R, \quad \text{where} \quad J_i{}^* = (q \in Q \mid J_i q \subseteq R).$$

So
$$R = T^{-1}T \supseteq J_1{}^*T \supseteq \ldots \supseteq T.$$

Eventually, $J_m{}^*T = J^*_{m+1}T = \ldots$, for some $m > 0$, and $J_m{}^* = J_m{}^*TT^{-1} = J^*_{m+1} = \ldots$.

However $J_m = J_m{}^{**} = J^{**}_{m+1} = J_{m+1}$, etc, where $J_m{}^{**} = (q \in Q \mid qJ_i{}^* \subseteq R)$. This completes the proof of max-ℓ, but its proof really requires the following.

(3.22). Let J be a projective left ideal of R and $J^* = (q \mid Jq \subseteq R)$, $J^{**} = (q \mid qJ^* \subseteq R)$, then $J = J^{**}$.

<u>Proof</u>. $J \subseteq J^{**}$ is clear and

$$J = RJ \supseteq J^{**}J^*J \supseteq J^{**}, \quad \text{since} \quad J^*J = O_r(J) \ni 1,$$

by (3.11) for left ideals.

<u>Krull dimension</u>. See P. Gabriel and R. Rentschler [8].

(3.23). Let S be a set of R-modules partially ordered by inclusion, $A \supseteq B$; $A,B \in S$. The interval (A,B) is the set $A \supseteq X \supseteq B$.

Define   K-dim S = $-\infty$, when S is discrete ($A \supsetneq B$ does not occur)

K-dim S = 0 , when S is Artinian (d.c.c. on chains $A_1 \supsetneq A_2 \supsetneq \ldots$).

K-dim S = n , when K-dim S $\neq -\infty,0,1,\ldots,n-1$ and every chain $A_1 \supsetneq A_2 \supsetneq \ldots$ with K-dim$(A_i, A_{i+1}) > n-1$ for $i = 1,2,\ldots$ is finite.

K-dim S = ∞ , when greater than any positive n.

In particular, the Krull-dimension of an R-module M is obtained by taking that of the set of its submodules and the right Krull dimension of a ring R is that of the module R_R. Denote by k(R) or k(M).

ex). Let R be a ring with 1 then k(R) = 0 if and only if R is right Artinian.

(3.24). Let R be a semi-prime ring with right quotient ring Q, and E a proper essential right ideal. Then

$$k(R) \geq 1 + k(^R/E).$$

Proof. $E \supseteq cR$, where c is a regular element and the chain $R \supset cR \supset c^2R \supset \ldots$ is infinite with distinct terms and isomorphic factors. Then $k\left(\dfrac{c^mR}{c^{m+1}R}\right) = k\left(\dfrac{R}{cR}\right) \geq k\left(\dfrac{R}{E}\right)$ and k(R) must be $\geq 1 + k(^R/E)$.

The following result is due to Michler [30].

(3.30). Let R be a prime ring with 1 and $k(R/E) = 0$ for all essential right ideals E. The following are equivalent

(1) R has finite right rank, see (1.28).

(2) $k(R) \le 1$.

Proof. (1) → (2). Let rank $R < \infty$. If $k(R) > 1$ there is an infinite chain $A_1 \supsetneq A_2 \supsetneq \dots$ of right ideals with $k(A_i/A_{i+1}) > 0$ for $i = 1,2,\dots$. Now

$$0 \le \operatorname{rank} A_i \le \operatorname{rank} R$$

so rank A_i = rank A_s (hence $A_i \sim A_s$ see (1.15)) for $i \ge s$, and a suitably large s. Let C be a complement of A_s then it is so for all A_i, $i \ge s$, and $A_i \oplus C$ is essential in R. Now

$$0 \le k\left(\frac{A_s \oplus C}{A_{s+1} \oplus C}\right) \le k\left(\frac{R}{A_{s+1} \oplus C}\right) = 0 \, .$$

Then $\frac{A_s}{A_{s+1}} \approx \frac{A_s \oplus C}{A_{s+1} \oplus C}$ implies that $k\left(\frac{A_s}{A_{s+1}}\right) = 0$.

(2) → (1). Let $k(R) \le 1$; if $k(R) = 0$ then R is an Artinian simple ring so that (1) holds. In this case, of course, $R = E$ and $k(R/E) = -\infty$; not quite the object of study.

Let $k(R) = 1$ and $A_1 \oplus A_2 \oplus \dots$ a sum of right ideals. Set $B_i = A_i \oplus A_{i+1} \oplus \dots$ for $i = 1,2,\dots$; then $B_1 \supsetneq B_2 \supsetneq \dots$. There has to be $s > 0$ with $k\left(B_s/B_{s+1}\right) = 0$ and, as $B_s/B_{s+1} \approx A_s$, this means that A_s is an Artinian R-module.

Hence R has minimal right ideals and their sum H is the direct sum of some of them. If $H = R$ then R is Artinian, certainly rank $R < \infty$ (but the case is not being considered). If H is a finite direct sum and $H \neq R$ we still have rank $R < \infty$. Hence there only remains to examine the case where there is an infinite direct sum $M_1 \oplus M_2 \oplus \ldots$ in H. Let $p_1, p_2, \ldots$ be the primes $2,3,5,\ldots$ and

$$D_o = \sum_1^\infty M_j, \quad D_1 = \sum_{2 \nmid j} M_j, \quad D_2 = \sum_{2,3 \nmid j} M_j \ \ldots$$

Certainly $D_o \supsetneq D_1 \supsetneq D_2 \ldots$ and, moreover, $k\left({}^{D_n}/D_{n+1}\right) > 0$ because ${}^{D_n}/D_{n+1}$ is not Artinian. It follows that this case does not occur.

(3.31). Let M, N be R-modules and M N, then

$$k(M) = \max(k(N), k({}^M/N)).$$

Proof. It follows from the easy case $M = N \oplus {}^M/N$.

(3.32). An R-module M is k-critical if $k(M) = k$ and $k({}^M/N) < k$ for all proper factor modules. See also R. Hart [16] who uses the term restricted for critical.

ex) M is 0-critical means that M is simple.

M is 1-critical and Noetherian means that all factor modules have finite length (composition series) but M does not. This is often called the restricted minimum condition.

Let M be a Noetherian module with $k(M) = k$, then a chain

$$M \supset N_o \supset N_1 \supset \dots \supset N_k \supset 0$$

exists such that M/N_s is s-critical for $s = 0,\dots,k$. Choose N_o to be a maximal submodule of M. Since $k(N_o) = k$ and letting $k \geqslant 1$, we can choose N_1 maximal in the set of submodules N of N_o for which $k(\frac{N_o}{N}) = 1$. Then N_o/N_1 is 1-critical, hence M/N_1 is 1-critical, and so on.
Note that $k(N_k) \leq k$, but no more than that can be said about N_k in general.

(3.32). A k-critical module is uniform.
Every non-zero submodule of M is k-critical.
Every endomorphism of M has zero kernel, so $\mathrm{End}_R M$ is an integral domain.

Proof. Let $N \cap N' = 0$; $N, N' \subseteq M$ and $\neq 0$. Since $N \approx \frac{N \oplus N'}{N'}$ we have $k(N) < k$, and as $k(M/N) < k$, this gives a contradiction.

Let $\theta \in \mathrm{End}_R M$ then $\theta M \approx \frac{M}{\ker\theta}$. Now the left hand side has Krull dimension k or $-\infty$ and the right hand side $< k$. Hence $\theta M = 0$ or $\ker \theta = 0$.

(3.33). A right ideal V of R is a k-critical right ideal if the module R/V is k-critical.

(3.34). Let V be k-critical in R and $V_a = (x \in R \mid ax \in V)$, where $a \notin V$. Then V_a is a k-critical right ideal.

Proof. The map $[r + V_a] \to [ar + V]$, $r \in R$ is an $\approx$ of R/V_a onto $aR+V/V$ and the latter is k-critical.

(3.35). Let R be a ring with max-r and $P_0 \subsetneqq P_1 \subsetneqq \cdots$ be a chain of prime ideals of R then $n \leq k(R)$.

Proof. We need only discuss the case of primes 0 and $P \neq 0,R$. Now $P \supseteq cR$, where c is a regular element and the chain $R \supset cR \supset c^2R \supset \ldots$ is infinite where the factors are isomorphic and of dimension $\geq k(R/P)$. Hence $k(R) > k(R/P)$.

ex) In a commutative ring R we have $k(R) = \sup n$, where n is the length of a chain of primes, but $k(R)$ may be infinite. In a non-commutative ring we can have $k(R) > \sup n$, as for example, in a simple Noetherian ring R, where $n = 0$ but $k(R) = 0$ only when R is Artinian.

(3.36). Let R be a commutative ring then the k-critical ideals are prime ideals and conversely, provided that the factor ring has finite Krull dimension.

Proof. Let V be k-critical, let $ab \in V$ and $a \notin V$, $b \notin V$. Then $V_a \supsetneqq V$ and $R/V_a \approx \frac{aR+V}{V}$ has dimension k, but as a factor, has dimension $< k$; contradiction.

Conversely, let P be a prime and $a \notin P$ then $P_a = P$ hence $R/P \approx \frac{aR+P}{P}$ which induces isomorphisms

$$\frac{R}{aR+P} \longleftrightarrow \frac{aR+P}{a^2R+P} \longleftrightarrow \frac{a^2R+P}{a^3R+P} \longleftrightarrow \cdots .$$

If $k(\frac{R}{aR+P}) = s$ then $k(\frac{R}{P}) \geq s+1$; hence all proper factor modules of R/P have lower dimension than that of R/P, so R/P is k-critical for some k.

(3.37). A module M is <u>compressible</u> if there is an R-isomorphism of M into any non-zero submodule of M.

When R has max-r, a cyclic submodule of M is Noetherian, hence any compressible module over R is Noetherian. Similarly, it is a uniform module. If $k(M) < \infty$ then M is k-critical. For if $0 \subsetneq N \subset M$ and $k(M/N) = s$, say, then the injection of M into N carried out repeatedly produces an infinite chain

$$M \supset N \supset N' \supset N'' \supset P' \supset P'' \supset Q' \supset Q'' \supset \ldots$$

where $M/N \approx N'/N'' \approx P'/P'' \approx Q'/Q'' \ldots$. Then $k(M) \geq s+1$. Thus M is k-critical.

(3.38). Let R be a commutative ring and T an ideal of R. Then T is a prime ideal if and only if R/T is a compressible R-module.

<u>Proof</u>. Let $a \notin T$, then $R/T_a \approx \frac{aR+T}{T}$. If T is prime, then $T_a = T$ and R/T is isomorphic to an arbitrary cyclic submodule, so R/T is compressible. Conversely, if $ab \in T$, there is an isomorphism of R/T into $\frac{aR+T}{T}$ and $[b + T] \to 0$, hence $b \in T$, so T is prime.

Note that compressible modules have stronger properties than k-critical modules and with the added advantage that they are not tied to a finite Krull dimension. But do they exist in general settings?

(3.39). Let R be a semi-prime ring with right quotient ring Q and U be a uniform right ideal of R. Then U is a compressible R-module.

Proof. Let $V \trianglelefteq_r R$, $0 \neq V \subseteq U$. Now $VU \neq 0$ because $V^2 \neq 0$. Now if $r(x) \cap U \neq 0$ then $xU = 0$. See Goldie [10]. Hence $v \in V$ exists with $r(v) \cap U = 0$ and then $U \approx vU \subseteq V$.

(3.40). Let V be a right ideal of R then the idealizer $\mathbb{I}\, V = (x \in R \mid xV \subseteq V)$ and the eigen-ring is $\mathbb{I}V/V$.

It is well-known that $\mathbb{I}\,V/V \approx \mathrm{End}_R(R/V)$ under the map $\theta \longleftrightarrow [a + V]$ given by

$$(r \in R) : \quad \theta[r + V] = [ar + V] \quad (\text{it needs } a \in \mathbb{I}V).$$

and $\theta[1 + V] = [a + V]$ in particular.

(3.41). Let R/V be compressible and $k(R/V) < \infty$, then $\mathbb{I}\,V/V$ is a right Ore domain.

Proof. Let $s_1, s_2 \in \mathbb{I}\,V - V$. Now R/V is k-critical and can be injected into $\dfrac{(s_1R+V) \cap (s_2R+V)}{V}$ so an element $s \in \mathbb{I}V - V$ exists with

$$s = s_1a_1 + v_1 = s_2a_2 + v_2 \;;\; a_i \in R, \;\; v_i \in V.$$

Now $s_i(a_iV + V) \subseteq V$ and $V_{s_i} \supseteq V$ and R/V_{s_i} is k-critical, so $V_{s_i} = V$. Hence $a_iV \subseteq V$, $a_i \in \mathbb{I}V$. Thus the common right multiple property holds in the eigen ring, which we know to be a domain from (3.32).

(3.42) Theorem. Let R have max-r and V be a right ideal such that R/V is compressible and k-critical. Let T = bound V and $L = (r \in R \mid rV \subseteq T)$. Then

(1) R/T is a prime ring,

(2) If $T = L \neq \mathbb{I}V$ then $\mathbb{I}V/T$ is a prime ring with the same right quotient ring as R/T.

(3) If $T \neq L \neq \mathbb{I}V$ then $L \nsubseteq V$ and $V \nsubseteq L$. The ring $\mathbb{I}V/T$ has two minimal primes L/T and V/T, with $(L \cap V)^2 \subseteq T$, moreover T = bound L in R and $\mathbb{I}V$ is the idealizer $\mathbb{I}'L = (r \in R \mid Lr \subseteq L)$.

Proof. Let $A, B \triangleleft R$ and $AB \subseteq T$. If $A \nsubseteq V$ then $A + V$ meets $\mathbb{I}V - V$ since R/V is compressible. Hence $B \subseteq V$, as V is a prime ideal of $\mathbb{I}V$. Thus T is a prime ideal of R.

Let $A, B \triangleleft \mathbb{I}V$ and $AB \subseteq L$, then $(RAV)(RBV) \subseteq LV \subseteq T$ so either $RAV \subseteq T$ or $RBV \subseteq T$, so either A or B is in L. Then L is prime in $\mathbb{I}V$. Clearly $(L \cap V)^2 \subseteq LV \subseteq T$ and it follows that L, V are the only minimal primes of the ideal T in the ring $\mathbb{I}V$.

When $L \subseteq V$ then $LR \subseteq V$ and $LR \subseteq T$, so $L = T$. When $V \subseteq L$ then $V^2 \subseteq T$, hence $(RV)^2 \subseteq T$ and $RV \subseteq T$. Then

$V = RV = T$ and $R = \mathbb{I}V$. This possibility is trivial but is covered by the theorem.

In case (3), L and V are distinct primes of $\mathbb{I}V$. Let $rV \subseteq V$ then $LrV \subseteq T$ so $Lr \subseteq L$. Let $Lx \subseteq L$ then $LxV \subseteq T$ and, as $L \nsubseteq V$ there is $r \in L - V$, $r \in \mathbb{I}V$ with $r(xV+V) \subseteq V$. Hence $xV \subseteq V$, $x \in \mathbb{I}V$. Thus $\mathbb{I}'L = \mathbb{I}V$.

Finally, let $T \subseteq T' \subseteq L$, where $T' \triangleleft R$. Then $T'V \subseteq T$ hence $T' \subseteq T$ or $V \subseteq T$. The latter means that $V = T \subseteq L = \mathbb{I}V$, which is not allowed, so $T = \text{bound } L$.

ex) Both cases of the theorem occur. When R is a simple artin ring and V is a maximal right ideal, $V = eR$, say, with $e^2 = e$. Then $L = R(1 - e)$, $T = 0$ and

$$\mathbb{I}V = R(1 - e) + eR.$$

This is an example of (3). In fact (3) is the generic situation for V/T non-essential in R/T as right ideal. Then taking $T = 0$ for convenience, we have $L = \ell(V)$ and $V = r(u)$, where u is some uniform element of R. For $R/V \approx uR$, proves that uR is uniform, since R/V is compressible.

Now (2) occurs whenever V is an essential right ideal of R, as for example when $R = A_2$ (see later).

A natural question is whether $\mathbb{I}V$ has max-r when R has max-r. That this is so for the case when V is a maximal right ideal or, indeed a semi-maximal (finite intersection of maximals) right ideal is due to Robson [32].

(3.43) Lemma. Let R be a ring, V be a semi-maximal right ideal, and $S = \mathbb{I}V$. Let U be a simple R-module, $U \approx {}^{R}/_{M}$ where M is a maximal right ideal. Let $B = (r \in R \mid rV \subseteq M)$. Then (1) $\mathrm{Hom}({}^{R}/_{V}, {}^{R}/_{M}) = 0 \rightarrow U$ is a simple R-module,

(2) $RV \subseteq M \rightarrow U$ is a simple R-module,

(3) $RV \not\subseteq M$ and $V \subseteq M \rightarrow U$ has a unique S-composition series of length 2 given by $R \supset B = S + M \supset M$.

Proof. If U is not S-simple then it has a submodule $0 \neq uS \neq U$ and then $uV \subseteq uS$ is an R-submodule, so $uV = 0$. The map $[r + V] \rightarrow ur$ $(r \in R)$ is in $\mathrm{Hom}({}^{R}/_{V}, U) \approx \mathrm{Hom}({}^{R}/_{V}, {}^{R}/_{M})$ and is not zero, which contradicts (1).

In (2) and (3) we have $M \supseteq V$ and ${}^{B}/_{M} \approx \mathrm{Hom}({}^{R}/_{V}, {}^{R}/_{M})$ under the map $[1 + V]r \rightarrow [b + M]r$, $(r \in R)$, for assigned $b \in B$. Let $0 \neq \alpha, \beta \in \mathrm{Hom}({}^{R}/_{V}, {}^{R}/_{M})$. As ${}^{R}/_{V}$ is a semi-simple module, there exists $\gamma \in \mathrm{End}({}^{R}/_{V})$ with $\alpha\gamma = \beta$. Then ${}^{B}/_{M}$ is a simple S-module.

When $RV \subseteq M$, we have $R = B$ and ${}^{R}/_{M} \approx U$ is a simple S-module.

When $RV \not\subseteq M$, then ${}^{B}/_{M}$ is a simple S-module and ${}^{R}/_{B}$ is likewise. For, let $r \in R$, $r \notin B$, $rV \not\subseteq M$ then $rV + M = R$ and ${}^{R}/_{B}$ is S-generated by the coset $[r + B]$. Finally, $B = S + M$, because that is the only object that it can be.

In order to prove that $\mathbb{I}V$ is Noetherian when R is, it is necessary to transfer V into V', where $RV' = R$.

(3.44) Lemma. Let V be a semi-maximal right ideal of R.

There exists $V' \underset{r}{\triangleleft} R$ with $V' \supseteq V$, $RV' = R$ and $\mathbb{I}V' = \mathbb{I}V = S$, say.

Proof. R/V is a semi-simple R-module, so RV/V has a lattice complement V'/V. Thus $RV + V' = R$ and $RV \cap V' = V$. Then $RV' = RV + V' = R$. Let $xV' \subseteq V'$ then $xV \subseteq RV \cap V' = V$ so that $\mathbb{I}V' \subseteq \mathbb{I}V$. Let $xV \subseteq V$ then $xV'V \subseteq V$, because $V'V \subseteq V$. Set $I = (a \in R \mid aRV \subseteq V)$. Now $I \supseteq V'$ and we have $I = V'$, provided that $I \cap RV = V$. This is because lattice complements are unique. Let $I \cap RV = J$ and take $K \underset{r}{\triangleleft} R$ such that $K \cap J = V$ and $K + J = R$. Then $JV = V$ and

$$RV = JV + KV = V + KV \subseteq K$$

so that $J = I \cap RV \subseteq J \cap K = V$. Hence $xV \subseteq V$ implies that $xV' \subseteq V'$, so that now we have $\mathbb{I}V = \mathbb{I}V'$.

(3.45) Theorem. Let V be a semi-maximal right ideal of R. Then R has max-r (r. Artinian) if and only if $S = \mathbb{I}V$ has max-r (r. Artinian).

Proof. We can assume $RV = R$ for convenience. Let R have max-r and let $A \underset{r}{\triangleleft} S$, then $AR \geq A \geq AV$. Now AR, AV are f.g. over R and hence over S. We are using the fact that $RV = R$ and R has 1 means that R is f.g. over S (because f.g. over V). Then AR/AV is an image of a finite number of copies of R/V and hence is a Noetherian S-module, since $|R/V| < \infty$ over S by (3.43). In fact $|\frac{AR}{AV}| < \infty$. Then

$|\frac{A}{AV}| < \infty$, hence A is f.g. modulo AV and AV is f.g. over R, hence over S. Thus S has max-r. The converse is trivial, because R is a f.g. S-module, hence Noetherian S-module, hence certainly has max. for R-submodules (namely right ideals).

The argument in the first part makes it plain that R r. Artinian implies that S is r. Artinian. The converse is equally obvious.

ex). (Björk [2], Robson). This example shows that (3.45) need not hold when $|^R/V| < \infty$ but V is not semi-maximal.

Let K be a field with derivation' and $K_o = (b \in K \mid k' = 0)$ be such that $|K : K_o| = \infty$. Let $T = K[y]$ and define a derivation ∂ on T by

$$\partial(k_n y^n + \ldots + k_o) = (k_n' y^n + \ldots + k_o')y.$$

Set $R = T[x]$ with the law $tx - xt = \partial t$; $t \in T$. Then R is a Noetherian integral domain. Let $V = xR + y^2R$, $W = xR + yR$. Now $^R/V$ has a unique composition series $R \supset W \supset V$ but $^{\Pi V}/V \approx K_o + \bar{y}K$, $\bar{y}^2 = 0$, so even the eigen ring is not Noetherian. Also $^R/W^2$ is Artinian but $^{\mathbb{I}(V)}/W^2$ is not.

Now $RV = W$ but on passing to $R' = \begin{pmatrix} R & R \\ R & R \end{pmatrix}$ and $V' = \begin{pmatrix} R & R \\ V & V \end{pmatrix}$, $W' = \begin{pmatrix} R & R \\ W & W \end{pmatrix}$ then $R' \supset W' \supset V'$ is an R'-composition series and $R'V' = R'$. Now $\mathbb{I}(V') = \begin{pmatrix} R & R \\ V & \mathbb{I}(V) \end{pmatrix}$ is not Noetherian because $\mathbb{I}(V) \approx e_{22}\, \mathbb{I}(V')e_{22}$ is not.

ex). When (3.45) is applied to $A_1 = F[x,y]$, where $xy-yx = 1$

and char $F = 0$, using $V = xA_1$ we obtain $\mathbb{I}V = F + xA_1$. This is a hereditary Noetherian domain with a single two-sided ideal $P \neq 0$, $P^2 = P$, where $P = aX_1$. This example is needed later.

(3.45) <u>Alternative partial proof</u>. Suppose that R has max-r then $S = \mathbb{I}V$ has max-r. Let $A \ntriangleleft_r S$ and $a_1,\ldots,a_k \in A$ with

$$AR = a_1R + \ldots + a_kR + AV.$$

Let $J_1 = (r \in R \mid a_1r \in a_2R + \ldots a_kR + AV)$ and let K_1 be a complement of J_1 wrt V; then $J_1 \cap K_1 = V$ and $J_1 + K_1 = R$. Then

$$AR = a_1K_1 + a_2R + \ldots + a_kR + AV.$$

Let $J_2 = (r \in R \mid a_2r \in a_1K_1 + a_3R + \ldots + a_kR + AV)$ and K_2 be a complement of J_2 wrt V. Then

$$AR = a_1K_1 + a_2K_2 + a_3R + \ldots + a_kR + AV.$$

Repeat until we have

$$AR = a_1K_1 + \ldots + a_kK_k + AV.$$

Let $a \in A$ and $a = a_1x_1 + \ldots + a_kx_k \pmod{AV}$, $x_i \in K_i$. Then $x_iV + V \subseteq J_i \cap K_i = V$ so $x_i \in \mathbb{I}V$, hence

$$A = a_1S + \ldots + a_kS + AV.$$

Thus A is f.g. over S because AV is.

(3.46). Let $S = \mathbb{I}V$ where $|{}^R/V| < \infty$ and $RV = R$. Then $k(R) = k(S)$.

<u>Proof</u>. R_S is f.g. over S, as $RV = R$. Then $k(R_S) \leq k(S) \leq k(R_S)$, so $k(R_S) = k(S)$. No need to compare $K(R_S)$ and $k(R_R)$. Suppose that f.g. R-modules M have $k(M_R) = k(M_S)$ up to $k(M_R) = k(M_S) = n-1$. Take $M = M_R = M_S$ with K - dim = n with respect to S. Then M has an infinite chain

$$M \supset M_1 \supset M_2 \supset \dots$$

of f.g. S-submodules with $k_S({}^{M_i}/M_{i+1}) = n - 1$.
Now $\left|\frac{M_i}{M_i V}\right| < \infty$ wrt both R and S, because M_i is f.g. over R. Hence $n-1 = k_S\left(\frac{M_i}{M_{i+1}}\right) = k_S\left(\frac{M_i V}{M_{i+1}V}\right) = k_R\left(\frac{M_i V}{M_{i+1}V}\right)$. Hence $k_R(M_i V) \geq n$ and so $k_R(M) \geq n$. But $k_R(M) \leq k_R(M) = n$, hence the two are equal.

(3.47) (ex). We discuss here a particular idealizer of A_2 which provides a good example of a compressible module.

Let $R = A_2 = F[x,y,x_1,y_1]$ $xy - yx = 1$, $xx_1 = x_1x$, etc. F of characteristic 0.
Let $V = xA_2$. Then elements of R may be written

$$x^n\phi_n(y) + \dots + \phi_o(y) : \phi_i(y) \in A_1[y].$$

and $x\phi_i(y) - \phi_i(y)x = \phi_i'(y)$, usual derivative.

This element is in $S = \mathbb{I}V$ if and only if $\phi_o(y)$ is a constant in A_1. So we have

$$S = A_1 \oplus xR : R = A_1[y] \oplus xR.$$

The eigen-ring $\approx A_1$, hence has max-r.

Any right S-module W in R with $W \supsetneq V$ has an element $\phi(y)$ and $\phi'(y) = \phi(y)x - x\phi(y) \in W$. Repeat this and see that the coefficients of $\phi(y) \in W$. Hence $W \cap A_1[y]$ is a homogeneous right ideal in $A_1[y]$. Thus the right ideals W_R of R which contain V are (1-1) with right ideals of A_1 under $W_R \longleftrightarrow W_R \cap A_1$ and $W_R = (W_R \cap A_1)R$. Hence V is the intersection of the maximal right ideals of R which contain V and R/V is compressible and 1-critical.

Moreover, S has max-r. This follows from the discussion of the 'finite-down' case (3.45) provided we can prove that simple R-modules are Noetherian over S. Now R/M, M maximal, is a simple S-module if $V \not\subset M$. Now let $V \subset M$. Then R/M has a unique maximal S-submodule B/M, where $B = (b \in R \mid bV \subset M)$. B/M is oetherian as an R-module if it is f.g. as an S/V-module, as S/V has max-r. Now verify that $B = M + A_1$, so that B/M is cyclic over S/V.

Whether any 1-critical right ideal of A_2 is compressible with Noetherian idealizer is not yet settled.

<u>Dedekind prime rings</u>

Refer especially to Eisenbud and Robson [5], [6].

(3.48). A Dedekind prime ring R is a hereditary, Noetherian prime ring HNPR over which the fractional ideals form a group.

An equivalent formulation is that R is HNPR and is a maximal (~) order. See (3.15).

Another is, that R is HNPR with no proper idempotent ideals.

<u>Proof</u>. Clearly, if R has a proper idempotent ideal then this cannot belong to the ideal group, so that does not exist.

Conversely, let $T \triangleleft R$, $T^* = (q \mid qT \subseteq R)$, $T^+ = (q \mid Tq \subseteq R)$ then T^*T and TT^+ are idempotent ideals in R, because $TT^* = O_\ell(T)$.

In fact we have

(3.49). Let R be a right order in a simple Artinian ring Q and I be a projective fractional right R-ideal. Then

(i) I^*I is an idempotent ideal of R,

(ii) $I = I^* \setminus R = I^{**}$.

<u>Proof</u>. Done essentially in (3.22) for left ideals. However, we repeat again. $II^* = O_\ell(I)$, hence $(I^*I)(I^*I) = I^*O_\ell(I)I = I^*I$. Also $I^{**} = I^*\setminus R = (q \in Q \mid I^*q \subseteq R) \subseteq O_\ell(I)I^{**} = I(I^*I^{**}) \subseteq IR = I$. Then $I^{**} \supseteq I$ is clear and implies that $I^{**} = I$.

The following case has to be eliminated before considering Dedekind prime rings properly.

(3.50). A Noetherian prime ring with a minimal right ideal is simple Artinian.

<u>Proof</u>. Let M be a minimal right ideal, it is uniform and hence contains an isomorphic copy of any other uniform right

ideal. It follows that all uniform right ideals are minimal, their sum is essential and contains cR, c regular, and $cR \approx R$. But cR is a direct sum of minimal right ideals, hence so is R and the result follows.

(3.51) Lemma. (See Webber [36]; also [5])

Let R be HNPR, not simple Artinian, and $J \subseteq I$ be right ideals, then $|I/J| < \infty$ if and only if J is essential in I.

Proof. If J be not essential in I and $J \oplus K \subseteq I$ and $J \oplus K \sim I$, then $I/J \supseteq J\oplus K/J \approx K$ cannot have a composition series, unless K contains a minimal right ideal, which is not allowed.

If J is essential in I then $I \oplus H$, $J \oplus H$ are essential in R for some $H \triangleleft_r R$. Now $\frac{I \oplus H}{J \oplus H} \approx \frac{I}{J}$, so we may suppose I, J were essential in R to begin with. Let

$$I \supseteq I_1 \supseteq \dots \supseteq J$$

be a chain of essential right ideals (hence integral). Then $I_1{}^* \subseteq I_1{}^* \subseteq \dots \subseteq J^*$ are a f.g. left module $(J \supseteq cR \to J^* \subseteq Rc^{-1})$ and we are supposing that conditions on R hold on the left so $I_n{}^* = I_{n+1}^*$ for some n. Then $I_n = I_n{}^{**} = I_{n+1}^{**} = I_{n+1}$ and the result follows.

(3.52). If R is a non-Artinian HNPR then its left and right Krull dimensions are 1.

(3.53). Let R be HNPR, then any projective module is a direct sum of uniform right ideals.

Proof. Refer to Cartan and Eilenberg [4] for the result that if $R \approx I_1 \oplus \ldots \oplus I_n$ with $I_k \triangleleft_r R$, then any projective module P is $\approx J_1 \oplus \ldots \oplus J_t$ with J's in I's.
Now $0 \to r(a) \to R \to aR \to 0$ exact with aR projective, means that $R \approx r(a) \oplus ?$ We can take r(a) to be a uniform right ideal (minimal right annihilator) and R is now split. The process is repeated and gives $R \approx I_1 \oplus \ldots \oplus I_n$ with all I_k uniform right ideals. The result follows.

(3.54). Following chapter 1 we say that $a \in A$, a module over R, is a torsion element if $ac = 0$ for some regular element c. When R is a prime ring the set of torsion elements is a submodule t(A) of A and $A/t(A)$ is torsion-free.

(3.55). A f.g. module over HNPR is a torsion module if and only if it is Artinian.

Proof. Let $M = m_1R + \ldots + m_sR$ be torsion. Now $mc = 0$ implies that $mR \approx \frac{R}{\text{ann}(m)}$ = Artinian module, as ann(m) is an essential right ideal. Hence M is an Artinian module.

Conversely, given mR is Artinian then m is a torsion element by (3.51).

(3.56). A f.g. torsion-free module N is a submodule of a free module and hence is projective.

Proof. Being torsion-free N can be embedded in $N \otimes_R Q$ which is a f.g. Q-module and is a submodule of a free Q-module of rank n, say. A generating set for N can be regarded as a finite set of n-tuples of elements of Q. Thus $N \approx \sum_{i=1}^{k} (c^{-1}a_1^i, \ldots, c^{-1}a_n^i)R$ where $c \in R$ is a common denominator. Then $cN \approx \sum_{i=1}^{k} (a_1^i, \ldots, a_n^i)R$ and $N \approx cN$. Hence N meets the requirements.

(3.57) Theorem. Let R be HNPR, a a f.g. right R-module. Then ${}^{A}/t(A)$ is projective and $A \approx t(A) \oplus {}^{A}/t(A)$.

Proof. $0 \to t(A) \to A \to {}^{A}/t(A) \to 0$ splits by (3.56).

For further results we need to look at Dedekind prime rings because there are HNPR which provide counter-examples.

(3.58). Let $I \triangleleft_r R$ then $I^*I = R$.

Proof. Let $I^*I = T$ then $T \triangleleft R$, T is integral, hence is invertible. Then $T^{-1}T = R = T^{-1}I^*I$. Thus $T^{-1}I^* \subseteq I^*$ and $T^{-1}I^*I \subseteq I^*I$ so $T^{-1}T = R \subseteq I^*I$, hence $I^*I = R$.

The following theorem was proved by D. Webber [36] for simple Noetherian domains and extended in Eisenbud and Robson [5].

(3.59) Theorem. Let R be a Dedekind prime ring, let $I,J,K \triangleleft_r R$ and I,K be related as right ideals (see chapter 1). There exists $L \triangleleft_r R$, $L \subseteq J$, L essential in J and

$$I \oplus J \approx K \oplus L.$$

Proof. We can assume that $I \subseteq K$ and $|\frac{K}{I}| < \infty$. For by Goldie [10], Th. (3.7) there exists $a \varepsilon K$ with $r(a) \cap K = 0$, hence $r(a) \cap I = 0$ and $I \approx aI \subseteq K$. We prove the theorem by induction on $n = |\frac{K}{I}|$. If $n = 1$, ${}^{K}/_{I}$ is simple, there is a non-zero map of J onto ${}^{K}/_{I}$, let this be θ, say. Choose $L = \ker \theta$ then ${}^{J}/_{L} \approx {}^{K}/_{I}$. By Schanuel's lemma $I \oplus J \approx K \oplus L$ and L essential in J is clear. Let $n > 1$ and $K' \underset{r}{\triangleleft} R$ with $K \supseteq K' \supseteq I$ and $|\frac{K'}{I}| = 1$. The inductive hypothesis gives a right ideal L' essential in J and $I \oplus J \approx K' \oplus L'$. As $|{}^{K}/_{K'}| = n - 1$ there is an L essential in L' with $K' \oplus L' \approx K \oplus L$. So $I \oplus J \approx K \oplus L$ and L is essential in J.

A property used in the proof is the following.

(3.60). J is a generator in the categorical sense, in particular, J can be mapped onto any simple R-module.

Proof. The former is really $J^*J = R$, but we give a short proof of the latter. Let M be a maximal right ideal of R. If $qJ \subseteq M$ for all $q \varepsilon J^*$ then $J^*J \subseteq M$ which contradicts $J^*J = R$. Hence we can take $q \varepsilon Q$ with $qJ \nsubseteq M$. Now

$$J \to qJ \to \frac{qJ}{qJ \cap M} \to \frac{qJ+M}{M} = \frac{R}{M}$$

and each map is the natural one and is onto. Remember that $qJ \neq 0$ since J has a regular element.

(3.61). Any right ideal of R is generated by two elements.

Proof. Let K be any essential right ideal, $I = J = R$ then

$R \oplus R \approx K \oplus L$ for some L essential in R. Thus K has two generators, hence so does any direct summand of K, which is enough.

(3.62) Theorem. A f.g. projective module A over a Dedekind prime ring R has the form $R \oplus \dots \oplus R \oplus I$, where I is a right ideal. A projective module, which is not f.g., is free.

Proof. We know that A is isomorphic to a direct sum of uniform right ideals of R. If $\operatorname{rank} R = n$ then any direct sum of $k \leq n$ uniform right ideals is isomorphic to a right ideal, because any uniform right ideal can be injected into any other. If $k = n$, the sum is essential. Thus forming groups of n uniform summands of A gives $A \approx I_1 \oplus \dots \oplus I_t \oplus J$, where $J \triangleleft_r R$ and $I_1, \dots, I_t$ are essential right ideals. Now $I_1 \oplus I_2 \approx R \oplus L_1$, $L_1 \oplus I_3 \approx R \oplus L_2$, etc, hence $I_1 \oplus \dots \oplus I_t \approx R \oplus \dots \oplus R \oplus L_{t-1}$. Now $L_{t-1} \oplus J \approx R \oplus I$, say, and we are finished.

When A is not f.g. the argument above still shows that A is a direct sum of essential right ideals and to finish we need only consider a countable sum of these, say $A = I_1 \oplus I_2 \oplus \dots$. Now $I_1 \oplus I_2 \approx R \oplus J_1$; $I_3 \oplus I_4 \approx J_1' \oplus J_2$; where J_1' is chosen by asking that $J_1 \oplus J_1' \approx R \oplus R$, etc. Then

$$A \approx (I_1 \oplus I_2) \oplus (I_3 \oplus I_4) \oplus \dots \approx (R \oplus J_1) \oplus (J_1' \oplus J_2) \oplus (J_2' \oplus J_3) \oplus \dots$$
$$\approx R \oplus (J_1 \oplus J_1') \oplus (J_2 \oplus J_2') \dots .$$

So A is an ascending union of free modules each a summand of the next sum, hence A is free. The above proof follows a recasting by L. Levy of the proof for the commutative case due to I. Kaplansky [22].

(3.63). A module is <u>completely faithful</u> if every submodule of every factor module is faithful.

A module is <u>unfaithful</u> if it has a non-zero right annihilator.

(3.64). Let A be a module and $|A| < \infty$; let C be a submodule with A/C cyclic. Then

(1) If C is simple and $0 \to C \to A \to A/C \to 0$ does not split then A is cyclic.

(2) If C is completely faithful then A is cyclic.

<u>Proof</u>. (1) Let $A/C = [a+c]R$ then $aR = A$. For aR has a representative of each coset of A over C. If $aR \not\supseteq C$ then aR is mapped isomorphically onto A/C and the inverse of this map splits the exact sequence, which is not allowed.

(2) Use induction on $|C|$. For $C = 0$ it is clear. Now choose a simple submodule $S \subseteq C$. By hypothesis A/S is cyclic, so if $0 \to S \to A \to A/S \to 0$ does not split, then A is cyclic.

On the other hand, suppose that it splits, then $A \approx S \oplus A/S$. S being simple, then ann S is an intersection of maximal right ideals (annihilating individual elements), and ann S = 0.

Write $^A/_S \approx {}^R/_K$, where $K \underset{r}{\vartriangleleft} R$. Note $K \neq 0$ as $|^A/_S| < \infty$. There is a maximal right ideal M with $^R/_M \approx S$ and $M \nsupseteq K$. Then we have

$$\frac{R}{M \cap K} \approx \frac{K}{M \cap K} \oplus \frac{M}{M \cap K} \approx \frac{M+K}{M} \oplus \frac{M+K}{K}$$

$$\approx \frac{R}{M} \oplus \frac{R}{K} \approx S \oplus \frac{A}{S} .$$

and $\frac{R}{M \cap K}$ is clearly cyclic.

The following result is assumed; see W. D. Gwynne and J. C. Robson [15].

(3.65). Let $I \underset{r}{\vartriangleleft} R$ and $T \vartriangleleft R$, $T \neq 0$, then $^I/_{IT}$ is a cyclic R-module.

(3.66) Theorem. Let $J \subseteq I$ be right ideals of R such that J is essential in I. Then $^I/_J$ is a cyclic R-module.

Proof. This is by induction on $|^I/_J| = n < \infty$. The case $n = 1$ is trivial. For $n > 1$ choose a simple submodule $S \subset {}^I/_J$ and write $0 \to S \to {}^I/_J \to {}^I/_{J'} \to 0$. The induction hypothesis says that $^I/_{J'}$ is cyclic and hence $^I/_J$ is cyclic unless S is unfaithful and the sequence splits. As S was arbitrary we conclude that, if $^I/_J$ is not cyclic, then it is a finite direct sum of unfaithful simple modules, say, $^I/_J \approx \amalg S_i' \oplus \amalg S_i'' \oplus \ldots$ where the isomorphic simple parts are grouped together. We need only show that the sum in each group is cyclic, for ${}^{R}/K_j \approx \amalg S_i^{(j)}$ implies that ${}^{I}/J \approx \frac{R}{\cap K_j}$ (check the length of composition series). So we only need to consider the single case ${}^{I}/J \approx \amalg S_i$.

Let $T = \text{ann } S_i$, $T \triangleleft R$, $T \neq 0$ and ${}^{I}/IT$ is cyclic. Now $({}^{I}/J)T = 0$ so $IT \subseteq J$, hence ${}^{I}/J$ is cyclic.

(3.67). Any submodule B of an Artinian cyclic module A is cyclic.

Proof. $A \approx {}^{R}/K$, where K is an essential right ideal. Then $B \approx {}^{L}/K$ for some right ideal $L \supset K$.

(3.68). Let I be an essential right ideal of R and $c \in I$ be a regular element, then $I = cR + iR$ for some $i \in R$.

Proof. Let $J = cR$ in (3.66).

References

1. S. A. Amitsur, Prime rings having polynomial identities with arbitrary coefficients, Proc. London Math. Soc., (3), 17, 1967, 470-486.

2. J. E. Björk, Conditions which imply that subrings of Artinian rings are Artinian, J. Algebra (to appear).

3. N. Bourbaki Algèbre commutative XXVII, Hermann, Paris.

4. H. Cartan and S. Eilenberg, Homological Algebra, Princeton, 1956.

5. D. Eisenbud and J. C. Robson, Modules over Dedekind prime rings, J. Algebra 16(1970), 67-85.

6. ______, Hereditary Noetherian prime rings, J. Algebra, 16(1970), 86-104.

7. P. Gabriel, Des catégories abéliennes, Thèse 1962.

8. P. Gabriel and R. Rentschler, Sur la dimension des anneaux er ensembles ordonnés, C. R. Acad. Sci., Paris, 265(1967), 712-715.

9. A. W. Goldie, The structure of prime rings under ascending chain conditions, Proc London Math. Soc., 8(1958), 589-608.

10. A. W. Goldie, Semi-prime rings with maximum condition, Proc. London Math. Soc.,10(1960), 201-220.

11. A. W. Goldie, Torsion-free modules and rings, J. Algebra (1964), 268-287.

12. __________, Localization in non-commutative Noetherian rings, J Algebra 5(1967), 89-105.

13. __________, A note on non-commutative localization, J. Algebra 8(1968), 41-44.

14. __________, Some aspects of ring theory, Bull. London Math. Soc., 1(1969), 129-154.

15. W. D. Gwynne and J. C. Robson, Completions of Dedekind prime rings, (to appear).

16. R. Hart, Krull dimension and global dimension of simple Ore-extensions, Math. Zeit. (to appear).

17. C. R. Hajamavis and T. H. Lenagan, Localization in asano orders (to appear).

18. Y. Hinohari, Notes on non-commutative local rings, Nagoya Math. J., 17(1960), 161-166.

19. N. Jacobson, The theory of rings, Amer. Math. Soc. Surveys, VI(New York, 1943).

20. __________, Structure of rings, Amer Math. colloq. Publ., 37(1956, 1964).

21. __________, Lie algebras, Interscience (New York, 1962).

22. I. Kaplansky, Fields and Rings, Univ. of Chicago Press, 1969.

23. J. Lambek, Lectures on rings and modules, Blaisdell, 1966.

24. ________, Torsion theories, Springer lecture notes 177.

25. T. H. Lenagan, Bounded asano orders are hereditary, (to appear).

26. J. C. McConnell, The intersection theorem for a class of non-commutative rings, (3) 17(1967), 487-498.

27. ________, Localization in enveloping rings, J. London Math. Soc., 43(1968), 421-428.

28. A. Malcev, On the immersion of an algebraic ring into a field, Math. Ann., 113(1936), 686-691.

29. G. O. Michler, Maximal asano orders, Proc. London Math. Soc., (3), 19(1969), 421-443.

30. ________, Primringe mit Krull-dimension eins, J. für. r.u. angewandte Math., 239 (240), (1970), 366-381.

31. J. C. Robson, Non-commutative Dedekind rings, J. Algebra, 9(1968), 249-265.

32. ________, Idealizers and hereditary Noetherian prime rings, J. of Algebra, (to appear).

33. L. W. Small, On some questions in Noetherian rings, Bull. Amer. Math. Soc., 72(1966), 853-857.

34. L. W. Small, Orders in Artinian rings, J. Algebra 4(1966), 13-41.

35. __________, Semi-hereditary rings, Bull. Amer. Math. Soc., 73(1967), 656-658.

36. D. Webber, Ideals and modules of simple Noetherian hereditary rings, (to appear).

37. O. Zariski and P. Samuel, Commutative algebra I, II, Van Nostrand, 1960.

QUASISIMPLE MODULES AND OTHER TOPICS IN RING THEORY

by

Kwangil Koh

Department of Mathematics
North Carolina State University
Raleigh, North Carolina

Introduction

This lecture note is based on a course on topics in the non-commutative ring theories given to a group of graduate students at Tulane University in the fall of 1970. Topics which are covered in this lecture note are the simple modules, the injective hulls of modules and rings and the quasi-simple modules. An associative ring which does not necessarily contain the multiplicative identity is the main concern of this note. Throughout the note, the density theory of N. Jacobson plays a central role. In fact, we prove through a generalized density theorem, that if R is a prime ring with a uniform right ideal U which has the zero singular submodule, then for any finite subset $u_1, u_2, \ldots, u_n$ in U which is linearly independent over the endomorphism ring D of the quasi-injective hull of U, the endomorphism ring of $R/\bigcap_{i=1}^{n}(u_i)^{\perp}$, ($(u_i)^{\perp}$ is the annihilators of u_i in R), is a right order in the $n \times n$ matrix ring over D.

I am deeply indebted to Professors K. H. Hofmann and L. Fuchs at Tulane University for their stimulating interest and encouragement. I am particularly grateful to J Dauns, Hans H. Storrer, Mike Mislove, William Nico, Jo Ledhetter and J. Luh for catching a number of slips and also for some valuable suggestions which have improved the accuracy or the clarity at several points in the manuscript.

Chapter 1

If R is a ring and M is a faithful simple (right) R-module then M is a vector space over its endomorphism ring and R becomes naturally a subring of the ring of linear transformations on M. If $x \in M$, let $(x)^{\perp}$ be the set of annihilators of x in R. Main theorems in this chapter are that $x_1, \dots, x_n$ of M are linearly independent if and only if $(x_j)^{\perp} \not\supseteq \bigcap_{\substack{i=1 \\ i \neq j}}^{n} (x_i)^{\perp}$ for every $1 \leq j \leq n$ and that the set F of elements in R which are of a finite rank is a minimal two-sided ideal which is a regular ring and F is also the sum of all minimal right ideals and the sum of all minimal left ideals of R.

1.0 Definition. Let R be a ring. An (right) R-module M is called simple provided that $MR \neq 0$ and 0 and M are the only submodules of M.

1.1 Proposition. Let M be a simple R-module for some ring R. Then

(i) every non-zero element of M is a generator of M

(ii) there is a maximal modular right ideal I in R such that $M \cong R/I$.

Proof. Let $0 \neq m$ be an element of M. It suffices to show that $mR \neq 0$. If $mR = 0$ then the set $R^{\top} = \{m \in M \mid mR = 0\}$ is a non-zero submodule of M and hence $R^{\top} = M$. Therefore $MR = 0$ and this is a contradiction. To see (ii), let

$I_m = \{r \in R \mid mr = 0\}$. Then I_m is a maximal right ideal of R since $M = mR \cong R/I_m$ by a mapping: $mr \longmapsto r+I_m$, $r \in R$. Since $mR = M$, $m = ma$ for some $a \in R$ and $r-ar \in I_m$ for every $r \in R$. Thus I_m is a maximal modular right ideal of R.

1.2 Proposition. If I is a maximal modular right ideal of a ring R then R/I is a simple R-module.

Proof. Since the only submodules of R/I are zero and R/I itself, we only need to show that $R^2 \not\subseteq I$. Since I is a modular right ideal, there is $a \in R$ such that $r-ar \in I$ for every $r \in R$. If $R^2 \subseteq I$ then $a \in I$ since $a-aa \in I$ and $a^2 \in I$. Therefore $R = I$ and this is a contradiction.

1.3. Example. There is a ring R and a maximal right ideal I in R such that R/I is a simple R-module but I is not modular. Let $R = \begin{pmatrix} \mathbb{Z}_4 & 2\mathbb{Z}_4 \\ \mathbb{Z}_4 & 2\mathbb{Z}_4 \end{pmatrix}$, $I = \begin{pmatrix} \mathbb{Z}_4 & 2\mathbb{Z}_4 \\ 2\mathbb{Z}_4 & 2\mathbb{Z}_4 \end{pmatrix}$ where $\mathbb{Z}_4$ is the integers modulo (4). Then R/I is a simple R-module but I is not modular since $R/I = \{\begin{pmatrix} 0 & 0 \\ 0 & 0 \end{pmatrix} + I, \begin{pmatrix} 0 & 0 \\ 1_4 & 0 \end{pmatrix} + I\}$ where 1_4 is the multiplicative identity of the ring $\mathbb{Z}_4$ and
$\begin{pmatrix} 0 & 0 \\ 1_4 & 0 \end{pmatrix}\begin{pmatrix} 0 & 0 \\ 1_4 & 0 \end{pmatrix} = \begin{pmatrix} 0 & 0 \\ 0 & 0 \end{pmatrix}$.

1.4 Proposition. Let I be a maximal right ideal of a ring R. Then R/I is simple if and only if $R^2 \not\subseteq I$.

Proof. If I is a maximal right ideal of R then the module R/I has only two submodules. Hence if $R^2 \not\subseteq I$ then R/I is simple. If R/I is simple then by 1.0 $R^2 \not\subseteq I$.

1.5 <u>Proposition</u>. Let R be a ring and let $\sigma(R) = \{I \mid I$ is a maximal right ideal such that $R^2 \not\subseteq I\}$. Every member of $\sigma(R)$ is also a left ideal if and only if $R/I \cong R/J$, as R-modules $I,J \in \sigma(R)$, implies that $I = J$.

<u>Proof</u>. Suppose every member of $\sigma(R)$ is also a left ideal. Assume $I,J \in \sigma(R)$ and there exists an R-isomorphism $\phi : R/I \to R/J$. Let $a \in R$ such that $aR \not\subseteq J$. Then $\phi(b+I) = a+J$ for some b. Now $aJ \subseteq J$ implies $bJ \subseteq I$ since ϕ is a monomorphism. Now $(b+I)^{\perp} = \{r \in R \mid (b+I)r = 0\}$ $= I$ since $(b+I)I = 0$ and $(b+I)R \neq 0$ (for if $(b+I)R = 0$, then $(a+J)R = 0$ implies $aR \subseteq J$ which is a contradiction). Thus $I = J$. Conversely, assume that if $R/I \cong R/J$ as R-modules where $I,J \in \sigma(R)$, then $I = J$. Let $I \in \sigma(R)$ and $x \in R\backslash I$. It suffices to show that $xI \subseteq I$. Now $R/I = (x+I)R$ by 1.1 and $(x+I)R \cong R/(x+I)^{\perp}$. As R-modules $(x+I)^{\perp} \in \sigma(R)$, therefore $(x+I)^{\perp} = I$ and $xI \subseteq I$.

1.6 <u>Corollary</u>. If R is a commutative ring with 1 and I,J are two different maximal ideals, then $R/I \not\cong R/J$ (module isomorphic).

1.7 <u>Proposition</u>. Let R be a ring and I be a modular right ideal $(I \neq R)$ such that $xR \subseteq I$ implies $x \in I$ for any $x \in R$. Then $\mathrm{Hom}_R(R/I,R/I) \cong N(I)/I$ as rings where $N(I)$ is the largest subring of R in which I is a two-sided ideal.

Proof. First observe that $N(I) = \{r \in R \mid rI \subseteq I\}$.

Let $a \in R$ such that $ar - r \in I$ for every $r \in R$.

Then $R/I = (a + I)R$. Let $f \in \mathrm{Hom}_R(R/I, R/I)$. Define: $f \longmapsto f(a + I) = b_f + I$. Then $b_f + I \in N(I)/I$. Also $f \circ g \longmapsto f(b_g + I) = f(a+I)b_g = b_f \cdot b_g + I = (b_f + I)(b_g + I)$. The mapping is a monomorphism. To see that it is also an epimorphism, let $t + I \in N(I)/I$, where $t \in N(I)$. Define $h_t(x+I) = tx + I$ for every $x \in R$. Then $h_t \in \mathrm{Hom}_R(R/I,R/I)$ and $ta + I = t + I$ since $tar-tr = t(ar-r) \in I$ for every $r \in I$, i.e. $(ta-t)R \subseteq I$ implies $ta-t \in I$. Thus $h_t \to t+I$.

1.8 Proposition. (Schur's Lemma). If M is a simple R-module, then $\mathrm{Hom}_R(M,M)$ is a division ring.

Proof. Clear.

1.9 Example. There exists a ring R and a cyclic R-module M such that $\mathrm{Hom}_R(M,M)$ is a field but M is not simple. Let $R = \begin{pmatrix} \mathbb{Z}_2 & 0 \\ \mathbb{Z}_2 & \mathbb{Z}_2 \end{pmatrix}$, $M = R/\begin{pmatrix} \mathbb{Z}_2 & 0 \\ 0 & 0 \end{pmatrix}$. Then $\begin{pmatrix} \mathbb{Z}_2 & 0 \\ \mathbb{Z}_2 & 0 \end{pmatrix}/\begin{pmatrix} \mathbb{Z}_2 & 0 \\ 0 & 0 \end{pmatrix} \subsetneq M$ but $\mathrm{Hom}_R(M,M) \cong \begin{pmatrix} 0 & 0 \\ 0 & \mathbb{Z}_2 \end{pmatrix}$.

1.10 Note. If M is a simple R-module then M is a left vector space over $\mathbb{D} = \mathrm{Hom}_R(M,M)$. Note also that if $0 \neq x \in M$, then $(x)^{\perp}$ is a maximal right ideal of R. If S is a nonempty subset of R, let $S^{\top} = \{m \in M \mid ms = 0$ for every $s \in S\}$.

1.11 Proposition. Let M be a simple R-module. Then for any finite subset $\{x_1,\ldots,x_n\}$ of M, $(\mathbb{D}x_1+\mathbb{D}x_2+\ldots+\mathbb{D}x_n)^{\perp\top} = \mathbb{D}x_1+\mathbb{D}x_2+\ldots+\mathbb{D}x_n$.

Proof. Proceed by induction on n. For $n = 1$, clearly $(\mathbb{D}x_1)^{\perp\top} \supseteq \mathbb{D}x_1$. Let $y \in (\mathbb{D}x_1)^{\perp\top}$. Define $f:x_1R \to yR$ by $f(x_1r) = yr$ for every $r \in R$. If $x_1r = 0$, then $\mathbb{D}x_1r = 0$ and $r \in (\mathbb{D}x_1)^{\perp}$. Hence $yr = 0$ and $f \in \mathbb{D}$. Since $x_1 \in x_1R$, $f(x_1r) = f(x_1)r$ ($x_1 = x_1a$ for some $a \in R$, so $f(x_1) = f(x_1a) = ya$ and $f(x_1r) = f(x_1ar) = yar = f(x_1)r$), and $(f(x_1)-y)R = 0$. Therefore $f(x_1) = y$ and $y \in \mathbb{D}x_1$, so $(\mathbb{D}x_1)^{\perp\top} = \mathbb{D}x_1$.

Now assume $(\mathbb{D}x_1+\ldots+\mathbb{D}x_{n-1})^{\perp\top} = \sum_{i=1}^{n-1}\mathbb{D}x_i$. Let $J = (\sum_{i=1}^{n-1}\mathbb{D}x_i)^{\perp}$. Then $J^{\top} = \sum_{i=1}^{n-1}\mathbb{D}x_1$ by the inductive hypothesis. Now suppose $\sum_{i=1}^{n}\mathbb{D}x_i \subsetneq (\sum_{i=1}^{n}\mathbb{D}x_i)^{\perp\top}$ and let $v \in (\sum_{i=1}^{n}\mathbb{D}x_i)^{\perp\top}$ such that $v \notin \sum_{i=1}^{n}\mathbb{D}x_i$. Define $f:x_nj \to vj$ for all $j \in J$. If $x_nJ = 0$, then $x_n \in \sum_{i=1}^{n-1}\mathbb{D}x_i$ and $\sum_{i=1}^{n-1}\mathbb{D}x_i = \sum_{i=1}^{n}\mathbb{D}x_i$ which is a contradiction. So $x_nJ \neq 0$ and $x_nJ = M$. Therefore, $f \in \mathbb{D}$ and $(f(x_n)-v)J = 0$. Hence $f(x_n)-v \in J^{\top} = \sum_{i=1}^{n-1}\mathbb{D}x_i$ and $\sum_{i=1}^{n}\mathbb{D}x_i$, which is a contradiction.

1.11 Proposition. Let M be a simple R-module. Then for any finite subset $\{x_1,\ldots,x_n\}$ of M, $(\mathbb{D}x_1+\mathbb{D}x_2+\ldots+\mathbb{D}x_n)^{\perp\top} = \mathbb{D}x_1+\mathbb{D}x_2+\ldots+\mathbb{D}x_n$.

Proof. Proceed by induction on n. For n = 1, clearly $(\mathbb{D}x_1)^{\perp\top} \supseteq \mathbb{D}x_1$. Let $y \in (\mathbb{D}x_1)^{\perp\top}$. Define $f: x_1R \to yR$ by $f(x_1r) = yr$ for every $r \in R$. If $x_1r = 0$, then $\mathbb{D}x_1r = 0$ and $r \in (\mathbb{D}x_1)^{\perp}$. Hence $yr = 0$ and $f \in \mathbb{D}$. Since $x_1 \in x_1R$, $f(x_1r) = f(x_1)r$ ($x_1 = x_1a$ for some $a \in R$, so $f(x_1) = f(x_1a) = ya$ and $f(x_1r) = f(x_1ar) = yar = f(x_1)r$), and $(f(x_1)-y)R = 0$. Therefore $f(x_1) = y$ and $y \in \mathbb{D}x_1$, so $(\mathbb{D}x_1)^{\perp\top} = \mathbb{D}x_1$.

Now assume $(\mathbb{D}x_1+\ldots+\mathbb{D}x_{n-1})^{\perp\top} = \sum_{i=1}^{n-1}\mathbb{D}x_i$. Let $J = (\sum_{i=1}^{n-1}\mathbb{D}x_i)^{\perp}$. Then $J^{\top} = \sum_{i=1}^{n-1}\mathbb{D}x_1$ by the inductive hypothesis. Now suppose $\sum_{i=1}^{n}\mathbb{D}x_i \subsetneq (\sum_{i=1}^{n}\mathbb{D}x_i)^{\perp\top}$ and let $v \in (\sum_{i=1}^{n}\mathbb{D}x_i)^{\perp\top}$ such that $v \notin \sum_{i=1}^{n}\mathbb{D}x_i$. Define $f: x_nj \to vj$ for all $j \in J$. If $x_nJ = 0$, then $x_n \in \sum_{i=1}^{n-1}\mathbb{D}x_i$ and $\sum_{i=1}^{n-1}\mathbb{D}x_i = \sum_{i=1}^{n}\mathbb{D}x_i$ which is a contradiction. So $x_nJ \neq 0$ and $x_nJ = M$. Therefore, $f \in \mathbb{D}$ and $(f(x_n)-v)J = 0$. Hence $f(x_n)-v \in J^{\top} = \sum_{i=1}^{n-1}\mathbb{D}x_i$ and $v \in \sum_{i=1}^{n}\mathbb{D}x_i$, which is a contradiction.

Proof. Let $r_j \in \bigcap_{\substack{i=1\\ i\neq j}}^{n} (x_i)^{\perp}$ such that $r_j \notin (x_j)^{\perp}$. This we can do by 1.12 since $(x_j)^{\perp} \not\supseteq \bigcap_{\substack{i=1\\ i\neq j}}^{n} (x_i)^{\perp}$. Then $y_j \in x_j r_j R$ and $y_j = x_j r_j a_j$ for some a_j. Let $r = \sum_{j=1}^{n} r_j a_j$. Then $x_i r = y_i$, $i = 1,2,\ldots,n$.

1.15 Remark. Let $\mathcal{L} = \mathrm{Hom}_{\mathbb{D}}(M,M)$. Then R can be embedded into $\mathcal{L}$ and M is an $\mathcal{L}$-module. Let F be the collection of all finite subsets of M and consider $\tau_o = \{X^{\perp} \mid X \in F\}$ where $X^{\perp} = \{f \in \mathcal{L} \mid xf = 0$ for every $x \in X\}$. Then τ_o forms a neighborhood system of 0 in $\mathcal{L}$. Let $\tau = \{f + X^{\perp} \mid f \in \mathcal{L}, X \in F\}$, and let τ_1 be a topology generated by τ. Then $(\mathcal{L}, \tau_1)$ becomes a topological abelian group. Here the closure of R, $\bar{R} = \mathcal{L}$. To see this, let $\ell \in \mathcal{L}$ and $X \in F$. To show $(\ell + X^{\perp}) \cap R \neq \emptyset$, let $X = \{m_1, m_2, \ldots, m_t\}$ and let $X_1 = \{m_{i_1}, m_{i_2}, \ldots, m_{i_n}\}$ be a maximal linearly independent subset of X. Then $X_1^{\perp} = X^{\perp}$. Now let $y_{i_1} = m_{i_1}\ell$, $y_{i_2} = m_{i_2}\ell, \ldots, y_{i_n} = m_{i_n}\ell$. Then there exists $a \in R$ such that $m_{i_1}a = y_{i_1}$, $m_{i_2}a = y_{i_2}, \ldots, m_{i_n}a = y_{i_n}$; $a - \ell \in X_1^{\perp}$ and $a \in (\ell + X^{\perp}) \cap R$.

1.16 Corollary. If $\dim_{\mathbb{D}} M$ is finite, say $\dim_{\mathbb{D}} M = n$, then $R \cong \mathbb{D}_n$.

1.17 Proposition. Let M be a simple R-module and $\{m_1, m_2, \ldots, m_t\}$ be linearly independent elements over $\mathbb{D}$ in M where $\mathbb{D} = \mathrm{Hom}_R(M,M)$. Set $(m_i)^{\perp} = \{r \in R \mid m_i r = 0\}$.

Then $\operatorname{Hom}_R(R/\bigcap_{i=1}^{t}(m_i)^{\perp}, R/\bigcap_{i=1}^{t}(m_i)^{\perp}) \cong \mathbb{D}_t$.

Proof. Let $I = \bigcap_{i=1}^{t}(m_i)^{\perp}$. Let $e \in R$ such that $m_i e = m_i$ for every $i = 1,2,\ldots,t$; then I is modular. Furthermore if $aR \subseteq I$ for some $a \in R$, then $m_i aR = 0$ which implies $m_i a = 0$ for every $i = 1,2,\ldots,t$ and $a \in I$. Therefore by 1.7 $\operatorname{Hom}_R(R/I,R/I) \cong N(I)/I$. Let $M_t = \mathbb{D}m_1 + \mathbb{D}m_2 + \ldots + \mathbb{D}m_t$. Then $M_t^{\perp} = I$. Consider the set $R^{(t)} = \{r \in R \mid M_t r \subseteq M_t\}$. Then $I \subseteq R^{(t)}$ and $R^{(t)}/I \cong \mathbb{D}_t$ by 1.16. Now if $a \in R$ such that $aI \subseteq I$, then $M_t a \subseteq I^{T} = M_t$ which implies $N(I) \subseteq R^{(t)}$. If $a \in R^{(t)}$ then $M_t a \subseteq M_t$, therefore $aI \subseteq M_t^{\perp} = I$. Hence $N(I) = R^{(t)}$ and $N(I)/I \cong \mathbb{D}_t$.

1.18 Remark. Let M be a faithful simple R-module such that $\dim_{\mathbb{D}}M$ is countable where $\mathbb{D} = \operatorname{Hom}_R(M,M)$. Let $\{m_1,m_2,\ldots\}$ be a basis for M and let $G_n = \bigcap_{i=1}^{n}(m_i)^{\perp}$ for each positive integer n. Topologize R by taking $\{G_n\}_{n=1}^{\infty}$ as a fundamental system of neighborhoods of 0. Since $\bigcap_{i=1}^{\infty} G_n = 0$, this topology is Hausdorff and R becomes a topological abelian group. Let $\check{R}$ be the completion of R, that is $\check{R} = \{$all Cauchy sequences in $R\}/\sim$ where $(a_i)\sim(b_i)$ if and only if $a_i - b_i$ approaches 0 as i grows large. It is easy to see that $\sim$ is a congruence relation with respect to addition. Define $[(a_i)]\cdot r = [(a_i \cdot r)]$. It is easy to see that this is well defined and $\check{R}$ becomes an R-module.

1.19 <u>Theorem</u>. $\mathcal{L}_R \cong R_R$ where $\mathcal{L} = \text{Hom}_D(M,M)$.

<u>Proof</u>. Define $\theta: \mathcal{L} \to R$ as follows: Let $x \in \mathcal{L}$ and

$$m_1 r_1 = m_1 x \qquad m_1 r_2 = m_1 x \qquad m_1 r_3 = m_1 x$$
$$m_2 r_2 = m_2 x \qquad m_2 r_3 = m_2 x$$
$$m_3 r_3 = m_3 x \qquad \text{etc.}$$

Consider (r_i). To show (r_i) is Cauchy observe that given
any positive integer k, $r_i - r_j \in G_k$ for $i,j \geq k$ since
$m_t r_i = m_t r_j$ for all $1 \leq t \leq k$. Now set $\theta(x) = [(r_i)]$.
Suppose (r_i') is another choice such that $m_j r_k' = m_j x$ for
$j = 1,2,..,k$. Then $m_j(r_k' - r_k) = 0$ for $1 \leq j \leq k$ and
$r_k' - r_k \in G_k$. Since $G_j \subseteq G_k$ if $j \geq k$, then $r_j' - r_j \in G_j$
and $r_j' - r_j \in G_k$ for all $j \geq k$. Hence $[(r_i')] = [(r_i)]$.
It is easy to see that θ is an R-homomorphism. We claim θ
is a monomorphism. Suppose $\theta(x) = [(r_i)] = [(0)]$. If
$x \neq 0$, then there exists m_t such that $m_t x \neq 0$. Then for
G_t there exists a positive integer N such that for every
$i \geq N$, $r_i \in G_t$. Take $j = t + N$, then $r_j \in G_t$ and
$0 = m_1 r_j = m_1 x$, $0 = m_2 r_j = m_2 x,..., 0 = m_t r_j = m_t x$ which
is a contradiction. Finally we must show that θ is an
epimorphism. Let $[(s_i)]$ be an element of $\check{R}$. Given G_1
there exists $n_1 \geq 1$ such that $s_i - s_j \in G_1$ for every
$i,j \geq n_1$. If n_{k-1} is given, then there exists an integer
$n_k > n_{k-1}$ such that $s_i - s_j \in G_k$ for $i,j \geq n_k$. Thus we
obtain a subsequence (s_{n_i}) of (s_i) with $s_{n_k} - s_{n_j} \in G_k$
for all $j \geq k$. Now define $x \in \mathcal{L}$ such that $m_k x = m_k s_{n_k}$
for every $k = 1,2,...$. Then $\theta(x) = [(s_{n_i})]$. However

$[(s_{n_i})] = [(s_n)]$. This completes the proof.

1.20 Proposition. Let M be a faithful simple R-module. If $a \in R$, $a \neq 0$ is of finite rank then there exists $b \in R$ such that abR is a minimal right ideal. Conversely, if $c \in R$ such that cR is a minimal right ideal then Mc is a one-dimensional space.

Proof. Suppose $Ma = \mathbb{D}x_1 + \mathbb{D}x_2 + \ldots + \mathbb{D}x_n$ (where the x_j's are independent). Then there exists $b \in R$ such that $x_j b = x_1$ for all $j = 1,2,\ldots,n$ and there exists $m \in M$ such that $mab = x_1$. Let J be a right ideal where $0 \neq J \subseteq abR$. If $abr \in abR$, then there exists $abi \in J$ such that $mabi = x_1 r$. Hence $m(abi - abr) = 0$. Also, if $s \in R$, then $ms(abi-abr) = (msab)(i-r) = f(mab)(i-r) = 0$ where f is some element of $\mathbb{D}$. Then $M(abi-abr) = 0$ and $abi = abr$. Conversely, let cR be a minimal right ideal and suppose Mc is not one-dimensional. Then there exists m_1c and m_2c in Mc such that m_1c and m_2c are independent. Hence, let $d \in R$ such that $m_1cd = m_1c$ and $m_2cd = m_1c$. Then $(m_1-m_2)cdR = 0$. But $cdR = cR$ and $(m_1-m_2)\,c\,R = 0$ implies $m_1c = m_2c$, which is a contradiction.

1.21 Proposition. Let M be a faithful simple R-module and let F be a non-zero two sided ideal of R. Then M is a faithful simple F-module and $\mathrm{Hom}_F(M,M) = \mathrm{Hom}_R(M,M)$.

Proof. Clearly M is a faithful F-module. Suppose N is a nonzero F-submodule of M. Then $NF \subseteq N$ and $NF \neq 0$

(for if $NF = 0$, then $0 = NRF = MF$, which is a contradiction). Furthermore NF is an R-module. Thus $M = NF \subseteq N$ and $M = N$. Suppose there exists $f \in Hom_F(M,M)$ such that $f \notin Hom_R(M,M)$. Then there exists $m \in M$ and $r \in R$ such that $f(mr) - f(m)r = m_0 \neq 0 \in M$. Let $s \in F$, then $m_0 s = f(mr)s - f(m)rs = f(mrs) - f(m(rs)) = 0$. Hence $m_0 F = 0$. Let $X = \{m \in M \mid mF = 0\}$. Then X is a nonzero R-submodule of M; hence $X = M$ and $MF = 0$, which is a contradiction.

1.22 Theorem. The set $F = \{a \in R \mid a$ is of finite rank$\}$ is a two-sided ideal of R and F is a regular ring. F is the sum of all minimal right ideals of R and the sum of all minimal left ideals of R.

Proof. Let F_1 be the sum of all minimal right ideals of R. Then $F_1 \subseteq F$. Let $a \in F$. Then $M = \ker a \oplus X$ (†) and $X = \langle x_1, x_2, \ldots, x_n \rangle$ is finite dimensional since Xa is of finite dimension. For each $x_i a$ there exists $b_i \in R$ such that $x_i ab_i = x_i$ and $x_j ab_i = 0$ if $i \neq j$. Then $Mab_i = \mathbb{D}x_i$, and hence ab_iR is a minimal right ideal of R. Let $c = \sum_{i=1}^{n} ab_i a$. Then $c \in F_1$ and $M(a-c) = 0$. Therefore $a = c \in F_1$ and $F = F_1$. Now let F_2 be the sum of all minimal left ideals of R. We claim that F_2 is a two-sided ideal and $F_2 \subseteq F$. Let L be a minimal left ideal. Then $L^2 \neq 0$ and there exists $d \in L$ such that $Ld \neq 0$ and $Ld = L$. Hence there exists $e \in L$ such that $ed = d$. Therefore $(e^2-e)d = 0$. If $e^2-e \neq 0$, then

---

(†) $\ker a = \ker f$, where $f(m) = ma$, $\forall\, m \in M$.

$L \cap \{r\epsilon d \mid rd = 0\} \neq 0$ and $L \subseteq \{r\epsilon R \mid rd = 0\}$, which is a contradiction. Hence $e^2 = e$ and $Re = L$. By Schur's Lemma, eRe is a division ring. Now suppose $L \nsubseteq F$. Then there exists $a \in L$ such that $a \notin F$ and $a = ae$. Now eR is a minimal right ideal; for if $0 \neq I$ is a right ideal such that $I \subseteq eR$, then there exists $i \in I$ such that $ie \neq 0$ and $ie = eie$. Hence there exists $eye \in eRe$ such that $ieye = e$, $e \in I$ and $I = eR$. Therefore $a = ae$ is of finite rank and $a \in F$, which is a contradiction. Furthermore, F_2 is a two-sided ideal since $a \in R$ and J a minimal left ideal imply Ja is a minimal left ideal or (0). Now we have already shown that $F_1 = F \supseteq F_2$. If I is a minimal right ideal of R, then $I \cap F_2 \neq 0$ which implies $I \subseteq F_2$. Therefore $F_1 \subseteq F_2$ and $F_1 = F = F_2$.

1.23 <u>Proposition</u>. Let $F = \{a\epsilon R \mid a$ is of finite rank$\}$. Then F is a simple ring.

<u>Proof</u>. Let S be an ideal of F such that $S \neq 0$. Then $0 \neq FSF \subseteq S$ ($FSF \neq 0$ since F is a primitive ring in its own right). Now FSF is a nonzero ideal of R. Hence FSF must contain all minimal right ideals of R and $FSF = F$. Therefore $F = S$.

1.24 Example. Let V be the set of all finite real sequences. Then V is a vector space over $\mathbb{R}$ (where $\mathbb{R}$ denotes the real numbers). Let $R = \mathrm{Hom}_{\mathbb{R}}(V,V)$. Then V is a faithful simple R-module and F is a subring of R without unity. Also every right ideal of F is a direct summand of F. Therefore, if I is a right ideal of F and ϕ is an F-homomorphism from I_F to F, then ϕ can be extended to an F-homomorphism from F to F. F has neither a right identity nor a left identity.

1.25 Proposition. If R is a subdirect sum of a finite number of simple rings $R_1, R_2, \ldots, R_n$, then R is isomorphic to the direct sum of some of the R_i's.

Proof. By definition, there exists a ring monomorphism μ from R into $\sum_{i=1}^{n} \oplus R_i$ such that $R \xrightarrow{\mu} \sum_{i=1}^{n} \oplus R_i \xrightarrow{\pi_i} R_i$ and $\pi_i \circ \mu$ is an epimorphism for all $1 \leq i \leq n$. Proceed by induction on n. If $n = 1$, then $R \cong R_1$. Assume the statement is true for $1 \leq k < n$. If for each i there exists $0 \neq x_i \in R_i$ and $(0,0,\ldots,0,x_i,0,\ldots,0) \in \mu(R)$, then $R \cong \sum_{i=1}^{n} \oplus R_i$. Indeed, $R_i x_i R_i = R_i$ and for any $r_i \in R_i$, there exists $a_i \in R$ such that $r_i = \pi_i \circ \mu(a_i)$ and μ is a ring isomorphism. So suppose there exists i such that for any non-zero $x_i \in R_i$, $(0,\ldots,0_{i-1},x_i,0_{i+1},\ldots,0_n) \notin \mu(R)$. Define $f : \mu(R) \to R_1 \oplus \ldots \oplus R_{i-1} \oplus R_{i+1} \oplus \ldots \oplus R_n$ by $f(x_1,x_2,\ldots,x_{i-1},x_i,x_{i+1},\ldots,x_n) = (x_1,x_2,\ldots,x_{i-1},x_{i+1},\ldots,x_n)$. Then f is a monomorphism. Therefore $R \xrightarrow{f\circ\mu} \sum_{\substack{j=1\\ j\neq i}}^{n} \oplus R_j \xrightarrow{\pi_j} R_j$, where $f \circ \mu$ is a monomorphism and $\pi_j \circ f \circ \mu$ is an epimorphism for each j. Hence by the induction hypothesis, the assertion is true.

1.26 Proposition. If an R-module M is a subdirect sum of a finite number of simple modules $M_1, M_2, \ldots, M_n$, then M is isomorphic to the direct sum of some of the M_i's.

Proof. There exists a monic R-homomorphism μ such that $M \xrightarrow{\mu} \sum_{i=1}^{n} \oplus M_i \xrightarrow{\pi_i} M_i$ and such that $\pi_i \circ \mu$ is an epimorphism for each $i = 1,2,\ldots,n$. Proceed by induction on n. If $n = 1$, the proof is clear. Assume the statement is true for $1 \leq k < n$. If for each i, there exists $0 \neq x_i \in M_i$ and $(0_1,\ldots,0_{i-1},x_i,0_{i+1},\ldots,0_n) \in \mu(M)$, then $M \cong \sum_{i=1}^{n} \oplus R_i$. If not, let i be such that for any non-zero $x_i \in M_i$, $(0_1,\ldots,0_{i-1},x_i,0_{i+1},\ldots,0_n) \notin \mu(M)$. Then $f : \mu(M) \to M_1 \oplus \ldots \oplus M_{i-1} \oplus M_{i+1} \oplus \ldots \oplus M_n$ defined by $f((x_1,\ldots,x_{i-1},x_i,x_{i+1},\ldots,x_n)) = (x_1,\ldots,x_{i-1},x_{i+1},\ldots,x_n)$ is a monomorphism. Therefore $f \circ \mu$ is a monomorphism from M into $\sum_{\substack{i=1\\ i\neq j}}^{n} \oplus M_i$ and $\pi_i \circ f \circ \mu$ is an epimorphism for each $i = 1,2,\ldots,n$, $i \neq j$. Hence M is a subdirect sum of $\sum_{\substack{i=1\\ i\neq j}}^{n} \oplus M_i$, and by the induction hypothesis the conclusion follows.

1.27 Theorem. The following statements are equivalent:

(i) $1 \in R$ and R is a direct sum of a finite number of minimal right ideals.

(ii) $1 \in R$ and R is a direct sum of a finite number of minimal left ideals.

(iii) R is semi-simple and R satisfies the minimum condition on right ideals.

(iv) R is semi-simple and R satisfies the minimum condition on left ideals.

(v) R is a direct sum of a finite number of simple rings $R_1, R_2, \ldots, R_n$ such that for each i $R_i \cong D_{n_i}^{(i)}$ for some division ring $D^{(i)}$ and positive integer n_i.

Proof. We shall show that (i) implies (iii) implies (v) implies (i) and that (ii) implies (iv) implies (v) implies (ii). Assume (i) holds. If I is a minimal right ideal and $J(R)$ is the Jacobson radical of R, then $IJ(R) = 0$. Hence, since $1 \in R$, R must be semi-simple. The minimum condition on right ideals follows from the fact that R has a composition series. Therefore (i) implies (iii). Now assume (iii) holds. Note that $J(R) = \bigcap_{i \in \Lambda} P_i$ where P_i is a primitive ideal for each $i \in \Lambda$. Hence $\bigcap_{i \in \Lambda} P_i = (0)$, and the minimum condition on right ideals implies that $\bigcap_{i=1}^{n} P_i = 0$ for some integer n. Also since R satisfies the minimum condition on right ideals, then so does R/P_i for each $i = 1,2,\ldots,n$ Thus by corollary 1.16, $R/P_i = D_{n_i}^{(i)}$ for some division ring $D^{(i)}$ and positive integer n_i. In particular, R/P_i is a simple ring. Since $\bigcap_{i=1}^{n} P_i = 0$, R is a subdirect sum of simple rings $R/P_1, R/P_2, \ldots, R/P_n$. Hence (by 1.25), R is a direct sum of some of the $R/P_1, R/P_2, \ldots, R/P_n$. Clearly (v) implies (i). In a similar manner we can prove the remaining implications.

1.28 Proposition. Let R be a simple ring with a minimal right ideal. Then the following hold:

(i) R is the sum of all minimal right ideals and all minimal left ideals.

(ii) R is the direct sum of minimal right ideals and minimal left ideals.

(iii) Every pair of simple modules are isomorphic to each other.

(iv) R is a regular ring.

Proof. (i) If I is a minimal right ideal, then I is a faithful simple R-module. Hence $F = \{a \in R \mid Ia$ has finite dimension$\} \neq 0$, and by 1.22 F is a two-sided ideal. Thus R = F = the sum of all minimal right ideals of R = the sum of all minimal left ideals of R.

(ii) Let $\Sigma = \{ \sum_{\alpha \in \Lambda} I_\alpha \mid I_\alpha$ is a minimal right ideal for every $\alpha \in \Lambda$ and such that $I_\alpha \cap \sum_{\beta \neq \alpha} I_\beta = 0\}$. Then $\Sigma \neq \emptyset$ and Σ may be ordered by set inclusion. Let $\mathcal{C}$ be a chain in Σ. Then $\sum_{\ell \in \mathcal{C}} \ell$ is an upper bound for $\mathcal{C}$ in Σ and Σ

has a maximal member $\mathcal{M}$. If J is a minimal right ideal, then either $J \subseteq \mathcal{M}$ or $J \cap \mathcal{M} = 0$. The latter is impossible, since $\mathcal{M}$ is maximal. Hence $R = \mathcal{M}$ and R is a direct sum of minimal right ideals. The proof that R is a direct sum of minimal left ideals is similar.

(iii) Let M and N be simple R-modules. If $mR = M$ and $nR = N$, then $(m)^{\perp}$ and $(n)^{\perp}$ are maximal right ideals (by 1.1). Hence by (ii) there exist minimal right ideals eR and fR with $R = eR \oplus R = fR \oplus R$ and $meR = M$, $nfR = N$. Note that $fR\cdot eR = f(ReR) = fR \neq 0$, so let $s_0 \in R$ such that $0 \neq fs_0eR = fR$. Define $\phi : M \to N$ by $\phi(mer) = nfs_0er$. Then ϕ is an R-homomorphism and $\phi(M) = N$. Hence $\ker \phi = 0$ and ϕ is an isomorphism.

(iv) Let $a \in R$ and let I be a minimal right ideal of R. By (i), a is in the sum of minimal right ideals of R. Hence as a corollary to the proof of 1.22 (setting $M = I$), we can write $a = c = \sum_{i=1}^{n} ab_ia = a(\sum_{i=1}^{n} b_i)a$ for some choice of b_i's. Therefore a is a regular element and R is a regular ring.

1.29 <u>Proposition</u>. Let R be a regular ring (not necessarily with an identity). Then for any $a,b \in R$, there exists $g = g^2$ such that $aR + bR = gR$.

Proof. Let $aR = eR$ for some $e = e^2$. Consider $B_1 = (1-e)bR = \{br-ebr \mid r \in R\}$. Note that $aR + B_1 \subseteq aR + bR$. If $x \in aR + bR$, then $x = er_1+br_2 = er_1+br_2-ebr_2+ebr_2 \in aR + B_1$. Therefore $aR + B_1 = aR + bR$. Now there exists

$f_1 \in R$ such that $f_1{}^2 = f_1$ and $(1-e)bR = f_1R$. Then $f_1 = (1-e)w$ and $ef_1 = 0$. Let $f = f_1(1-e)$. Then $ff_1 = (f_1-f_1e)f_1 = f_1 - f_1ef_1 = f_1$. Therefore $f_1 \in fR$, $f \in f_1R$ and $f_1R = fR = B_1$. Hence $aR + bR = aR + B_1 = eR + fR$. Now let $g = e+f$. Since $fe = 0 = ef$, then $g = g^2$ and $(e+f)e = e$ and $(e+f)f = f$. Therefore $(e+f)R = gR = aR + bR$.

1.30 Theorem (Litoff's Theorem). Let R be a simple ring with a minimal right ideal. If B is a finite subset of R, then there exist a subring R' and a division ring $\mathbb{D}$ such that $B \subseteq R'$ and $R' \cong \mathbb{D}_n$ for some positive integer n.

Proof. Let $B_0 = \{b_1, b_2, \ldots, b_n\}$ be a finite subset of R. Let B be the least right ideal which contains B_0, (by 1.28). Since R is regular, $B = b_1R + b_2R + \ldots + b_nR$ and there exists $f = f^2$ such that $fR = B$ (by 1.29). If B' is the least left ideal containing B_0 and f, then $B' = Rb_1 + Rb_2 + \ldots + Rb_n + Rf$. Hence there exists $g = g^2$ such that $B' = Rg$ and $fg = f$. Let $e = f+g - gf$, so $ef = f$, $ge = g$, and $e^2 = e$. Therefore, $B_0 = fB_0g = efB_0ge \subseteq eRe$. Now eRe is a simple ring, for if $S \neq 0$ is an ideal of eRe, then $S \supseteq eReSeRe = eR(eSe)Re = eRSRe = eRe$. Note that $e = 1_S$ (the identity on S). Now if e_iR is a minimal right ideal of R such that $ee_iR \neq 0$, then ee_iRe is a minimal right ideal of eRe. Indeed if $0 \neq I$ is a right ideal of eRe such that $I \subseteq ee_iRe$, then $eIe = I$ and

$Iee_iRe \subseteq I$ which implies $eIe_iRe \subseteq I$. But $eIe_iRe = ee_iRe$ since $e_ire \neq 0$ implies $e_ire \cdot ee_iRe = e_iRe$ and $Ie_iR = ee_iR$. Therefore eRe is a simple ring with identity and contains a minimal right ideal. Therefore $eRe \cong \mathbb{D}_n$ for some division ring, where $\mathbb{D} \cong e_1Re_1$ and e_1R is a minimal right ideal of R.

Chapter 2

An Injective test theorem and an injective producing lemma are given in this chapter. An attempt is made to give a general theorem of embedding a not necessarily unitary module into its injective hull. A theorem of Villamayor that if $1 \in R$ then every simple (right) R-module is injective if and only if every right ideal of R is the intersection of maximal right ideals is given without assuming that $1 \in R$.

2.0 Theorem (Injective Test Theorem). Let R be a ring with identity and let M be a right R-module such that $M^0 = \{m \in M \mid mR = 0\}$ is zero. Then M is injective if and only if for each right ideal I of R, every R-homomorphism of I to M can be extended to a homomorphism of R to M.

Proof. If M is injective, then the result follows immediately. Now assume that for each right ideal I of R, every homomorphism of I to M can be extended to a homomorphism of R to M. Let $f: X \to M$ be an R-homomorphism and let X be a submodule of Y. Consider a family $\Sigma = \{f_\alpha \mid f_\alpha$ is an R-homomorphism of some submodule of Y which contains X into $M\}$. Then $\Sigma \neq \emptyset$ since $f \in \Sigma$. If $\mathcal{L}$ is any linearly ordered subcollection of Σ, then $\bigcup_{f_\alpha \in \mathcal{L}} f_\alpha \in \Sigma$. Hence by Zorn's Lemma, there exists a maximal member f_0 of Σ. Let Y_0 be the domain of f_0. If $Y_0 = Y$, then we are done. So suppose $Y_0 \subsetneqq Y$, and let $y \in Y$ such that $y \notin Y_0$. If $I = \{r \in R \mid yr \in Y_0\}$, then I is a right ideal of R.

Consider the homomorphism g from I into M defined by $g(r) = f_0(yr)$. By hypothesis there exists a homomorphism $\bar{g}$ from R into M such that $\bar{g}(1) = q$ for some $q \in M$ and $\bar{g}(r) = qr = f_0(yr)$ for all $r \in I$. Let Y_1 be the submodule of Y generated by $\{y\} \cup Y_0$. If $x \in Y_1$, then there exist $r \in R$, $y_0 \in Y_0$, and integer n such that $x = yr + ny + y_0$. Define h from Y_1 into Y by $h(x) = h(yr + ng + y_0) = qr + nq + f_0(y_0)$. Clearly h is additive. To show that it is an R-homomorphism, it suffices to show that $h(0) = 0$. Let $yr + ny + y_0 = 0$. Then $yr + ny = -y_0$ and $f_0(yr + ny) = -f_0(y_0)$. Now $[f_0(yr+ng) - (qr+nq)]a = f_0(yra + nya) - (qra + nqa) = f_0(y(ra+na)) - (qra+nra) = g(ra+na) - (qra+nqa) = (qra+nqa) - (qra+nqa) = 0$, for all $a \in R$. Therefore $f_0(yr+ny) = qr+nq$ and $h(0) = 0$. Hence we have a contradiction to the hypothesis that $Y_0 \subsetneq Y$.

2.1 Example. If $1 \notin R$, then the Injective Test Theorem is no longer valid. For example, take $R = F$ in 1.24. Then R satisfies the conditions of 2.0 except for the fact that R has an identity. So if R is injective, then the identity mapping 1_R from R onto R can be lifted to an R-homomorphism $\bar{I}_R$ from R_1 into R (where R_1 denotes the ring with unity containing a subring isomorphic to R). But $\bar{I}_R(1) = a \in R$, and hence R has a left identity namely a. Note that a is also a right identity since $(ra-r)R = 0$ hence $ar = r$ for every $r \in R$. This clearly cannot occur.

2.2 <u>Proposition</u>. R is a semi-simple artinian ring if and only if every R-module M is injective.

<u>Proof</u>. If R is semi-simple and artinian, then R has an identity and R is a finite direct sum of minimal right ideals (by 1.27). Hence every right ideal of R is a direct summand of R, and by 2.0, every (unitary) module is injective. Conversely assume every R-module is injective. Then in particular R is injective and the identity homomorphism 1_R on R can be lifted to a homomorphism $\bar{1}_R$ from R_1 into R (where R_1 is a ring with unity containing R as a subring). If $\bar{1}_R(1) = a \in R$, then $ar = r$ for every $r \in R$. Let I be a right ideal of R such that $a \notin I$. Then I is a subset of a maximal right ideal M. Since every right ideal of R is a direct summand of R, there exists a minimal right ideal K such that $M \oplus K = R$. Let S be the sum of all minimal right ideals. If $a \notin S$, then S can be embedded into a maximal right ideal M_0. Thus $M_0 \oplus J = R$ for some minimal right ideal J and $J \cap S = 0$, which is a contradiction. Hence $a \in S$ and $S = R$. By the proof of 1.28(ii), R is a direct sum of minimal right ideals, say $R = \sum_{\alpha} \oplus I_\alpha$. Let $a = i_{\alpha_1} + i_{\alpha_2} + \ldots + i_{\alpha_n}$. Then $aR = R = i_{\alpha_1}R + i_{\alpha_2}R + \ldots + i_{\alpha_n}R = I_{\alpha_1} \oplus I_{\alpha_2} \oplus \ldots \oplus I_{\alpha_n}$. If $T = \{r \in R \mid ra = 0\}$, then T is a right ideal. But $I_{\alpha_j} a \neq 0$ for every $j = 1,2,\ldots,n$. Hence $T = 0$ and $ra = r$ for every $r \in R$ (since $ra - r \in T$ for every $r \in R$). Therefore by 1.27, R is a semi-simple artinian ring.

2.3 <u>Proposition</u>. R is a semi-simple artinian ring if and only if every R-module M is projective.

<u>Proof</u>. If R is a semi-simple artinian ring, then every R-module is injective by 2.2. Thus every exact sequence $N \to M \to 0$ of R-modules splits. Therefore every R-module is projective. Conversely, if every R-module is projective, consider an exact sequence $0 \to M \to N$ of R-modules. This sequence splits since the sequence $N \to N/M \to 0$ splits. Therefore, every R-module is injective and by 2.2, R is a semi-simple artinian ring.

2.4 <u>Corollary</u> (to Injective Test Theorem). An abelian group G is divisible if and only if G is $\mathbb{Z}$-injective.

<u>Proof</u>. Suppose G is divisible. Let I be an ideal of $\mathbb{Z}$ and let f be a $\mathbb{Z}$-homomorphism of I into G. Then there exists a positive integer p such that $I = (p)$. Let $f(p) = g$. Since G is divisible, there exists $g_0 \in G$ such that $pg_0 = g$. Let $\bar{f}(n) = ng_0$. Then $\bar{f}(p) = g$. Hence $\bar{f}|I = f$ and we are done. Conversely, assume G is $\mathbb{Z}$-injective. Let $g \in G$ and n be an integer. Define $f(nm) = mg$. Then f is a $\mathbb{Z}$-homomorphism from (n) into G. Hence there exists $\bar{f} : \mathbb{Z} \to G$ such that $\bar{f}|_{(n)} = f$. Let $\bar{f}(1) = g_0$. Then $g = \bar{f}(n) = ng_0$.

2.5 <u>Proposition</u> (Injective Producing Lemma). If G is a divisible abelian group and R is a ring with identity, then the R-module $\mathrm{Hom}_{\mathbb{Z}}(R,G)$ is R-injective (R-module in the sense that $h\circ r(x) = h(rx)$, $h \in \mathrm{Hom}_{\mathbb{Z}}(R,G)$ and $r \in R$).

Proof. First note that $h \cdot r(1) = h(r)$ for any $h \in \text{Hom}_{\mathbb{Z}}(R,G)$.
Let I be a right ideal of R and let f be an R-homomor-
phism of I into $\text{Hom}_{\mathbb{Z}}(R,G)$, that is

$$0 \longrightarrow I \longrightarrow R$$
$$f \downarrow$$
$$\text{Hom}_{\mathbb{Z}}(R,G).$$

If $i \in I$, let $f(i) = f_i$. Then $f(i_1 + i_2) = f_{i_1+i_2} =$
$f_{i_1} + f_{i_2}$ and $f_{ir}(1) = f_i r(1) = f_i(r)$. Define $g(i) = f_i(1)$.
Then g is a $\mathbb{Z}$-homomorphism of $I_{\mathbb{Z}}$ into $G_{\mathbb{Z}}$. Since $G_{\mathbb{Z}}$ is
injective, g extends to $\bar{g}$ of $\text{Hom}_{\mathbb{Z}}(R,G)$. Define $\bar{f}(r) =$
f'_r such that $f'_r(x) = g(rx)$ for every $x \in R$. Then
$\bar{f}(r_1+r_2) = f'_{r_1+r_2}$ and $\bar{f}(r_1+r_2)(x) = \bar{g}((r_1+r_2)x) = \bar{g}(r_1 x) +$
$\bar{g}(r_2 x)$; thus $\bar{f}(r_1+r_2) = \bar{f}(r_1) + \bar{f}(r_2)$. Also, $\bar{f}(r_1 r) =$
$f'_{r_1 r}$, $f'_{r_1 r}(x) = \bar{g}(r_1 rx) = f'_{r_1}(rx) = f'_{r_1} r(x)$, and
$\bar{f}(r_1) \cdot r = f'_{r_1} \cdot$. Hence $\bar{f}(r_1 r) = \bar{f}(r_1)r$. Now $\bar{f}(i) = f'i$
and $f'_i(x) = \bar{g}(ix) = g(ix) = f_{ix}(1) = f_i(x)$; so $\bar{f}|_I = f$
and we are done.

2.6 Corollary. Let R be a ring with identity and let M
be a unitary R-module. Then there exists an injective
R-module M_0 such that $M \subseteq M_0$.
Proof. Let $\bar{M}$ be a divisible group containing M. Then
$M_0 = \text{Hom}_{\mathbb{Z}}(R,\bar{M})$ is injective and M can be embedded into M_0.

2.7 Proposition. Let R be a ring and let M be an
R-module. Then there exists an injective module M_0 such
that $M \subseteq M_0$.

Proof. Let $R^* = \mathbb{Z} \times R$. Then M becomes a unitary R^* module by defining $m \cdot (n,r) = nm + mr$. Also, M can be embedded into $M_0 = \mathrm{Hom}_{\mathbb{Z}}(R^*, \overline{M})$, which is an injective R^*-module. However, M_0 is also injective as an R-module. To see this, consider the diagram

$$\begin{array}{ccccc} 0 & \longrightarrow & A & \xrightarrow{i} & B \\ & & {\scriptstyle f}\big\downarrow & & \\ & & M_0 & & \end{array}$$

where A and B are R-modules. Then A and B are also R^* modules and hence there exists an R^*-homomorphism $\bar{f}: B \to M_0$ such that $i \circ \bar{f} = f$. Since $\bar{f}$ is also an R-homomorphism, we are done.

2.8 Definition. An R-module M_1 is said to be an essential extension of a submodule M if and only if $H \cap M = 0$ implies $H = 0$ for every submodule H of M_1. In this case, we call M an essential submodule of M_1.

2.9 Proposition. Let M be an R-module and let $M \subset M_1 \subset M_2$ where M_1 and M are submodules of M_2. Then M_2 is an essential extension of M if and only if M_2 is an essential extension of M_1 and M_1 is an essential extension of M.

Proof. Suppose M_2 is an essential extension of M. If $H \neq 0$ is a submodule of M_2, then $M_1 \cap H \supseteq M \cap H \neq 0$. If $K \neq 0$ is a submodule of M_1, then K is a submodule of M_2 and $K \cap M \neq 0$ and we are done. Conversely, if M_2 is an essential extension of M, and M_1 is an essential extension of M_2, let $H \neq 0$ be a submodule of M_2. Then $H \cap M_1 \neq 0$ and $0 \neq M \cap (H \cap M_1) \leq M \cap H$.

2.10 Proposition. Let M be an R-module. If M is injective and $M \subset M_1$, then M_1 is an essential extension of M is and only if $M_1 = M$.

Proof. Consider the following diagram:

$$\begin{array}{ccc} M_1 & & \\ \uparrow & \searrow^{h} & \\ M & \xrightarrow{1_M} & M. \end{array}$$

Then $\ker h \cap M = 0$ and $M_1 = \ker h \oplus M$ since $h(h(m_1)-m_1)=0$ for every $m_1 \in M_1$. Therefore $\ker h = 0$ and $M_1 = M$.

2.11 Proposition. If M is an R-module then M is contained in a maximal essential extension.

Proof. Let M'' be an injective R-module such that $M \leq M''$ and let $\Sigma(M) = \{M_\alpha \leq M'' \mid M_\alpha$ is an essential extension of $M\}$. Then $M \in \Sigma(M)$, $\Sigma(M)$ is a partially ordered set by "inclusion," and $\Sigma(M)$ is inductive. If $\mathcal{L}$ is a linearly ordered subset of $\Sigma(M)$, then $M' = \bigcup_{M_\alpha \in \mathcal{L}} M_\alpha$ is an R-module and an essential extension of M. Indeed, if $H \leq M'$ such that $H \cap M = 0$, then $H \cap M_\alpha = 0$ for every $M_\alpha \in \mathcal{L}$. Therefore $H \cap M' = 0$ and $H = 0$.

2.12 Proposition. Let M be an R-module. If M' is an extension of M, then there exists a submodule H of M' such that $H \cap M = 0$ and M'/H is an essential extension of $(M \oplus H)/H$.

Proof. Let H be a largest submodule of M' such that $H \cap M = 0$. Then M' is an essential extension of $H \oplus M$. If there exists a submodule N of M' such that $H \subsetneq N$ and $N/H \cap (M \oplus H)/H = 0$, then $H = N \cap (M \oplus H)$ implying $0 = M \cap N$, which is a contradiction.

2.13 Proposition. Let M be an R-module. Then the following statements are equivalent:

(i) M is injective.

(ii) M has no proper essential extension.

(iii) M is a direct summand of every extension of M.

Proof. (i) implies (ii) is a result of 2.10. Assume (ii). To show (iii), let M' be an extension of M. If $M' \neq M$, then there exists a nonzero submodule H of M' such that $H \cap M = 0$. Let H be the largest such submodule such that $H \oplus M/H$ has an essential extension M'/H. Then $M'/H = M \oplus H/H$ and $M' = M \oplus H$. Finally, assume (iii) holds. Then M is a direct summand of an injective module M_0 containing M. Since a direct summand of an injective module is injective, then M is injective.

2.14 Corollary. Let M be an R-module. Any two minimal injective extensions (maximal essential extensions) of M are isomorphic.

Proof. Let M_1 and M_2 be two minimal injective extensions of M and consider the diagram

$$\begin{array}{ccc} M_1 & & \\ \uparrow & \searrow^{\phi} & \\ M & \xrightarrow{1_M} & M_2. \end{array}$$

Then $\ker \phi = 0$. Now if $\phi(M_1) \subsetneq M_2$, then $\phi(M_1)$ is a direct summand of M_2. Since $M \leq \phi(M_1)$, this means M_2 is not an essential extension of M, which is a contradiction.

2.15 Proposition. Let M be an R-module. Then there exists a minimal injective extension $\hat{M}$ of M which is unique up to an isomorphism. $\hat{M}$ is called the injective hull of M.*

Proof. By 2.11, M is contained in a maximal essential extension $\hat{M}$. Also, $\hat{M}$ is injective by 2.13, and $\hat{M}$ is unique up to isomorphism by 2.14.

2.16 Definition. Let Λ be a directed set and let $M = \{M_\alpha\}_{\alpha\in\Lambda}$ be a set of R-modules. Let $\sigma = \{\Pi_{\alpha\beta} : M_\alpha \to M_\beta \mid \Pi_{\alpha\beta}$ is an R-homomorphism $\alpha \leq \beta$ and $\alpha,\beta \in \Lambda\}$. Then (Λ, M, σ) is called a directed system of modules over Λ provided that

(i) $\Pi_{\alpha\alpha} = 1_{M_\alpha}$ for every $\alpha \in \Lambda$

(ii) $\Pi_{\delta\beta}\Pi_{\gamma\delta} = \Pi_{\gamma\beta}$ for every $\gamma,\delta,\beta \in \Lambda$ such that $\gamma \leq \delta \leq \beta$.

Let X be an R-module and let $d = \{\theta_\alpha : M_\alpha \to X \mid \theta_\alpha$ is an R-homomorphism and $\alpha \in \Lambda\}$. We say that d is a compatible family with respect to σ provided that

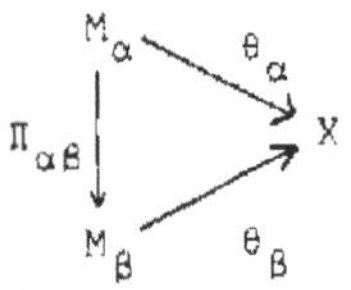

is a commutative diagram for every $\alpha,\beta \in \Lambda$.

(X,d) is called a <u>direct</u> <u>limit</u> of (Λ, M, σ) provided that for every compatible family $\{\psi_\alpha : M_\alpha \to Y\}_{\alpha\in\Lambda}$ there exists a unique morphism $\theta : X \to Y$ such that for every $\alpha \in \Lambda$,

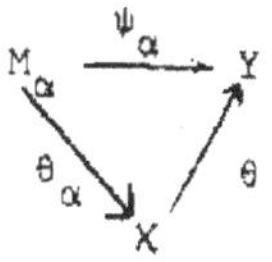

is a commutative diagram. X is usually denoted by $\varinjlim_{\alpha} M_\alpha$.

2.17 <u>Proposition</u>. Let I be an essential right ideal of a ring R and let $r \in R$. Then $r^{-1}I = \{x\in R \mid rx \in I\}$ is an essential right ideal.

<u>Proof</u>. Suppose $r^{-1}I$ is not an essential right ideal. Then there exists a right ideal $K \neq 0$ such that $r^{-1}I \cap K = 0$. However, $rK \neq 0$, hence $rK \cap I \neq 0$, which is a contradiction.

2.18 <u>Proposition</u>. Let $L_r^\Delta = \{A_\alpha \mid A_\alpha$ is an essential right ideal of $R\}$, which is indexed by Λ. Order Λ such that $\alpha \le \beta$ if and only if $A_\alpha \supseteq A_\beta$. If $f \in \mathrm{Hom}_R(A_\alpha, M)$ where M is an R-module, let $\Pi_{\alpha\beta}(f) = f|_{A_\beta}$. Then $(\Lambda, \{\mathrm{Hom}_R(A_\alpha,M) \mid \alpha\in\Lambda\}, \{\Pi_{\alpha\beta} \mid \alpha \le \beta$ and $(\alpha,\beta) \in \Lambda\times\Lambda\})$ is a directed system. Consider a set $\bigcup_{\alpha\in\Lambda} \mathrm{Hom}_R(A_\alpha,M)/\sim$ where $f \sim g$ if and only if there exists $A_\gamma \in L_r^\Delta(R)$ such that $f(x) = g(x)$ for every $x \in A_\gamma$. $f + g$ is defined on $\mathrm{dom}\, f \cap \mathrm{dom}\, g$ and $f\cdot r(x) = f(rx)$ for every $x \in r^{-1}\mathrm{dom}\, f$. Then $\bigcup_{\alpha\in\Lambda} \mathrm{Hom}_R(A_\alpha,M)/\sim$ is an R-module and $\varinjlim_{\alpha} \mathrm{Hom}_R(A_\alpha,M) = \bigcup_{\alpha\in\Lambda} \mathrm{Hom}_R(A_\alpha,M)/\sim$, where $\theta_\alpha : \mathrm{Hom}_R(A_\alpha,M) \to \bigcup_{\alpha\in\Lambda} \mathrm{Hom}_R(A_\alpha,M)/\sim$

and $\theta_\alpha(f_\alpha) = [f_\alpha]$.

<u>Proof</u>. Set $X = \bigcup_{\alpha\in\Lambda} \mathrm{Hom}_R(A_\alpha,M)/\sim$ and let $\{\psi_\alpha : \mathrm{Hom}_R(A_\alpha,M) \xrightarrow{\psi_\alpha} Y\}$ be a compatible family. Define $\theta : X \to Y$ by $\theta([f_\alpha]) = \psi_\alpha(f_\alpha)$. If $[f_\alpha] = [f_\beta]$, then there exists $\gamma \geq \alpha$ and $\gamma \geq \beta$ such that $A_\gamma \subseteq A_\alpha \cap A_\beta$ and $f_\beta(x) = f_\alpha(x) = f_\gamma(x)$ for every $x \in A_\gamma$.

Note that the diagram

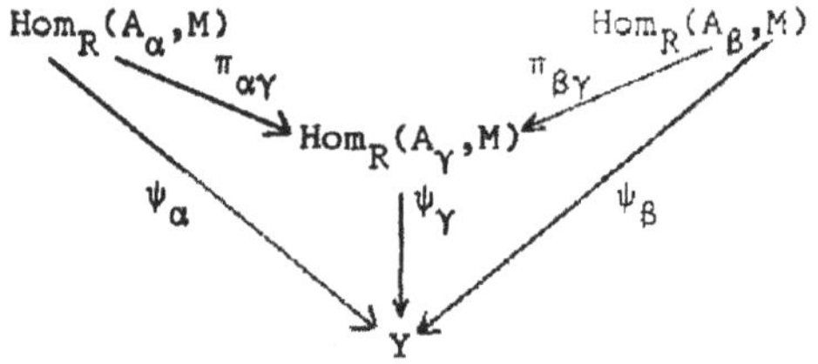

is commutative. Hence $\psi_\beta(f_\beta) = \psi_\gamma\Pi_{\beta\gamma}(f_\beta) = \psi_\gamma(f_\gamma) = \psi_\gamma\Pi_{\alpha\gamma}(f_\alpha) = \psi_\alpha(f_\alpha)$. Therefore θ is a mapping and the diagram

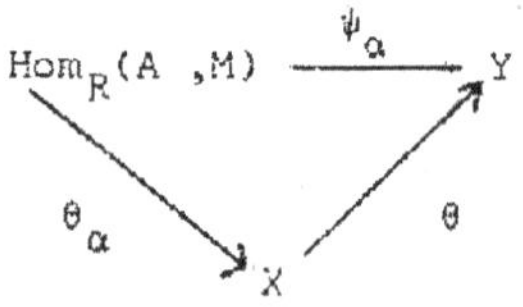

is commutative.

The uniqueness of θ is clear since for any $[f] \in X$ there exists f_α such that $\theta_\alpha(f_\alpha) = [f]$.

2.19 <u>Proposition.</u> Let R be a primitive ring with a minimal right ideal M. Then $\hat{M} = \varinjlim_\alpha \mathrm{Hom}_R(A_\alpha,M) \cong \mathrm{Hom}_R(F,M)$ where F is the socle of R (i.e. the sum of all minimal right ideals and the sum of all minimal left ideals).

<u>Proof</u>. First we will show that $X = \bigcup_{\alpha\in\Lambda} \mathrm{Hom}_R(A_\alpha,M)/\sim$ is a maximal essential extension of M. If $m \in M$, define $T_m(x) = mx$ for every $x \in R$. Then $m \overset{\phi}{\longmapsto} [T_m]$ is a monomorphism of M into X. Hence M is isomorphically contained in X and certainly X is an essential extension of M as an R-module (for if $[f] \in X$ and $[f] \neq 0$, then $[f]R \cap M \neq 0$). Suppose X' is an essential extension of X. Let $t \in X'$ such that $t \neq 0$. Then $tR \neq 0$. If $A = \{a\in R \mid ta \in M\}$, then A is an essential right ideal of R since X' is an essential extension of M. Therefore $X' = X$. Now to see that $\varinjlim_\alpha \mathrm{Hom}_R(A_\alpha,X) \cong \mathrm{Hom}_R(F,M)$, observe that $F \subseteq A_\alpha$ for every $\alpha \in \Lambda$ and F is an essential right ideal, since F is an ideal. Let $\eta_\alpha\colon \mathrm{Hom}_R(A_\alpha,M) \to \mathrm{Hom}_R(F,M)$ be a restriction map, i.e. $\eta_\alpha(f_\alpha) = f_\alpha|_F$. Let $\eta\colon \mathrm{Hom}_R(F,M) \to X$ such that $\eta(f) = [f]$. Then η factors θ_α for each $\alpha \in \Lambda$. Furthermore, the diagram

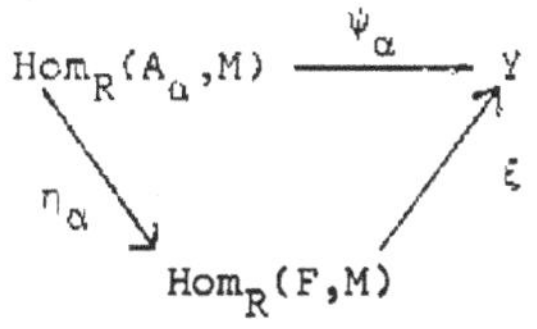

commutes for every compatible family $\{\psi_\alpha : \mathrm{Hom}(A_\alpha,M) \to Y\}$, and ξ is unique as in the proof of 2.18. Thus $\mathrm{Hom}_R(F,M) \cong X$.

2.20 <u>Theorem</u> (R. E. Johnson). Let R be a primitive ring with minimal right ideal M. Let $M = eR$ where $e = e^2$. Then $\hat{M} = \mathrm{Hom}_{eRe}(Re, eRe)$.

<u>Proof</u>. Let F be the set of all elements of finite rank. Then F is a direct sum of minimal left ideals and minimal right ideals. By 2.20, $\hat{M} = \mathrm{Hom}_R(F,M)$. Let $x \in \hat{M}$. Since $Re \subseteq F$, $x(re) = ea$ for some a and $x(re) = x(re^2) = eae$. Therefore $ea = eae$. Furthermore, if $x(Re) = 0$, then $x(ReR) = 0$ implying $x(F) = 0$ and $x = 0$. For each $x \in \hat{M}$, let $T_x(re) = x(re)$ for every $re \in Re$. Thus $T_x \in \mathrm{Hom}_{eRe}(Re, eRe)$ and $x \longmapsto T_x$ is a monomorphism. Now let $f \in \mathrm{Hom}_{eRe}(Re, eRe)$. Since $F = \sum_i \oplus A_i$ where the A_i's are minimal right ideals, then $Re \cap A_i \neq 0$ for each i. If $0 \neq a_i \in Re \cap A_i$, then $F = \sum_i \oplus a_i R$. Let $b \in F$. Then $b = \sum_i a_i r$ (where the sum is finite). Define $x'(b) = \sum_i f(a_i) r_i$. If $\sum_i a_i r_i = \sum_i a_i t_i$, then $a_i (r_i - t_i) = 0$ for each i. Now $a_i e = a_i$ and $(a_i)^{\perp} = (e)^{\perp}$ for each i (for if $a_i s = 0$ and $es \neq 0$, then $a_i \in J = \{r \in R \mid res = 0\}$ and $Re \subseteq J$; hence $0 = e \cdot e\,(es) = es$ which is a contradiction). Hence $r_i - t_i = d_i - ed_i$ for some $d_i \in R$. Thus $x'(b-b) = \sum_i f(a_i)(r_i - t_i) = \sum_i f(a_i)(d_i - ed_i) = 0$ and $x' \in \mathrm{Hom}_R(F,M)$. Therefore $f = T_{x'}$ and $\hat{M} \cong \mathrm{Hom}_{eRe}(Re, eRe)$.

2.21 <u>Proposition</u>. Let R be a ring such that if A is a non-zero right ideal of R then $AR = A$. Then the following statements are equivalent:

(i) Every proper right ideal of R is an intersection of maximal right ideals.

(ii) If M is a simple R-module, then $M = \hat{M}R$ where $\hat{M}$ is the injective hull of M.

<u>Proof</u>. Suppose (i) holds. Let M be a simple R-module and let $0 \neq x \in \hat{M}$. Then $(x)^{\perp} = \bigcap_{\alpha\in\Lambda} I_\alpha$, where $\{I_\alpha \mid \alpha\in\Lambda\}$ is a set of maximal right ideals. Let $M_\alpha = R/I_\alpha$. Then there exists a monomorphism $\mu : xR \to \prod_{\alpha\in\Lambda} M_\alpha$ and $xR \xrightarrow{\mu} \prod_{\alpha\in\Lambda} M_\alpha \to M_\alpha$. If there exists $\alpha \in \Lambda$ such that $\Pi_\alpha \circ \mu$ is a monomorphism, then xR is simple and $xR = M$. Now if for each $\alpha \in \Lambda$, there exists a $y_\alpha \in \hat{M}$ such that $\Pi_\alpha \circ \mu(y_\alpha) \neq 0$, then the kernel of $\Pi_\alpha \circ \mu$ must contain M (since $y_\alpha R$ is an essential submodule of $\hat{M}$). But $\bigcap_{\alpha\in\Lambda} \ker(\Pi_\alpha \circ \mu) \neq 0$ implying $M = 0$, which is a contradiction. Hence $xR = M$ for every $x \in \hat{M}$ and $M = \hat{M}R$.

Conversely, assume $A \neq R$ is a right ideal and that $\{I_\alpha \mid \alpha\in\Lambda\}$ is the family consisting of R and the maximal right ideals containing A. Suppose there exists $x \in \bigcap_{\alpha\in\Lambda} I_\alpha$ but $x \notin A$. Let A_0 be a right ideal of R maximal with respect to the property that $A \subseteq A_0$ but $x \notin A_0$. Then the submodule J/A_0 of R/A_0 generated by $x + A_0$ is a simple R-module. Note that $(\widehat{J/A_0})R = J/A_0$. Consider the diagram

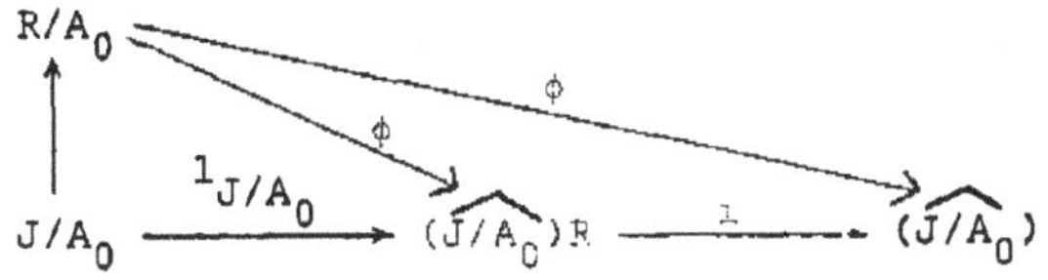

such that $\phi|_{J/A_0} = i \circ 1_{J/A_0}$. Now $\phi(R/A_0)R = \phi((R/A_0)R) = \phi(R/A_0) \subseteq (\hat{J}/A_0)R$. Thus $R/A_0 = J/A_0 \oplus \ker \phi$ and $\ker \phi = 0$ (by the maximal property of A_0). Hence $R/A_0 = J/A_0$ and A_0 is a maximal right ideal, which is a contradiction.

2.22 Proposition. Let R be a ring. If every simple R-module is injective, then for any simple module M the R-homomorphism $\phi: M \to \mathrm{Hom}_R(R,M)$ where $\phi(m)(r) = mr$ for every $r \in R$ is an isomorphism.

Proof. Clearly ϕ is a monomorphism. Let $f \in \mathrm{Hom}_R(R,M)$ and consider the diagram

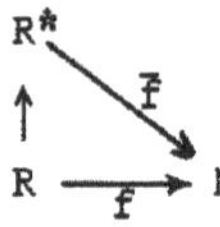

If $\bar{f}(1) = m_0$, then $f(r) = m_0 r$ for every $r \in R$. Therefore $f = \phi(m_0)$ and ϕ is an isomorphism.

2.23 Proposition. Let R be a ring such that $R^2 = R$. If every simple R-module is injective then every maximal right ideal I is modular.

Proof. Let $M = R/I$ where I is a maximal right ideal of R. Then by hypothesis M is injective and $\phi(M) = \mathrm{Hom}_R(R,M)$ (where ϕ is the isomorphism of 2.22). Let $g(r) = r+I$ for each $r \in R$. Then $g \in \mathrm{Hom}_R(R,M)$ and there exists $m \in M$ such that $g = \phi(m)$, so $g(r) = r + I = \phi(m)(r) = mr$ for every $r \in R$. Therefore $\ker g = (m)^{\perp} = I$. Since $m \in mR$, $m = me$ for some $e \in R$. Hence $r - er \in (m)^{\perp} = I$ for every $r \in R$.

2.24 **Theorem** (Villamayor). Let R be a ring such that if A is a non-zero right ideal of R then $AR = A$. Then the following statements are equivalent:

   (i) Every proper right ideal of R is the intersection of maximal modular right ideals.

   (ii) If M is a simple R-module, then M is injective.

Proof. Note that (ii) implies (i) by 2.21 and 2.23. Now assume (i) holds. Since $\hat{M}R = M$ (by 2.21), it suffices to show that $\hat{M}R = \hat{M}$. Consider $\phi:\hat{M} \to \operatorname{Hom}_R(R,\hat{M}R)$ as in 2.22. Clearly ϕ is a monomorphism. Now let $f \in \operatorname{Hom}_R(R,\hat{M}R)$ and consider the diagram

$$
\begin{array}{ccc}
R^* & & \\
\uparrow & \searrow{\scriptstyle \bar{f}} & \\
R \xrightarrow{\;f\;} \hat{M}R & \longrightarrow & \hat{M}
\end{array}
$$

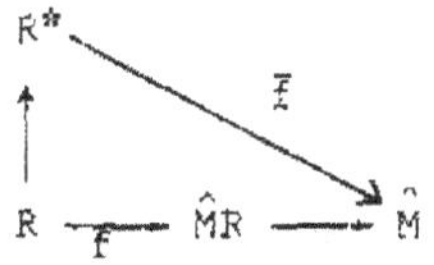

Then $\bar{f}(1) = \hat{m}$ for some $\hat{m} \in \hat{M}$ and $f = \phi(\hat{m})$. Therefore $\phi(\hat{M}) = \operatorname{Hom}_R(R,\hat{M}R)$. To see that $\phi(M) = \operatorname{Hom}_R(R,\hat{M}R)$, let $f \in \operatorname{Hom}_R(R,\hat{M}R)$. If $f \neq 0$, then $R/\ker f \cong \hat{M}R = M$. Hence there exists $e \in R$ such that $er-r \in \ker f$. So $f(e)r = f(r)$ for every $r \in R$ and $f = \phi(f(e))$. Therefore $\phi(M) = \operatorname{Hom}_R(R,\hat{M}R)$.

2.25 **Corollary** (Villamayor). Let R be a ring with identity. Then every proper right ideal of R is an intersection of maximal right ideals if and only if every simple R-module is injective.

2.26 **Proposition.** Let R be a ring. Suppose I is a right ideal of R such that $N(I)\backslash I$ is a semigroup (with respect to ring multiplication) and such that if J is a right ideal of R with $I \subsetneqq J$, then $J \cap (N(I)\setminus I) \neq \emptyset$. Then $\{m \in \widehat{(R/I)} \mid mR = 0\} = 0$.

Proof. Let $N = \{m \in \widehat{(R/I)} \mid mR = 0\}$ and suppose $N \neq 0$. Then N is a submodule of $\widehat{(R/I)}$ and $N \cap R/I \neq 0$. Hence $N \cap R/I = J/I$ for some right ideal J with $I \subsetneqq J$. Let $a \in J \cap (N(I)\setminus I)$. Then $aR \subseteq I$ and in particular, $a \cdot a \notin N(I)\setminus I$, which is a contradiction.

2.27 Proposition. Let R be a ring such that if $a \in R$ then there exists $c \in R$ such that $(ca)^{\perp} \subseteq (a)^{\perp}$. Suppose I is a right ideal of R such that $N(I)\setminus I$ is a semigroup (with respect to ring multiplication) and such that if J is a right ideal of R with $I \subsetneqq J$, then $J \cap (N(I)\setminus I) \neq \emptyset$. Then $\widehat{(R/I)}^{\perp} = \{r \in R \mid$ for each $x \in R$, there exists $y_x \notin I$ such that $rxy_x = 0\}$.

Proof. Let $0_I = \{r \in R \mid$ for each $x \in R$, there exists $y_x \notin I$ such that $rxy_x = 0\}$. Let $a \in \widehat{(R/I)}^{\perp}$. If $a \notin 0_I$, then there exists $x_0 \in R$ such that $(ax_0)^{\perp} \subseteq I$. Let $c \in R$ such that $(cax_0)^{\perp} \subseteq (ax_0)^{\perp}$. If $m \in T = \{m \in \widehat{(R/I)} \mid m(cax_0)^{\perp} = 0\}$, define $f_m\colon cax_0R \to \widehat{(R/I)}$ by $f_m(cax_0r) = mr$ for every $r \in R$. Let $\bar{f}_m$ be an extension of f_m to R. Then $\bar{f}_m(c) = m_0$ for some $m_0 \in \widehat{(R/I)}$ and $\bar{f}_m(cax_0r) = m_0ax_0r = mr$. Therefore $(m_0ax_0-m)R = 0$ and by 2.26, $m = m_0ax_0$. Since $\widehat{(R/I)}a = 0$, then $m = 0$ and $T = 0$. Let $b \in N(I)\setminus I$. Then $(b+I)((cax_0)^{\perp}) \subseteq b(cax_0)^{\perp} + I \subseteq bI + I = 0$. Hence $b+I \in T$ and $b \in I$, which is a contradiction. Therefore $\widehat{(R/I)}^{\perp} \subseteq 0_I$.

Conversely, if $0_I \not\subseteq (\hat{R}/I)^1$, then there exists $\hat{m} \in \hat{R}/I$ and $b \in 0_I$ such that $\hat{m}b \neq 0$. Also, there exists $r \in R$ with $0 \neq \hat{m}br \in R/I$. Then $\hat{m}br = a+I$ for some $a \notin I$. Consider the right ideal $aR + I$. If $aR + I = I$, then $(a+I)R = 0$ and $a \in I$, which is a contradiction. Therefore $I \subsetneq aR+I$ and there exists $t \in (aR+I)\cap(N(I)\setminus I)$ and $t+I = ar' + I$ for some $r' \in R$. Then $(\hat{m}br)r' = ar' + I = t + I$. Since $b \in 0_I$, there exists $y_{rr'} \notin I$ such that $brr'y_{rr'} = 0$. Hence $ty_{rr'} \in I$. However $y_{rr'}R \not\subseteq I$, and thus $I \subsetneq y_{rr'}R + I$. Let $y_{rr'}s + j \in y_{rr'}R + I \cap N(I)\setminus I$. Then $t(y_{rr'}s + j) \in I$, which is a contradiction (since $N(I)\setminus I$ is closed under multiplication).

2.28 <u>Proposition</u>. If R is a regular ring without nilpotent elements, then every one-sided ideal of R is two-sided.

<u>Proof</u>. Let $a \in R$. Then there exists $x \in R$ such that $axa = a$. If $e = ax$, then for any $r \in R$, $(er-ere)(er-ere) = 0$. Hence $er = ere$. Similarly $(re-ere)(re-ere) = 0$ and $re = ere$. Since $aR = eR$ and $Re = Ra$, then aR is a left ideal and Ra is a right ideal. Thus every one-sided ideal of R is two-sided.

2.29 <u>Proposition</u>. If I is a maximal right ideal of a regular ring R, then I is modular if and only if $N(I)\setminus I \neq \emptyset$.

<u>Proof</u>. Let $a \in N(I)\setminus I$. Note that $I = \{r \in R \mid ar \in I\}$. If $x \in R$ with $axa = a$, then $a(xar-r) = 0$ for every $r \in R$. Hence $xar-r \in I$ and I is modular.

2.30 <u>Proposition</u>. Let R be a regular ring without nonzero nilpotent elements. Then every simple R-module is injective.

<u>Proof</u>. Let R/I be a simple R-module. Then (by 2.27), $0_I \subseteq I$. Let $x \in I$. If $x \notin 0_I$ then there exists $r_0 \in R$ such that $(xr_0)^{\perp} \subseteq I$. Note that $xr_0yxr_0 = xr_0$ for some $y \in R$. Let $e = yxr_0$. Then $e \in I$ (since I is an ideal) and $xr_0e = xr_0$. Hence $\{ea-a \mid a \in R\} \subseteq (xr_0)^{\perp} \subseteq I$ and $R \subseteq I$, which is a contradiction. Therefore $0_I = I$. Note that $0_I = (\widehat{R/I})^{\perp}$. Thus if $0 \neq m \in (\widehat{R/I})$, then mR is a simple module and therefore $(\widehat{R/I})R = R/I$. Since every maximal right ideal is modular, then $R/I = \widehat{R/I}$ by 2.21 and 2.24.

2.31 <u>Theorem</u> (Kaplansky). Let R be a commutative ring with identity. Then R is regular if and only if every simple R-module is injective.

<u>Proof</u>. If R is regular, then R has no non-zero nilpotent elements since R is commutative. Hence every simple R-module is injective by 2.30. Conversely, if $a \in R$ such that $a^2R \neq R$, let I be a maximal ideal containing a^2R. If $a \notin I$, then $R = aR + I$. Hence, there exists $i \in I$ and $r \in R$ with $1 = ar + i$. Then $a = aar + ai \in I$, which is a contradiction. Hence by 2.25, $a \in a^2R$ and R is regular.

2.32 <u>Definition</u>. Let R be a ring and let M be an R-module. If M' is an extension of M, then M' has <u>finite length</u> if there exists a set $\{M_1, M_2, \ldots, M_n\}$ of R-modules such that $M' = M_1 \supset M_2 \supset \ldots \supset M_{n-1} \supset M_n = M$ and

M_i/M_{i+1} is simple for every $i = 1,2,\ldots,n-1$.

2.33 Theorem (Rosenberg and Zelinsky). Let R be a ring with identity and let $J(R)$ be the Jocobson radical of R. If the injective hull of every simple R-module has finite length, then $J(R)$ is a nil ideal and $\bigcap_{i=1}^{\infty} J(R)^i = 0$.

Proof. Let $x \in J(R)$ and suppose $x^i \neq 0$ for every $i = 1,2,\ldots$. Consider the right ideal $K = \bigcup_{i=1}^{\infty} (x^i)^{\perp}$. Note that $x \notin K$, so there exists a maximal right ideal I such that $K \subseteq I$. Then for each $i = 1,2,\ldots$, $x^i \notin 0_I$. Hence by 2.27, $(\widehat{R/I})x^i \neq 0$. There exist R-modules $B_1, B_2, \ldots, B_n$ such that $(\widehat{R/I}) = B_1 \supset B_2 \supset \ldots \supset B_{n-1} \supset B_n = R/I$ and B_i/B_{i+1} is simple for each $i = 1,2,\ldots,n-1$. Therefore, $B_1 J(R) \subseteq B_2$, $B_1 J(R)^2 \subseteq B_2 J(R) \subseteq B_3$ etc. Hence $(\widehat{R/I}) J(R)^k = 0$ for some positive integer k and $(\widehat{R/I})x^k = 0$, which is a contradiction. Now let $x \in \bigcap_{i=1}^{\infty} J(R)^i$. Then $(\widehat{R/I_\alpha})x = 0$ for each maximal right ideal I_α. Now if $x \neq 0$, then $(x)^{\perp} \subsetneq R$ and there exists a maximal right ideal I_β such that $(x)^{\perp} \subseteq I_\beta$. Hence $x \notin 0_I$, which is a contradiction.

Chapter 3

The singular ideal of a ring and the singular submodule of a module are investigated and B. Osofsky's example that an injective hull of a ring may not be a ring in case the singular ideal of the ring is not zero is included here. Let $\{A_\alpha \mid \alpha\in\Lambda\}$ be the family of essential right ideals of a ring R. A detailed account of embedding a ring R with zero singular ideal into a direct limit of $\{\mathrm{Hom}_R(A_\alpha,R) \mid \alpha\in\Lambda\}$ which turns out to be the injective hull of R as a right regular R-module is also given.

3.0 Definition. If R is a ring, let $Z_r(R) = \{a\in R \mid (a)^r$ is an essential right ideal of $R\}$ where $(a)^r = \{r\in R \mid ar = 0\}$ and let $Z_\ell(R) = \{a\in R \mid (a)^\ell$ is an essential left ideal of $R\}$. $Z_r(R)$ is called a right singular ideal of R and $Z_\ell(R)$ is called a left singular ideal of R.

3.1 Proposition. Let R be a ring. Then $Z_r(R)$ and $Z_\ell(R)$ are ideals of R.

Proof. If $a,b \in Z_r(R)$, then $(a)^r \cap (b)^r \subseteq (a-b)^r$ and since $(a)^r \cap (b)^r$ is essential, $a-b \in Z_r(R)$. Also, if $x \in R$, then $x^{-1}((a)^r)$ is essential by 2.17. Hence $(ax)^r$ is essential. Clearly $xa \in Z_r(R)$ since $(a)^r \subseteq (xa)^r$. Similarly one can show that $Z_\ell(R)$ is an ideal.

3.2 Example. Let $R = \{\begin{pmatrix}0&0\\0&0\end{pmatrix},\begin{pmatrix}1&1\\0&0\end{pmatrix},\begin{pmatrix}0&0\\1&1\end{pmatrix},\begin{pmatrix}1&0\\1&0\end{pmatrix},\begin{pmatrix}0&1\\0&1\end{pmatrix},\begin{pmatrix}1&1\\1&1\end{pmatrix}\}$ where $\{0,1\} = \mathbb{Z}_2$. Then $Z_r(R) = \{\begin{pmatrix}0&0\\0&0\end{pmatrix}\}$ and $Z_\ell(R) = \{\begin{pmatrix}0&0\\0&0\end{pmatrix},\begin{pmatrix}1&1\\1&1\end{pmatrix}\}$.

3.3 Example (Osofsky). $\hat{R}_R$ is not necessarily a ring which contains R as a subring.

Let $R = \begin{pmatrix} \mathbb{Z}_4 & 2\mathbb{Z}_4 \\ 0 & \mathbb{Z}_4 \end{pmatrix}$, $I = \begin{pmatrix} 0 & 0 \\ 0 & 2 \end{pmatrix}R$, $J = \begin{pmatrix} 0 & 2 \\ 0 & 0 \end{pmatrix}R$. Define

$f: I \to J$ by $f(\begin{pmatrix} 0 & 0 \\ 0 & 2 \end{pmatrix}r) = \begin{pmatrix} 0 & 2 \\ 0 & 0 \end{pmatrix}r$. Note that f is a monomorphism since $\begin{pmatrix} 0 & 0 \\ 0 & 2 \end{pmatrix}^{\perp} = \begin{pmatrix} \mathbb{Z}_4 & 2\mathbb{Z}_4 \\ 0 & 2\mathbb{Z}_4 \end{pmatrix}$ and $\begin{pmatrix} 0 & 2 \\ 0 & 0 \end{pmatrix}^{\perp} = \begin{pmatrix} \mathbb{Z}_4 & 2\mathbb{Z}_4 \\ 0 & 2\mathbb{Z}_4 \end{pmatrix}$.

Consider $f^{-1}: J \to I$. Then there exists $m \in \hat{I}$ such that $f^{-1}(j) = mj$ for every $j \in J$. For $m' = m\begin{pmatrix} 1 & 0 \\ 0 & 0 \end{pmatrix} \in \hat{I}$, $m'\begin{pmatrix} 1 & 0 \\ 0 & 0 \end{pmatrix} = m\begin{pmatrix} 1 & 0 \\ 0 & 0 \end{pmatrix}\begin{pmatrix} 1 & 0 \\ 0 & 0 \end{pmatrix} = m\begin{pmatrix} 1 & 0 \\ 0 & 0 \end{pmatrix} = m'$. Now if $2m' \neq 0$, then there exists $r \in R$ such that $2m'r \neq 0 \in I$. Let $r = \begin{pmatrix} x & y \\ 0 & w \end{pmatrix}$. Then $2m'r = m\begin{pmatrix} 2 & 0 \\ 0 & 0 \end{pmatrix}\begin{pmatrix} x & y \\ 0 & w \end{pmatrix} = m\begin{pmatrix} 2x & 0 \\ 0 & 0 \end{pmatrix}$. Since $m\begin{pmatrix} 2n & 0 \\ 0 & 0 \end{pmatrix} \in I$, $2m'r = \begin{pmatrix} 0 & 0 \\ 0 & w' \end{pmatrix}$ for some $w' \in \mathbb{Z}_4$. Therefore $2m'r\begin{pmatrix} 1 & 0 \\ 0 & 0 \end{pmatrix} = 2m'r \neq \begin{pmatrix} 0 & 0 \\ 0 & 0 \end{pmatrix}$ and $\begin{pmatrix} 0 & 0 \\ 0 & w' \end{pmatrix}\begin{pmatrix} 1 & 0 \\ 0 & 0 \end{pmatrix} = \begin{pmatrix} 0 & 0 \\ 0 & 0 \end{pmatrix}$, which is a contradiction. Thus $2m' = 0$. Now assume $\hat{R}_R$ is a ring. Then if $\hat{f}$ is the extension of f^{-1} from R to $\hat{I}$,

$$\begin{pmatrix} 0 & 0 \\ 0 & 0 \end{pmatrix} = 2m'\hat{f}(\begin{pmatrix} 0 & 0 \\ 0 & 1 \end{pmatrix}) = m'\hat{f}\begin{pmatrix} 0 & 0 \\ 0 & 2 \end{pmatrix} = m'\begin{pmatrix} 0 & 2 \\ 0 & 0 \end{pmatrix} = m\begin{pmatrix} 1 & 0 \\ 0 & 0 \end{pmatrix}\begin{pmatrix} 0 & 2 \\ 0 & 0 \end{pmatrix} =$$

$m\begin{pmatrix} 0 & 2 \\ 0 & 0 \end{pmatrix} = \begin{pmatrix} 0 & 0 \\ 0 & 2 \end{pmatrix}$. Hence $\hat{R}_R$ cannot contain R as a subring.

3.4 Proposition. Let R be a ring and M be an R-module. If $Z(M) = \{m \in M \mid (m)^{\perp}$ is an essential right ideal of $R\}$, then Z(M) is a submodule of M. Z(M) is called the singular submodule of M.

Proof. Let $m \in Z(M)$ and $r \in R$. Then $r^{-1}(m)^{\perp}$ is an essential right ideal since $(m)^{\perp}$ is essential. Hence

$(mr)^{1}$ is an essential right ideal and $mr \in Z(M)$. Clearly if $m_1, m_2 \in Z(M)$, then $m_1 - m_2 \in Z(M)$.

3.5 <u>Proposition</u>. Let M be an R-module such that $Z(M) = 0$. Then $M = \varinjlim_{\alpha \in \Lambda} \mathrm{Hom}_R(I_\alpha, M)$, where $\{I_\alpha \mid \alpha \in \Lambda\}$ is the set of essential right ideals of R.

<u>Proof</u>. Let $X = \varinjlim_{\alpha \in \Lambda} \mathrm{Hom}_R(I_\alpha, M)$. Consider $\phi: M \to X$ defined by $\phi(m) = [t_m]$ where $t_m(r) = mr$ for every $r \in R$. Certainly ϕ is an R-homomorphism of M into X. Also if $[t_m] = 0$, then there exists an essential right ideal I_α such that $t_m(I_\alpha) = 0$. Hence $m \in Z(M)$ and $m = 0$. Thus ϕ is a monomorphism. Now let $0 \neq [f] \in X$. Then there exists an essential right ideal I_β such that $f(I_\beta) \neq 0$ and $f(i) = m \neq 0$ for some $i \in I_\beta$ and $m \in M$. Hence $[f]i = [t_m] \neq 0$. Therefore X is an essential extension of $\phi(M)$. We will show that any essential extension of X is isomorphic to X. Let X' be an essential extension of X. If $0 \neq x' \in X'$, then there exists $r \in R$ such that $0 \neq x'r \in X$. Otherwise the set $\Gamma = \{x' \in X' \mid x'R = 0\}$ is a non-zero submodule of X' and $0 \neq \Gamma \cap M \subseteq Z(M)$, which is impossible. Hence $x'R \cap \phi(M) \neq 0$. Let $I_{x'} = \{r \in R \mid x'r \in \phi(M)\}$. Then $I_{x'}$ is an essential right ideal since $\phi(M)$ is essential in X'. Define $\psi(x')(r) = m$ where $r \in I_{x'}$ and $x'r = \phi(m) = [t_m]$. We claim that $x' \longmapsto [\psi(x')]$ is an isomorphism of X' onto X. First note that $\psi(x') \in \mathrm{Hom}_R(I_{x'}, M)$ since ϕ is an R-homomorphism and $\psi(x')(rr_1) = \phi(x'(rr_1))$. Also, if $x_1' = x_2'$, then certainly $\psi(x_1') = \psi(x_2')$. Now if

$[\psi(x_1')] = [\psi(x_2')]$, then there exists an essential right

ideal $I \subseteq \operatorname{dom} \psi(x_1') \cap \operatorname{dom} \psi(x_2')$ such that

$(\psi(x_1') - \psi(x_2'))(i) = 0$ for every $i \in I$. Therefore

$x_1'i = x_2'i$ for every $i \in I$ and $x_1' - x_2' \in Z(X')$.

Furthermore, $x_1' - x_2' = 0$ since $Z(X') \cap M \subseteq Z(M)$ and

$Z(M) = 0$. Thus $X' \cong X$.

3.6 <u>Proposition</u>. Let R be a ring and let $\{I_\alpha \mid \alpha \epsilon \Lambda\}$ be

the family of essential right ideals of R. Then

$\varinjlim_{\alpha} \operatorname{Hom}_R(I_\alpha, R)$ is a regular ring with identity. If $R^\ell = 0$

(where $R^\ell = \{r \epsilon R \mid rR = 0\}$), then there is a bijection from

$\hat{R}/Z(\hat{R}_R)$ onto $\varinjlim_{\alpha} \operatorname{Hom}_R(I_\alpha, R)$.

<u>Proof</u>. By 2.19, $X = \varinjlim_{\alpha} \operatorname{Hom}_R(I_\alpha, R) = \bigcup_{\alpha \epsilon \Lambda} \operatorname{Hom}_R(I_\alpha, R/\sim)$

is an R-module. If $[f], [g] \epsilon X$, define $[f] \cdot [g] = [f \cdot g]$.

If $[f] = [f']$ and $[g] = [g']$ then there exist essential

right ideals I_1 and I_2 such that $f(a) = f'(a)$ for every

$a \epsilon I_1$ and $g(b) = g'(b)$ for every $b \epsilon I_2$. Let

$K_1 = \{r \epsilon R \mid g(r) \epsilon \operatorname{dom} f\}$ and $K_2 = \{r \epsilon R \mid g'(r) \epsilon \operatorname{dom} f'\}$.

If $K = K_1 \cap K_2 \cap I_2$, then K is an essential right ideal

and $f \cdot g(k) = f' \circ g'(k)$ for every $k \epsilon K$. Thus $\cdot$ is well

defined and $(X, +, \cdot)$ is a ring. Now let $0 \neq [f] \epsilon X$ and

let $I_\alpha = \operatorname{dom} f$. Since $[f] \neq 0$, ker f is not an essential

right ideal. Let J be a non-zero right ideal of R such

that $J \cap \ker f = 0$ and $J \oplus \ker f$ is an essential right

ideal. Then $(J \oplus \ker f) \cap I_\alpha = (J \cap I_\alpha) \oplus \ker f$, which is

an essential right ideal. Let $S_\alpha = f(J \cap I_\alpha)$ and let K

be a right ideal such that $S_\alpha \oplus K$ is an essential right ideal. Now define $g: S_\alpha \oplus K \to R$ by $g(f(t) \dot{+} k) = t$ for $f(t) \in S_\alpha$ and $k \in K$. Then $[g] \in X$ and $(f \circ g \circ f - f)(t \dot{+} i) = 0$ for every $t \dot{+} i \in J \cap I_\alpha \oplus \text{Ker } f$. Hence $[f][g][f] = [f]$. Finally, to show that there is a bijection from $\hat{R}/Z(\hat{R}_R)$ onto X, for each $q \in \hat{R}$, define $I(q) = \{r \in R \mid qr \in R\}$. Then $I(q)$ is an essential right ideal of R. Define $Tq(r) = qr$ for every $r \in I(q)$. Then $[Tq] \in X$. Now define ψ from $\hat{R}/Z(\hat{R}_R)$ to X by $\psi(q + Z(\hat{R})) = [Tq]$. If $q_1 - q_2 \in Z(\hat{R})$ then there exists an essential right ideal I_α such that $(q_1 - q_2)I_\alpha = 0$. Hence $[Tq_1] = [Tq_2]$. Conversely, if $[Tq_1] = [Tq_2]$, then there exists an essential right ideal I_β such that $Tq_1(r) = Tq_2(r)$ for every $r \in I_\beta$. Hence $q_1 + Z(\hat{R}_R) = q_2 + Z(\hat{R}_R)$ and ψ is an injection. Now let $[f] \in X$. Then $f \in \text{Hom}_R(I_\alpha, R)$ for some essential right ideal I_α. Hence there exists $\hat{f} \in \text{Hom}_R(\hat{R}, \hat{R})$ such that $\hat{f}|_{I_\alpha} = f$. Since $R^\ell = 0$, R is isomorphically contained in $\text{Hom}_R(R,R)$, and since $\text{Hom}_R(R,R)$ is an essential extension of R_R, $\text{Hom}_R(R,R)$ can be embedded into $\hat{R}$. Hence there exists $a \in \hat{R}$ such that $ar = r$ for every $r \in R$. Let $\hat{f}(a) = q$. Then $[f] = \psi(q + Z(R))$.

3.7 <u>Theorem</u> (R. E. Johnson). Let R be a ring. Then $Z(R_R) = 0$ if and only if $\hat{R}_R$ is a regular ring with identity.

<u>Proof.</u> If $Z(R_R) = 0$, then $R^\ell = 0$ and $Z(\hat{R}) = 0$. Thus by 3.5 and 3.6, R is a regular ring with identity. Conversely if $\hat{R}$ is a regular ring, then $Z(\hat{R}) = 0$. Otherwise, if

$0 \neq q \in Z(\hat{R})$, then there exists $q' \in \hat{R}$ such that $q'q$ is a nonzero idempotent in $Z(\hat{R})$. Hence $(q'q)^{\perp} \cap (q'q\hat{R} \cap R) = 0$, which is impossible. Therefore $Z(R_R) \subseteq Z(\hat{R}) = 0$.

3.8 Definition. A right ideal U of a ring R is called uniform if and only if every nonzero submodule of U_R is essential in U_R.

3.9 Proposition. If R is a regular ring and U is a uniform right ideal of R then U is a minimal right ideal of R.

Proof. Let A be a nonzero right ideal such that $A \subseteq U$. Then there exists $a^2 = a \in A$ such that $aR \oplus \{r-ar \mid r \in R\} = R$. Let $B = \{r-ar \mid r \in R\}$. If $aR \neq U$, then $U \cap B \neq 0$ since $aR \subseteq U$. Then $aR \cap (U \cap B) = 0$ and U is not uniform, which is a contradiction.

3.10 Proposition. If R is a ring with a uniform right ideal U and such that $Z_r(R) = 0$, then $\hat{R}$ is a regular ring with a minimal right ideal.

Proof. Let $A = U\hat{R}$. We claim that A is a uniform right ideal of $\hat{R}$. Let a and b be nonzero elements of A. Then $a = \sum_{i=1}^{n} u_i\hat{r}_i$, $b = \sum_{i=1}^{m} v_i\hat{q}_i$ for some $u_i, v_i \in U$ and $\hat{r}_i, \hat{q}_i \in \hat{R}$. Let $K(\hat{r}_i) = \{x \in R \mid \hat{r}_i x \in R\}$. Then $K(\hat{r}_i)$ is an essential right ideal of R since $\hat{R}$ is an essential extension of R. Let $K_0 = \bigcap_{i=1}^{n} K(\hat{r}_i)$, $K_1 = \bigcap_{i=1}^{m} K(\hat{q}_i)$. Then $K = K_0 \cap K_1$ is an essential right ideal. Also, $aK \neq 0$ and

$bK \neq 0$ since $Z(\hat{R}_R) = 0$. Now aK, bK are submodules of U and $aK \cap bK \neq 0$. Thus A is a uniform right ideal of $\hat{R}$. By 3.7, $\hat{R}$ is a regular ring and hence $\hat{R}$ has a minimal right ideal by 3.9.

3.11 Corollary. Let U be a uniform right ideal of a ring R with $Z_r(R) = 0$. Then $\hat{U} = U\hat{R}$.

Proof. Since $U \subseteq U\hat{R}$, $\hat{U} \subseteq \widehat{(U\hat{R})}$. Now $U\hat{R}$ is a direct summand of $\hat{R}$ since $U\hat{R}$ is a minimal right ideal of a regular ring $\hat{R}$. Therefore $U\hat{R}$ is injective. Since U is essential in $U\hat{R}$, then $\hat{U} = U\hat{R} = \widehat{(U\hat{R})}$.

Chapter 4

Main concerns of this chapter are the structures of the injective hull $\hat{R}$ of any R as a regular right R-module. Major theorems in this chapter are that if R is a regular ring such that R_R is injective then there exist two ideals A and B in R such that $R = A \oplus B$ where A is a strongly regular ring, every non-zero ideal of B contains a non-zero nilpotent element, and B is generated by idempotents and that $\hat{R}_R$ is a strongly regular ring if and only if the right singular ideal of R is zero and R contains no pair of isomorphic non-zero right ideals whose intersection is zero.

4.0   Definition. A ring R is called a semi-prime ring if and only if aRa = 0 implies a = 0 for any a ϵ R. A ring R is called a prime ring if and only if aRb = 0 implies that either a = 0 or b = 0 for any a, b ϵ R. An ideal P of R is called a prime ideal if and only if $^R/P$ is a prime ring.

4.1   Definition. A subset S of a ring R is called an m-system if and only if for any a, b ϵ S there exists x ϵ R such that axb ϵ S.

4.2   Definition. If R is a ring, let rad R = {x ϵ R| if S is an m-system such that x ϵ S, then 0 ϵ S}.

4.3 Proposition. Let R be a ring. Then $\text{rad } R = \bigcap_{\alpha \in \Lambda} P_\alpha$ where $\{P_\alpha | \alpha \in \Lambda\}$ is the only set of prime ideals of R.

Proof. Clearly $\text{rad } R \subseteq P_\alpha$ for every $\alpha \in \Lambda$ since $R \backslash P_\alpha$ is an m-system and $0 \notin R \backslash P_\alpha$. Now if $x \in \bigcap_{\alpha \in \Lambda} P_\alpha$ and $x \notin \text{rad } R$, then there exists an m-system S such that $x \in S$ but $0 \notin S$. Let P be an ideal maximal with respect to the property that $P \cap S = \phi$. Note that P is a prime ideal since, for $a, b \in R$, $aRb \subseteq P$ implies $(a)(b) \subseteq P$ and $a \in P$ or $b \in P$. This is a contradiction.

4.4 Proposition. Let R be a ring. If A is an ideal of R such that $A \subseteq \text{rad } R$ and R/A is a semi-prime ring, then $A = \text{rad } R$.

Proof. If there exists $x \in \text{rad } R$ and $x \notin A$, then $xRx \nsubseteq A$. Let $N = \text{rad } R \backslash A$. Then $xRx \cap N \neq \phi$. Construct a set $X = \{x_1 = x, x_2 = x_1 r_1 x_1, x_3 = x_2 r_2 x_2, \ldots\}$ where $x_1 r_1 x_1 \in N$, $x_2 r_2 x_2 \in N$, etc. Let $x_i, x_j \in X$ (say $i \leq j$). Then $x_{j+1} \in x_j R x_j \subseteq x_i R x_j$. So X is an m-system and $0 \notin X$, which is a contradiction.

4.5 Corollary. R is a semi-prime ring if and only if $\text{rad } R = 0$.

Proof. If R is a semi-prime ring, then $0 = \text{rad } R$ by 4.4. Suppose $\text{rad } R = 0$ but R is not semi-prime. Then there exists $a \neq 0$ in R such that $aRa = 0$. Since $a \notin \text{rad } R$, there exists an m-system S with $a \in S$ but $0 \notin S$. This is impossible since $aRa = 0$.

4.6 Proposition. R is a semi-prime ring if and only if R contains no non-zero nilpotent ideals.

Proof. If R is a semi-prime ring, then certainly R contains no non-zero nilpotent ideals. Suppose now that R contains no non-zero nilpotent ideals. If R were not semi-prime, then there would exist $a \neq 0$ such that $aRa = 0$. If $aR \neq 0$, then $aR + RaR$ is a non-zero nilpotent ideal. Hence $aR = 0$. Let $N = \{a \in R | aR = 0\}$. Then N is a nilpotent ideal and $N \neq 0$. This is a contradiction.

4.7 Proposition. If N is a maximal nilpotent ideal of a ring R, then $N = \text{rad } R$.

Proof. It suffices to show that R/N is semi-prime. Let A/N be an ideal of R/N such that $A^n \subseteq N$. Then A is nilpotent and $A \subseteq N$. So by 4.6, R/N is semi-prime and $N = \text{rad } R$ by 4.4.

4.8 Definition. A ring R is said to be right noetherian if and only if R satisfies the ascending chain condition on right ideals.

4.9 Theorem (Utumi). If R is a semi-prime ring which satisfies the ascending chain condition on annihilator right ideals, then any nil left ideal is zero.

Proof. Let N be a nil left ideal. If $N \neq 0$, consider $F = \{(n)^r \mid 0 \neq n \in N\}$. Let $(n_o)^r$ be a maximal member of F. Then $n_oRn_o \neq 0$ since R is a semi-prime ring. If $x \in R$ such that $n_oxn_o \neq 0$, let k be the index of nilpotency of xn_o (i.e. $(xn_o)^k = 0$ but $(xn_o)^{k-1} \neq 0$). Note that $((xn_o)^{k-1})^r = (n_o)^r$ by the maximality of $(n_o)^r$. Hence $xn_o \in (n_o)^r$ and $n_oxn_o = 0$, which is a contradiction.

4.10 Corollary (Levitzki). If R is a right noetherian ring then any nil ideal is nilpotent.

Proof. Let N be a maximal nilpotent ideal (such exists by the noetherian condition). Then $N = \text{rad}\, R$ by 4.7. Hence R/N is a semi-prime ring with ascending chain condition. Thus if H is a nil ideal, then $H \subseteq N$ and H is nilpotent.

4.11 Proposition. Let R be a ring such that $R = \sum_{i=1}^{n} \oplus I_i$, where I_i is a minimal right ideal for each $i = 1,2,\ldots,n$ and such that if M is a minimal right ideal of R then $M^2 \neq 0$. Then R is a semi-simple ring which satisfies the descending chain condition on right ideals.

Proof. In view of 1.27, it suffices to show that R has an identity. Note that $I_i^2 = I_i$, so there exists $a_i \in I_i$ such that $a_iI_i = I_i$. Therefore there exists $e_i \in I_i$ with

$a_ie_i = a_i$ and $a_i(e_i^2 - e_i) = 0$. Since $(a_i)^r \cap I_i = 0$, $e_i^2 = e_i$. Hence we may write $R = e_1R \oplus e_2R \oplus \ldots \oplus e_nR$ where e_i is idempotent for each $i = 1,2,\ldots,n$. Consider $e_1R \oplus e_2R$. Let $B_1 = \{e_2r - e_1e_2r \mid r \in R\}$. Then $e_1R \oplus e_2R = e_1R \oplus B_1$. Hence $e_2R \cong B_1$ and B_1 is a minimal right ideal. Now $B_1^2 \neq 0$, so there exists $f_1 = f_1^2$ such that $f_1R = B_1$ and $f_1 = e_2r - e_1e_2r$ for some $r \in R$. Let $f = f_1 - f_1e_1$. Then $ff_1 = f_1 - f_1e_1f_1 = f_1$. Hence $f_1 \in fR$ and $fR = f_1R$. Note that $fe_1 = 0 = e_1f$ and $f^2 = (f_1 - f_1e_1)(f_1 - f_1e_1) = f_1-f_1e_1 = f$. Let $g = e_1 + f$. Then $g^2 = g$ and $gR = e_1R \oplus e_2R$. By continuing this process, we can find $e^2 = e$ such that $eR = R$. Clearly $ea = a$ for each $a \in R$. If $ae - a \neq 0$, for some $a \in R$ then $(ae - a)R \neq 0$, which is a contradiction.

4.12 Proposition. Let R be a ring which satisfies the ascending chain condition on annihilator right ideals. Then $Z_r(R)$ is nilpotent.

Proof. Note first that if $a,b \in Z_r(R)$ such that $ba \neq 0$ and $aR \neq 0$, then $(ba)^r \supsetneq (a)^r$. Indeed, if $(ba)^r = (a)^r$, then $(b)^r \cap aR \neq 0$ and there exists $ar \neq 0$ such that $bar = 0$, which is a contradiction. Now suppose $Z_r(R)$ is not nilpotent. Let $N = \{x \in R \mid xR = 0\}$. Then N is an ideal of R and $N^2 = 0$. By the ascending chain condition on annihilator right ideals, there exists a positive integer m such that $(Z_r^m(R))^r = (Z_r^{m+k}(R))^r$ for every positive integer k. Since $Z_r(R)$ is not nilpotent, there exists $a \in Z_r(R)$ such that $a \notin N$ and $Z_r^{m+1}(R)\, a \neq 0$. Hence there exists $b \in Z_r(R)$ such that $Z_r{}^m(R)ba \neq 0$. Therefore $(a)^r \subsetneq (ba)^r$. Since $Z_r{}^m(R)\, ba \neq 0$, $Z_r{}^m(R)Z_r(R)ba \neq 0$ and there exists $c \in Z_r(R)$ such that $Z_r{}^m(R)cba \neq 0$. Thus $(a)^r \subsetneq (ba)^r \subsetneq (cba)^r$. We can continue this process and obtain a chain of annihilator right ideals which does not terminate. This is a contradiction.

4.13 <u>Corollary</u>. If R is a commutative ring such that the ascending chain condition holds on annihilator ideals, then $(\text{rad } R)^n = 0$ for some positive integer n.

<u>Proof</u>. Note that $\text{rad } R \subseteq Z_r(R)$. Hence $(\text{rad } R)^n = 0$ for some positive integer n.

4.14 <u>Definition</u>. A right ideal C of a ring R is <u>closed</u> if and only if C has no proper essential extension in R as a right R-module.

4.15 <u>Proposition</u>. Let R be a ring with a zero right singular ideal. If D is a closed right ideal of R, then $\hat{D} = e\hat{R}$ for some $e = e^2 \in \hat{R}$ and $\hat{D} \cap R = D$.

<u>Proof</u>. Let $S = \{q \in \hat{R} \mid q^{-1} D \cap R$ is an essential right ideal of $R\}$. Then $\hat{D} \subseteq S$. If $q \in S$ such that $q \notin \hat{D}$, then $\hat{D} \subsetneq q\hat{R} + \hat{D}$ (by 3.7). Hence $q\hat{R} + \hat{D} = \hat{D} \oplus T$ for some submodule T of $\hat{R}_R$. Then $q = \hat{d} + t$ for some $\hat{d} \in D$ and $t \in T$ where $t \neq 0$. Let $K = q^{-1}D \cap R$. Then $tK = 0$ since $(q-\hat{d})K \subseteq \hat{D}$.

This is impossible since $Z(\hat{R}_R) = 0$. Therefore $\hat{D} = S$. If $q \in S$ and $\hat{r} \in \hat{R}$ then $(q\hat{r})^{-1}D \cap R$ is essential since $\hat{r}^{-1}(q^{-1}D \cap R) \cap R \subseteq (q\hat{r})^{-1}D \cap R$ and $\hat{r}^{-1}(q^{-1}D \cap R) \cap R$ is an essential right ideal of R. Hence $\hat{D}$ is a right ideal of $\hat{R}$ and $\hat{R} = \hat{D} \oplus F$ for some right ideal F of $\hat{R}$. By 3.7, there exists $e = e^2 \in \hat{R}$ such that $\hat{D} = e\hat{R}$. Clearly $D \subseteq \hat{D} \cap R$. If $D \subsetneq \hat{D} \cap R$, then $\hat{D} \cap R$ is not an essential extension of D since D is closed. Hence there exists a right ideal $H \subseteq \hat{D} \cap R$ such that $H \cap D = 0$.

4.16 Proposition. Let R be a ring such that $Z_r(R) = 0$. If e is an idempotent of $\hat{R}$ then $e\hat{R} \cap R$ is a closed right ideal of R.

Proof. If $e\hat{R} \cap R$ is not a closed right ideal of R, then there exists a right ideal C of R such that $e\hat{R} \cap R \subsetneq C$ and C is an essential extension of $e\hat{R} \cap R$. Let $c \in C \setminus e\hat{R} \cap R$ and let $K = c^{-1}(e\hat{R} \cap R) \cap R$. Then K is an essential right ideal of R and $(1-e)cK = 0$. Since $Z_r(R) = 0$, $(1-e)c = 0$ and $c = ec \in e\hat{R} \cap R$, which is a contradiction.

4.17 Proposition (R. E. Johnson). Let R be a ring. If $Z_r(R) = 0$, then $\hat{R}_R$ is a right self injective ring.

Proof. Let $S = \hat{R}_R$. Let A be a right ideal of S and let ϕ be an S-homomorphism of A into S. Since ϕ is also an R-homomorphism, there exists an R-homomorphism $\overline{\phi}: S_R \to S_R$.

It is clear that $\overline{\Phi}|_A = \phi$. Suppose there exist $s, t \in S$ such that $\phi(s)t - \phi(st) = q \neq 0$. Let $K = \{r \in R | tr \in R\}$. Then K is an essential right ideal of R. For $k \in K$, $qk = \phi(s)tk - \phi(st)k = \phi(stk) - \phi(stk) = 0$. Hence $q \in Z_r(R)$, which is a contradiction.

4.18 Proposition. Let R be a ring such that a right ideal I is essential if and only if I^{ℓ} (the left annihilator of I) is zero. If L is a non-zero left ideal of $\hat{R}_R$ then $L \cap R \neq 0$. (Assume $Z_r(R) = 0$.)

Proof. Since L is a non-zero left ideal of $\hat{R}_R$, there exist $e^2 = e \neq 0$ such that $\hat{R}e \subseteq L$ (by 3.7). Note that $A = e\hat{R} \cap R$ and $B = (1-e)\hat{R} \cap R$ are non-zero right ideals of R (if L is a proper left ideal). Since $A \cap B = 0$, $B^{\ell} \neq 0$. Let $r_o \in R$ such that $r_oB = 0$ and $r_o \neq 0$. Let $K = \{r \in R | (1-e)r \in R\}$. Then K is an essential right ideal of R and $r_o(1-e)K = 0$. Hence $r_o(1-e) = 0$ and $r_o = r_oe \in L \cap R$.

4.19 Proposition. Let R be a ring such that a right ideal I is essential if and only if $I^{\ell} = 0$. If A is a closed right ideal of R then $A^{\ell r} = A$. (Assume $Z_r(R) = 0$.)

Proof. Let $B = A^{\ell r}$. Then clearly $A \subseteq B$. If $A \neq B$, there exists a right ideal $C \neq 0$ such that $C \subseteq B$ and $A \cap C = 0$ since A is closed. Let D be a right ideal which is maximal with respect to the property that $A \subseteq D$ and $D \cap C = 0$. Then $D \oplus C$ is an essential right ideal. Let D' be an essential extension of D in R. If $D \subsetneqq D'$, let $0 \neq d' \in D' \cap C$. Since $d'R \neq 0$, $d'R \cap D \neq 0$. Let $x \in R$ such that $d'x \neq 0$ and $d'x \in D$. Then $d'x \in D' \cap C \cap D$, which is a contradiction. Hence D is closed. By 4.15, $D = e\hat{R} \cap R$ for some $e = e^2 \in \hat{R}$. Since $D \neq R$, $e \neq 1$ and $R(1-e) \cap R \neq 0$ by 4.18. Let $0 \neq x \in \hat{R}(1-e) \cap R$. Then $xD = 0$ and $xA = 0$ since $A \subseteq D$. Therefore $x \in A^{\ell}$ and $B \subseteq (x)^r$. Thus $xB = 0$ and $xC = 0$ since $C \subseteq B$. Therefore $x(D \oplus C) = 0$, which is a contradiction.

4.20 Proposition. Let R be a ring such that $Z_r(R) = 0$. If $0 \neq e^2 = e \in \hat{R}_R$, let $(e\hat{R} \cap R)^{\ell} = \{a \in R | a(e\hat{R} \cap R) = 0\}$ and let $(e\hat{R})^{\ell} = \{a \in R | ae\hat{R} = 0\}$. Then $(e\hat{R} \cap R)^{\ell} = (e\hat{R})^{\ell}$.

Proof. Clearly $(e\hat{R})^{\ell} \subseteq (e\hat{R} \cap R)^{\ell}$. Let $a(e\hat{R} \cap R) = 0$. If $ae\hat{R} \neq 0$, then $ae \neq 0$. Let $K = \{r \in R | er \in R\}$. Then K is an essential right ideal and $aeK = 0$. Hence $ae \in Z_r(\hat{R})$, which is a contradiction.

4.21 Proposition. Let R be a ring such that a right ideal I is essential if and only if $I^{\ell} = 0$. Then for any $e^2 = e \in \hat{R}_R$, $(\hat{R}e \cap R)^r = (1-e)\hat{R} \cap R$ and $(\hat{R}(1-e) \cap R)^r = e\hat{R} \cap R$.

Proof. By 4.20, $(e\hat{R} \cap R)^{\ell} = (e\hat{R})^{\ell} = \hat{R}(1-e) \cap R$. Now $(e\hat{R} \cap R)^{\ell r} = (\hat{R}(1-e) \cap R)^r$. By 4.19, $(e\hat{R} \cap R)^{\ell r} = e\hat{R} \cap R$. Hence $e\hat{R} \cap R = (\hat{R}(1-e) \cap R)^r$. Similarly, $(1-e)\hat{R} \cap R = (\hat{R}e \cap R)^r$.

4.22 Proposition. (Utumi). Let R be a ring such that if $I(L)$ is a right (left) ideal then $I^{\ell} = 0$ ($L^r = 0$) if and only if $I(L)$ is an essential right (left) ideal. Let $R^o = \{q \in \hat{R}_R \mid Rq^{-1}$ is an essential left ideal of $R\}$ and ${}^oR = \{q \in {}_R\hat{R} \mid q^{-1}R$ is an essential right ideal of $R\}$. Then R^o (${}^oR$) is a subring of $\hat{R}_R$ (${}_R\hat{R}$) such that if e is an idempotent in $\hat{R}_R$ (${}_R\hat{R}$), then $e \in R^o$ (${}^oR$) and $R \subseteq R^o$ ($R \subseteq {}^oR$).

Proof. If $q_1, q_2 \in R^o$, then $R(q_1 - q_2)^{-1} \supseteq Rq_1^{-1} \cap Rq_2^{-1}$ and $R(q_1 \cdot q_2)^{-1} \supseteq R(q_2^{-1})q_1^{-1}$. Hence R^o is a subring of $\hat{R}_R$. Let e be an idempotent of $\hat{R}_R$. If $1 = e$, then $e \in R^o$. So assume $1 - e \neq 0$. By 4.18, $L_1 = \hat{R}e \cap R \neq 0$ and $L_2 = \hat{R}(1-e) \cap R \neq 0$. We claim that $L_1 \oplus L_2$ is an essential right ideal. Observe that $(L_1 \oplus L_2)^r = (L_1)^r \cap (L_2)^r$ and by 4.21, $(L_1)^r = (1-e)R \cap R$ and $(L_2)^r = eR \cap R$. Hence $(L_1)^r \cap (L_2)^r = ((1-e)\hat{R} \cap R) \cap (e\hat{R} \cap R) = 0$ and $L_1 \oplus L_2$ is essential. Since $L_1 \oplus L_2 \subseteq Re^{-1}$, then $e \in R^o$. Similarly, ${}^oR$ contains all idempotents of ${}_R\hat{R}$.

4.23 Proposition. Let R be a regular ring such that R_R is injective. If A is an annihilator right ideal of R, then $A = eR$ for some $e = e^2$ in R.

Proof. Since A is an annihilator right ideal, there exists a non-empty subset S of R such that $A = \{r \in R \mid sr = 0\}$ for every $s \in S$. Since $Z(R) = 0$ (by 3.7), A is not an essential right ideal of R. Hence there exists a non-zero right ideal B of R such that $A \oplus B$ is an essential right ideal of R. Define an R-homomorphism $p: A \oplus B \to R$ by $p(a + b) = a$ for any $a \in A$, $b \in B$. Since R is R-injective, there exists $q \in R$ such that $qa = a$ for every $a \in A$ and $qB = 0$. Hence $A \subseteq qR$. Now if $q \notin A$, there exists $s \in S$ such that $0 \neq sq$ and $sq(A \oplus B) = 0$. Then $sq \in Z(R)$, which is a contradiction. Therefore $A = qR = eR$ where $e = qx$ and $qxq = q$.

4.24 Proposition. Let R be a regular ring such that R_R is injective. If e and f are idempotents of R such that $eRf \neq 0$, then there exist e' and f' in R such that $0 \neq e'R \subseteq eR$, $0 \neq f'R \subseteq fR$, and $e'R \cong f'R$.

Proof. Let $0 \neq x = eaf \in eRf$. Then $(x)^r = gR$ for some $g^2 = g$ in R. Define $T_x: fR \to eR$ by $T_x(fr) = xfr = eafr$. Now $xfgr = xgr = 0$ for any $fgr \in fgR$ and if $xfr = 0$, then $xr = 0$. Hence $r = gr'$ for some $r' \in R$ and $fr = fgr'$, so $\ker T_x = fgR$. Note that there exists $f_1 = f^2$ in R such that $f_1R = fgR$ and $fR = (1-f_1)fR \oplus f_1R$. Then $0 \neq xfR \subseteq eR$, $0 \neq (1-f_1)fR \subseteq fR$, and $xfR \cong (1-f_1)fR$.

4.25 Proposition. (Utumi). Let R be a regular ring such that R_R is injective. Let $0 \neq x \in R$ such that xR contains no direct sum of a pair of mutually isomorphic non-zero right ideals of R. If $e^2 = e \in R$ such that eR does not contain any right ideal which is isomorphic to xR, then $(1-e)R$ contains a non-zero central idempotent.

Proof. Let $\Sigma = \{\theta_\alpha \subseteq xR \times eR \mid \theta_\alpha$ is a monomorphism from some submodule of xR into $eR\}$. Note that Σ is partially ordered by set inclusion and Σ is inductive. Let θ be a maximal member of Σ, let D be the domain of θ, and let $E = \theta(D)$. Since xR and eR are direct summands of R, they are injective as R-modules. Hence $\hat{D} \subseteq xR$, and $\hat{E} \subseteq eR$. If $D \neq \hat{D}$, then θ extends to $\overline{\theta}$ which is a monomorphism from D into E, contradicting the maximality of θ. Since the isomorphic image of an injective module is injective, $E = \hat{E}$. Thus D and E are direct summands of xR and eR respectively. Let $E = e'R$ for some idempotent e' in R such that $eR = e'R \oplus e''R$ for some idempotent e'' in R, and let $xR = D \oplus fR$ for some idempotent f. We claim that $eRf = 0$. First note that if $e'Rf \neq 0$, then by 4.24 there exist $a,b \in R$ such that $aR \subseteq e'R$, $bR \subseteq fR$ and $aR \cong bR$. Hence $\theta^{-1}(aR) \oplus bR \subseteq xR$ and $\theta^{-1}(aR) \cong bR$, which is a contradiction. If $e''Rf \neq 0$, then there exists $c,d \in R$ such that $cR \subseteq e''R$, $dR \subseteq fR$ and $cR \cong dR$. Let η denote the isomorphism from cR onto dR. Define $\overline{\theta}: D \oplus dR \to E \oplus cR$ by $\overline{\theta}(y + dr) = \theta(y) + \eta^{-1}(dr)$. This violates the maximality of θ. Thus $eRf = 0$. Let $S = (eR)^r$. Then S is a non-zero two sided right ideal which is an annihilator right ideal. By 4.23, S is generated by a central idempotent element $e_o \neq 0$. Since $ee_o = 0$, $e_o \in (1-e)R$.

4.26 Proposition. Let R be a regular ring such that R_R is injective. Let $A = aR$ for some $a \in R$. Then $A \doteq B \oplus C$, for some right ideals B and C of R such that B is a direct sum of a pair of right ideals B_1 and B_2 with $B_1 \cong B_2$ and C contains no direct sum of mutually isomorphic non-zero right ideals.

Proof. Let $\Sigma = \{f_\alpha \mid f_\alpha$ is a monomorphism of a submodule of A into A such that $\mathrm{dom}\, f_\alpha \cap \mathrm{Im} f_\alpha = 0\}$. Then Σ is partially ordered by $f_\alpha \leq f_\beta$ if $\mathrm{dom}\, f_\alpha \subseteq \mathrm{dom}\, f_\beta$ and $f_\beta|\mathrm{dom}\, f_\alpha = f_\alpha$ and Σ is inductive. Let f be a maximal member of Σ. If $B_1 = \mathrm{dom}\, f$ and $B_2 = f(B_1)$, then $B = B_1 \oplus B_2 \subseteq A$. Since $A = aR$, A is a direct summand of R and A is injective. Note that $\hat{B}_1 \cap \hat{B}_2 = 0$. Indeed if $x \in \hat{B}_1 \cap \hat{B}_2$, then $x(x^{-1}B_1 \cap x^{-1}B_2) = 0$ and $x \in Z(R) = 0$. So $B_1 \oplus B_2 \subseteq \hat{B}_1 \oplus \hat{B}_2 \subseteq A$. If $B_1 \oplus B_2 \neq \hat{B}_1 \oplus \hat{B}_2$, then f can be extended to $\bar{f}$, an isomorphism between $\hat{B}_1$ and $\hat{B}_2$. This contradicts the maximality of f. Hence $B = \hat{B}_1 \oplus \hat{B}_2$ and $A = B \oplus C$ for some right ideal C. If C contains a pair of mutually isomorphic non-zero right ideals C_1 and C_2 such that $C_1 \cap C_2 = 0$, then $\hat{B}_1 \oplus C_1 \cong \hat{B}_2 \oplus C_2$. Therefore f can again be extended, contradicting the maximality of f as a member of Σ.

4.27 Proposition (Utumi). Let R be a ring and let θ be an element of $\mathrm{Hom}_R(R,R)$. Suppose that there are mutually isomorphic right ideals A_1 and A_2 such that $R = A_1 \oplus \ker \theta$

and A_2 is a direct summand of $\ker\theta$. Then θ is a sum of products of idempotents in $\mathrm{Hom}_R(R,R)$.

Proof. Let $R = A_1 \oplus A_2 \oplus A_3$ where $\ker\theta = A_2 \oplus A_3$ and $A_1 \cong A_2$ (by the isomorphism ω). Denote the projection mappings of $R = A_1 \oplus \ker\theta$ onto A_1 and $R = A_1 \oplus \ker\theta$ onto $\ker\theta$ by ε_1 and ε_2 respectively. Let $A_1' = \{x + \omega(x) \mid x \in A_1\}$. Then $A_1' \cap A_2 \oplus A_3 = 0$ since $\theta(x) + \theta(\omega(x)) = 0$ implies that $\theta(x) = 0$ and $x \in A_1 \cap \ker\theta$. Let $A_2' = \{\varepsilon_1\theta\omega^{-1}(y) + y \mid y \in A_2\}$. If $t \in A_2' \cap A_1 \oplus A_3$, $t = \varepsilon_1\theta\omega^{-1}(y) + y = a_1 + a_3$ for some $y \in A_2$, $a_1 \in A_1$, $a_3 \in A_3$ and $\varepsilon_1\theta\omega^{-1}(y) - a_1 = -y + a_3 \in A_1 \cap A_2 \oplus A_3 = 0$. Hence $A_2' \cap A_1 \oplus A_3 = 0$. Now let $A_1'' = \{x - \varepsilon_2\theta(x) | x \in A_1\}$. Then $A_1'' \cap \ker\theta = 0$. Denote the projection of $R = A_1' \oplus A_2 \oplus A_3$ onto A_2 by ε_3, $R = A_1 \oplus A_2' \oplus A_3$ onto A_1 by ε_4, and $R = A_1'' \oplus \ker\theta$ onto $\ker\theta$ by ε_5. For any $a_1 \in A_1$, $b \in \ker\theta$, consider $(\varepsilon_4\varepsilon_3\varepsilon_1 + \varepsilon_5\varepsilon_1)(a_1 + b) = \varepsilon_4\varepsilon_3(a_1) + \varepsilon_5(a_1)$. Also, $\varepsilon_3(a_1) = \varepsilon_3(a_1 + \omega(a_1) - \omega(a_1)) = -\omega(a_1) = \varepsilon_1\theta\omega^{-1}(\omega(a_1)) - \varepsilon_1\theta\omega^{-1}(\omega(a_1)) - \omega(a_1) = \varepsilon_1\theta\omega^{-1}(\omega(a_1)) - (\varepsilon_1\theta\omega^{-1}(\omega(a_1)) + \omega(a_1))$. Therefore $\varepsilon_4(-\omega(a_1)) = \varepsilon_1\theta\omega^{-1}(\omega(a_1)) = \varepsilon_1\theta(a_1)$. Now $\varepsilon_5(a_1) = \varepsilon_5(a_1 - \varepsilon_2\theta(a_1) + \varepsilon_2\theta(a_1)) = \varepsilon_2\theta(a_1)$.

Hence we have $(\varepsilon_4\varepsilon_3\varepsilon_1 + \varepsilon_5\varepsilon_1)(a_1 + b) = \varepsilon_1\theta(a_1) + \varepsilon_2\theta(a_1) = (\varepsilon_1 + \varepsilon_2)(\theta(a_1)) = \theta(a_1 + b)$ and $\varepsilon_4\varepsilon_3\varepsilon_1 + \varepsilon_5\varepsilon_1 = \theta$.

4.28 Proposition. (Utumi). Let R be a regular ring such that R_R is injective. Suppose every non-zero ideal of R contains a non-zero nilpotent element. Then R is generated by idempotent elements.

<u>Proof</u>. Let $x \in R$. Then $(x)^r = e_1R$ for some idempotent e_1 (by 4.23). Hence there exists an idempotent e_1' such that $R = e_1R \oplus e_1'R$. By 4.26, $e_1'R = e_2R \oplus e_3R \oplus e_4R$ where $e_2R \cong e_3R$ and e_4R contains no direct sum of mutually isomorphic non-zero right ideals. Without loss of generality, we may assume that e_1, e_2, e_3, and e_4 are orthogonal idempotents since $R = e_1R \oplus e_2R \oplus e_3R \oplus e_4R$ and $1 = e_1 + e_2 + e_3 + e_4$. Clearly $e_1R \oplus e_3R \oplus e_4R \subseteq (xe_2)^r$. If $xe_2y = 0$, then $e_2y \in (x)^r = e_1R$ and $e_2y = e_1r'$ for some r'. Hence $e_1e_2y = e_2y = 0$ since $e_1e_2 = 0$. Now $y = (e_1 + e_2 + e_3 + e_4)y = e_1y + e_3y + e_4y \in e_1R \oplus e_3R \oplus e_4R$. Therefore, $(xe_2)^r = e_1R \oplus e_3R \oplus e_4R$. Similarly $(xe_3)^r = e_1R \oplus e_2R \oplus e_4R$ and $(xe_4)^r = e_1R \oplus e_2R \oplus e_3R$. Let T be the subring of R which is generated by all idempotents of R. Note that $R = e_2R \oplus (xe_2)^r$, and since $(xe_2)^r$ contains a direct summand $e_3R \cong e_2R$, by 4.27, $xe_2 \in T$. Similarly, $xe_3 \in T$ since $R = e_3R \oplus (xe_3)^r$. Now $e_1R \oplus e_2R \oplus e_3R = (e_1 + e_2 + e_3)R$. Suppose $e_1R \oplus e_2R \oplus e_3R$ does not contain a right ideal which is isomorphic to e_4R. Then $e_4 \neq 0$ and e_4R contains no mutually isomorphic non-zero right ideals whose direct sum is contained in e_4R. Hence by 4.25, $[1-(e_1+e_2+e_3)]R = e_4R$ contains a non-zero central idempotent c. Since $cR \subseteq e_4R$, cR contains no mutually isomorphic pair of non-zero right

ideals whose direct sum is in cR. We claim that cR contains no non-zero nilpotent elements. Let $S = cR$. Then S is a regular ring. Suppose $0 \neq a \in S$ and $a^2 = o$. Then $S = (a)^r \oplus B$ for some right ideal B of S and $aS \subseteq (a)^r$. Now $B \cong S/(a)^r$ and $aS \cong S/(a)^r$. Note that every right ideal of S is also a right ideal of R. Hence cR contains mutually isomorphic right ideals aS and B such that $aS \oplus B \subseteq e_4R$, which is a contradiction. So cR contains no non-zero nilpotent elements and cR is an ideal, which is contrary to our hypothesis. Thus $e_1R \oplus e_2R \oplus e_3R$ contains a right ideal A which is isomorphic to e_4R. Since e_4R is injective, A is a direct summand of $e_1R \oplus e_2R \oplus e_3R$. Since $(xe_4)^r = e_1R \oplus e_2R \oplus e_3R$, then $xe_4 \in T$. Now $x = x(e_1+e_2+e_3+e_4) = xe_1 + xe_2 + xe_3 + xe_4 \in T$.

4.29 <u>Proposition</u>. Let R be a ring with identity. If R is a direct sum of right ideals $I_1, I_2, \ldots, I_n$, then there exist idempotents $e_1, e_2, \ldots, e_n$ such that $1 = e_1 + e_2 + \ldots + e_n$ and $e_ie_j = 0$ for $i \neq j$.

<u>Proof</u>. Let $R = \sum_{i=1}^{n} \oplus I_i$. Then $1 = e_1 + e_2 + \ldots + e_n$ for some $e_i \in I_i$, $i = 1,2,\ldots,n$. If $a \in R$, then $a = a_1 + a_2 + \ldots + a_n$ uniquely for some $a_i \in I_i$, $i = 1,2,\ldots, n$. Hence $a_i = e_ia$. In particular, $e_i = e_ie_i$ and $0 = e_ie_j$ if $i \neq j$.

4.30 <u>Proposition</u>. Let R be a ring and let e_1 and e_2 be idempotents in R. Then $e_1R \cong e_2R$ (as modules) if and only if there exist elements e_{12} and e_{21} in R such that $e_1e_{12}e_2 = e_{12}$, $e_2e_{21}e_1 = e_{21}$, $e_{12}e_{21} = e_1$ and $e_{21}e_{12} = e_2$.

Proof. Assume f is an isomorphism of e_1R onto e_2R. Let $f(e_1) = e_{21}$ and $f^{-1}(e_2) = e_{12}$. Now $f(e_1e_{12}e_2) = f(e_{12}e_2)$ $= f(e_{12})e_2 = e_2e_2 = e_2 = f(e_{12})$. Therefore $e_1e_{12}e_2 = e_{12}$. Similarly $f^{-1}(e_2e_{21}e_1) = f^{-1}(e_21)$ and $e_2e_{21}e_1 = e_{21}$. Also, $e_1 = f^{-1}f(e_1) = f^{-1}(e_{21}) = f^{-1}(e_2e_{21}) = f^{-1}(e_2)e_{21} = e_{12}e_2$, and $e_2 = ff^{-1}(e_2)= f(e_{12}) = f(e_1e_{12}) = f(e_1)e_{12} = e_{21}e_{12}$.

Conversely, if e_{12} and e_{21} are given such that $e_1e_{12}e_2 = e_{12}$, $e_2e_{21}e_1 = e_{21}$, $e_{12}e_{21} = e_1$ and $e_{21}e_{12} = e_2$, then define $g:e_1R \to e_2R$ by $g(e_1x) = e_{21}x$. If $e_1x = 0$, then $e_2e_{21}e_1x = e_{21}x = 0$ and g is well defined. Furthermore, if $y \in e_2R$ then $y = e_2r$ for some $r \in R$ and $e_2r = e_{21}e_{12}r$. Hence $e_2r = g(e_1e_{12}r)$ and g is an epimorphism. Now define $h:e_2R \to e_1R$ by $h(e_2x) = e_{12}x$. Thus h is also an R-homomorphism from e_2R onto e_1R. Also, $g \circ h(e_2r) = g(e_{12}r) =$ $g(e_1)e_{12}r = e_{21}e_{12}r = e_2r$. Similarly $h \circ g(e_1r) = e_1r$ and $e_1R \cong e_2R$.

4.31 Definition. Let R be a ring with identity. A finite subset $\{e_{ij} \mid i,j = 1,2,\ldots,n\}$ of R is called a set of matrix units in R if and only if $\sum_{i=1}^{n} e_{ii} = 1$ and $e_{ij}e_{k\ell} = \delta_{jk}e_{i\ell}$ where $\delta_{jk} = 1$ if $j = k$ and $\delta_{jk} = 0$ otherwise.

4.32 Proposition. Let R be a regular ring containing an element $a \neq 0$ such that $a^2 = 0$. Then there exists $u^2 = u \neq 0$

in R such that $uRu \subseteq (a)$ and uRu contains a set of matrix units which consists of four elements.

Proof. Since $aR \neq 0$, aR contains a non-zero idempotent element e_1. Hence $e_1 = ar$ for some $r \in R$ and $e_1 = are_1$. Let $b = re_1$ and $e_2 = ba$. Then $e_1 = ab$, $be_1 = b$, $e_2e_2 = baba = be_1a = ba = e_2$, and $e_2e_1 = baab = 0$. Let $u = e_1 + e_2 - e_1e_2$. Then $u^2 = (e_1+e_2 - e_1e_2)(e_1+e_2-e_1e_2) = e_1+e_2 - e_1e_2 = u$, $e_1u = e_1 = ue_1$, and $e_2u = e_2 = ue_2$. Hence $\{e_1, e_2\} \subseteq uRu \subseteq (a)$. Now let $B = uRu$. Then $e_1B \cap e_2B = 0$, for if $x \in e_1B \cap e_2B$, then $x = e_1b_1 = e_2b_2$ and $x = 0$ since $e_2e_1 = 0$. If $x \in uRu$ then $x = uru$ for some $r \in R$ and $x = (e_1 + e_2 - e_1e_2)ru$. Hence $uRu = e_1B \oplus e_2B$. Let $e_{12} = aba$ and $e_{21} = bab$. Then $e_{12}e_{21} = ababab = e_1$ and $e_{21}e_{12} = bababa = ba = e_2$. Furthermore, $e_1e_{12}e_2 = abababa = aba = e_{12}$ and $e_2e_{21}e_1 = bababab = bab = e_{21}$. Thus by 4.30, $e_1B \cong e_2B$. Let f_{11} and f_{12} be orthogonal idempotents such that $e_1B = f_{11}B$ and $e_2B = f_{22}B$. This is possible by 4.29. Then by 4.30, there exist f_{12}, f_{21} in R such that $f_{12}f_{21} = f_{11}$, $f_{21}f_{12} = f_{22}$, $f_{11}f_{12}f_{22} = f_{12}$, and $f_{22}f_{21}f_{11} = f_{21}$. Then $f_{12}f_{12} = f_{11}f_{12}f_{22}f_{11}f_{12}f_{22} = 0$, $f_{12}f_{11} = 0$, $f_{21}f_{21} = f_{22}f_{21}f_{11}f_{22}f_{21}f_{11} = 0 = f_{21}f_{22}$. Therefore $\{f_{11}, f_{12}, f_{21} f_{22}\}$ is a set of matrix units.

4.33 Proposition. Let R be a regular ring. Let $S = \sum_{\alpha \in \Lambda} S_\alpha$ where $\{S_\alpha \mid \alpha \in \Lambda\}$ is the family of all ideals of R, each of which contains no non-zero nilpotent elements Then $S \in \{S_\alpha \mid \alpha \in \Lambda\}$ and $S^{\ell r} = S$.

Proof. It is clear that $S \subseteq S^{\ell r}$. We shall prove that $S^{\ell r}$ does not contain any non-zero nilpotent elements (note that $S^{\ell r}$ is a two sided ideal). Suppose $0 \neq a \in S^{\ell r}$ such that $a^2 = 0$. By 4.32 there exists a set of matrix units $e_{11}, e_{12}, e_{21}, e_{22}$ in $S^{\ell r}$. We claim that $e_{11}R \cap S_\alpha \neq 0$ for some $\alpha \in \Lambda$. If $e_{11}R \cap S_\alpha = 0$ for every $\alpha \in \Lambda$, then $e_{11}R\,S_\alpha = 0$ for every $\alpha \in \Lambda$ and $(e_{11}R)S = 0$. Therefore $(e_{11}R)(e_{11}R) \subseteq (S^\ell)(S^{\ell r}) = 0$. This means $e_{11}{}^2 = 0$, which is a contradiction. Hence $e_{11}R \cap S_\alpha \neq 0$ for some $\alpha \in \Lambda$. Let $0 \neq e \in e_{11}R \cap S_\alpha$, and let $f_{11} = e_{11}ee_{11}$, $f_{12} = e_{11}ee_{12}$. Then $f_{11} \neq 0$, for if $f_{11} = 0$ then $e_{11}ee_{11}e = 0$ and $e = e_{11}e = 0$ since S_α contains no non-zero nilpotent elements. Also $f_{12} \neq 0$, for if $f_{12} = 0 = e_{11}ee_{12}$ then $e_{11}ee_{11} = e_{11}ee_{12}e_{21} = 0$, and $f_{11} = 0$. However, $0 \neq f_{12} = e_{11}ee_{12} = ee_{12} \in S_\alpha$ and $f_{12}f_{12} = e_{11}ee_{12}e_{11}ee_{12} = 0$. This is a contradiction.

4.34 Proposition. (Utumi) Let R be a regular ring such that R_R is injective. Then there exist two ideals A and B in R such that $R = A \oplus B$. Here A is a stron ly regular ring, every non-zero ideal of B contains a non-zero nilpotent element, and B is generated by idempotent elements.

Proof. Let $A = S$, where S is the sum of all ideals of R containing no non-zero nilpotent elements. Since $A = S^{\ell r}$, by

4.23 A is a direct summand of R. Let $R = A \oplus B$ for some right ideal B of R. By 4.33, A is strongly regular. Since every ideal of B is also an ideal of R, then every non-zero ideal of B contains a non-zero nilpotent element. By 4.28, B is generated by idempotent elements.

4.35 Proposition. Let R be a ring such that $Z_r(R) = 0$. If A is a right ideal of R then $\hat{A}$ is a right ideal of $\hat{R}_R$. If B is a right ideal of R such that $A \cap B = 0$ then $\hat{A} \cap \hat{B} = 0$.

Proof. Suppose $\hat{A} \subsetneqq \hat{A}\hat{R}$. Then there exists a non-zero submodule H of $\hat{A}\hat{R}$ such that $H \cap \hat{A} = 0$. Let $0 \neq h \in H$ anc write $h = \sum_{i=1}^{n} a_i q_i$ where $a_i \in \hat{A}$ and $q_i \in \hat{R}$ for $i = 1,2, \ldots,n$. Let $K = \bigcap_{i=1}^{n} q_i^{-1}R$. Then $hK \subseteq \hat{A} \cap H = 0$, which is a contradiction. Hence $\hat{A}$ is a right ideal of $\hat{R}$. Now if $A \cap B = 0$, suppose $x \in \hat{A} \cap \hat{B}$. Let $I_1 = x^{-1}A$ and $I_2 = x^{-1}B$. Then $I = I_1 \cap I_2$ is an essential right ideal of R and $xI \subseteq A \cap B = 0$. Hence $x = 0$.

4.36 Definition. Let R be a ring. The least upper bound of all integers n such that R contains n mutually isomorphic non-zero right ideals of R is denoted by $m(R)$.

4.37 Proposition. Let R be a ring. Then the following statements are equivalent.

(i) R_R is a strongly regular ring

(ii $Z_r(R) = 0$ and $m(R) = 1$.

Proof. Assume (i) holds. Then $Z_r(R) = 0$ by 3.7. If $m(R) \neq 1$, then there exist right ideals $A \neq 0$ and $B \neq 0$ of R such that $A \cong B$ and $A \cap B = 0$. By 4.35, $\hat{A} \cap \hat{B} = 0$ and $\hat{A}$ and $\hat{B}$ are right ideals of $\hat{R}$. Furthermore, $\hat{A} \cong \hat{B}$ as $\hat{R}$-modules. Hence $\hat{A}\hat{A} = 0$ since $\hat{A}$ and $\hat{B}$ are two-sided ideals and $\hat{A} \cap \hat{B} = 0$. This is a contradiction. Conversely, assume (ii) holds. By 3.7, $\hat{R}_R$ is a regular ring. If there exists $0 \neq a \in \hat{R}_R$ such that $a^2 = 0$, let $(a)^r = g\hat{R}$ for some $g = g^2$ in R. Then $a\hat{R} \subseteq g\hat{R}$. Let $\hat{R} = g\hat{R} \oplus X$ where X is a non-zero right ideal of $\hat{R}$ and let $K = a^{-1}R$. Then K is an essential right ideal of R. Let $K_o = K \cap X$. Then K_o is a right ideal of R, $aK_o \cong K_o$, and $aK_o \cap K_o \subseteq g\hat{R} \cap X = 0$. This means that $m(R) \neq 1$. Therefore $\hat{R}$ contains no non-zero nilpotent elements and $\hat{R}$ is a strongly regular ring.

4.38 Corollary. If R is a strongly regular ring, then $\hat{R}_R$ and ${}_R\hat{R}$ are strongly regular rings.

4.39 Proposition. Let R be a ring such that $R^\ell = 0$. If R_R is injective then R has an identity.

Proof. Consider the diagram

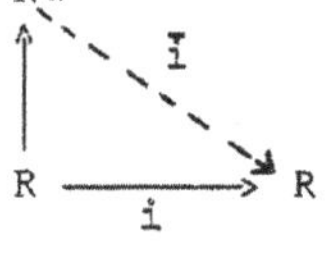

where R^* is the ring with identity

containing R, $\bar{i}(x) = x$ for every $x \in R$, and $\bar{i}$ is the extension of i to R^*. Let $\bar{i}(1) = a_o$. Then $a_o r = r$ for every $r \in R$ and $(ra_o - r)R = 0$ for every $r \in R$. Hence a_o is the identity for R.

4.40 Proposition. Let R be a ring. If there exist strongly regular rings P and Q such that P and Q are overrings of R (i.e. R is a subring of P and R is a subring of Q) such that if $0 \neq p \in P$ there exist $0 \neq a \in R$ and $0 \neq b \in R$ such that $ap = b$ and if $0 \neq q \in Q$ there exist $0 \neq a' \in R$ and $0 \neq b' \in R$ such that $qa' = b'$, then ${}_R\hat{R} = \hat{R}_R$.

Proof. First observe that the hypothesis implies that the right singular ideal of R and the left singular ideal of R are 0. Furthermore ${}_RP$ is an essential extension of ${}_RR$ and Q_R is an essential extension of R_R. Hence by 4.38, ${}_R\hat{R}$ and $\hat{R}_R$ are both strongly regular rings. Let $p \in {}_R\hat{R}$ such that $p \neq 0$. Then there exist $a \neq 0$ and $b \neq 0$ in R such that $ap = b \neq 0$. Since $a \in \hat{R}_R$, there exists $q \in \hat{R}_R$ such that $aqa = a = a^2q = qa^2$ and aq and qa are in the center of $\hat{R}_R$. Note that $b^2 \neq 0$ hence $bap \neq 0$ and $ba \neq 0$. Hence $ab \neq 0$ and $0 \neq a^2p = ab = a^2qb$. There exists $k \in R$ such that qbk is a non-zero element of R and $a^2qbk \neq 0$. Then $a^2pk - a^2qbk = 0 = a^2(pk-qbk)$. Since $pk - qbk \in {}_R\hat{R}$ and ${}_R\hat{R}$ is strongly regular, $(pk - qbk)a^2 = 0$. Hence $p(ka^2) = (pk)a^2 = qbka^2 \neq 0$. This means that $pR \cap R \neq 0$. Therefore ${}_R\hat{R}$ is an essential extension of R as a right R-module. Hence ${}_R\hat{R} \subseteq \hat{R}_R$. Similarly $\hat{R}_R \subseteq {}_R\hat{R}$ and $\hat{R}_R = {}_R\hat{R}$.

Chapter 5

If M is a right R-module for some ring R then we say M is <u>quasi-simple</u> provided that the endomorphism of its quasi-injective hull is a division ring and every non-zero submodule of M contains an isomorphic image of M. In this chapter we develop theories on quasi-simple modules along lines which are analogous to the theories on simple R-modules. A major theorem here is that if R is a prime ring with a uniform right ideal U such that the singular submodule of U is zero then for any linearly independent elements $u_1, u_2, \ldots, u_n$ of U over the endomorphism ring of the quasi-injective hull of U, $\mathrm{Hom}_R(R/\bigcap_{i=1}^{n}(u_i)^{\perp}, R/\bigcap_{i=1}^{n}(u_i)^{\perp})$ is a right order in the $n\times n$ matrix ring over the endomorphism ring of the quasi-injective hull of U.

5.0 <u>Definition.</u> Let R be a ring and let M be a right R-module. M is called <u>quasi-injective</u> if and only if every R-homomorphism from a submodule of M into M can be extended to an endomorphism of M.

5.1 <u>Proposition.</u> Let R be a ring. If M is an R-module and $\Lambda = \mathrm{Hom}_R(\hat{M},\hat{M})$, then M is quasi-injective if and only if M is a (Λ,R)-submodule of $\hat{M}$.

<u>Proof.</u> Let M be a (Λ,R)-submodule of M and let f be an R-homomorphism from a submodule of M into M. Then f can

be extended to an R-homomorphism $\bar{f}$ of M into $\hat{M}$. Since $\bar{f} \in \Lambda$, $\bar{f}(M) \subseteq M$ and M is quasi-injective. Conversely, assume that M is quasi-injective and let $f \in \mathrm{Hom}_R(\hat{M},\hat{M})$. Let $N = f^{-1}(M) \cap M$. Then there exists $f_1 \in \mathrm{Hom}_R(M,M)$ such that $f_1|_N = f|_N$ and there exists $f_2 \in \mathrm{Hom}_R(\hat{M},\hat{M})$ such that $f_2|_M = f_1$. If $f_2(m) = f(m)$ for every $m \in M$ then certainly $f(M) \subseteq M$ and M is a (Λ,R)-module. So suppose $(f_2 - f)(m) \neq 0$ for some $m \in M$. Then $(f_2 - f)(M) \cap M \neq 0$ and there exists $m_o \in M$ such that $(f_2 - f)(m_o) = m_1 \neq 0$ in M. Then $f(m_o) = f_2(m_o) - m_1 \in M$. Since $f_2(m_o) \in M$, $m_o \in N$ and $f(m_o) = f_1(m_o) = f_2(m_o)$, which is a contradiction.

5.2 Proposition. Let R be a ring. Each right R-module M has a unique minimal quasi-injective essential extension.

Proof. The unique minimal quasi-injective essential extension of M is the intersection of all (Λ,R)-submodules of $\hat{M}$ containing M.

5.3 Proposition. Let R be a ring and let M be a right R-module such that $Z(M) = 0$. Then M is quasi-injective if and only if $\mathrm{Hom}_R(M,M) \cong \mathrm{Hom}_R(\hat{M},\hat{M})$.

Proof. Let $K = \mathrm{Hom}_R(M,M)$ and let $\Lambda = \mathrm{Hom}_R(\hat{M},\hat{M})$. Define a mapping ϕ from K into Λ by $\phi(f) = \bar{f}$, where $\bar{f}$ is an extension of f to $\hat{M}$. Suppose there exist $\bar{f}_1, \bar{f}_2 \in \Lambda$ such that $\bar{f}_1|_M = \bar{f}_2|_M$ but $\bar{f}_1 \neq \bar{f}_2$. Then $(\bar{f}_1 - \bar{f}_2)(\hat{M}) \neq 0$ and $(\bar{f}_1 - \bar{f}_2)(\hat{M}) \cap M \neq 0$. Let $(\bar{f}_1 - \bar{f}_2)(m) = m_o \neq 0$ for some

$m \in M$ and $m_o \in M$. Let $I = m^{-1}M$. Then I is an essential right ideal of R and $(\bar{f}_1 - \bar{f}_2)(mi) = m_o i = 0$ for every $i \in I$. Hence $m_o \in Z(M)$, which is a contradiction. Clearly ϕ is a monomorphism. Hence K is isomorphic to a subring of Λ. If $\phi(K) = \Lambda$, then M is quasi-injective by 5.1. If M is quasi-injective then $\Lambda M \subseteq M$ and hence $\phi(K) = \Lambda$.

5.4 Proposition. Let R be a ring and let M be a quasi-injective R-module such that $mR = 0$ implies $m = 0$ for $m \in M$. Let $\mathbb{D} = \mathrm{Hom}_R(M,M)$. If $\{x_1, x_2, \ldots, x_n\}$ is a finite subset of M then $(\mathbb{D}x_1 + \mathbb{D}x_2 + \ldots + \mathbb{D}x_n)^{\perp\top} = \mathbb{D}x_1 + \mathbb{D}x_2 + \ldots + \mathbb{D}x_n$.

Proof. Proceed by induction on n. Clearly $\mathbb{D}x_1 \subseteq (\mathbb{D}x_1)^{\perp\top}$. Let $y \in (\mathbb{D}x_1)^{\perp\top}$ and define a mapping $f: x_1R \to yR$ by $f(x_1r) = yr$ for every $r \in R$. If $x_1r = 0$ then $\mathbb{D}x_1r = 0$. Hence $yr = 0$. Let $\bar{f}$ be an extension of f to M. Then $\bar{f} \in \mathbb{D}$ and $(\bar{f}(x_1) - y)R = 0$. Therefore $\bar{f}(x_1) = y$ and $y \in \mathbb{D}x_1$. Now assume $(\mathbb{D}x_1 + \mathbb{D}x_2 + \ldots + \mathbb{D}x_{n-1})^{\perp\top} = \sum_{i=1}^{n-1} \mathbb{D}x_i$. Let $J = (\sum_{i=1}^{n-1} \mathbb{D}x_i)^{\perp}$. Then $J^{\top} = \sum_{i=1}^{n-1} \mathbb{D}x_i$ by the induction hypothesis. Now suppose $\sum_{i=1}^{n} \mathbb{D}x_i \subsetneq (\sum_{i=1}^{n} \mathbb{D}x_i)^{\perp\top}$ and let $v \in (\sum_{i=1}^{n} \mathbb{D}x_i)^{\perp\top} \setminus \sum_{i=1}^{n} \mathbb{D}x_i$. Define a mapping $f: x_nJ \to vJ$ by $x_nj \to vj$ and let $\bar{f}$ be an extension of f to M. Then $\bar{f} \in \mathbb{D}$ and $(\bar{f}(x_n) - v)J = 0$. Therefore $\bar{f}(x_n) - v \in J^{\top}$ and $v \in \sum_{i=1}^{n} \mathbb{D}x_i$, which is a contradiction.

5.5 Definition. Let R be a ring and let M be a (right)

R-module such that $MR \neq 0$. Then M is called quasi-simple if and only if the following two conditions hold:

(Q1) $\mathrm{Hom}_R(\tilde{M},\tilde{M})$ is a division ring where $\tilde{M}$ is the quasi-injective hull of M.

(Q2) Every non-zero submodule of M contains an isomorphic image of M.

5.6 Proposition. Let R be a ring and let I be a right ideal of R such that the following conditions hold:

(i) $R^2 \nsubseteq I$ and every non-zero submodule of R/I is essential.

(ii) If $0 \neq \bar{a} \in R/I$ then there exists $r \in R$ such that $(\overline{ar})^{\perp} = I$.

(iii) If $\bar{a} \in R/I$ then either $I \subsetneqq (\bar{a})^{\perp}$ or $(\bar{a})^{\perp} \nsubseteq I$.

If $a^{-1}J \supseteq I$ for some $a \in R$ and right ideal J of R, then $I \subsetneqq a^{-1}J$ if $I \subsetneqq J$.

Proof. By (iii), either $(\bar{a})^{\perp} \supseteq I$ or $(\bar{a})^{\perp} \nsubseteq I$. If $I \subsetneqq (\bar{a})^{\perp}$, then $\phi \neq (\bar{a})^{\perp} \setminus I \subseteq a^{-1}J \setminus I$. If $(\bar{a})^{\perp} = I$ then $a \notin I$ and by (ii), there exists $r \in R$ such that $(ar)^{\perp} = I$. Now $rI \subseteq I$ since $(\bar{a})^{\perp} = I$. By (i), $arJ+I/I \cap J/I \neq 0$ and $arj = j_1 \neq 0$ for some $j, j_1 \in J$. Thus $j \in J \setminus I$, $rj \in R \setminus I$ and $rj \in a^{-1}J$. If $(a)^{\perp} \nsubseteq I$ then there exists $r \in (\bar{a})^{\perp}$ but $r \notin I$. This means that $I \subsetneqq a^{-1}J$.

5.7 Proposition. Let R be a ring and let I be a right ideal of R such that the following conditions hold:

(i) $R^2 \nsubseteq I$ and every non-zero submodule of R/I is essential.

(ii) If $0 \neq \bar{a} \in R/I$ then there exists $r \in R$ such that $(\overline{ar})^{\perp} = I$

(iii) If $\bar{a} \in R/I$ then either $I \subseteq (\bar{a})^{\perp}$ or $(\bar{a})^{\perp} \nsubseteq I$.

If f is a non-zero R-homomorphism from a submodule of R/I into R/I, then $\ker f = 0$.

Proof. If $\ker f \neq 0$, let $\ker f = J/I$ for some right ideal J of R. Let $\operatorname{dom} f = J'/I$ for some right ideal J' of R, and let $f(J'/I) = K/I$ for some right ideal K of R. By (ii) there exists $k \in K$ such that $(\bar{k})^{\perp} = I$. Then $f(\bar{b}) = \bar{k}$ for some $b \in J'$ and $bI \subseteq J$. Hence $I \subseteq b^{-1}J$ and by 5.6 $I \subsetneqq b^{-1}J$. Let $r \in b^{-1}J$ such that $r \notin I$. Then $f(\overline{br}) = \overline{kr} = 0$ and $(\bar{k})^{\perp} \neq I$, which is a contradiction.

5.8 Proposition. Let R be a ring and let I be a right ideal of R. The right R-module R/I is quasi-simple and $N(I) \setminus I \neq \phi$ if and only if the following conditions hold:

(i) $R^2 \nsubseteq I$ and every non-zero submodule of R/I is essential.

(ii) If $0 \neq \bar{a} \in R/I$ then there exists $r \in R$ such that $(\overline{ar})^{\perp} = I$.

(iii) If $\bar{a} \in R/I$ then either $I \subseteq (\bar{a})^{\perp}$ or $(\bar{a})^{\perp} \nsubseteq I$.

Proof. Assume R/I is quasi-simple and $N(I) \setminus I \neq \phi$. Since $(R/I)R \neq 0$, $R^2 \nsubseteq I$. Also $\operatorname{Hom}_R(\widetilde{R/I}, \widetilde{R/I})$ is a division ring implies that every non-zero submodule of R/I is essential. We first prove that the set $R^{\tau} = \{a\varepsilon R | aR \subseteq I\} = I$. If $I \subsetneq R^{\tau}$, then R^{τ}/I is a non-zero submodule of R/I and by (Q2), $f(R/I) \subseteq R^{\tau}/I$ for some non-zero R-homomorphism f of R/I. Note that $f(R^2/I) = 0$ and $R^2 \subseteq I$, which is a contradiction. Now let $a \in R \setminus I$. Then $aR + I / I$ is a non-zero submodule of R/I. Hence there exists $0 \neq f \in \mathrm{Hom}_R(R/I, R/I)$ such that $f(R/I) \subseteq {aR+I}/I$. Let $b \in N(I) \setminus I$. Then $f(\bar{b}) = \overline{ar}$ for some $r \in R$. Note that $I \subseteq (ar)^{\perp}$ since $\ker f = 0$ and $bI \subseteq I$. If $I \subsetneq (\overline{ar})^{\perp}$ then the kernel of the R-homomorphism $T_b: R/I \to R/I$ defined by $T_b(\bar{x}) = \overline{bx}$ for every $x \in R$ is not zero. But T_b extends to $\tilde{T}_b: \widetilde{R/I} \to \widetilde{R/I}$ and $\ker \tilde{T}_b \neq 0$, which is a contradiction of (Q1). To see (iii), let $a \in R \setminus I$. Suppose $I \nsubseteq (\bar{a})^{\perp}$. Then $aI \nsubseteq I$ and we must show $(\bar{a})^{\perp} \nsubseteq I$. Assume $(\bar{a})^{\perp} \subseteq I$. Define $g: {aR+I}/I \to R/I$ by $g(\overline{ar}) = \bar{r}$. Let h be the extension of g to $\widetilde{R/I}$. Then $h^{-1}(\bar{r}) = \overline{ar}$. Since $(\bar{a})^{\perp} \supsetneq I$, there exists $i \in I$ such that $ai \notin I$. Hence $h^{-1}(\bar{I}) = \overline{aI} = 0$, which is a contradiction.

Conversely, assume (i), (ii), and (iii) hold. By (i), $(R/I)R \neq 0$ and $I \neq R$. Let $a \in R$ such that $a \notin I$. Then by (ii), there exists $r \in R$ such that $(\overline{ar})^{\perp} = I$. This means that $ar \in N(I)$ and $ar \notin I$. Now let $0 \neq g \in \mathrm{Hom}_R(\tilde{R}/I, \tilde{R}/I)$. Since $g \neq 0$, there exists $f \in \mathrm{Hom}_R(\widehat{R/I}, \widehat{R/I})$ such that $gf(m) \neq 0$ for some $m \in R/I$. Hence $R/I \nsubseteq \ker gf$. Now if $\ker gf \neq 0$, then $gf(R/I) = 0$ by 5.7, which is a contradiction.

Since $f(R/I) \neq 0$, $f(R/I) \cap R/I \neq 0$ by (i). Let $m_o \in R/I$ such that $0 \neq f(m_o) \in R/I$. Then $gf(m_o) \neq 0$ since $\ker gf = 0$. This means that $g(R/I) \neq 0$. Hence $\ker g = 0$. To prove (Q2), let J/I be a non-zero submodule of R/I and let $a \in J \setminus I$. Then by (ii), there exists $r \in R$ such that $(\overline{ar})^{\perp} = I$ the mapping T_{ar} defined by $T_{ar}(\bar{x}) = \overline{arx}$ is a monomorphism of R/I into J/I.

5.9 Definition. Let I be a right ideal of a ring R. A multiplicative system S is a complemented multiplicative system with respect to I provided that $S \cap I = \phi$ and if J is a right ideal of R such that $I \subsetneq J$, then $J \cap S \neq \phi$.

5.10 Proposition. Let R be a ring and let I be a right ideal of R. Then the following hold if and only if (i), (ii), and (iii) of 5.8 hold.

(a) $N(I) \setminus I \neq \phi$ is a complemented multiplicative system.

(b) If $a^{-1}J \supseteq I$ for some $a \in R$ and right ideal J of R, then $I \subsetneq a^{-1}J$ if $I \subsetneq J$.

Proof. If (i), (ii), and (iii) are given, then (a) and (b) follow as a consequence of 5.8 provided that $N(I) \setminus I$ is a multiplicative system . Suppose $a, b \in N(I) \setminus I$ and $ab \in I$. Consider the mapping $T_b : R/I \to (\bar{a})^{\perp}/I$ defined by $T_b(\bar{x}) = \overline{bx}$. Note that T_b is not the zero homomorphism since $bR \nsubseteq I$. Hence $\ker T_b = 0$ and $ba \notin I$. Now consider the mapping $T_{ba} : R/I \to \overline{ba}R$ defined by $T_{ba}(\bar{x}) = \overline{bax}$. Again T_{ba} is not the zero homomorphism since $baR \nsubseteq I$. Therefore $\ker T_b = 0$ and $a \notin \ker T_b$. Hence $bab \notin I$ and $b \notin N(I)$, which is a contradiction.

Conversely, assume (a) and (b) hold. Then certainly $R^2 \nsubseteq I$ since $N(I) \setminus I$ is a nonempty complemented multiplicative system. Let A and B be right ideals of R such that $A \cap B = I$. If $I \subsetneq A$ and $I \subsetneq B$, let $a \in A \cap N(I) \setminus I$. Then $I \subseteq a^{-1}B$ and since $I \subsetneq B$, $I \subsetneq a^{-1}B$ by (b). Hence if $b \in (a^{-1}B \setminus I) \cap N(I)$ then $ab \in A \cap B \setminus I$, which is a contradiction. To see (ii), let $a \in R$ such that $a \notin I$. Then $I \subsetneq I + aR$, for if $aR + I = I$ then $aR \subseteq I$ and $a \in N(I) \setminus I$. Since $N(I) \setminus I$ is a multiplicative system, $a^2 \notin I$ and $aR \nsubseteq I$, which is a contradiction. So let $ar + i \in N(I) \setminus I$. Then $ar \in N(I) \setminus I$. Suppose $art \in I$ and $t \notin I$. Then $I \subsetneq J = \{s \in R | ars \in I\}$ and there exists $b \in J \cap N(I) \setminus I$. Then $arb \in I$, which is a contradiction to the fact that $N(I) \setminus I$ is a multiplicative system. Hence $(\overline{ar})^{\perp} = I$. To see (iii), let $a \in R$ such that $I \nsubseteq (\bar{a})^{\perp}$. To prove $(\bar{a})^{\perp} \nsubseteq I$, assume $(\bar{a})^{\perp} \subseteq I$. Note that $aI \nsubseteq I$, so let $J = aI + I$. Then $I \subsetneq J$ and $I \subsetneq a^{-1}J$ by (b). Hence there exists $r \notin I$ such that $ar \in J$, that is $ar = ai_1 + i_2$ for some $i_1, i_2 \in I$. Thus $a(r-i_1) = i_2$, $r-i_1 \in (\bar{a})^{\perp} \subseteq I$, and $r \in I$ which is a contradiction.

5.11 Definition. A right ideal I of a ring R is almost maximal if and only if I satisfies (a) and (b) of 5.10.

5.12 **Example.** Every maximal modular ideal is almost maximal. There exists a right ideal I of a ring R such that R/I is quasi-simple but $N(I) \setminus I = \not{c}$. Let $R = \begin{pmatrix} \mathbb{Z}_4 & 2\mathbb{Z}_4 \\ \mathbb{Z}_4 & 2\mathbb{Z}_4 \end{pmatrix}$, $I = \begin{pmatrix} \mathbb{Z}_4 & 2\mathbb{Z}_4 \\ 2\mathbb{Z}_4 & 2\mathbb{Z}_4 \end{pmatrix}$. Then R/I is simple but $N(I) \setminus I = \phi$.

5.13 **Example.** Note that (a) does not imply (b) in 5.10. Let $R = F[x,y]$ where F is a field. Then $xR \cap yR = 0$. Let $I = (0)$. Then $N(I) = R$ and $R\setminus\{0\}$ is a complemented multiplicative system. However $y^{-1}(xR) = 0$.

5.14 **Proposition.** Let I be a right ideal of a ring R such that $N(I) \setminus I$ is a nonempty complemented multiplicative system. Then $\{y \in R/I \mid yR = 0\} = \{0\}$.

Proof. Suppose there is a non-zero element y in R/I such that $yR = 0$. Thus $N = \{y \in R/I \mid yR = 0\}$ is a non-zero submodule of R/I. So $N = J/I$ for some right ideal J of R and $J \cap (N(I) \setminus I) \neq \phi$. Let $a \in J \cap N(I) \setminus I$. Then $(a+I)R = 0$ (mod I) and $a^2 \notin N(I) \setminus I$. This contradicts the fact that $N(I) \setminus I$ is a multiplicative system.

5.15 **Proposition.** Let I be a right ideal of a ring R such that $N(I) \setminus I$ is a complemented multiplicative system. Let $H = \{r \in R \mid$ if $x \in R$ there exists $b \in R \setminus I$ such that $rxb = 0\}$. Then H is an ideal of R.

Proof. Let $a, b \in H$ and let x_0 be an arbitrary element of R.

Consider the right ideal $(ax_o)^{\perp} + I$. Since $a \in H$, I is a proper subset of $(ax_o)^{\perp} + I$. Hence $[(ax_o)^{\perp} + I] \cap N(I)\backslash I \neq \phi$. Let $w \in [(ax_o)^{\perp} + I] \cap N(I) \setminus I$. Then $w = s + i$ for some $s \in (ax_o)^{\perp}$ and $i \in I$ and $s \in N(I) \setminus I$ since $w - i \in N(I) \setminus I$. Now consider the right ideal $(bx_o s)^{\perp} + I$. Since $b \in H$, I is also a proper subset of $(bx_o s)^{\perp} + I$ and one can find $z \in [(bx_o s)^{\perp} + I] \cap N(I) \setminus I$. Let $z = t + i_1$ for some $t \in (bx_o s)^{\perp}$ and $i_1 \in I$. Then $st \in N(I) \setminus I$ and $(a-b)x_o st = 0$. Hence $a-b \in H$. If $r \in R$ and $a \in H$, then clearly $ra \in H$ and $ar \in H$.

5.16 Proposition. Let R be a ring and let H and I be as in 5.15. Then $H \subseteq I$.

Proof. If $H \nsubseteq I$, then $(H+I) \cap N(I) \setminus I \neq \phi$. Let $a \in (H+I) \cap N(I) \setminus I$. Then $a = h + i$ for some $h \in H \cap N(I) \setminus I$ and $i \in I$. Now I is a proper subset of $(h)^{\perp} + I$; hence $[(h)^{\perp} + I] \cap N(I) \setminus I \neq \phi$. Let $t \in [(h)^{\perp} + I] \cap N(I) \setminus I$. Then $t = s + i'$ for some $s \in (h)^{\perp}$, $i' \in I$ and $t \in N(I) \setminus I$. Now $ht = hi' \in I$, which contradicts the fact that $N(I) \setminus I$ is a complemented multiplicative system.

5.17 Theorem. Let R be a ring with a regular element. If I is a right ideal of R such that $N(I) \setminus I$ is a nonempty complemented multiplicative system, then $(\widehat{R/I})^{\perp} = H$ (where H is as in 5.15).

Proof. If $(\widehat{R/I})^{\perp} \nsubseteq H$ then there exists $r_o \in (\widehat{R/I})^{\perp}$ such that for some $x_o \in R$, $r_o x_o b \neq 0$ for every $b \in R \setminus I$. Let c be a regular element in R. Then $cr_o x_o b \neq 0$ for every $b \in R \setminus I$. Let $T = \{x \in \widehat{R/I} \mid xa = 0$ for every $a \in R$ such that $cr_o x_o a = 0\}$. If $y \in T$, define an R-homomorphism f from $cr_o x_o R$ into $\widehat{R/I}$ by $f(cr_o x_o r) = yr$ for every $r \in R$. Let $\bar{f}$ be an extension of f to R. Since $(y-\bar{f}(cr_o x_o))R = 0$, by 5.14 $\bar{f}(cr_o x_o) = y$. Hence $y = \bar{f}(cr_o x_o)$ and $y \in (\widehat{R/I})r_o x_o = 0$. Hence $T = \{0\}$. However if $b \in N(I) \setminus I$ then $b + I \in T$ since $ba \in I$ for every $a \in R$ such that $cr_o x_o a = 0$. Thus the assumption that $(\widehat{R/I})^{\perp} \not\subseteq H$ is false.

Conversely suppose $H \not\subseteq (\widehat{R/I})^{\perp}$. Then there exist $x \in \widehat{R/I}$ and $b \in H$ such that $xb \neq 0$. Since $\widehat{R/I}$ is an essential extension of R/I, there is $r_o \in R$ such that $xbr_o \neq 0$ and $xbr_o \in R/I$. By 5.14, xbr_oR is a non-zero submodule of R/I. Hence there is a right ideal J in R such that $I \subsetneq J$ and $J/I = xbr_oR$. Let $t \in J \cap N(I) \setminus I$. Then $t + I = xbr_o r$ for some $r \in R$ and $br_o ra = 0$ for some $a \in R \setminus I$ since $b \in H$. This means that $ta \in I$ and the right ideal $t^{-1}I = \{a \in R \mid ta \in I\}$ properly contains I. Then $t^{-1}I \cap N(I) \setminus I \neq \phi$ and this is impossible since $N(I) \setminus I$ is a multiplicative system.

5.18 <u>Proposition</u>. Let M be an R-module for some ring R. If $\mathbb{D} = \mathrm{Hom}_R(\tilde{M},\tilde{M})$, then $\tilde{M} = \mathbb{D}M$.

<u>Proof</u>. Clearly $\mathbb{D}M \subseteq \tilde{M}$ since $\mathbb{D}\tilde{M} \subseteq \tilde{M}$. Let $\Lambda = \mathrm{Hom}_R(\hat{M},\hat{M})$. Then by 5.1, $\Lambda M = \tilde{M}$. If $x \in \tilde{M}$ then $x = \sum_{i=1}^{n} f_i(m_i)$ for some

positive integer n where $f_i \in \Lambda$ and $m_i \in M$ for all $1 \leq i \leq n$. Let $d_i = f_i|_{\tilde{M}}$. Since $f_i(\tilde{M}) = f_i(\Lambda M)$, $f_i(\tilde{M}) \subseteq \tilde{M}$. Hence $d_i \in \mathbb{D}$ for each $i = 1,2,\ldots,n$. Therefore $x \in \mathbb{D}M$ and $\mathbb{D}M = \tilde{M}$.

5.19 Proposition. Let R be a ring and let M be a quasi-simple R-module. Then for any non-zero submodule N of M, $\tilde{N} = \tilde{M} = \mathbb{D}N$ where $\mathbb{D} = \mathrm{Hom}_R(\tilde{M},\tilde{M})$.

Proof. We know that $\hat{M} = \hat{N}$. Let $\Lambda = \mathrm{Hom}_R(\hat{M},\hat{M})$. Then $\Lambda N = \tilde{N}$ by 5.2. Let $d \in \mathbb{D}$ and let $\hat{d}$ be an extension of d to $\hat{M}$. Then $d(N) = \hat{d}(N) \subseteq \Lambda N = \tilde{N}$. Hence $\mathbb{D}N \subseteq \tilde{N}$. By (Q2) there exists $0 \neq d_0 \in \mathbb{D}$ such that $d_0(M) \subseteq N$. Hence $\mathbb{D}d_0(M) \subseteq \tilde{N}$ and since $\mathbb{D}d_0 = \mathbb{D}$, $\mathbb{D}M \subseteq \mathbb{D}N$. Therefore $\mathbb{D}M = \mathbb{D}N$. Hence $\tilde{M} = \Lambda M = \mathbb{D}M = \mathbb{D}N = \Lambda N = \tilde{N}$.

5.20 Proposition. Let R be a ring and let M be a quasi-simple R-module. If $m \in M$ such that $m \neq 0$, then $mR \neq 0$ and $mR \cong R/I$ for some almost maximal right ideal I of R where $I = \{r \in R | mr = 0\}$.

Proof. Let $N_0 = \{m \in M | mR = 0\}$. If $N_0 \neq 0$ then N_0 is a non-zero submodule of M. Hence by (Q2), $0 \neq f(M) \subseteq N_0$ for some R-homomorphism f of M. This means $f(M)R \subseteq N_0R = 0$ and $f(MR) = 0$. Since the kernel of f is zero, $MR = 0$ and this is a contradiction. Now let $I = (m)^{\perp}$. Since $mR \neq 0$, by (Q2), there exists a non-zero homomorphism g of M such that $0 \neq g(M) \subseteq mR$. Hence $0 \neq g(m) = mr$ for some $r \in R$. Thus $mrI = 0$ since $mI = 0$ and $rI \subseteq I$. Since $g(m) \neq 0$, $r \notin I$ and $N(I) \setminus I \neq \phi$. Now by 5.19, $\mathrm{Hom}_R(\widetilde{mR},\widetilde{mR})$ is a division ring. Any non-zero submodule of mR is also a non-zero submodule of M, thus every non-zero submodule of mR contains an isomorphic image of mR.

5.21 __Proposition__. Let R be a ring. Let Σ be the set of quasi-simple R-modules and let σ be the set of almost maximal right ideals of R. Then $\bigcap_{M_\alpha \in \Sigma} M_\alpha^{\perp} = \bigcap_{I_\alpha \in \sigma} I_\alpha$. (We write $\bigcap_{I_\alpha \in \sigma} I_\alpha = W(R)$ if $\sigma \neq \phi$ and $W(R) = R$ if $\sigma = \phi$).

__Proof__. If $M_\alpha \in \Sigma$, then $M_\alpha^{\perp} = \bigcap_{m_\alpha \in M_\alpha} (m_\alpha)^{\perp}$. Since $(m_\alpha)^{\perp} \in \sigma$ by 5.20, $\bigcap_{M_\alpha \in \Sigma} M_\alpha^{\perp} \subseteq \bigcap_{I_\alpha \in \sigma} I_\alpha$. Let $a \in \bigcap_{M_\alpha \in \Sigma} M_\alpha^{\perp}$ and suppose $a \notin I_\alpha$ for some $I_\alpha \in \sigma$. By 5.8, $R/I_\alpha \in \Sigma$ and hence $Ra \subseteq I_\alpha$. Let $t \in N(I_\alpha) \setminus I_\alpha$. Then $ta \in I_\alpha$ and $a \in I_\alpha$, which is a contradiction.

5.22 Proposition. Let L be a nil one-sided ideal of a ring R. Then $L \subseteq W(R)$.

__Proof__. If L is a right ideal then certainly $L \subseteq I_\alpha$ for every $I_\alpha \in \sigma$ (where σ is as in 5.21). Suppose L is a left ideal and $L \nsubseteq I_\alpha$ for some $I_\alpha \in \sigma$. Let $a \in L$ such that $a \notin I_\alpha$. Then there exists $r \in R$ such that $ar \in N(I_\alpha) \setminus I_\alpha$. Now if $L^n = 0$, then $(ar)^{n+1} = 0$. But $(ar)^{n+1}$

is in $N(I_\alpha) \setminus I_\alpha$, which is a contradiction.

5.23 Proposition. Let R be a ring and let S be an ideal of R. If I is an almost maximal right ideal of R such that $S \subseteq I$, then I/S is an almost maximal right ideal of R/S. Conversely, if I/S is an almost maximal right ideal of R/S for some right ideal I of R, then I is almost maximal.

Proof. Assume I is an almost maximal right ideal and that $S \subseteq I$. Since $N(I) \setminus I$ is a complemented semigroup, so is $N(I/S) \setminus I/S$. If $I/S \subseteq (a+S)^{-1}(J/S)$ and $I/S \subsetneq J/S$ for some right ideal J/S of R/S then $I \subsetneq a^{-1}J$. Thus I/S is an almost maximal right ideal of R/S. Conversely, assume I/S is an almost maximal right ideal of R/S. Since $N(I/S) \setminus I/S$ is a complemented semigroup in R/S, $N(I) \setminus I$ is a complemented semigroup in R. If $I \subseteq a^{-1}J$ for some $a \in R$ and right ideal J such that $I \subsetneq J$, then $(a+S)^{-1}(J/S) \supsetneq I/S$. Thus $I \subsetneq a^{-1}J$.

5.24 Corollary. Let R be a ring. Then $W(R/W(R)) = 0$.

Proof. This follows immediately from 5.23 since $W(R)$ is an ideal by 5.10.

5.25 Proposition. Let R be a ring. Then (0) is an almost maximal right ideal if and only if R is a right Ore-domain.

Proof. If (0) is an almost maximal right ideal then $N(0) = R$.

Hence R has no non-zero divisors except 0 itself. If $a \neq 0$ and $b \neq 0$ in R, then $(0) \subsetneq b^{-1}(aR)$ and $bR \cap aR \neq 0$. Conversely, if R is a right Ore-domain then certainly $R\backslash(0)$ is a complemented semigroup. Let $a \in R$ and let J be a non-zero right ideal such that $(0) \subseteq a^{-1}J$. Then $J \cap aR \neq 0$ if $a \neq 0$. Hence $(0) \subsetneq a^{-1}J$ and (0) is almost maximal.

5.26 Proposition. Let R be a ring. Every proper right ideal of R is almost maximal if and only if R is a division ring.

Proof. Suppose that every right ideal of R is almost maximal. Then in particular, (0) is almost maximal. Hence R is a right Ore-domain by 5.25. Let $a \in R$ such that $a \neq 0$. Suppose $a^3R \subsetneq a^2R$. Then $a \in N(a^3R) \setminus a^3R$ and $N(a^3R) \setminus a^3R$ is not a semigroup. Thus $a^3R = a^2R$. If $x \in R$ then $a^2x = a^3y$ for some $y \in R$ and $x = ay$. Thus $R = aR$ and R is a division ring.

5.27 Corollary. If R is a commutative ring then R is a field if and only if every proper ideal of R is a prime ideal.

Proof. Note that every prime ideal here is almost maximal, and the result follows immediately from 5.26.

5.28 Proposition. Let R be a ring. Every non-zero proper right ideal oi R is almost maximal if and only if one of the following holds.

(i) R is a zero ring of a prime order.

(ii) R is a division ring

(iii) $R/J(R)$ is a division ring and $J(R)$ is the only proper right ideal of R.

(iv) R has an identity and R is a direct sum of two minimal right ideals.

Proof. Suppose every non-zero proper right ideal of R is almost maximal. Note that if I is a proper right ideal of R and $a \in I$ such that $aI \neq 0$ then $a \in aI$. To verify this note that $a \in N(aI)$. Hence if $a \notin aI$ then $a \in N(aI) \setminus aI$ and $a^2 \in N(aI) \setminus aI$, which is a contradiction. Now assume R is not a division ring and R is not a zero ring of a prime order. First we shall show that $R \neq J(R)$. If $R = J(R)$ then there exists $a \in J(R)$ such that $aJ(R) \neq 0$ since R is not a zero ring. If $aJ(R) = J(R)$ then $a = ab$ for some $b \in J(R)$ and $br = b+r$ for some $r \in R$. Hence $ar = abr = a(b+r) = ab + ar$ and $a = 0 = ab$, which is a contradiction. So $aJ(R)$ is a proper non-zero right ideal of R. Hence $a \in aJ(R)$, which is impossible. Thus $R \neq J(R)$ and there exists a maximal modular right ideal I of R. If $I = J(R)$ then $R/J(R)$ is a division ring. Suppose there exists a right ideal A of R such that $(0) \subsetneq A \subsetneq J(R)$. Note that $(J(R))^2 = 0$ by the first part of the proof. Thus A is not almost maximal, and this is a contradiction. If $I \neq J(R)$ then there exists a modular maximal right ideal I_1 of R such that $I \neq I_1$. If $I_1 \cap I_2 = A \neq 0$ then $R/A = I/A \oplus I_1/A$ and A is not almost maximal. So $0 = I \cap I_1 = J(R)$ and $R = I \oplus I_1$. Also I and I_1 are minimal right ideals of R. Hence by 4.11, R has an identity. The converse is clear.

5.29 Proposition. Let R be a ring and let I be an almost maximal right ideal of R. Let n be a positive integer and let R_n be the $n{\times}n$ matrix ring over R. If k is an integer such that $1 \leq k \leq n$ then define $I_k^* = \{(x_{ij}) \in R_n | x_{kj} \in I$ for every $1 \leq j \leq n\}$. Then I_k^* is an almost maximal right ideal of R_n.

Proof. Let $\delta \in N(I) \setminus I$. Then the matrix (x_{ij}), where $x_{kk} \in \delta$ and $x_{ij} = 0$ if $i \neq k$ or $j \neq k$ is a member of $N(I_k^*) \setminus I_k^*$. Let (b_{ij}) and (d_{ij}) be in $N(I_k^*) \setminus I_k^*$. Let $\alpha = (\alpha_{ij})$ such that the $k^{\underline{th}}$ row of α is the $k^{\underline{th}}$ row of (b_{ij}) and $\alpha_{ij} = 0$ if $i \neq k$. Let $\beta = (\beta_{ij})$ such that the $k^{\underline{th}}$ row of β is the $k^{\underline{th}}$ row of (d_{ij}) and $\beta_{ij} = 0$ if $i \neq k$. If $\alpha' = (b_{ij}) - \alpha$ and $\beta' = (\alpha_{ij}) - \beta$, then α' and β' are members of I_k^*. Note that b_{kk} and d_{kk} are members of $N(I) \setminus I$ and b_{kj} and d_{kj} are members of I if $k \neq j$. Hence if $(b_{ij}) \cdot (d_{ij}) \in I_k^*$, then $b_{kk}d_{kk} \in I$, which is a contradiction Now let J^* be a right ideal of R_n such that $I_k^* \subsetneqq J^*$. Then there exists $(c_{ij}) \in J^*$ such that $c_{ij} = 0$ if

$i \neq k$ and $c_{k\ell} \notin I$ for some $1 \leq \ell \leq n$. Since I is almost maximal and $c_{k\ell} \notin I$, there exists $a \in R$ such that $c_{k\ell}a \in N(I) \setminus I$. Let (a_{ij}) be a matrix such that $a_{\ell k} = a$ and $a_{ij} = 0$ if $i \neq \ell$ or $j \neq k$. Then $(c_{ij})(a_{ij}) \in J^*$ and $(c_{ij})(a_{ij}) \in N(I_k^*) \setminus I_k^*$. Now suppose K^* is a right ideal of R_n such that $I_k^* \subsetneqq K^*$ and let $\beta = (b_{ij}) \in R_n$ such that $\beta I_k^* \subseteq K^*$. If $b_{kk} \in I$, let $\gamma = (c_{ij}) \in R_n$ such that $\gamma \notin I_k^*$ and $c_{ij} = 0$ if $i \neq k$. Then $\beta \cdot \gamma \in I_k^*$. Now suppose $b_{kk} \notin I$. Let $K = \{r \in R \mid (c_{ij}) \in K^*$ with $c_{kk} = r$, $c_{ij} = 0$ if $i \neq k\}$. Then $I \subsetneqq K$ since there exists $(x_{ij}) \in K^*$ such that $x_{k\ell} \notin I$ for some $1 \leq \ell \leq n$ and $x_{ij} = 0$ if $i \neq k$. Hence there exists $a \in R$ such that $x_{k\ell}a \in N(I) \setminus I$. Let (a_{ij}) be a matrix such that $a_{\ell k} = a$ and $a_{ij} = 0$ if $i \neq k$ or $j \neq \ell$. Then $(x_{ij})(a_{ij}) \in K^*$ and $x_{k\ell}a_{\ell k} \in K$. Hence $I \subsetneqq K$. For any $a \in I$, the matrix (a_{ij}) such that $a_{kk} = a$ and $a_{ij} = 0$ if $i \neq k$ or $j \neq k$ is a member of I_k^*. Hence $\beta(a_{ij}) \in K^*$. Therefore $b_{kk}I \subseteq K$ and $I \subsetneqq b_{kk}^{-1}K$. Let $c \in R$ such that $b_{kk}c \in K$ but $c \notin I$. There exists a matrix $(x_{ij}) \in K^*$ such that $x_{kk} = b_{kk}c$ and $x_{ij} = 0$ if $i \neq k$. Let $\gamma = (c_{ij})$ be a matrix such that $c_{kk} = c$ and $c_{ij} = 0$ if $i \neq k$ or $j \neq k$. Then $\beta\gamma \in K^*$ and $\gamma \notin I_k^*$.

5.30 Proposition. If R is a ring then $W(R_n) \subseteq (W(R))_n$.

Proof. Let $\alpha \in W(R_n)$, say $\alpha = (\alpha_{ij})$. If $\alpha \notin (W(R))_n$ then $\alpha_{k\ell} \notin W(R)$ for some $1 \leq k,\ell \leq n$. Hence there exists an almost maximal right ideal I such that $\alpha_{k\ell} \notin I$. Let $I_k^* = \{(b_{ij}) \in R_n \mid b_{kj} \in I$ for $1 \le j \le n\}$. Then I_k^* is an almost maximal right ideal by 5.29 and $\alpha \notin I_k^*$. Thus $\alpha \notin W(R_n)$, and this is a contradiction.

5.31 Proposition. Let J^* be an almost maximal right ideal of R_n for some ring R. Suppose there exists $\beta = (b_{ij}) \in N(J^*) \setminus J^*$ such that $b_{ij} = 0$ if $i \ne k$ and let $\gamma = (c_{ij})$ be a matrix such that $c_{kk} = b_{kk} \ne 0$ and $c_{ij} = 0$ if $i \ne k$ or $j \ne k$. Then $I^* = \{\rho \in R_n | \gamma\rho \in J^*\}$ is almost maximal and $\gamma \in N(I^*) \setminus I^*$.

Proof. Note that $\gamma \notin J^*$ since $\beta^2 = \gamma\beta$ and $\beta^2 \in N(J^*) \setminus J^*$. Since J^* is almost maximal, by 5.20 I^* is almost maximal. Let $\rho \in I^*$. Then $\gamma\rho \in J^*$. Hence $\beta\gamma\rho \in J^*$ since $\beta \in N(J^*)$ and $\beta\gamma\rho = \gamma^2\rho = \gamma(\gamma\rho) \in J^*$. Therefore $\gamma\rho \in I^*$ and $\gamma \in N(I^*)$. Now if $\gamma \in I^*$ then $\gamma^2 \in J^*$. Since $\gamma \notin J^*$ and J^* is almost maximal there exists $\alpha \in R_n$ such that $\gamma\alpha \in N(J^*) \setminus J^*$. Hence $\gamma^2\alpha = \beta\gamma\alpha \in N(J^*) \setminus J^*$, which is a contradiction.

5.32 Proposition. Let J^* be an almost maximal right ideal of R_n for some ring R. Let β, γ, and I^* be as in 5.31, and let $I = \{r \in R | r = r_{kk}$ for some $(r_{ij}) \in I^*\}$. If $a \in R \setminus I$ then there exists $r \in R$ such that $ar \in N(I) \setminus I$.

Proof. Let $\rho = (r_{ij})$ with $r_{kk} = a$ and $r_{ij} = 0$ if $i \ne k$ or $j \ne k$. Then $\rho \notin I^*$. Hence there exists $\tau = (t_{ij}) \in R_n$

such that $\rho\tau \in N(I^*) \setminus I^*$. We claim that $at_{kk} \in N(I) \setminus I$. Note that if $(x_{ij}) \in R_n$ and $x_{kj} = 0$ for every $1 \le j \le n$ then $(x_{ij}) \in I^*$. Suppose $\rho\tau \in N(I^*) \setminus I^*$, the matrix α whose k^{th} row is the kth row of $\rho\tau$ and all other rows are zero is a member of $N(I^*) \setminus I^*$. Suppose there exists $c \in I$ such that $at_{kk}c \notin I$. Since $c \in I$ there exists a matrix $(c_{ij}) \in I^*$ such that $c_{kk} = c$. Hence the matrix (d_{ij}) whose k^{th} row is the kth row of (c_{ij}) and all other rows are zero is a member of I^*. Now $\alpha(d_{ij}) \in I^*$ and the element of the kth row and k^{th} column of $\alpha(d_{ij})$ is $at_{kk}c$. Hence $at_{kk}c \in I$, which is a contradiction. If $at_{kk} \in I$, there exists $\mu = (m_{ij}) \in I^*$ with $m_{kk} = at_{kk}$ and $m_{ij} = 0$ for $i \ne k$. Then $\mu\gamma = \rho\tau\gamma \in I^*$ which implies $\gamma \in I^*$. This is a contradiction.

5.33 Proposition. Let R be a ring and let J^*, β, γ, I^*, and I be as in 5.32. If $(x_{ij}) \in N(I^*) \setminus I^*$ then $x_{kk} \in N(I) \setminus I$, and if $a \in N(I) \setminus I$ then the matrix (a_{ij}) where $a_{kk} = a$ and $a_{ij} = 0$ if $i \ne k$ or $j \ne k$ is a member of $N(I^*) \setminus I^*$.

Proof. Let $x = (x_{ij}) \in N(I^*) \setminus I^*$. Then the matrix (d_{ij}) where the kth row of (d_{ij}) is the kth row of (x_{ij}) and all other rows are zero is a member of $N(I^*) \setminus I^*$. Hence, as in 5.32, $x_{kk} \in N(I)$. If $x_{kk} \in I$ then there exists $\mu = (m_{ij}) \in I^*$ such that $m_{kk} = x_{kk}$ and $m_{ij} = 0$ if $i \ne k$. Then $\mu\gamma = (x_{ij}')\gamma \in I^*$, where the kth row of (x_{ij}') is the kth row of (x_{ij}) and all other rows are zero. This is a contradiction since I^* is almost maximal. Now let $a \in N(I) \setminus I$. First note that the set $\{a \in R \mid aR \subseteq I\}$ is I by 5.32. Consider the matrix (a_{ij}) where $a_{kk} = a$ and $a_{ij} = 0$ if $i \neq k$ or $j \neq k$. Then $(a_{ij}) \notin I^*$. Let $(c_{ij}) \in I^*$. Without loss of generality, we may assume that every row except the $k^{\underline{th}}$ row of (c_{ij}) is zero. Furthermore, $c_{k\ell}R \subseteq I$ for every $1 \leq \ell \leq n$ since $s \in R$, $(c_{ij})(s_{ij}) \in I^*$ where $s_{\ell k} = s$ and $s_{ij} = 0$ if $i \neq \ell$ or $j \neq k$. Thus $c_{k\ell} \in I$ and $ac_{k\ell} \in I$ for every $1 \leq \ell \leq n$. Let $\rho = (r_{ij})$ be a matrix such that $r_{kj} = ac_{kj}$. If $\rho \notin I^*$ then there exists $\tau = (t_{ij}) \in R_n$ such that $\rho\tau \in N(I^*) \setminus I^*$. Let $\rho\tau = \sigma = (s_{ij})$. We may assume $s_{ij} = 0$ if $i \neq k$. Let $\mu = (m_{ij}) \in I^*$ such that $m_{kk} = ac_{kk}t_{kk} \in I$ and $m_{ij} = 0$ if $i \neq k$. Then $\rho\tau\gamma = \mu\gamma \in I^*$, which is a contradiction.

5.34 <u>Proposition</u>. Let R be a ring and let J^*, β, γ, I^*, and I be as in 5.32. Then $I^* = \{(r_{ij}) \mid r_{kj} \in I$ for every $1 \leq j \leq n\}$.

<u>Proof</u>. Note that $\{(r_{ij}) \mid r_{kj} \in I$ for every $1 \leq j \leq n\}$ is contained in I^* as a result of the proof of 5.33. Suppose $\rho = (r_{ij}) \in I^*$. Then $r_{kj}R \subseteq I$ for every $1 \leq j \leq n$. Hence by 5.32, $r_{kj} \in I$ for every $1 \leq j \leq n$.

5.35 <u>Proposition</u>. Let R be a ring and let J^*, β, γ, I^*, and I be as in 5.32. Then I is an almost maximal right ideal of R.

Proof. Let a and b belong to $N(I) \setminus I$. Then (a_{ij}) and (b_{ij}), where $a_{kk} = a$, $b_{kk} = b$ and $a_{ij} = 0 = b_{ij}$ if $i \neq k$ or $j \neq k$, are members of $N(I^*) \setminus I^*$ by 5.33. Hence $ab \in N(I) \setminus I$ by 5.33. If K is a right ideal of R such that $I \subsetneqq K$ then $K \cap N(I) \setminus I \neq \phi$ by 5.32. Now let $I \subseteq r^{-1}K$ for some $r \in R$ and for some right ideal J of R such that $I \subsetneqq K$. Let $\sigma = (s_{ij})$ be a matrix such that $s_{kk} = r$ and $s_{ij} = 0$ if $i \neq k$ or $j \neq k$. Let $K^* = \{(r_{ij}) \mid r_{kj} \in K$ for every $1 \leq j \leq n\}$. Then $\sigma I^* \subseteq K^*$. Since $I^* \subsetneqq K^*$ by 5.34, $I^* \subsetneqq \sigma^{-1}K^*$. Therefore $I \subsetneqq r^{-1}K$.

5.36 Theorem. If R is a ring then $(W(R))_n \subseteq W(R_n)$.

Proof. Suppose there exists a matrix $\alpha = (\alpha_{ij})$ such that $\alpha \in (W(R))_n$ but $\alpha \notin W(R_n)$. Let $\alpha^{(i)} = (a_{ij})$ be such that the $i^{\underline{th}}$ row of $\alpha^{(i)}$ is the $i^{\underline{th}}$ row of α and all other rows are zero. Then $\alpha = \alpha^{(1)} + \alpha^{(2)} + \ldots + \alpha^{(n)}$. Since $\alpha \notin W(R_n)$, there exists $1 \leq k \leq n$ such that $\alpha^{(k)} \notin W(R_n)$. Let J^* be an almost maximal right ideal of R_n such that $\alpha^{(k)} \notin J^*$. Hence there exists $\beta = (b_{ij}) \in (N(J^*) \setminus J^*) \cap \alpha^{(k)}R_n$. Then $b_{ij} = 0$ if $i \neq k$. Observe that $b_{kk} \in W(R)$. Let I be an almost maximal right ideal of R as in 5.32. Then $b_{kk} \in I$. Hence there exists $\rho = (r_{ij}) \in I^*$, where I^* is as in 5.32, with $r_{ij} = 0$ if $i \neq k$ and $r_{kk} = b_{kk}$. Let $\gamma = (c_{ij})$ be a matrix such that $c_{kk} = b_{kk}$ and $c_{ij} = 0$ if $i \neq k$ of $j \neq k$. Then $\gamma^2 = \rho\gamma \in I^*$. But this is impossible since $\gamma \in N(I^*) \setminus I^*$ by 5.31.

5.37 Proposition. Let M be a quasi-simple R-module for some ring R. If N is any non-zero submodule of M then $\tilde{M}^{\perp} = N^{\perp}$.

Proof. By 5.19, $\tilde{N} = \tilde{M} = \mathbb{D}N$ where $\mathbb{D} = \mathrm{Hom}_R(\tilde{M},\tilde{M})$. Since $(\mathbb{D}N)^{\perp} = N^{\perp}$, $\tilde{M}^{\perp} = N^{\perp}$.

5.38 Definition. Let Q be a ring with identity and let R be a subring of Q. Then R is a right order in Q provided that for any $q \in Q$ there exist $a,b \in R$ where b is regular such that $q = ab^{-1}$ and every regular element of R has an inverse in Q.

5.39 Proposition. Let M be a quasi-simple R-module for some ring R and let $\mathbb{D} = \mathrm{Hom}_R(\tilde{M},\tilde{M})$. If N is a non-zero submodule of M and if $K(N) = \mathrm{Hom}_R(N,N)$ then $K(N)$ is a right order in $\mathbb{D}$.

Proof. Let $d \neq 0$ be an element of $\mathbb{D}$ and let $N(d) = \{m \in M \mid d(m) \in N\}$. Then $N(d)$ is a non-zero submodule of M. Let $N_1 = N \cap N(d)$. Since $N_1 \neq 0$, there exists $f \neq 0$ in $\mathrm{Hom}_R(M,M)$ such that $f(M) \subseteq N_1$. Hence $df(M) \subseteq d(N_1) \subseteq N$ and $df|_N \in K(N)$. This proves that $K(N)$ is a right order in $\mathbb{D}$.

5.40 Proposition. Let M be a quasi-simple R-module for some ring R and let $\mathbb{D} = \mathrm{Hom}_R(\tilde{M},\tilde{M})$. If $\{x_i\}_{i=1}^{n}$ is a finite subset of $\tilde{M}$ and $y \in \tilde{M}$ then y is a $\mathbb{D}$-linear combination of $\{x_i\}_{i=1}^{n}$ if and only if $\bigcap_{i=1}^{n}(x_i)^{\perp} \subseteq (y)^{\perp}$.

Proof. If y is a $\mathbb{D}$-linear combination of $\{x_i\}_{i=1}^{n}$ then clearly $\bigcap_{i=1}^{n}(x_i)^{\perp} \subseteq (y)^{\perp}$. Now by 5.4, $(\mathbb{D}x_1 + \mathbb{D}x_2 + \ldots \mathbb{D}x_n)^{\perp T} = \sum_{i=1}^{n}\mathbb{D}x_i$. Since $(\sum_{i=1}^{n}\mathbb{D}x_i)^{\perp} = \bigcap_{i=1}^{n}(x_i)^{\perp}$, if $\bigcap_{i=1}^{n}(x_i)^{\perp} \subseteq (y)^{\perp}$ then $(y)^{\perp T} \subseteq (\sum_{i=1}^{n}\mathbb{D}x_i)^{\perp T}$. Since $y \in (y)^{\perp T}$ and $(\sum_{i=1}^{n}\mathbb{D}x_i)^{\perp T} = \sum_{i=1}^{n}\mathbb{D}x_i$, y is a $\mathbb{D}$-linear combination of $x_1, x_2, \ldots, x_n$.

5.41 Proposition. Let M be a quasi-simple R-module for some ring R. Let $\mathbb{D} = \mathrm{Hom}_R(\tilde{M},\tilde{M})$ and let $K(N) = \mathrm{Hom}_R(N,N)$ where N is a non-zero submodule of M. If $\{x_i\}_{i=1}^{n}$ is a finite $\mathbb{D}$-linearly independent subset of $\tilde{M}$ and $\{y_i\}_{i=1}^{n}$ is a sequence in M then there exists $r \in R$ and $0 \neq k \in \mathrm{Hom}_R(M,M)$ such that $k(N) \subseteq N$ and $x_i r = ky_i$ for all $1 \leq i \leq n$.

Proof. Let $I_j = \bigcap_{\substack{i=1 \\ i\neq j}}^{n}(x_i)^{\perp}$ for $1 \leq j \leq n$. Then $x_j I_j \neq 0$ by 5.40 for any $1 \leq j \leq n$. Hence $x_j I_j \cap N$ is a non-zero submodule of M. Since M is quasi-simple there is $0 \neq f_j \in \mathrm{Hom}_R(M,M)$ such that $f_j(M) \subseteq x_j I_j \cap N$. Hence if $y_j \neq 0$ then $f_j(y_j) = x_j a_j \neq 0$ for some $a_j \in I_j$. Let $N_o = [\bigcap_{y_j \neq 0} f_j(y_j)R] \cap N$. Then N_o is a non-zero submodule of M and there exists $0 \neq k_o \in \mathrm{Hom}_R(M,M)$ such that $k_o(M) \subseteq N_o$. Hence $k_o(y_j) = f_j y_j r_j \neq 0$ for some $r_j \in R$ if $y_j \neq 0$. Note that $k_o(N) \subseteq N$. Let $r = \sum_{j=1}^{n} a_j r_j$ where $a_j = 0$ if $y_j = 0$ and $f(y_j) = x_j a_j$ if

$y_j \neq 0$. Then $x_i r = x_i \sum_{j=1}^{n} a_j r_j = x_i a_i r_i = f_i(y_i) r_i = k_o y_i$ for all $1 \leq i \leq n$.

5.42 <u>Proposition.</u> Let M be a quasi-simple R-module for some ring R. If the singular submodule of M is zero then M is a quasi-simple B-module for any essential right ideal B of R.

<u>Proof</u>. Let $f \in \mathrm{Hom}_B(N,M)$ for a submodule N of M. If $f \notin \mathrm{Hom}_R(N,M)$ then $f(nr_o) - f(n)r_o \neq 0$ for some $r_o \in R$ and $n \in N$. Let $A = \{a \in R \mid r_o a \in B\}$. Then A is essential since B is essential. Hence $G = A \cap B$ is an essential right ideal of R and $(f(nr_o) - f(n)r_o)G = 0$. This contradicts the fact that the singular submodule of M is zero. Thus $f \in \mathrm{Hom}_R(N,M)$ and $\mathrm{Hom}_R(N,M) \subseteq \mathrm{Hom}_B(N,M) \subseteq \mathrm{Hom}_R(N,M)$. Likewise, since the singular submodule of $\tilde{M}$ is zero, $\mathrm{Hom}_R(\tilde{M},\tilde{M}) = \mathrm{Hom}_B(\tilde{M},\tilde{M})$. Now if N is a non-zero R-submodule of M then $0 \neq NB \subseteq N$. Since NB is a non-zero R-submodule of M, there exists $0 \neq f \in \mathrm{Hom}_R(M,M) = \mathrm{Hom}_B(M,M)$ such that $f(M) \subseteq NB \subseteq N$. Hence M is a quasi-simple B-module.

5.43 <u>Proposition.</u> Let R be a ring and let M be a faithful finite-dimensional quasi-simple R-module. Then every essential right ideal B of R contains a regular element.

<u>Proof</u>. Let $\hat{M} = \sum_{i=1}^{n} \mathbb{D} x_i$ for some $\mathbb{D}$-linearly independent subset $\{x_1, x_2, \ldots, x_n\}$ of M where $\mathbb{D} = \mathrm{Hom}_R(\hat{M},\hat{M})$. Let N

be the singular submodule of M. If $N \neq 0$, without loss of generality let $\{x_1, x_2, \ldots, x_t\} \subseteq N$ for some $1 \leq t \leq n$. Note that for each $1 \leq j \leq n$, $\bigcap_{\substack{i=1 \\ i \neq j}}^{n} (x_i)^{\perp} = 0$. Indeed, if $\bigcap_{\substack{i=1 \\ i \neq j}}^{n} (x_i)^{\perp} \neq 0$ for some j, then $\bigcap_{\substack{i=1 \\ i \neq j}}^{n} (x_i)^{\perp} \subseteq (x_j)^{\perp}$ and by 5.40, x_j is a linear combination of $\{x_i\}_{i \neq j}$, contradicting the $\mathbb{D}$-linear independence of $\{x_1, x_2, \ldots, x_n\}$. But $x_1 \in N$ implies $(x_1)^{\perp}$ is essential and $(x_1)^{\perp} \cap (\bigcap_{i=2}^{n} (x_i)^{\perp}) \neq 0$, which is a contradiction to the fact that M is faithful. Hence $N = 0$. Now let B be an essential right ideal of R. Then by 5.41 and 5.42 there exists $b \in B$ and $0 \neq k \in \mathrm{Hom}_B(M,M)$ such that $x_i b = k(x_i)$ for each $i = 1, 2, \ldots, n$. Since R can be embedded into $\mathrm{Hom}_{\mathbb{D}}(\tilde{M}, \tilde{M}) \equiv \mathbb{D}_n$ and $\ker b = 0$, b^{-1} exists in $\mathbb{D}_n$. Thus b is regular.

5.44 <u>Proposition.</u> Let R be a ring such that the right singular ideal of R is zero. If U is a uniform right ideal of R then for any $u \in U$ either $uU = 0$ or $(u)^{\perp} \cap U = 0$.

<u>Proof</u>. Suppose there exists $u \in U$ such that $uU \neq 0$ and $(u)^{\perp} \cap U \neq 0$. Let $A = (u)^{\perp} \cap U$. If $x = uy \neq 0$ for some $y \in U$ then there exists a non-zero right ideal I in R such that $(uy)^{\perp} \cap I = 0$. Hence $(y)^{\perp} \cap I = 0$ and $yI \cap A \neq 0$. Let $0 \neq t = yi = a$ for some $0 \neq i \in I$ and $a \in A$. Then $ut = uyi = 0$ and $(uy)^{\perp} \cap I \neq 0$, which is a contradiction.

5.45 <u>Proposition.</u> Let R be a semi-prime ring such that the right singular ideal of R is zero. If U is a uniform right ideal of R then U is quasi-simple.

Proof. By hypothesis the singular submodule of U is zero. Hence the singular submodule of $\tilde{U}$ is zero. Let $0 \neq f \in \mathrm{Hom}_R(\tilde{U},\tilde{U})$. Let $M = \ker f$ and suppose $M \neq 0$. Since $f \neq 0$, $f(\tilde{U}) \neq 0$. Let $x \in \tilde{U}$ such that $f(x) \neq 0$. Let $K = \{r \in R \mid xr \in M\}$. Since M is essential in $\tilde{U}$, K is an essential right ideal and $f(x)K = 0$, which is impossible. Hence there exists $g \in \mathrm{Hom}_R(\tilde{U},\tilde{U})$ such that $g \circ f = 1_{\tilde{U}}$ (the identity mapping on $\tilde{U}$). If $f \circ g \neq 1_{\tilde{U}}$, then $\tilde{U} = (1_{\tilde{U}} - f \circ g)\tilde{U} \oplus (f \circ g)\tilde{U}$. This is impossible. Therefore $g = f^{-1}$. Now let N be a non-zero submodule of U. Then $NU \neq 0$ since R is semiprime. Hence there exists $0 \neq t \in N$ such that $0 \neq tU \subseteq N$ and $(t)^{\perp} \cap U = 0$ by 5.44. Thus U is a quasi-simple R-module.

Remark. In fact, we have shown that if R is a ring and U is a uniform right ideal of R such that $Z(U) = 0$ and $U \cap \mathrm{rad}\, R = 0$, then U is quasi-simple.

5.46 Proposition. If R is a prime ring with a maximal annihilator right ideal then $Z_r(R) = Z_\ell(R) = 0$.

Proof. Let $Z = Z_r(R)$ and let $(a)^{\perp}$ be a maximal annihilator right ideal for some $0 \neq a \in R$. Suppose $Z \neq 0$. Then $aZa \neq 0$ since R is a prime ring. Hence there exists $z \in Z$ such that $aza \neq 0$ and $(aza)^{\perp} = (a)^{\perp}$ since $(a)^{\perp}$ is a maximal annihilator right ideal and $(a)^{\perp}$ is essential. Thus $(a)^{\perp} \cap zaR \neq 0$.

Let $0 \neq t \in (a)^{\perp} \cap zaR$. Then $t = zar_o \neq 0$ for some $r_o \in R$ and $0 = at = azar_o$. Hence $r_o \in (a)^{\perp} = (aza)^{\perp}$ and $t = 0$, which is a contradiction. Now let $Z = Z_{\ell}(R)$ and suppose $Z \neq 0$. Again there exists $z \in Z$ such that $0 \neq aza \in Z$ and $(za)^{\ell}$ is essential since $za \in Z$. Hence $(za)^{\ell} \cap Ra \neq 0$. Let $t \in (za)^{\ell} \cap Ra$ such that $t \neq 0$. Then $t = r_o a$ for some $r_o \in R$ and $0 = tza = r_o aza$. Since $(r_o a)^{\perp} = (a)^{\perp}$ then $aza = 0$, which is a contradiction.

5.47 Proposition. Let R be a prime ring such that $Z_r(R) = 0$ and let U be a uniform right ideal of R. Then $\mathrm{Hom}_R(\tilde{U},\tilde{U}) \cong \mathrm{Hom}_R(\hat{U},\hat{U}) \cong \mathrm{Hom}_{\hat{R}}(\hat{U},\hat{U})$.

Proof. Define a mapping $\phi:\mathrm{Hom}_R(\tilde{U},\tilde{U}) \to \mathrm{Hom}_R(\hat{U},\hat{U})$ by $\phi(f) = \hat{f}$ where $\hat{f}$ is an extension of f to $\hat{U}$. Note that ϕ is well defined, for if $f = g$ but $\hat{f} \neq \hat{g}$ then there exists $x \in \hat{U}$ such that $(\hat{f} - \hat{g})(x) \neq 0$ and $(\hat{f} - \hat{g})(\tilde{U}) = 0$. Since $\tilde{U}$ is essential in $\hat{U}$, $K = \{r \in R \mid (\hat{f}-\hat{g})(x)r \in \tilde{U}\}$ is an essential right ideal of R and $(f-g)(x)K = 0$. This means that $Z(\hat{U}) \neq 0$, which is a contradiction. ϕ is clearly a monomorphism. Let $h \in \mathrm{Hom}_R(\hat{U},\hat{U})$. The above argument shows that if $h(\tilde{U}) = 0$ then $h(\hat{U}) = 0$. Hence if $h \neq 0$, then $h(\tilde{U}) \cap \tilde{U} \neq 0$. Let $U_1 = \{x \in \tilde{U} \mid h(x) \in \tilde{U}\}$. Then U_1 is an essential submodule since it is a non-zero submodule of $\tilde{U}$. Let $h_1 = h|_{U_1}$ and let $\tilde{h}_1$ be the extension of h_1 to $\tilde{U}$. Then $\phi(\tilde{h}_1) = h$. Thus $\mathrm{Hom}_R(\tilde{U},\tilde{U}) \cong \mathrm{Hom}_R(\hat{U},\hat{U})$. Now it is easy to see that $\mathrm{Hom}_R(\hat{U},\hat{U}) \cong \mathrm{Hom}_{\hat{R}}(\hat{U},\hat{U})$.

5.48 <u>Proposition</u>. Let R be a primitive ring and let M be a faithful simple R-module. Let $\{m_1, m_2, \ldots, m_n\}$ be a finite linearly independent subset of M over $\mathbb{D}$ where $\mathbb{D} = \mathrm{Hom}_R(M,M)$. Then $\mathrm{Hom}_R(R/\bigcap_{i=1}^n (m_i)^\perp, R/\bigcap_{i=1}^n (m_i)^\perp)$ is isomorphic to the ring of $n \times n$ matrices over $\mathbb{D}$.

<u>Proof</u>. Let $M_n = \sum_{i=1}^n \oplus \mathbb{D}m_i$ and let $R^{(n)} = \{r \in R \mid M_n r \subseteq M_n\}$. Since R is a dense ring of linear transformations over $\mathbb{D}$, $R^{(n)}/M_n^\perp = \mathbb{D}_n$. Note that $N(M_n^\perp) = \{r \in R \mid r(M_n^\perp) \subseteq M_n^\perp\}$, where $N(M_n^\perp)$ is the largest two sided ideal of R containing $M_n^\perp$. Let $a \in N(M_n^\perp)$. Then $M_n a(M_n^\perp) = 0$ since $a(M_n^\perp) \subseteq M_n^\perp$ and $M_n a \subseteq (M_n^\perp)^\ell$ where $(M_n^\perp)^\ell = \{m \in M \mid m(M_n^\perp) = 0\}$. Since $(M_n^\perp)^\ell = M_n$, $a \in R^{(n)}$. If $a \in R^{(n)}$ then $M_n a \subseteq M_n$. Hence if $a \notin N(M_n^\perp)$ then there exists $x \in M_n^\perp$ such that $ax \notin M_n^\perp$. Then $M_n ax \neq 0$, but $M_n ax \subset M_n x = 0$, which is a contradiction. Thus $N(M_n^\perp) = R^{(n)}$. Clearly $M_n^\perp = \bigcap_{i=1}^n (m_i)^\perp$. Now there exists $e \in R$ such that $m_i e = m_i$ for every $i = 1,2,\ldots,n$ since R is a dense ring of linear transformations of M over $\mathbb{D}$. Hence $er - r \in \bigcap_{i=1}^n (m_i)^\perp$ for any $r \in R$ and $(e+M_n^\perp)R = RM_n^\perp$. Note that if $r \notin M_n^\perp$, then $rR \not\subseteq M_n^\perp$, so by 1.7, $N(M_n^\perp)/M_n^\perp \cong \mathrm{Hom}_R(R/\bigcap_{i=1}^n (m_i)^\perp, R/\bigcap_{i=1}^n (m_i)^\perp)$.

5.49 <u>Theorem</u>. Let R be a prime ring and let U be a uniform right ideal of R such that the singular submodule of U is zero. Let $\mathbb{D} = \mathrm{Hom}_R(\hat{N}, \hat{N})$. If $\{u_1, u_2, \ldots, u_n\}$ is a linearly independent subset of U over D, then $\mathrm{Hom}_R(R/\bigcap_{i=1}^n (u_i)^\perp, R/\bigcap_{i=1}^n (u_i)^\perp)$ is a right order in the $n \times n$ matrix ring over $\mathbb{D}$.

Proof. Let $\hat{R}$ denote the injective hull of R_R and let $(u_i)^r = \{q \in \hat{R} \mid u_i q = 0\}$. Then $(u_i)^\perp = (u_i)^r \cap R$ for each $i = 1,2,\ldots,n$. By 4.35, $\hat{U}$ is a right ideal of $\hat{R}$. Since $\hat{R}$ is a prime regular ring, $\hat{U}$ is a minimal right ideal of $\hat{R}$ and $\hat{R}$ is a primitive ring with a minimal right ideal $\hat{U}$. Let $I = \bigcap_{i=1}^{n} (u_i)^r$. Then by 5.48, $N(I)/I \cong \mathbb{D}_n$ where $\mathbb{D} = \mathrm{Hom}_R(U,U) = \mathrm{Hom}_R(\hat{U},\hat{U}) = \mathrm{Hom}_{\hat{R}}(\hat{U},\hat{U}) = \mathrm{Hom}_R(\tilde{U},\hat{U})$. Let $\mathring{I} = \bigcap_{i=1}^{n} (u_i)^\perp$. Then $N(\mathring{I})/\mathring{I}$ can be embedded into $N(I)/I$, for if $a + \mathring{I} \in N(\mathring{I})/\mathring{I}$ for some $a \in N(\mathring{I})$ then $aI \subseteq I$. Otherwise, there exists $x \in I$ such that $ax \notin I$ and $ax \notin (u_i)^r$ for some $1 \le i \le n$. Since $u_i a\mathring{I} = 0$ for all i, by 5.4 $u_i a \in \sum_{i=1}^{n} \oplus \mathbb{D}u_i$. In particular, $u_i ax = 0$, which is a contradiction. Now let $q \in N(I) \setminus I$ and let $I_q = \{r \in R \mid qr \in R\}$. If $J = I_q + \mathring{I}$ then J is an essential right ideal of R. Hence by 5.41 and 5.42, there exist $a \in J$ and $0 \neq k \in \mathbb{D}$ such that $u_i a = ku_i$ for every $i = 1,2,\ldots,n$. Now $a = x+y$ for some $x \in I_q$ and $y \in \mathring{I}$. Therefore $qa = qx + qy$ where $qx \in R$ and $qy \in I$. Let $b = qx$. Since $q \in N(I)$, $qa = b \bmod I$ and $q + I = (b+I)(a+I)^{-1}$. Thus $N(\mathring{I})/\mathring{I}$ is a right order in $N(I)/I$.

5.50 Definition. Let R be a ring and let C be a right ideal of R. Then C is said to be a complement right ideal provided there exists a right ideal B of R such that $B \cap C = 0$ and if C' is a right ideal of R such that $C \subsetneq C'$ then $C' \cap B \neq 0$.

5.51 Definition. A ring R is called a right Goldie ring if and only if the following conditions hold.

(i) R satisfies the ascending chain condition on annihilator right ideals.

(ii) R satisfies the ascending chain condition on complement right ideals.

5.52 Proposition. Let R be a ring such that $Z_r(R)$ is zero. If A and B are closed right ideals of R then $A \cap B$ is a closed right ideal.

Proof. First we note that if B is closed, then $B = \{r \in R \mid r^{-1}B$ is an essential right ideal of $R\}$. Clearly if $b \in B$, then $b^{-1}B = R$ is essential. Also, if $x^{-1}B$ is essential and $r \in R$, then $(xr)^{-1}B = \{s \in R \mid xrs \in B\} = \{s \in R \mid rs \in x^{-1}B\} = r^{-1}(x^{-1}B)$, which is an essential right ideal. Suppose $x^{-1}B$ is essential and $x \notin B$. Then $B \subsetneq B + xR$. Hence there exists $b \in B$ and $x_o \in xR$ such that $(b+x_o)R \cap B = 0$ and $b + x_o \neq 0$. Note that $x_o^{-1}B$ is essential and $(b+x_o)(x_o^{-1}B) \subseteq B$. Since $Z_r(R) = 0$, $(b+x_o)(x_o^{-1}B) \neq 0$ and $(b+x_o)(x_o^{-1}B) \subseteq (b+x_o)R \cap B = 0$, which is a contradiction. Now suppose A and B are closed and $A \cap B \subseteq D$, where D is an essential extension of $A \cap B$ in R. Then for any $d \in D$, $d^{-1}(A \cap B)$ is an essential right ideal. Otherwise, there exists a right ideal $H \neq 0$ such that $d^{-1}(A \cap B) \cap H = 0$. But if $dH = 0$, then $H \subseteq d^{-1}(A \cap B)$, and if $dH \neq 0$ then $dH \cap (A \cap B) \neq 0$. Hence there exists

$0 \neq h \in H$ such that $dh \in A \cap B$ and $h \in d^{-1}(A \cap B)$, which is a contradiction. Now $d^{-1}(A \cap B) \subseteq d^{-1}A \cap d^{-1}B$. Thus $d^{-1}A$ and $d^{-1}B$ are both essential right ideals and $d \in A \cap B$. Therefore $D \subseteq A \cap B$ and $A \cap B$ is closed.

5.53 Proposition. Let R be a ring such that $Z_r(R) = 0$. R satisfies the ascending chain condition on complement right ideals if and only if R satisfies the descending chain condition on closed right ideals.

Proof. Suppose R satisfies the ascending chain condition on complement right ideals and let $B_1 \supseteq B_2 \supseteq B_3 \supseteq \ldots$ be a descending chain of closed right ideals. Let A_1 be a maximal right ideal such that $A_1 \cap B_1 = 0$. Select a sequence $\{A_i\}_{i=1}^{\infty}$ of right ideals such that for each $i = 1,2,\ldots$ $A_i \subseteq A_{i+1}$ and A_{i+1} is maximal with respect to the property that $A_{i+1} \cap B_{i+1} = 0$. Clearly each A_i is a complement right ideal. Hence the chain $A_1 \subseteq A_2 \subseteq A_3 \subseteq \ldots$ terminates. Since the B_i's are closed, $B_1 \supseteq B_2 \supseteq B_3 \supseteq \ldots$ also terminates.

Conversely let $A_1 \subseteq A_2 \subseteq A_3 \subseteq \ldots$ be a chain of complement right ideals of R. Select a sequence $\{C_i\}_{i=1}^{\infty}$ of right ideals such that $C_i \cap A_i = 0$ and C_i is maximal with respect to this property for each $i = 1,2,\ldots$. Note that each C_i is closed, so the chain $C_1 \supseteq C_1 \cap C_2 \supseteq C_1 \cap C_2 \cap C_3 \supseteq \ldots$ must terminate. Hence $A_1 \subseteq A_2 \subseteq A_3 \subseteq \ldots$ also terminates.

5.54 Proposition. Let R be a prime right Goldie ring. If

U is a uniform right ideal of R, then there exists a finite subset $\{u_1,u_2,\ldots,u_n\}$ of U such that $\bigcap_{i=1}^{n}(u_i)^{\perp} = 0$.

Proof. Let u be a non-zero element of U. We claim that $(u)^{\perp}$ is a closed right ideal. Note that $Z_r(R) = 0$ by 5.46. Hence $u \notin Z_r(R)$ and there exists a non-zero right ideal A of R maximal with respect to the property that $A \cap (u)^{\perp} = 0$. Suppose D is a proper essential extension of $(u)^{\perp}$ in R. Note that $D \cap A = 0$ since D is an essential extension of $(u)^{\perp}$. Also $uD \neq 0$ and $uA \neq 0$. Since U is uniform, there exists $d \in D$ and $a \in A$ such that $0 \neq ud = ua \in uD \cap uA$. Hence $d - a \in (u)^{\perp} \subseteq D$ and $a \in D$, which is a contradiction to $D \cap A = 0$. Hence $(u)^{\perp}$ is a closed right ideal of R. Now let C be the collection of all finite intersections of annihilators of elements of U. By 5.53, C has a minimal element $\bigcap_{i=1}^{n}(u_i)^{\perp}$. Let $s \in \bigcap_{i=1}^{n}(u_i)^{\perp}$. Since R is a prime ring, $(U)^{\perp} = 0$. Hence if $s \neq 0$, there exists $u \in U$ such that $us \neq 0$. Then $\bigcap_{i=1}^{n}(u_i)^{\perp} \cap (u)^{\perp} \subsetneq \bigcap_{i=1}^{n}(u_i)^{\perp}$, which is a contradiction.

5.55 Corollary (Goldie). If R is a right prime Goldie ring then R is a right order in the $n\times n$ matrix ring over a division ring for some positive integer n.

Proof. Suppose R has no uniform right ideal. For any right ideal I of R, let $\bar{I}$ denote the least closed right ideal of R containing I (by 5.53, $\bar{I}$ exists for every right ideal I).

Since R is not uniform, there exist non-zero right ideals I_1 and J_1 such that $I_1 \oplus J_1 \subsetneqq R$. Again I_1 is not uniform and there exist non-zero right ideals I_2 and J_2 such that $I_2 \oplus J_2 \subsetneqq I_1$. By continuing this process, we can obtain a sequence $\{I_i\}_{i=1}^{\infty}$ of right ideals such that $\bar{I}_1 \supsetneqq \bar{I}_2 \supsetneqq \bar{I}_3 \supsetneqq \cdots$ is an infinite chain. This is a contradiction. So let U be a uniform right ideal of R. Then there exists a finite subset $\{u_1, u_2, \ldots, u_n\}$ of u such that $\bigcap_{i=1}^{n} (u_i)^{\perp} = 0$. By 5.40 it suffices to choose $\{u_1, u_2, \ldots, u_n\}$ to be a linearly independent set over $\mathbb{D}$, where $\mathbb{D} = \mathrm{Hom}_R(\tilde{U}, \tilde{U}) = \mathrm{Hom}_R(\hat{U}, \hat{U})$. If R has an identity then $R \cong \mathrm{Hom}_R(R, R)$ and the assertion follows from 5.49. If R has no identity, for $q \in \mathbb{D}_n$ there exist $f, g \in \mathrm{Hom}_R(R, R)$ such that $q = fg^{-1}$ by 5.49. For each $a \in R$ define $T_a(x) = ax$ for all $x \in R$. Let a be a regular element of R. Then $qT_{g(a)} = qgT_a = fT_a = T_{f(a)}$. Hence $q = T_{f(a)} T^{-1}_{g(a)}$ and R is a right order in $\mathbb{D}_n$.

Bibliography

1. Faith, C., Lectures on Injective Modules and Quotient Rings, Springer-Verlag, 49 (1967).

2. Faith, C. and Y. Utumi, "Maximal Quotient Rings", Proc. Amer. Math. Soc., 16(1965), 1084-1089.

3. Goldie, A. W., "Semiprime Rings with Maximum Condition", Proc. London Math. Soc., 10(1960), 201-220.

4. Jacobson, N., Structure of Rings, Amer Math. Soc. Colloq. Pub. (1956).

5. Johnson, R. E., "The Extended Centralizer of a Ring Over a Module", Proc. Amer. Math. Soc., 2(1951) 891-895.

6. ____________, Structure Theory of Faithful Rings III, Irreducible Rings, Proc. Amer. Math. Soc., 2(1960), 710-717.

7. ____________, "Quotient Rings of Rings with Zero Singular Ideal", Pacific J. Math., II(1961), 1385-1392.

8. Johnson, R. E. and E. T. Wong, "Quasi-Injective Modules and Irreducible Rings", J. London Math. Soc., 36(1961), 260-268.

9. Koh, K., "On Almost Maximal Right Ideals, Proc. Amer. Math. Soc., 25(1970), 266-272.

10. ____________, "On the Annihilators of the Injective Hull of a Module", Canad. Math. Bull., vol. 12, No. 6, 1969.

11. Koh, K. and A. C. Mewborn, "Prime Rings with Maximal Annihilator and Maximal Complement Right ideals," Proc. Amer. Math. Soc., 16(1965), 1073-1076.

12. _______________, "A Class of Prime Rings", Can. Math. Bull., 9(1966), 63-72.

13. _______________, "The Weak Radical of a Ring", Proc. Amer. Math. Soc., 18(1967), 554-559.

14. Koh, K. and J. Luh, "On a Finite Dimensional Quasi-Simple Module", Proc. Amer. Math. Soc., 25(1970), 801-807.

15. Osofsky, B., "On Ring Properties of Unjective Hulls", Can. Math. Bull., 7(1964), 405-413.

16. Utumi, Y., "On Continuous Regular Rings and Semisimple Self Injective Rings", Can. J. Math., 12(1960), 597-605.

17. ________, "On Rings of Which Any One-Sided Quotient Rings are Two-Sided", Proc. Amer. Math. Soc., 14(1963), 141-147.

BLOCKS AND CENTERS OF GROUP ALGEBRAS

by

G. O. Michler

Department of Mathematics
University of Tübingen
Tübingen, Germany

Introduction. These lectures present a ringtheoretical approach to R. Brauer's theory of blocks of finite groups (see [4] till [7]) using A. Rosenberg's definition of the defect groups of a block and the methods of his paper [36]. The last two sections of these lecture notes contain applications of the theory to several questions in the study of group algebras of finite groups over arbitrary fields of characteristic $p > 0$.

In section 1 the definition and the main properties of a block ideal of an Artinian ring are given. Section 2 contains the definition of a block $B \iff e \iff \lambda$ of the group algebra FG of the finite group G over the (arbitrary) field F with characteristic $p > 0$ dividing the order $|G|$ of G. If p does not divide $|G|$, all the results of these lectures remain formally valid but become trivial. In section 3 the defect groups $\delta(B)$ of a block $B \iff e \iff \lambda$ are defined and shown to be uniquely determined by $B \iff e \iff \lambda$ up to conjugacy by elements of G. R. Brauer's first main theorem on blocks (Theorem 4.12) is proved in section 4 following closely A. Rosenberg's ringtheoretical arguments [36]. In section 5 we give D. S. Passman's proof [29] of Osima's theorem [27] asserting that every central idempotent e of the group algebra FG

is a linear combination of the class sums of the p-regular conjugacy classes of G. Section 6 contains W. Hamernik's and the author's generalization [18] of P. Fong's theorem [11] on blocks and normal subgroups to group algebras FG over arbitrary fields F (see Theorem 6.11). As an application of Fong's theorem in section 7 R. Brauer's generalized first main theorem ([6], Theorem 5C) is proved using D. S. Passman's paper [30]. Section 8 contains W. Hamernik's and the author's proof [18] of R. Brauer's main theorem on blocks with normal defect groups ([4], Theorem 12A) reducing the question of counting the blocks $B \leftrightarrow e \leftrightarrow \lambda$ of FG with a given p-subgroup D as one of their defect groups to the question of counting certain blocks of defect zero of a certain group of smaller order (see Theorem 8.3). In section 9 the author's relation [24] between the number b of all the blocks $B \leftrightarrow e \leftrightarrow \lambda$ of the group algebra FG with a given p-subgroup D as one of their defect groups $\delta(B)$ and the number c of all the p-regular conjugacy classes C with defect groups $\delta(C)$ G-conjugate to D is proved (see Theorem 9.4). As corollaries of it, we obtain all the results by R. Brauer [4] and R. Brauer and C. Nesbitt [3] estimating the number b of blocks of FG with a given defect d by the number c of conjugacy classes of G with defect d, including R. Brauer's Theorem 13A of [4] on the number of blocks with defect zero.

Therefore these notes contain all the results of R. Brauer's paper [4] on blocks of the group algebra FG which do not contain assertions on characters of the group G. Our proofs operate solely within the group algebra FG and do not use any result proved by means of the theory of characters of finite groups. Moreover, they do not use any result of modular representation theory dealing with fields of characteristic zero.

In order to keep these lectures as self-contained as possible, Brauer's theorem on the number of simple FG-modules is proved in section 10 and Clifford's theorem and the Frobenius reciprocity theorem are presented in section 11. Section 12 contains the author's application [25] of the theory of blocks developed in these notes so far to the study of the structure of the group algebras FG of finite p-nilpotent groups G over arbitrary fields F of characteristic $p > 0$. The main result (Theorem 12.17) of this section gives the construction of the (up to isomorphisms) unique indecomposable projective FG-module T of every block $B \leftrightarrow e \leftrightarrow \lambda$ and determines the Jacobson radical $J(FG)$ of FG. In the course of its proof using the Wedderburn-Malcev theorem an easy ringtheoretical proof is given for W. F. Reynolds' recent result [34] asserting that every block $B \leftrightarrow e \leftrightarrow \lambda$ of FG contains only one simple FG-module (Corollary 12.16). As another corollary we obtain D. A. R. Wallace's criterion [41] for the normality of the p-Sylow subgroups in a p-nilpotent group (Theorem 12.20). In the last section we present H. Spiegel's elementary proof [37] for D. A. R. Wallace's theorem [42] determining the structure of all finite groups G having the property that the Jacobson radical $J(FG)$ of the group algebra FG over any field F of characteristic $p > 0$ is contained in the center ZFG of FG (Theorem 13.8).

Throughout these lectures it is assumed that all rings R have an identity element and that R-modules are unital right R-modules. The reader is only assumed to know the basic results on Artinian rings, completely reducible modules and (only in section 12) on separable algebras. As a reference for these results we mention J. Lambek's book [23] and Curtis and Reiner [9], Chapter 10. From the theory of finite groups we only use well-known standard results which one can find in M. Hall [17] or B. Huppert [19].

The author is grateful to the Department of Mathematics of Tulane University for the financial support and all the help during his visit. In particular, I should like to thank Mrs. Meredith R. Mickel for her beautiful typing. I am also indebted to my student, Wolfgang Hamernik, for several suggestions and his valuable help with the proofs.

0. Notation

$\mathbb{Z}$	:	ring of rational integers
R	:	ring with identity
J(R)	:	Jacobson radical of R
M_R	:	unital right R-module M
$N \cong_R M$	:	R-module isomorphism
F	:	commutative field
$\|X\|$	:	number of elements of the finite set X.
G	:	finite group of order $\|G\| = p^a q$, where $(p,q)=1$
FG	:	group algebra of G over F
RG	:	group ring of G over R
ZFG	:	center of the group algebra FG
J(FG)	:	Jacobson radical of the group algebra FG
k	:	number of conjugacy classes K_i of G
c_i	:	$= \sum_{g \in K_i} g \in FG$ class sum of conjugacy class K_i
$N_G(X)$	:	normalizer of the subset X of G
$C_G(X)$	:	centralizer of the subset X of G
$U \leq V$	:	U is a subset of V
$U < V$	:	U is a proper subset of V
U^g	:	gUg^{-1} for $g \in G$ and $U \leq G$
U^D	:	$\{u^d \in G \mid u \in U, d \in D\}$, where $U,D \leq G$
$G' = [G,G]$	:	commutator subgroup of G
$U \underset{G}{\leq} V$	:	U is conjugate to a subset of V
$U \underset{G}{<} V$	:	U is conjugate to a proper subset of V

$G_1 \times G_2$	:	direct product of the groups G_1 and G_2
$\sum_{i=1}^{n} \oplus A_i$	:	direct sum of ideals or modules
$A \otimes_R B$	:	tensor product of A_R and ${}_RB$ over R
$a \mid b$	:	a divides b, $a,b \in \mathbb{Z}$
$a \nmid b$	:	a does not divide b, $a,b \in \mathbb{Z}$
$\dim_F A$	:	F-vector space dimension of A
$\emptyset$	:	empty set
$\nu(\lvert G\rvert)$	:	highest power of p dividing the order $\lvert G\rvert$ of G.
M^s	:	direct sum of s copies of M_R

§1. Block ideals

Throughout this section R is a right Artinian ring. Its Jacobson radical $J(R)$ is the intersection of all maximal right ideals. It is well known that $J(R)$ is also the intersection of all maximal left ideals of R, and that $J(R)$ is a nilpotent ideal, if R is right Artinian (see e.g. [23]).

Definition. The idempotent e of the ring R is primitive, if eR is not a direct sum of two nonzero proper R-submodules.

Without proof we state the following well known result on right Artinian rings (see e.g. [23], p.55-79).

Theorem 1.1. Let $1 = e_1 + e_2 + \dots + e_n$ be a decomposition of the identity 1 of the right Artinian ring R into primitive orthogonal idempotents $e_i \in R$. Suppose that $\{e_jR \mid j = 1,2,\dots,m \le n\}$ is a full set of non-isomorphic among the right ideals e_iR, $i = 1,2,\dots,n$. Then

a) $e_iJ(R)$ is the unique maximal R-submodule of e_iR, $i = 1,2,\dots,n$.

b) $e_jR/e_jJ(R)$, $j = 1,2,\dots,m$, is a full set of non-isomorphic simple right R-modules.

c) $e_iR \cong_R e_tR$ if and only if $e_iR/e_iJ(R) \cong_R e_tR/e_tJ(R)$.

d) The finitely generated right R-module M has a composition factor which is isomorphic to $e_jR/e_jJ(R)$ as a right R-module if and only if $Me_j \neq 0$.

Definition. The primitive idempotents e and f of the right Artinian R are linked, if there is a sequence

$$e = e_1, e_2, e_3, \ldots, e_{n-1}, e_n = f$$

of primitive idempotents e_i of R such that e_iR and $e_{i+1}R$ have a common composition factor for $i = 1,2,\ldots,n-1$.

Remark. "Linkedness" of primitive idempotents of R is an equivalence relation.

Definition. The idempotent e of the ring R is called centrally primitive, if e is a primitive idempotent of the center Z of the ring R.

Definition. The two-sided ideal $K \neq 0$ of the ring R is a block ideal of R, if

1) $R = K \oplus K'$ for some two-sided ideal K' of R.
2) K is not a direct sum of two proper ideals $K_i < K$, $i = 1,2$, of R.

Lemma 1.2. The two-sided ideal K of the ring R is a block ideal of R if and only if $K = eR$, where e is a centrally primitive idempotent of R.

Proof. If K is a block ideal, then $R = K \oplus K'$ for some two-sided ideal K' of R. Thus $1 = k + k'$. Since K and K' are two-sided ideals of R the idempotents k and k' belong to the center Z of R. If k were not primitive in Z, then K would not be a block ideal of R. The converse is trivial.

Lemma 1.3. Every right Artinian ring R has only finitely many block ideals B_i, $i = 1,2,\ldots,n$, and

$$R = B_1 \oplus B_2 \oplus \ldots \oplus B_n.$$

Proof follows at once by complete induction.

Lemma 1.4. Let e be a central idempotent of the ring R contained in the ideal A of R. Then the decompositions of e into centrally primitive idempotents of A and of R are the same.

Proof. Let $e = \sum_{i=1}^{s} f_i$ where the f_i are orthogonal centrally primitive idempotents of R. Then

$$f_j = f_j^2 = \sum_{i=1}^{s} f_i f_j = ef_j \in A \quad \text{for } j = 1,2,\ldots,s.$$

Conversely, if $e = \sum_{i=1}^{s} f_i$ where the f_i are orthogonal primitive idempotents of the center Z(A) of the ideal A of R, then B = eR is a two-sided ideal of R contained in A, and by the above argument f_j belongs to the center Z(B) of B for $j = 1,2,\ldots,s$. Since $R = B \oplus (1-e)R$ and $Z(R) = Z(B) \oplus Z[(1-e)R]$, the idempotents f_i are centrally primitive idempotents of R.

Lemma 1.5. Let f be a primitive idempotent and let e be a centrally primitive idempotent of the ring R. Then $f \in eR$

if and only if $ef \neq 0$.

Proof. $f \in eR$ implies $0 \neq f = ef$, because e is the identity of the ring eR. Suppose that $ef \neq 0$. Clearly, $fR = feR \oplus f(1-e)R = efR \oplus f(1-e)R$. Since fR is an indecomposable right R-module and since $0 \neq efR$, it follows that $f(1-e)R = 0$. Thus $f \in fR = efR \leq eR$.

Lemma 1.6. Every primitive idempotent f of the right Artinian ring R belongs to exactly one block ideal B_i of R.

Proof. The result is clear in view of Lemma 1.5 and Lemma 1.3.

Theorem 1.7. The primitive idempotents e and f of the right Artinian ring R are linked if and only if e and f belong to the same block ideal B of R.

Proof. If e and f are linked, then there exist primitive idempotents $e = e_1, e_2, \ldots, e_n = f$ of R such that e_iR and $e_{i+1}R$ have a common composition factor for $i = 1,2,\ldots,n-1$. By Lemma 1.6 there is a uniquely determined block ideal B of R such that $e \in B$. Since linkedness is an equivalence relation, we may assume that $n = 2$, because an inductive argument will then prove the general case. Therefore eR and fR have a composition factor tR/tJ in common, where t is a primitive idempotent of R and J is the Jacobson radical of R. Hence $eRt \neq 0 \neq fRt$ by Theorem 1.1. Let g be the centrally primitive idempotent of B. Then $e = eg$ by Lemma 1.5. Thus

$$0 \neq eRt = egRt = eRgt$$

implies $gt \neq 0$. Hence $t \in B$ by Lemma 1.5. Therefore $0 \neq fRt \leq B$, and $gf \neq 0$. Thus $f \in B$ by Lemma 1.5.

Conversely, suppose that e and f belong to the same block ideal B of R. Let $1 = e_1 + e_2 + \ldots + e_n$ be a decomposition of the identity 1 of R into orthogonal primitive idempotents e_i of R, where $e = e_1$. Suppose that e_1, $e_2, \ldots, e_k$ are linked to e_1, but e_j is not linked to e_1 for $j = k+1, k+2, \ldots, n$. Then for every $i \in \{1,2,\ldots,k\}$ and $j \in \{k+1, k+2, \ldots, n\}$, e_i is not linked to e_j, because linkedness is an equivalence relation. Hence $e_iRe_j = 0 = e_jRe_i$ by Theorem 1.1, because e_iR and e_jR do not have a common composition factor. If $T = \sum_{i=1}^{k} e_iR$ and $S = \sum_{j=k+1}^{n} e_jR$, then $R = T + S$, and $TS = 0 = ST$. Therefore R is the direct sum of the two-sided ideals S and T. Since $e_1 \in B$ and since $e_2, e_3, \ldots, e_k$ are linked to e_1, the proof of the first part of Theorem 1.7 implies that $e_i \in B$ for $i = 1, 2, \ldots, k$. Thus $T \leq B$, and $B = T \oplus (S \cap B)$. As B is a block ideal of R, it follows that $B = T$. Since $f \in B$, we therefore obtain elements $r_i \in R$ such that

$$f = e_1r_1 + e_2r_2 + \ldots + e_kr_k.$$

Thus $e_if \neq 0$ for at least one $i \in \{1,2,\ldots,k\}$, and $e_iRf \neq 0$. By Theorem 1.1 the right ideals e_iR and fR have a common composition factor. Hence e_iR and fR are linked. Since

e_1R is linked to e_iR and linkedness is an equivalence relation, the right ideals e_1R and fR are linked.

Definition. The simple right R-module M belongs to the block ideal B of R, if M is isomorphic (as a right R-module) to some composition factor of the right R-module B.

Corollary 1.8. Let R be a right Artinian ring with Jacobson radical J. Then the following properties of the primitive idempotents e and f of R are equivalent:

(1) e and f are linked.

(2) eR/eJ and fR/fJ belong to the same block ideal B of R.

(3) e and f are contained in the same block ideal B of R.

Proof. The equivalence of (1) and (3) follows from Theorem 1.7. Clearly (2) follows from (3). Let g be the centrally primitive idempotent of B. Since eR/eJ and fR/fJ are composition factors of the right R-module B, it follows from Theorem 1.1 that $Be \neq 0 \neq Bf$. Thus $fg \neq 0 \neq eg$ and $e,f \in B$ by Lemma 1.5.

Because of Theorem 1.1 and Corollary 1.8 two simple R-modules F_1 and F_2 are called linked, if F_1 and F_2 belong to the same block ideal B of R.

Corollary 1.9. For every right Artinian ring R there are the following one-to-one correspondences:

$$\left\{\begin{array}{c}\text{classes of linked}\\ \text{simple}\\ \text{right R-modules}\end{array}\right\} \longleftrightarrow \{\text{block ideals}\} \longleftrightarrow \left\{\begin{array}{c}\text{centrally primitive}\\ \text{idempotents}\end{array}\right\}$$

<u>Proof</u> is obvious in view of Corollary 1.8 and Lemma 1.2.

§2. Linear characters

In this section the definition of a block of the group algebra FG of the finite group over the (arbitrary) field F of characteristic $p > 0$ is given.

Since FG is a finite-dimensional F-algebra, all its F-subalgebras are Artinian rings. In particular, FG has only finitely many block ideals B_i, $i = 1,2,\dots,n$, by Lemma 1.3. Let e_i be the centrally primitive idempotent of B_i. Then the center ZFG of FG is decomposed into the direct sum

$$ZFG = \sum_{i=1}^{n} \oplus\ e_i ZFG,$$

and the rings $e_i ZFG$ are local finite-dimensional F-algebras with Jacobson radical $e_i J(ZFG)$, where $J(ZFG)$ denotes the Jacobson radical of ZFG. Thus $F_i = e_i ZFG/e_i J(ZFG)$ is a finite extension of the field F for $i = 1,2,\dots,n$.

Definition. The natural homomorphism λ_i from the center ZFG of the group algebra FG onto the finite extension $F_i = e_i ZFG/e_i J(ZFG)$ is called the linear character of the group algebra FG belonging to the centrally primitive idempotent e_i.

Remark 2.1. If $\{e_i \mid i = 1,2,\dots,n\}$ is the set of all centrally primitive idempotents of the group algebra FG, and if λ_i is the linear character of FG belonging to e_i, then

$$\lambda_i(e_j) = \begin{cases} 0 \in F_i, & \text{if } i \neq j \\ 1 \in F_i, & \text{if } i = j \end{cases}$$

This follows at once from the fact that $\{e_i \mid i = 1,2,\ldots,n\}$ consists of orthogonal primitive idempotents of ZFG.

More generally than the above definition is the following

Definition [36]. A linear character of the finite dimensional algebra R over the field F is an F-algebra homomorphism of the center $Z(R)$ of R into a finite extension field of F.

Proposition 2.2. If $\{e_i \mid i = 1,2,\ldots,n\}$ is the set of all centrally primitive idempotents of the group algebra FG, and if λ_i is the linear character of FG belonging to e_i, then for every linear character λ of the center ZFG of FG there is an F-algebra isomorphism ψ from $\lambda(ZFG)$ onto $\lambda_i(ZFG)$ for exactly one $i \in \{1,2,\ldots,n\}$ such that $\psi\lambda = \lambda_i$.

Proof. Let $\bar{F}$ be the algebraic closure of F. Then we may assume that $\lambda(ZFG) \leq \bar{F}$. Thus $\lambda(ZFG)$ is an Artinian ring without zero-divisors, and so it is a field. Hence $\ker \lambda$ is a maximal ideal of ZFG. Now $ZFG = \sum_{i=1}^{n} \oplus\, e_i ZFG$, and $e_i ZFG$ is a local Artinian ring with Jacobson radical $e_i J(ZFG)$, where $J(ZFG)$ is the Jacobson radical of ZFG. Hence ZFG has exactly n maximal ideals

$$M_i = e_i J(ZFG) \oplus \sum_{\substack{j=1 \\ j \neq i}}^{n} e_j ZFG$$

Hence $\ker \lambda = M_i$ for exactly one $i \in \{1,2,\ldots,n\}$. Clearly, $\lambda(ZFG) \cong ZFG/\ker \lambda \cong ZFG/M_i = \lambda_i(ZFG)$. Therefore, the result is now obvious in view of Remark 2.1.

Definition. Two linear characters λ and μ of the center ZFG of the group algebra FG are called equivalent, if there exists an F-algebra isomorphism ψ from $\lambda(ZFG)$ onto $\mu(ZFG)$ with $\psi\lambda = \mu$.

Because of Proposition 2.2 the center ZFG of the group algebra FG has only n equivalence classes of linear characters, and $\{\lambda_i \mid i = 1,2,\dots,n\}$ is a full set of representatives of them. Therefore Corollary 1.9 implies

Corollary 2.3. For every group algebra FG there are the following one-to-one correspondences:

{block ideals B}<->{centrally primitive idempotents e}<->{F-equivalence classes of linear characters λ}

Definition. A block B <-> e <-> λ of the group algebra FG of the finite group G over the field F is the triple B <-> e <-> λ, where B is a block ideal of FG, e is its centrally primitive idempotent and λ is a representative of the F-equivalence class of the linear character

$$\lambda(e) : ZFG \to e(ZFG)/eJ(ZFG).$$

of the center ZFG of FG.

By Proposition 2.2 and Corollary 2.3, every group algebra FG of a finite group G has only finitely many blocks B_i <-> e_i <-> λ_i, $i = 1,2,\dots,n$.

§3. Defect groups

In this section A. Rosenberg's definition (see [36]) of the defect groups $\delta(B)$ of a block $B \leftrightarrow e \leftrightarrow \lambda$ of the group algebra FG of the finite group G over the field F of characteristic $p > 0$ is given, and it is shown that they are uniquely determined by $B \leftrightarrow e \leftrightarrow \lambda$ up to conjugacy by elements g of G.

3.1 Notation. By K_i, $i = 1,2,\ldots,k$ we denote the k conjugacy classes of G, and $c_i = \sum_{g \in K_i} g \in FG$ is the class sum of the i-th conjugacy class of G.

Lemma 3.2. The class sums c_i of the k conjugacy classes K_i of G form an F-vector space basis of the center ZFG of the group algebra FG.

Proof. Clearly the elements c_i, $i = 1,2,\ldots,k$ are k linearly independent elements of ZFG over F. Let $z \in ZFG$, and $z = \sum_{g \in G} r_g g$, $r_g \in F$. Then $z = hzh^{-1}$ for every $h \in G$. Since FG is a free F-module with basis G, it follows that $r_g = r_{g'}$ whenever g and g' belong to the same conjugacy class K_i of G. Thus $z = \sum_{i=1}^{k} r_i c_i$, and Lemma 3.2 holds.

Remark. Lemma 3.2 and its proof hold for group rings over commutative rings.

Lemma 3.3. With the notation of 3.1 the following statement holds:

If g_t is a fixed element of the conjugacy class K_t for $t = 1,2,\ldots,k$, then

$$c_i c_j = \sum_{t=1}^{k} a_{ij}^{(t)} c_t,$$

where $a_{ij}^{(t)}$ is modulo p equal to the number of solutions $(x,y) \in K_i \times K_j$ of $xy = g_t$.

Proof is trivial.

Definition 3.4. Let g be a fixed element of the conjugacy class K of G. Then a defect group of K is a p-Sylow subgroup $\delta(K)$ of $C_G(g)$. If $p^{d(K)} = |\delta(K)|$, then $d(K)$ is the defect of K.

Since the p-Sylow subgroups are conjugate, and since $hC_G(g)h^{-1} = C_G(h_g h^{-1})$ for every $h \in G$, the defect groups $\delta(K)$ of K are uniquely determined by K up to G-conjugacy. Therefore the defect $d(K)$ of K is uniquely determined by K.

Definition: Let D be a subgroup of G, then the subset X of G is D-invariant, if $X^D = X$.

Lemma 3.5. Let D be a p-subgroup of G, and let X and Y be D-invariant subsets of G. For every $t \in C_G(D)$ let

$$S_t = \{(x,y) \in X \times Y \mid xy = t\}.$$

If $Y \cap C_G(D) = \phi$, then $|S_t| \equiv 0 \bmod p$.

Proof. Define an equivalence relation $\sim$ on S_t by $(x,y) \sim (a,b)$ if $(a,b) = (x^d,y^d)$ for some $d \in D$. For every pair $(x,y) \in X \times Y$ let $T(x,y) = \{d \in D \mid (x^d,y^d) = (x,y)\}$, and let $A(x,y)$ be the $\sim$ equivalence class of (x,y). Then $|A(x,y)| = |\{(x^d,y^d) \mid d \in D\}| = |D| : |T(x,y)|$. If p^b is the order of D then $|A(x,y)| = p^h$ for some integer $0 \leq h \leq b$, because $T(x,y) = D \cap C_G(x) \cap C_G(y)$. Since $Y \cap C_G(D) = \phi$, $D \nsubseteq C_G(y)$. Hence $T(x,y) < D$ and $1 \neq p^h = p^b : |T(x,y)|$. Thus $|A(x,y)| \equiv 0 \bmod p$. Therefore $|S_t| \equiv 0 \bmod p$, because S_t is the disjunct union of the equivalence classes $A(x,y)$.

Definition. Since by Lemma 3.2 every $z \in ZFG$ has the unique representation $z = \sum_{i=1}^{k} r_i c_i$, where $r_i \in F$, the set $\text{Sup } z = \{c_i \mid r_i \neq 0\}$ is called the central support of z.

If $x = \sum_{g \in G} x_g g \in FG$, then the support of x is the set $\text{sup } x = \{g \in G \mid x_g \neq 0\}$.

Lemma 3.6 (Osima [27]). For every p-subgroup D of G the sets $J_D = \{z \in ZFG \mid c_i \in \text{Sup } z \text{ implies } \delta(K_i) \leq_G D\}$, and $\hat{J}_D = \{z \in ZFG \mid c_i \in \text{sup } z \text{ implies } \delta(K_i) <_G D\}$ are ideals of the center ZFG of the group algebra FG.

Proof. Clearly J_D is an F-vector space. Let $z = \sum_{i=1}^{k} r_i c_i$, $r_i \in F$, be any element of J_D, and let c_j be any class sum of G. Then $zc_j \in J_D$ if $c_i c_j \in J_D$ for every $c_i \in \text{Sup } z$. By Lemma 3.3

$$c_i c_j = \sum_{t=1}^{k} a_{ij}^{(t)} c_t ,$$

where $a_{ij}^{(t)}$ is modulo p equal to the number of all ordered pairs $(x,y) \in K_i \times K_j$ such that $xy = g_t$ for some fixed element $g_t \in K_t$. If $c_t \in \text{Sup}\, c_i c_j$, then $a_{ij}^{(t)} \not\equiv 0 \bmod p$. Let $\delta(K_t) = H$ be a defect group of K_t. Then K_i and K_j are H-invariant sets and $g_t \in C_G(H)$. If $K_i \cap C_G(H) = \phi$, then $a_{ij}^{(t)} \equiv |S_{g_t}| = |\{(x,y) \in K_i \times K_j \mid xy = g_t\}| \equiv 0 \bmod p$, by Lemma 3.5, a contradiction! Therefore $K_i \cap C_G(\delta(K_t)) \neq \phi$. Let g be in this intersection. Then the p-subgroup $\delta(K_t) \leq C_G(g)$. Therefore $\delta(K_t) \leq \delta(K_i)$, where $\delta(K_i)$ is a p-Sylow subgroup of $C_G(g)$. Since $g \in K_i$, $\delta(K_i)$ is a defect group of K_i. Hence $\delta(K_i) \underset{G}{\leq} D$, because $c_i \in \text{Sup}\, z$ and $z \in J_D$. Thus also $\delta(K_t) \underset{G}{\leq} D$, and $c_t \in J_D$. Therefore J_D is an ideal of ZFG.

Similarly, it is proved that $\mathfrak{J}_D$ is an ideal of ZFG.

<u>Lemma 3.7</u>. If $B \longleftrightarrow e \longleftrightarrow \lambda$ is a block of the group algebra FG, and if I is an ideal in the center ZFG of FG such that $\lambda(I) \neq 0$, then $e \in I$.

<u>Proof</u> [36]. Since $\lambda(Z)$ is a field, and $0 \neq \lambda(I)$ is an ideal of $\lambda(Z)$, it follows $\lambda(Z) = \lambda(I)$. Thus $\lambda(e) = \lambda(i)$ for some $i \in I$. Therefore

$$e-i \in \ker \lambda = J(Z)e \oplus Z(1-e),$$

by the proof of Proposition 2.2. Hence

$$e-i = re + z(1-e), \quad \text{where } r \in J(Z),\ z \in Z.$$

Thus

$$e = ie + re, \quad \text{and} \quad e-ie = re \in J(Z).$$

Since Z is Artinian, $J(Z)$ is nilpotent. As Z is an algebra over the field F of characteristic p, it follows $0 = (re)^{p^s} = (e-ie)^{p^s} = e^{p^s} + (ie)^{p^s} = e+(ie)^{p^s}$ for some positive integer s. Therefore $e = -(ie)^{p^s} \in I$.

Theorem 3.8 [36]. For every block $B \leftrightarrow e \leftrightarrow \lambda$ of the group algebra FG there exists a p-subgroup D of G such that $e \in J_D$, but $e \notin \hat{J}_D$. D is uniquely determined by $B \leftrightarrow e \leftrightarrow \lambda$ up to G-conjugacy.

Proof. By Lemma 3.2

$$e = \sum_{i=1}^{k} r_i c_i \quad , \quad r_i \in F.$$

Thus

$$1 = \lambda(e) = \sum_{i=1}^{k} \lambda(r_i)\, \lambda(c_i).$$

Hence $\lambda(c_{io}) \neq 0$ for some $c_{io} \in \mathrm{Sup}\ e$. Let D be a defect group of K_{io}. Then $c_{io} \in J_D$ by Lemma 3.6. Thus $\lambda(J_D) \neq 0$, and $e \in J_D$ by Lemma 3.7, because J_D is an ideal of ZFG by Lemma 3.6.

If $e \in \hat{J}_D$, then $\delta(K_j) \underset{G}{\lneq} D$ for all $c_j \in \mathrm{Sup}\ e$, and $c_{io} \notin \mathrm{Sup}\ e$, a contradiction! Hence $e \notin \hat{J}_D$.

Let H be another p-subgroup of G such that $e \in J_H$ and $e \notin \hat{J}_H$. Then $\delta(K_t) \underset{G}{\leq} H$ for all $c_t \in \mathrm{Sup}\ e$. Since $c_{io} \in \mathrm{Sup}\ e$, it follows that $D = \delta(K_{io}) \underset{G}{\leq} H$.

As $e \in J_D$, Lemma 3.6 implies that $\delta(K_t) \underset{G}{\leq} D$ for all $c_t \in \mathrm{Sup}\ e$. Because of $e \notin \hat{J}_H$ there is a $c_{to} \in \mathrm{Sup}\ e$ such that

$\delta(K_{to}) \underset{G}{=} H$. Hence $H \underset{G}{\leq} D$, and $H \underset{G}{=} D$.

Definition 3.9 [36]. A p-subgroup D of G is a defect group $\delta(B)$ of the block $B \leftrightarrow e \leftrightarrow \lambda$ if $e \in J_D$ but $e \notin \hat{J}_D$.

By Theorem 3.8 $\delta(B)$ ($= \delta(e)$) is uniquely determined by the block $B \leftrightarrow e \leftrightarrow \lambda$ up to G-conjugacy. The order $d(e) = d(B) = |\delta(B)|$ of $\delta(B)$ is called the defect of the block $B \leftrightarrow e \leftrightarrow \lambda$.

§4. First Main Theorem on Blocks

The main result of this section is R. Brauer's first main theorem on blocks ([4], 10B) exhibiting a one-to-one correspondence of the blocks of the group algebra FG with a given p-subgroup D as one of their defect groups and the blocks of the group algebra $FN_G(D)$ with defect group D. The one-to-one correspondence is given by

4.1 Definition [4]. Let D be a p-subgroup of the finite group G. Then the map $\sigma : ZFG \to ZFC_G(D)$ defined by

$$\sigma(c_i) = \begin{cases} \sum_{g \in K_i \cap C_G(D)} g & , \quad \text{if } K_i \cap C_G(D) \neq \phi \\ 0 & , \quad \text{if } K_i \cap C_G(D) = \phi \end{cases}$$

is the Brauer homomorphism from the center ZFG of the group algebra FG into the center $ZFC_G(D)$ of the group algebra $FC_G(D)$.

It is clear that σ is an F-vector space homomorphism, but in fact it is an F-algebra homomorphism by the following

Theorem 4.2 [4]. Let D be a p-subgroup of G and let H be any subgroup of G satisfying $C_G(D) \leq H \leq N_G(D)$. Then the Brauer homomorphism σ is an F-algebra homomorphism from the center ZFG of FG into the center ZFH of the group algebra FH such that

$$\ker \sigma = \{z \in Z \mid c_i \in \text{Sup } z \Rightarrow K_i \cap C_G(D) = \phi\}.$$

Proof. Since $C_G(D)$ is a normal subgroup of H, $K_i \cap C_G(D)$ consists of full H-conjugacy classes whenever it is not empty. Thus σ is an F-vector space map from ZFG into ZFH.

If $K_i' = K_i - (K_i \cap C_G(D))$, and if $c_i' = \sum_{g \in K_i'} g$ for $i = 1,2,\ldots,k$, then

$$(1) \qquad c_i = \sigma(c_i) + c_i'$$

In order to show that σ is an F-algebra homomorphism it suffices to show that $\sigma(c_i c_j) = \sigma(c_i)\sigma(c_j)$ for $i,j = 1,2,\ldots,k$. By Lemma 3.3

$$(2) \qquad c_i c_j = \sum_{t=1}^{k} a_{ij}^{(t)} c_t ,$$

where $a_{ij}^{(t)}$ are uniquely determined integers mod p. Now (1) and (2) imply

$$(3) \quad \sigma(c_i)\sigma(c_j)+\sigma(c_i)c_j'+c_i'\sigma(c_j)+c_i'c_j' = \sum_{t=1}^{k} a_{ij}^{(t)}\sigma(c_t)+\sum_{t=1}^{k} a_{ij}^{(t)}c_t' .$$

Clearly $\sigma(c_i)\sigma(c_j)$ and $\sum_{t=1}^{k} a_{ij}^{(t)}\sigma(c_t)$ belong to $FC_G(D)$. Furthermore $\sigma(c_i)c_j'$, $c_i'\sigma(c_j)$ and $\sum_{t=1}^{k} a_{ij}^{(t)}c_t'$ do not belong to $FC_G(D)$, if they are non-zero. If $c \in C_G(D)$ belongs to the support of $c_i'c_j'$, then let $S_c = \{(x,y) \in K_i' \times K_j' \mid xy = c\}$. Since $K_j' \cap C_G(D) =$, and since K_i' and K_j' are D-invariant Lemma 3.5 implies that $|S_c| \equiv 0 \bmod p$. Hence Sup $(c_i'c_j') \cap C_G(D) = \phi$ by Lemma 3.3. Therefore

$$\sigma(c_i c_j) = \sum_{t=1}^{k} a_{ij}^{(t)}\sigma(c_t) = \sigma(c_i)\sigma(c_j) \quad \text{for } i,j = 1,2,\ldots,k,$$

and so σ is an F-algebra homomorphism from ZFG onto ZFH with

$$\ker \sigma = \{z \in ZFG \mid c_i \in \text{Sup } z \text{ implies } K_i \cap C_G(D) = \phi\},$$

because $\sigma(c_i) = 0$ if and only if $K_i \cap C_G(D) = \phi$. This completes the proof of Theorem 4.2.

Definition. If H is a subgroup of the group G, then $(\omega H)FG$ is the right ideal of the group algebra FG generated by $\{1-h \mid h \in H\}$. If $H = G$, then $\omega G = (\omega G)FG$ is called the augmentation ideal of FG.

If H is a normal subgroup of G, then $(\omega H)FG$ is a two-sided ideal of the group algebra FG, in fact it is the kernel of the canonical epimorphism

$$FG \to F(G/H) \to 0.$$

(See [23], p. 153).

The following Lemma is well-known; we include a proof of it for the sake of completeness.

Lemma 4.3. If G is a finite p-group, then the augmentation ideal ωG of the group algebra FG is nilpotent, and $FG/\omega G \cong F$.

Proof. It is obvious that ωG is the kernel of the map $\phi: FG \to F$ defined by

$$\phi\left(\sum_{g \in G} f_g g\right) = \sum_{g \in G} f_g \in F.$$

Hence $FG/\omega G \cong F$.

Suppose that the group G is of least order among all the groups G for which ωG is not nilpotent. Say $|G| = p^n$. Then $(1-g)^{p^n} = 1 - 1 = 0$ for all $g \in G$. Thus G is not abelian, because otherwise ωG would be a nil ideal and therefore a nilpotent ideal of FG as is easily seen. Since G is a p-group, it has a non trivial center $Z(G)$ (see [19], p. 31). Thus $|G/Z(G)| < |G| > |Z(G)|$. Hence $\omega(G/Z(G))$ and $\omega(Z(G))$ are nilpotent. Since the sequence

$$0 \to \omega(Z(G))FG \to FG \to F(G/Z(G)) \to 0$$

is exact, it follows that

$$\omega G/\omega(Z(G))FG \cong \omega(G/Z(G)).$$

Hence ωG is nilpotent, a contradiction!

Lemma 4.4. Let D be a normal subgroup and $K = x^G$ be a conjugacy class of the group G. Let τ be the canonical group algebra epimorphism $FG \to F(G/D)$. If $\bar{G} = G/D$, $\bar{x} = xD$, $\bar{K} = \bar{x}^{\bar{G}}$ and $n = \dfrac{|G:C_G(x)|}{|\bar{G}:C_{\bar{G}}(\bar{x})|}$, then $\tau(c) = n\bar{c}$, where $c = \sum_{g \in K} g$ and $\bar{c} = \sum_{\bar{g} \in \bar{K}} \bar{g}$.

Proof is trivial.

Lemma 4.5 [36]. If D is a normal p-subgroup of G, and if τ is the canonical ring epimorphism $FG \to F(G/D)$, then $\ker \tau = (\omega D)FG = FG(\omega D)$ is a nilpotent ideal of the group algebra FG. Furthermore, $\{c_i \in ZFG \mid K_i \cap C_G(D) = \phi\} \leq \ker \tau$.

Proof. Since D is a p-group ωD is nilpotent by Lemma 4.3. As D is a normal p-subgroup of G, $\ker \tau = (\omega D)FG = FG(\omega D)$. Hence $\ker \tau$ is a nilpotent ideal of FG.

Let $K = x^G$ be a conjugacy class of G such that $K \cap C_G(D) = \phi$. Let $\bar{x} = xD$, $\bar{G} = G/D$ and $\bar{K} = \bar{x}^{\bar{G}}$. If $c = \sum_{g \in K} g$ and $\bar{c} = \sum_{\bar{g} \in \bar{K}} \bar{g}$, then Lemma 4.4 asserts that $\tau(c) = n\bar{c}$, where $n = \dfrac{|G:C_G(x)|}{|\bar{G}:C_{\bar{G}}(\bar{x})|}$. Let y be a fixed element of K.

Then $Y = y^{-1}D$ is D-invariant, because D is normal in K. If $S_1 = \{(t,r) \in Y \times K \mid tr = 1\}$, then $|S_1| \equiv 0 \bmod p$ by Lemma 3.5, because $K \cap C_G(D) = \phi$. Since $t = y^{-1}d$ for some $d \in D$, we have

$$tr = 1 \iff y^{-1}dr = 1 \iff dr = y \iff rD = yD,$$

because D is normal in G. Thus

$$|S_1| = n = \frac{|G:C_G(x)|}{|\bar{G}:C_{\bar{G}}(\bar{x})|},$$

and $\tau(c) = n\bar{c} = 0$.

Proposition 4.6 [36]. Let D be a normal p-subgroup of the group G and D' be a p-subgroup of G such that $D \nleq D'$. Then $J_{D'} = \{z \in ZFG \mid c_i \in \text{Sup } z \text{ implies } \delta(K_i) \leq_G D'\}$ is a nilpotent ideal of the center ZFG of the group algebra FG.

Proof. Let τ be the canonical ring epimorphism $FG \to F(G/D)$. Then by Lemma 3.6 and Lemma 4.5 it suffices to show that

$J_{D'} \leq \ker \tau$. If $c_j \in J_{D'}$, then $K_j \cap C_G(D) = \phi$, because otherwise we would have $D \leq \delta(K_j) \underset{G}{\leq} D'$. Thus $c_j \in \ker \tau$ by Lemma 4.5.

Corollary 4.7 [36]. If D is a normal p-subgroup of the finite group G, then D is contained in every defect group $\delta(B)$ of every block $B \leftrightarrow e \leftrightarrow \lambda$ of FG.

Proof. Since $e \in J_{\delta(B)}$ by Definition 3.9, $J_{\delta(B)}$ is not nilpotent. Thus $D \leq \delta(B)$ by Proposition 4.6.

Lemma 4.8 [4]. Let D be a p-subgroup of the group G and $H = N_G(D)$.

a) If K is a conjugacy class of G with class sum c and having D as one of its defect groups, then $\sigma(c)$ is the class sum in FH of a single conjugacy class of H.

b) If L is a conjugacy class of H having defect group $\delta(L) = D$, then there is a conjugacy class K of G with class sum c and defect groups $\delta(K) \underset{G}{=} D$ such that $\sigma(c) = \sum_{h \in L} h$.

Proof. a) Since D is a defect group of K, there is a $g \in K \cap C_G(D)$ such that D is a p-Sylow subgroup of $C_G(g)$. Hence $g^H \leq K \cap C_G(D)$. Let $t \in C_G(D) \cap K$. Then $t = xgx^{-1}$ for some $x \in G$, and $D \leq C_G(t)$. Hence D is a p-Sylow subgroup of $C_G(t)$, and so is $xDx^{-1} \leq C_G(t)$. By Sylow's Theorem ([19], p.34) there is a $y \in C_G(t)$ such that $D = yxDx^{-1}y^{-1}$. Hence $yx \in H$, and $t = yty^{-1} = yxgx^{-1}y^{-1} \in g^H$.

Thus $K \cap C_G(D) = g^H$.

b) Let L be a conjugacy class of $H = N_G(D)$ with defect group D. Then D is a p-Sylow subgroup of $C_G(h) \cap H = C_H(h)$ for some $h \in L$. If D is not a p-Sylow subgroup of $C_G(h)$, then $D < D_1$ for some p-Sylow subgroup D_1 of $C_G(h)$. By [45], p.137 we know that $N_{D_1}(D) > D$. Thus there exists a $d_1 \in N_{D_1}(D)$ such that $d_1 \notin D$. Let U be the subgroup of G generated by D and d_1. Then $U \leq C_G(h) \cap H$, because $d_1 \in D_1 \leq C_G(h)$ and $d_1 \in N_{D_1}(D) \leq N_G(D) = H$. Therefore $D < U$ is not a p-Sylow subgroup of $C_H(h)$, a contradiction. Thus D is a defect group of $K = h^G$, and from Definition 4.1 and Lemma 4.8 a) follows

$$\sigma(c) = \sum_{h \in L} h \quad , \quad \text{where} \quad c = \sum_{g \in K} g.$$

<u>Lemma 4.9</u>. Let D be a p-subgroup of G and σ the Brauer homomorphism from ZFG into ZFH, where $H = N_G(D)$. Then:

a) $\sigma(c_i) \neq 0$ if and only if $D \underset{G}{\leq} \delta(K_i)$

b) If $B \leftrightarrow e \leftrightarrow \lambda$ is a block of FG, then $\sigma(e) \neq 0$ if and only if $D \underset{G}{\leq} \delta(B)$

c) If $B_i \leftrightarrow e_i \leftrightarrow \lambda_i$, $i = 1,2$, are blocks of FG, with $\delta(B_i) \underset{G}{\equiv} D$, then $\sigma(e_1) = \sigma(e_2)$ if and only if $e_1 = e_2$.

<u>Proof</u>. a) By Definition 4.1 $\sigma(c_i) \neq 0$ if and only if $K_i \cap C_G(D) \neq \phi$. There is $g_i \in K_i \cap C_G(D)$ if and only if $D \leq C_G(g_i)$, and $D \leq C_G(g_i)$ for some $g_i \in K_i \cap C_G(D)$ if

and only if $D \underset{G}{\leq} \delta(K_i)$, which proves a).

b) By Lemma 3.2 $e = \sum_{i=1}^{k} r_i c_i$. Now $\sigma(e) \neq 0$ if and only if $\sigma(c_i) \neq 0$ for some conjugacy class K_i. Thus $D \underset{G}{\leq} \delta(K_i)$ for all K_i such that $\sigma(c_i) \in \text{Sup}\,\sigma(e)$ by a). Hence $D \leq \delta(B)$, by the definition of $\delta(B)$. The converse is now trivial.

c) By Theorem 4.2 σ is an F-algebra homomorphism from ZFG into ZFH. If $\sigma(e_1) = \sigma(e_2)$ for the centrally primitive idempotents $e_1 \neq e_2$ of FG with defect group D, then

$$0 = \sigma(e_1\, e_2) = \sigma(e_1)\, \sigma(e_2) = \sigma(e_1)^2 = \sigma(e_1) \neq 0$$

by b), a contradiction. Hence c) holds.

The following Lemma is well-known in Ring Theory. It is included here for the sake of completeness.

Lemma 4.10. Let R be right Artinian. Then:

a) Every mininal non-nil right ideal L is generated by a primitive idempotent of R

b) If R is commutative and if $\sigma: R \to S$ is a ring epimorphism such that $\sigma(e) = 0$ for every primitive idempotent e of R, then every element of S is nilpotent.

Proof. a) Since L is not nil, $L = yL$ for some non nilpotent element $y \in L$. Thus $y = yx$, and also x is not nilpotent. Let $N = L \cap y_r$, $y_r = \{z \in R \mid yz = 0\}$. Then $N < L$, because $y \notin N$. Clearly $x - x^2 \in N$. As N is nil,

$(x-x^2)^k = 0$ for some $k \geq 1$. Therefore $x^k = x^{k+1}g(x)$, where $g = g(x)$ is a monic polynomial in x with integer coefficients. Let $e = x^k g^k$. Then

$$e^2 = x^{2k}g^{2k} = x^{k-1}(x^{k+1}g)g^{2k-1} = x^{k-1}x^k g^{2k-1} = x^{2k-1}g^{2k-1}$$

$$= x^{2k-2}g^{2k-2} = \dots = x^k g^k = e.$$

As $e \in L$, $L = eR$, and e is primitive, because L is minimal among the non-nil right ideals of R.

b) As R is right Artinian, there is a maximal set $\gamma = \{e_i \mid i = 1,2,\dots,r\}$ of orthogonal primitive idempotents of R. Let $f = e_1 + e_2 + \dots + e_r$. Then

$$R = e_1R + \dots + e_rR + (1-f)R,$$

where $(1-f)R = \{r - fr \mid r \in R\}$. If $(1-f)R$ is not nil, then $(1-f)R$ contains a minimal non nil right ideal L. By a) $L = e_{r+1}R$, where e_{r+1} is a primitive idempotent of R. Clearly $e_ie_{r+1} = 0$ for $i = 1,2,\dots,r$. As R is commutative the set $\{e_i \mid i = 1,2,\dots,r+1\}$ consists of $r+1$ orthogonal primitive idempotents of R, and γ is not maximal. Hence $(1-f)R$ is nil. Now

$$S = \sigma(R) = \sigma(e_1R \oplus e_2R \oplus \dots \oplus e_rR + (1-f)R) = \sum_{i=1}^{r} \sigma(e_iR) + \sigma(1-f)R),$$

and $S = \sigma[(1-f)R]$. Thus S is a nil ring.

<u>Lemma 4.11</u>. Let D be a p-subgroup of the finite group G, $H = N_G(D)$, and let σ be the Brauer homomorphism from ZFG

into ZFH. Let $\{H_j \mid i = 1,2,\ldots,r\}$ be the set of all conjugacy classes of H and let u_j be the class sum of H_j in FH. If $V_D = \{z \in ZFH \mid u_j \in \text{Sup } z \text{ implies } \delta(H_j) = D\}$, and $J_D = \{z \in ZFG \mid c_i \in \text{Sup } z \text{ implies } \delta(K_i) \underset{G}{\leq} D\}$, then the following assertions hold:

a) $\sigma(J_D) = V_D$

b) If $W_D = \{z \in ZFH \mid u_j \in \text{Sup } z \text{ implies } \delta(H_j) \underset{H}{\leq} D\}$, and if $\hat{W}_D = \{z \in ZFH \mid u_j \in \text{Sup } z \text{ implies } \delta(H_j) \underset{H}{<} D\}$, then $\hat{W}_D$ is a nilpotent ideal of ZFH satisfying

$$W_D = V_D + \hat{W}_D \text{ and } V_D \cap \hat{W}_D = 0 .$$

c) V_D contains all idempotents of W_D.

d) If $B_i \leftrightarrow e_i \leftrightarrow \lambda_i$, $i = 1,2,\ldots,r$ are all the blocks of FG with defect groups $\delta(B_i) \underset{G}{=} D$, then $\sigma(e_i)$ is a centrally primitive idempotent of FH with defect group D for $i = 1,2,\ldots,r$.

e) The set $\{\sigma(e_i) \mid i = 1,2,\ldots,r\}$ consists of all centrally primitive idempotents of FH with defect group D.

Proof. a) By Lemma 3.6 J_D is an ideal of the center ZFG of FG. Since σ is an F-algebra homomorphism by Theorem 4.2, it follows that $\sigma(J_D)$ is an ideal of $\sigma(ZFG) \leq ZFH$. By the definition of J_D, Definition 4.1, Lemma 4.9 a) and Lemma 4.8 it follows that

$$\sigma(J_D) = \{z \in ZFH \mid u_j \in \text{Sup } z \text{ implies } \delta(H_j) = D\} = V_D.$$

b) By a) V_D is a subring of W_D and $\hat{W}_D$ is a nilpotent ideal of ZFH in view of Proposition 4.6. Clearly $V_D \cap \hat{W}_D = 0$

and $W_D = V_D + \hat{W}_D$.

c) Let $g = g^2 \in W_D$. Then $g = w + v$ for some $w \in \hat{W}_D$ and $v \in V_D$ by b). Furthermore, w is nilpotent. Hence $w^{p^k} = 0$ for some integer $k > 0$. Therefore

$$g = g^{p^k} = w^{p^k} + v^{p^k} = v^{p^k} \in V_D, \text{ because } V_D \text{ is a ring.}$$

d) By Theorem 3.8 the ideal J_D of the center ZFG of the group algebra FG contains all centrally primitive idempotents e_i of FG, $i = 1,2,\ldots,r$, with defect groups $\delta(e_i) \underset{G}{=} D$. Hence $e_i ZFG \leq J_D$ for $i = 1,2,\ldots,r$. Thus

$$(*) \qquad J_D = \sum_{i=1}^{r} \oplus\, e_i ZFG \oplus (1-f)J_D,$$

where $f = e_1 + e_2 + \ldots + e_r$ and $(1-f)J_D = \{r - fr \mid r \in J_D\}$. By Lemma 4.9 b) $\sigma(e_i) \neq 0$ for $i = 1,2,\ldots,r$. Therefore $\sigma(e_i)\,\sigma(ZFG)$ is a local Artinian ring for $i = 1,2,\ldots,r$. Hence $\sigma(e_i)$ is a primitive idempotent of $\sigma(ZFG)$ for every i. By statement a) $V_D = \sigma(J_D) \leq \sigma(ZFG)$. Therefore Lemma 1.4 implies that $\sigma(e_i)$ is a primitive idempotent of V_D for $i = 1,2,\ldots,r$. Because of b) and c) it follows that $\sigma(e_i)$ is a primitive idempotent of the ideal W_D of ZFH for every i. Hence another application of Lemma 1.4 implies that $\sigma(e_i)$ is a primitive idempotent of ZFH for $i = 1,2,\ldots,r$. By Lemma 4.8, Lemma 4.9 a) and Corollary 4.7 the defect group of every idempotent $\sigma(e_i)$ is D, because D is a normal subgroup of H.

e) If g is a centrally primitive idempotent of the group algebra FH with defect group D, then $g \in W_D$ by Lemma 3.6. Hence $g \in V_D = \sigma(J_D)$ by a) and c). Now (*) implies that

$$V_D = \sigma(J_D) = \sum_{i=1}^{r} \sigma(e_i)\, \sigma(ZFG) + \sigma[(1-f)J_D].$$

If $q = q^2 \in (1-f)J_D$ is a primitive idempotent of $(1-f)J_D$, then it is a primitive idempotent of the center ZFG of the group algebra FG by Lemma 1.4. Hence its defect group $\delta(q) \underset{G}{<} D$ by the definition of J_D and because $q \notin \{e_i \mid i = 1,2,\dots,r\}$. Thus Lemma 4.9 b) asserts that $\sigma(q) = 0$. Therefore Lemma 4.10 implies that $\sigma[(1-f)J_D]$ is a nil ring, and so $g \notin \sigma[(1-f)J_D]$. As $\sigma[(1-f)J_D]$ is a nil ring, it follows that $g \in \sum_{i=1}^{r} \sigma(e_i)\, \sigma(ZFG)$. Thus $g \in \{\sigma(e_i) \mid i = 1,2,\dots,r\}$ by Lemma 1.6.

Lemma 4.9 c) and Lemma 4.11 together prove now

Theorem 4.12 (First Main Theorem on Blocks, [4]).

Let D be a p-subgroup of the finite group G with order $|D| = p^d$, let $H = N_G(D)$ and F be a field of characteristic $p > 0$.

Then the Brauer homomorphism σ induces a one-to-one correspondence between blocks $B \leftrightarrow e \leftrightarrow \lambda$ of the group algebra FG with defect groups $\delta(B) \underset{G}{=} D$ and the blocks $b \leftrightarrow f \leftrightarrow \mu$ of the group algebra FH with defect d.

§5. Osima's Theorem

For every element $x = \sum_{g \in G} r_g g$, $r_g \in F$, of the group algebra FG,

$$\sup x = \{g \in G \mid r_g \neq 0\}$$

denotes the support of x.

In this section it is shown that for every central idempotent e of FG the support $\sup e$ consists of p-regular elements of G.

Definition. Let p be a prime number. Then the element g of the finite group G is p-regular, if p does not divide the order of g.

$g \in G$ is called p-singular, if its order is a power of p.

A conjugacy class K of G is called p-regular, if K consists of p-regular elements.

Lemma 5.1. Let p be a prime number. Then for every element g of the finite group G there exists exactly one p-regular element $g_1 \in G$ and exactly one p-singular element $g_2 \in G$ such that

$$g = g_1 g_2 = g_2 g_1 .$$

Furthermore, g_1 and g_2 are powers of g.

Proof. Let $n = p^a q$ be the order of g, $(p,q) = 1$. Thus $1 = p^a r + qt$ for some $r,t \in \mathbb{Z}$. If $g_1 = g^{p^a r}$ and $g_2 = g^{qt}$, then

$$g = g_1 g_2 = g_2 g_1.$$

Clearly q is the order of g_1 and p^a is the order of g_2.

Suppose that $g = g_3 g_4 = g_4 g_3$, where g_3 is p-regular and g_4 is p-singular. Let p^b be the order of g_4 and u be the order of g_3. Then $n = p^a q = p^b u$. As $\mathbb{Z}$ is a unique factorization domain, $b = a$ and $q = u$. Hence $g_i^{-1} g = g g_i^{-1}$, for $i = 3,4$, implies that

$$g_3 = g_3^{p^a r+qt} = g_3^{p^a r} = (g g_4^{-1})^{p^a r} = g^{p^a r}(g_4^{-1})^{p^a r} = g_1.$$

Similarly one shows that $g_4 = g_2$.

<u>Lemma 5.2</u> (R. Brauer). Let F be a field of characteristic $p > 0$, and let R be an algebra over F.

If $S = \{ab - ba \mid a,b \in R\}$ denotes the F-subspace of R generated by all commutators $ab - ba$, then

$$(a_1 + a_2 + \ldots + a_m)^{p^n} \equiv a_1^{p^n} + a_2^{p^n} + \ldots + a_m^{p^n} \mod S$$

for every $n \geq 0$ and every m-tuple $a_1, a_2, \ldots, a_m \in R$.

<u>Proof</u>. We first assume that $n = 1$ and $m = 2$. Let $a,b \in R$. Then

$$(a+b)^p - a^p - b^p = \sum c_1 c_2 \ldots c_p,$$

where the sum is taken over all $2^p - 2$ products $c_1 c_2 \ldots c_p$, where $c_i = a$ or $c_i = b$, but not every $c_i = a$ and not every

$c_i = b$. Clearly $c_2c_3\dots c_pc_1 \neq c_1c_2\dots c_p$, and $c_1(c_2\dots c_p) - (c_2c_3\dots c_p)c_1 \in S$. Hence p summands of $\Sigma c_1c_2\dots c_p$ always lie in the same residue class of R/S. As $2^p - 2$ is divisible by p and as F has characteristic p, it follows that $\Sigma c_1c_2\dots c_p \equiv 0 \bmod S$. Therefore our claim holds for $n = 1$ and $m = 2$. The final assertion now follows easily by an inductive argument.

We now can give Passman's proof [29] of Osima's Theorem [27].

Theorem 5.3. Let F be a field of characteristic $p > 0$ and G be a finite group.

Then the support $\sup(e)$ of every central idempotent e of the group algebra FG consists of p-regular elements.

Proof. Let $z \in \sup(e)$. By Lemma 5.1 there is a p-regular element $y \in G$ and a p-singular element $x \in G$ such that $z = xy = yx$. Furthermore x and y are uniquely determined by z and both elements are powers of z. Thus $z \in C_G(x)$, and z is p-regular if and only if $x = 1$.

If z is not p-regular, then $D = \langle x \rangle$ is a non-trivial p-subgroup of G. Clearly $z \in C_G(D)$. Let σ be the Brauer homomorphism of $Z(FG)$ into $Z(FC_G(D))$. By Definition 4.1 the support of $\sigma(e)$ contains z. As σ is an F-algebra homomorphism by Theorem 4.2, $\sigma(e) \neq 0$ is a central idempotent of the group algebra $F(C_G(D))$. As $x \in C_G(D) \cap D$, we may assume that x is contained in the center $Z(G)$ of G, i.e. $G = C_G(D)$.

Let p^d be the order of x and q be the order of y. Then $(p,q) = 1$. Therefore $\bar{p} = p + q\mathbb{Z}$ is a non-zero divisor in the finite ring $\mathbb{Z}/q\mathbb{Z}$. Thus $\bar{p}$ is a unit in $\mathbb{Z}/q\mathbb{Z}$, and

$$(*) \qquad p^m \equiv 1 \mod q \quad \text{for some} \quad m \geq 1.$$

If the order $|G| = p^a r$, $(p,r) = 1$, then Lemma 5.1 implies that g^{p^m} is p-regular for every $g \in G$, and every $m \geq a$ satisfying (*). Let m_o be a fixed integer m satisfying these conditions and let $u = y^{-1}e$. Then

$$u = \sum_{g \in G} r_g g \quad , \quad r_g \in F.$$

As in Lemma 5.2 let $S = \{ab - ba \mid a,b \in FG\}$ be the subspace of FG generated by all commutators $ab - ba$ of FG. Then

$$(**) \qquad u^{p^{m_o}} \equiv \sum_{g \in G} r_g^{p^{m_o}} g^{p^{m_o}} \mod S$$

by Lemma 5.2. Since e is a central idempotent of FG,

$$u^{p^{m_o}} = (y^{-1}e)^{p^{m_o}} = (y^{-1})^{p^{m_o}} e^{p^{m_o}} = y^{-1}e = u$$

by (*). As $z = xy \in \sup (e)$, $x \in \sup (y^{-1}e) = \sup (u)$. Since x is p-singular, and since $g^{p^{m_o}}$ is p-regular for every $g \in G$, it follows from (**) that $x \in \sup s$ for some $s \in S$. Hence there are $h,k \in G$ such that $hk \neq kh$, and $x \in \sup (hk-kh)$. Without loss in generality we may assume that $x = hk$. As $x \in Z(G)$, $x = h^{-1}xh$, and $hk \neq kh = h^{-1}(hk)h = h^{-1}xh = x = hk$. This contradiction proves that z is p-regular.

§6. Blocks and Normal Subgroups

Throughout this section K denotes a normal subgroup of the finite group G. If $b \leftrightarrow f \leftrightarrow \mu$ is a block of the group algebra FK, then $f^x = xfx^{-1}$ is a centrally primitive idempotent of FK for every $x \in G$; its linear character μ^x is then defined by

$$\mu^x(a) = \mu(x^{-1}ax) \qquad a \in ZFK.$$

Clearly $b^x = xbx^{-1}$ is the block ideal of f^x.

6.1 Definition. Let K be a normal subgroup of the finite group G. If $b \leftrightarrow f \leftrightarrow \mu$ is a block of the group algebra FK, then

$$T_G(b) = T_G(f) = \{x \in G \mid xfx^{-1} = f\}$$

is the inertia group of the block $b \leftrightarrow f \leftrightarrow \mu$ in G.

Let $b_j \leftrightarrow f_j \leftrightarrow \mu_j$, $j = 1,2,\dots,t$, be the distinct conjugates of $b \leftrightarrow f \leftrightarrow \mu$. Then $x = \sum_{j=1}^{t} f_j$ is a central idempotent of FG. Therefore

$$x = \sum_{j=1}^{t} f_j = \sum_{i=1}^{r} e_i ,$$

where every e_i is a centrally primitive idempotent of the group algebra FG. The idempotents e_i are uniquely determined by x and hence by f. Every block $B_i \leftrightarrow e_i \leftrightarrow \lambda_i$, $i = 1,2,\dots,r$, of FG covers each block $b_j \leftrightarrow f_j \leftrightarrow \lambda_j$, $j = 1,2,\dots,t$ of FK.

Lemma 6.2. Let K be a normal subgroup of the finite group G, and let $T = ZFG \cap ZFK$. Then:

a) If $b \leftrightarrow g \leftrightarrow \mu$ is a block of FK, and if $\{g_j \mid j = 1,2,\ldots,t\}$ is a transversal of $T_G(f)$ in G, then $x = \sum_{j=1}^{t} g_j fg_j^{-1}$ is a primitive idempotent of T.

b) If x is a primitive idempotent of T, then there is a block $b \leftrightarrow f \leftrightarrow \mu$ of FK such that

$$x = \sum_{j=1}^{t} g_j fg_j^{-1},$$

where $\{g_j \mid j = 1,2,\ldots,t\}$ is a transversal of $T_G(f)$ in G.

Proof. a) Let $Z = ZFG$, $Z_1 = ZFK$, and let $g_1 = 1$. If $f_j = g_j fg_j^{-1}$ for $j = 1,2,\ldots,t$, then $x = \sum_{j=1}^{t} f_j$ is a central idempotent of FG. Hence $x \in Z \cap Z_1 = T$. Suppose that x is not a primitive idempotent of T. Then $x = y + z$, where y and z are orthogonal non-zero idempotents of T. Since y and z belong to Z_1, there are centrally primitive idempotents $y_h \in FK$, $h = 1,2,\ldots,s$ and $z_q \in FK$, $q = 1,2,\ldots,u$, such that

$$y = \sum_{h=1}^{s} y_h \quad \text{and} \quad z = \sum_{q=1}^{u} z_q.$$

Hence $x = \sum_{j=1}^{t} f_j = \sum_{h=1}^{s} y_h + \sum_{q=1}^{u} z_q$.

Since every central idempotent of FK is a unique sum of the finitely many centrally primitive idempotents of FK, we may assume that $f = f_1 = y_1$. Thus $f = y_1 = yy_1 \in yZ_1$. As $(Z_1)^{g_j} = Z$ for $j = 1,2,\dots,t$, it follows that

$$f_j = f^{g_j} \in (yZ_1)^{g_j} = y^{g_j}(Z_1)^{g_j} = yZ_1 \quad \text{for all } j,$$

because $y \in Z$. Hence

$$x = f_1 + f_2 + \dots + f_t \in yZ_1.$$

Thus $x = uz_1$ for some $z_1 \in Z_1$. From $x = y+z = yz_1$ follows $z = xz = (yz_1)z = z_1yz = 0$, a contradiction! Hence x is a primitive idempotent of T.

b) Let x be a primitive idempotent of T. Then

$$(*) \qquad x = f_1 + f_2 + \dots + f_s,$$

where the f_i, $i = 1,2,\dots,s$, are uniquely determined primitive idempotents of the center of FK. Set $f = f_1$ and

$$T_G(f) = \{g \in G \mid gfg^{-1} = f\}.$$

Let $\{g_j \mid j = 1,2,\dots,t\}$ be a transversal of $T_G(f)$. Then $x = x^{g_j} = f^{g_j} + f_2^{g_j} + \dots + f_s^{g_j}$ for $j = 1,2,\dots,t$. Since every f^{g_j} is a primitive idempotent of the center of FK, the uniqueness of (*) implies

$$f^{g_j} \in \{f_i \mid i = 1,2,\dots,s\}.$$

If $y = \sum_{j=1}^{t} f^{g_j}$, then

$$x = y + z, \quad yz = zy, \quad z^2 = z \in Z(FK).$$

As y is a primitive idempotent of T, by a) it follows that $x = y$, because x is primitive in T.

For the sake of completeness we now prove I. S. Cohen's structure theorem for complete local rings R (see [26]) in the case where R is an Artinian local ring of characteristic $p > 0$ which we have to apply later. Its proof is based on the following lemma by A. Geddes [12].

Lemma 6.3. Let R be a commutative local ring of characteristic $p > 0$ with Jacobson radical J. If $J^2 = 0$, then there is a subfield W of R such that $R = W + J$ and $W \cap J = 0$.

Proof. Let $R^p = \{x \in R \mid x = y^p$ for some $y \in R\}$. If $0 \neq x \in R^p$ and $x = y^p$, then $y \notin J$, because $J^2 = 0$. Thus y and x are units in R. If $yz = 1$, then $x^{-1} = z^p \in R^p$. Therefore R^p is a subfield of R. By Zorn's Lemma R^p is contained in a maximal subfield W of R. Clearly $W \cap J = 0$.

If $\bar{W} = W + J/J \neq R/J = F$, then there is an element $\bar{a} = a + J \in F$ which is not contained in $\bar{W}$. As $R^p \leq W$, $\bar{a}^p \in \bar{W}$, and so $a^p \in W$. If X is an indeterminate over W, and if $X^p - a^p \in W[X]$ is not irreducible, then $X^p - a^p$ has an irreducible factor $(X-a)^r \in W[X]$, where $1 < r < p$, because $a \notin W$. Since

$$(X-a)^r = X^r - raX^{r-1}+\ldots+(-1)^r a^r \in W[X],$$

it follows that $ra \in W$. As $(r,p) = 1$, $1 = sr + tp$ for some $t,s \in \mathbb{Z}$. Since W has characteristic p, we obtain $a = ars + tpa = sra \in W$, a contradiction! Hence $X^p - a^p$ is irreducible over W. Therefore $W[a]$ is a subfield of R properly containing W. This contradiction proves that $R = W + J$.

Theorem 6.4. Let R be a finite-dimensional local commutative algebra over the field F with characteristic $p > 0$. If J is the Jacobson radical of R, then there is a subfield $W \geq F$ of R such that $R = W + J$ and $W \cap J = 0$.

Proof. Suppose that Theorem 6.4 is false, and that R is of least dimension over F among the algebras R over F for which Theorem 6.4 does not hold. By Lemma 6.3 $J^2 \neq 0$. Thus $R_1 = R/J^2$ has smaller dimension than R. Clearly $N_1 = J/J^2$ is the radical of R_1, and $(N_1)^2 = 0$. By Lemma 6.3 there is a subfield W_1 of R_1 such that (*) $R_1 = W_1 + N_1$ and $W_1 \cap N_1 = 0$. As $R_1/N_1 \cong R/J \cong W_1$, W_1 is an F-algebra. Thus $F \leq W_1$. Let $W_1 = B/J^2$ for some F-subalgebra B of R. Then (*) implies

$$(**) \qquad R = B + J \text{ and } B \cap J = J^2.$$

As B is a local proper F-subalgebra of R with Jacobson radical J^2, there is a subfield W of B such that

$$(***) \qquad B = W + J^2 \text{ and } W \cap J^2 = 0,$$

because $\dim_F B < \dim_F R$. Since W is an F-algebra $F \leq W$. From (**) and (***) follows

$$R = W + J \text{ and } W \cap J = W \cap B \cap J = W \cap J^2 = 0$$

which finishes the proof of Theorem 6.4.

Lemma 6.5. Let K be a normal subgroup of G. If $B \leftrightarrow e \leftrightarrow \lambda$ is a block of FG, and if $B \leftrightarrow f \leftrightarrow \mu$ is a block of FK, then the restrictions λ_1 and μ_1 of λ and μ to $T = ZFG \cap ZFK$ are non-trivial F-algebra homomorphisms from T into a finite extension field of F.

Proof. As G is finite, T is a commutative Artinian ring having the same identity element as FG and FK. Thus $T \nsubseteq \ker \lambda$ and $T \nsubseteq \ker \mu$. Therefore $M = T \cap \ker \lambda < T$. Since $\ker \lambda$ is a maximal ideal of ZFG, it follows that M is a maximal ideal of T, because every prime ideal of the Artinian ring T is maximal. Clearly $\lambda_1(T) \cong T/M$ which is a finite extension of F. Similarly it is shown that μ_1 is an F-algebra homomorphism from T into a finite extension field of F.

With these subsidiary results from commutative algebra we can now prove

Theorem 6.6 [18]. Let $B \leftrightarrow e \leftrightarrow \lambda$ be a block of the group algebra FG and let $b \leftrightarrow f \leftrightarrow \mu$ be a block of FK where K is a normal subgroup of G. Then B covers b if and only if there exists an F-algebra isomorphism π from $\mu(T)$ onto $\lambda(T)$ such that

$$\pi\mu(a) = \lambda(a) \quad \text{for all } a \in T = ZFG \cap ZFK.$$

Proof. Suppose that B covers b. Then there are centrally primitive idempotents $e = e_1, e_2, \ldots, e_r \in FG$ such that

$$x = \sum_{j=1}^{t} f^{g_j} = e_1 + e_2 + \ldots + e_r$$

where $\{g_j \mid j = 1,2,\ldots,t\}$ is a transversal of the inertia subgroup $T_G(f) = \{g \in G \mid gfg^{-1} = f\}$ of $b \iff f \iff \mu$ in G, and where $g_1 = 1$. By Lemma 6.2 x is a primitive idempotent of the commutative finite-dimensional F-algebra $T = ZFG \cap ZFK$. If $J(T)$ denotes the Jacobson radical of T, then xT is a finite-dimensional commutative local algebra over the field F with Jacobson radical $M = xJ(T)$. By Theorem 6.4 there exists a subfield W of xT such that $F \leq W$, and

$$xT = W + xJ(T) \quad \text{and} \quad W \cap xJ(T) = 0.$$

Thus for every $a \in T$ there exists a unique $w \in W$ and a unique $n \in xJ(T)$ such that

$$(*) \qquad xa = w + n.$$

Since $x = e + \sum_{i=2}^{r} e_i$ is the identity element of xT, it follows from (*) that

$$ae = we + ne.$$

As $n \in xJ(T) \leq J(ZFG)$ and as λ is the linear character of ZFG belonging to e, $\lambda(n) = \lambda(ne) = 0$. Hence

$$(**) \qquad \lambda(a) = \lambda(ae) = \lambda(we) + \lambda(ne) = \lambda(w).$$

By Lemma 6.5 the restriction λ_1 of λ to T is a non-trivial linear character of T. Therefore the restriction α of λ to W is an F-algebra isomorphism from W onto $\lambda(T)$ by (**). Hence $\alpha(W) = \lambda_1(T) = \lambda(T)$, and $\alpha(w) = \lambda(w)\ \forall\ w \in W$.

Since $x = f + \sum_{j=2}^{t} f^{g_2}$ is the identity of xT, it follows similarly from (*) that the restriction β of μ to W is an F-algebra isomorphism from W onto $\mu(T)$ such that

$$(***) \qquad \mu(a) = \mu(w) = \beta(w),$$

where a and w are chosen as in (*). Therefore, if $\pi = \alpha\beta^{-1}$, then π is an F-algebra isomorphism from $\mu(T)$ onto $\lambda(T)$. Furthermore, (**) and (***) imply that

$$\pi\mu(a) = \lambda(a) \quad \text{for every} \quad a \in T.$$

Conversely, let π be an F-algebra isomorphism from $\mu(T)$ onto $\lambda(T)$ satisfying this condition. Since $x = f + \sum_{j=2}^{t} f^{g_j} \in T$ is a central idempotent of FG by Lemma 6.2, there are centrally primitive idempotents $e_i \in FG$, $i = 1,2,\dots,r$, such that $x = \sum_{i=1}^{r} e_i$. If the block idempotent $e \notin \{e_i \mid i = 1,2,\dots,r\}$, then $\lambda(e_i) = 0$ for all i by Proposition 2.2. Hence

$$0 = \lambda(x) = \pi\mu(x) = \pi\mu(f+\sum_{j=2}^{t} f^{g_j}) = \pi\mu(f) = \pi(1) = 1,$$

a contradiction. Hence B covers b.

Lemma 6.7. Let K be a normal subgroup of G. Let $L = x^K$ be a conjugacy class of K. Then

$$N_G(L) = KC_G(x).$$

Proof. If $g \in N_G(L)$, then for every $h \in K$ exists $h_i \in K$ such that $ghxh^{-1}g^{-1} = h_i x h^{-1}$. Hence $h_1^{-1}gh \in C_G(x)$, and $g \in KC_G(x)$, because K is a normal subgroup of G.

Conversely, if $g \in KC_G(x)$, then $g = h_1 y$ for some $h_1 \in K$ and $y \in C_G(x)$. Thus $h_1^{-1}gxg^{-1}h_1 = x$. Hence $gxg^{-1} = h_1 x h_1^{-1}$. As K is a normal subgroup of G, for every $h \in K$ there is $h_2 = ghg^{-1} \in K$. Clearly $gh = h_2 g$, and

$$ghxh^{-1}g^{-1} = h_2 gxg^{-1}h_2^{-1} = h_2 h_1 x h_1^{-1} h_2^{-1} \in L.$$

Therefore $g \in N_G(L)$.

Lemma 6.8. Let K be a normal subgroup of G. Then every block $B \leftrightarrow e \leftrightarrow \lambda$ of FG covers exactly one conjugacy class of blocks $b \leftrightarrow f \leftrightarrow \mu$ of FK.

Proof. Let $T = ZFG \cap ZFK$. Decompose the identity element 1 of T into primitive idempotents $t_j \in T$, $j = 1,2,\ldots,k$. Then

$$1 = t_1 + t_2 + \ldots + t_k = e_1 + e_2 + \ldots + e_n,$$

where $\{e_i \mid i = 1,2,\ldots,n\}$ is the set of all primitive idempotents of ZFG. As every t_j is a unique sum of some of the idempotents e_i, we may assume that $t_1 = e + e_2 + \ldots + e_s$.

Since the t_j are orthogonal idempotents of T, t_1 is uniquely determined by e. Therefore e covers exactly one conjugacy class of blocks of FK by Lemma 6.2.

Lemma 6.9. Let K be a normal subgroup of G, and let $b \longleftrightarrow f \longleftrightarrow \mu$ be a block of FK. Among the blocks $B_i \longleftrightarrow \lambda_i \longleftrightarrow e_i$, $i = 1,2,\ldots,r$, of FG covering b let B_1 be of maximal defect. If $a = \sum_{i=1}^{r} e_i = \sum_{s=1}^{k} q_s c_s$, where $q_s \in F$, and where c_s is the class sum of the conjugacy class K_s of G, then there is a p-regular conjugacy class, K_1 say, of G such that:

a) $K_1 \leq K$

b) $q_1 \neq 0$ and $\lambda_1(c_1) \neq 0$

c) $\delta(B_1) \underset{G}{=} \delta(K_1)$

d) $\delta(B_i) \underset{G}{\leq} \delta(B_1)$ for $i = 1,2,\ldots,r$.

Proof ([18] and [30]). Let $\{b_j \longleftrightarrow f_j \longleftrightarrow \mu_j \mid j = 1,2,\ldots,t\}$ be the set of blocks of FK which are conjugate to $b \longleftrightarrow f \longleftrightarrow \mu$ under G. As $\{B_i \longleftrightarrow e_i \longleftrightarrow \lambda_i \mid i = 1,2,\ldots,r\}$ is the set of all blocks covering b, Lemma 6.2 implies

$$(*) \qquad a = \sum_{i=1}^{r} e_i = \sum_{j=1}^{t} f_i = \sum_{s=1}^{k} q_s c_s .$$

Since $\lambda_1(a) = \lambda_1(e_1) = 1$, it follows that $\lambda_1(c_s) \neq 0$ for at least one $s \in \{1,2,\ldots,k\}$, $s = 1$ say. Thus $q_1 \neq 0$, and $\lambda_1(c_1) \neq 0$. Hence b) holds.

As $a \in T = ZFG \cap ZFK$, the conjugacy class K_1 of G is contained in K. From (*) and Theorem 5.3 follows that K_1

consists of p-regular elements.

By Theorem 6.6 there are F-algebra isomorphisms π_i from $\mu(T)$ onto $\lambda_i(T)$ such that $\mu(a) = \pi_i^{-1}\lambda_i(a)$ for all $a \in T$, because the blocks B_i cover b. Since $c_1 \in T$, it follows $0 \neq \lambda_1(c_1) = \pi_1\mu(c_1) = \pi_1\pi_i^{-1}\lambda_i(c_1)$. Hence $\lambda_i(c_1) \neq 0$ for $i = 1,2,\ldots,r$. Let D be a defect group of K_1. Then $c_1 \in J_D$. Hence $\lambda_i(J_D) \neq 0$ for $i = 1,2,\ldots,r$. Therefore $e_i \in J_D$ by Lemma 3.7 for all i. Hence

$$(**) \qquad \delta(B_i) \underset{G}{\leq} D \underset{G}{=} \delta(K_1)\,.$$

By (*) $c_1 \in \operatorname{Sup} e_{i_o}$ for at least one $i_o \in \{i = 1,2,\ldots,r\}$. Thus $\delta(K_1) \underset{G}{\leq} \delta(B_{i_o}) \underset{G}{\leq} \delta(k_1)$, and $\delta(B_{i_o}) \underset{G}{=} \delta(K_1)$. Since $d(B_{i_o}) \leq d(B_1)$ it follows from (**) that $\delta(B_1) \underset{G}{=} \delta(K_1)$, and $\delta(B_i) \underset{G}{\leq} \delta(B_1)$ for $i = 1,2,\ldots,r$ which finishes the proof of Lemma 6.9.

For the proof of Fong's Theorem we need one further lemma.

Lemma 6.10. Let FG be the group algebra of the finite group G over the field F and let α be an automorphism of the group G. If α^* is the algebra isomorphism of FG induced by α, and if $B \leftrightarrow e \leftrightarrow \lambda$ is a block of FG with defect groups $\delta(B) \underset{G}{=} D$, then $\alpha^*(B) \leftrightarrow \alpha^*(e) \leftrightarrow \alpha^*(\lambda)$ is a block of FG with defect groups $\delta(\alpha^*(B)) \underset{G}{=} \alpha(D)$, where $[\alpha^*(\lambda)](z) = \lambda\alpha^*(z)$ for every $z \in ZFG$.

Proof is an easy consequence of Theorem 3.8, the definition of a defect group of a conjugacy class of G and the definition of a defect group of a block FG.

After all these preparations we now can prove the main result of this section which in case F is a splitting field for the finite group G was proved first by P. Fong in [11] using characters. The proof presented is due to W. Hamernik and the author [18] using interesting counting techniques of W. F. Reynolds [32] which are also applied by D. S. Passman [30].

<u>Theorem 6.11</u>. Let K be a normal subgroup of G, and let $b \leftrightarrow f \leftrightarrow \mu$ be a block of the group algebra FK. If the block $B_1 \leftrightarrow e_1 \leftrightarrow \lambda_1$ of the group algebra FG is of maximal defect among the blocks $B_i \leftrightarrow e_i \leftrightarrow \lambda_i$, $i = 1, 2,\dots,r$, of FG covering the block $b \leftrightarrow f \leftrightarrow \mu$ of FK, then:

a) $\delta(B_i) \underset{G}{\leq} \delta(B_1)$ for $i = 1,2,\dots,r$

b) $\delta(B_i) \underset{G}{\leq} T_G(b)$ for $i = 1,2,\dots,r$

c) $\delta(B_1) \cap K \underset{G}{=} \delta(b)$

d) $\delta(|K\ \delta(B_1)|) \leq \nu(|T_G(b)|)$

e) Equality holds in d), if F is a splitting field for K.

<u>Proof</u>. a) follows at once from Lemma 6.9.

b) By Lemma 6.9 there is a p-regular conjugacy class K_1 of G such that $K_1 = x^G \leq K$ and $\delta(B_1) \underset{G}{=} \delta(K_1)$. Furthermore, if $\{e_i \mid i = 1,2,\dots,r\}$ is the set of all centrally primitive idempotents of FG covering the block $b \leftrightarrow f \leftrightarrow \mu$ of FK, then $x \in \sup a$, where $a = \sum_{i=1}^{r} e_i$. Let $\{g_i \in G \mid i = 1,2,\ldots,q\}$ be a complete system of representatives of the double cosets $T_G(b)y\, C_G(x)$ in G, and choose $c_{ij} \in C_G(x)$ such that $\{g_i\, c_{ij} \mid i = 1,2,\ldots,q;\ j = 1,2,\ldots,r_i\}$ is a right transversal of $T_G(b)$ in $T_G(b)g_i\, C_G(x)$. Then

$$a = \sum_{i=1}^{q} \sum_{j=1}^{r_i} f^{g_i c_{ij}} .$$

If $f = \sum_{g \in G} \nu_g g$, where $\nu_g \in F$, then

$$a = \sum_{i=1}^{q} \sum_{j=1}^{r_i} \sum_{g \in G} \nu_g g^{g_i c_{ij}} ,$$

$$a = \sum_{i=1}^{q} \sum_{j=1}^{r_i} \sum_{\substack{g \in G \\ g^{g_i} = x}} \nu_g g^{g_i c_{ij}} + \sum_{i=1}^{q} \sum_{j=1}^{r_i} \sum_{\substack{g \in G \\ g^{g_i} \neq x}} \nu_g g^{g_i c_{ij}} .$$

Since $c_{ij} \in C_G(x)$ it follows that $x = g^{g_i c_{ij}}$ whenever $x = g^{g_i}$. Thus

$$a = \Big(\sum_{i=1}^{q} r_i\, \nu_{x^{g_i^{-1}}}\Big)x + \sum_{i=1}^{q} \sum_{j=1}^{r_i} \sum_{\substack{g \in G \\ g^{g_i} \neq x}} \nu_g\, g^{g_i c_{ij}} .$$

Hence there is at least one integer $i_o \in \{1,2,\ldots,q\}$ such that $p \nmid r_{i_o}$, because $x \in \sup a$. Since

$$r_{i_o} = \frac{|T_G(b)g_{i_o}C_G(x)|}{|T_G(b)|} = \frac{|T_G(b)g_{i_o}C_G(x)g_{i_o}^{-1}|}{|T_G(b)|} = \frac{|C_G(x^{g_{i_o}})|}{|T_G(b)\cap C_G(x^{g_{i_o}})|} ,$$

it follows therefore that $T_G(b)$ contains a p-Sylow subgroup of $C_G(x^{g_{i_o}})$. As $x^{g_{i_o}} \in K_1$, $\delta(K_1) \underset{G}{\leq} T_G(b)$, and b) holds.

d) Clearly $K \leq T_G(b)$. By b) we may assume that $T_G(b)$ contains a defect group $\delta(B_1)$ of B_1. Thus $K\,\delta(B_1) \leq T_G(b)$, and

(1) $$\nu(|K\,\delta(B_1)|) \leq \nu(|T_G(b)|).$$

e) Let $L = x^K$, and $\{z_i \in G \mid i = 1,2,\ldots,m\}$ be a system of representatives of the left cosets of $N_G(L)$ in G. If $\ell = \sum_{g \in L} g \in FG$, then $c_1 = \sum_{g \in K_1} g = \sum_{i=1}^{m} \ell^{z_i^{-1}}$. By Lemma 6.9 we have $\lambda_1(c_1) \neq 0$, and by Theorem 6.6 there is an F-algebra isomorphism π_1 from $\mu(T)$ onto $\lambda_1(T)$ such that $\pi_1\mu(a) = \lambda_1(a)\ \forall\, a \in T = ZFK \cap ZFG$. Thus

(2) $$0 \neq \lambda_1 c_1 = \pi_1\mu c_1 = \sum_{i=1}^{m} \pi_1\mu(b^{z_i^{-1}}).$$

Therefore (*) $\sum_{i=1}^{m} \mu^{z_i}(b) \neq 0$, if F is a splitting field for K.

Let $\{w_j \in G \mid j = 1,2,\ldots,s\}$ be a system of representatives of the double-cosets $T_G(b)w\,N_G(L)$ in G. Choose $t_{ij} \in T_G(b)$ such that $\{t_{ij}w_i \mid j = 1,2,\ldots,n_i\}$ is a system of representatives of the left cosets of $N_G(L)$ in $T_G(b)w_i\,N_G(L)$ Then

$$\sum_{i=1}^{m} \mu^{z_i}(\ell) = \sum_{i=1}^{s}\sum_{j=1}^{n_i} \mu^{t_{ij}w}(\ell) = \sum_{i=1}^{s} n_i\mu^{w_i}(\ell).$$

Therefore it follows from (2) that

(3) $$\sum_{i=1}^{s} n_i\,\mu^{w_i}(\ell) \neq 0.$$

Hence there is an integer $j_o \in \{1,2,\ldots,s\}$ such that $p \nmid n_{j_o}$. Since

$$\frac{|T_G(b)w_{j_o}N_G(L)|}{|T_G(b)|} = \frac{|T_G(b)N_G(w_{j_o}Lw_{j_o}^{-1})|}{|T_G(b)|} = \frac{|T_G(b)|}{|T_G(b)\cap N_G(L^{w_{j_o}})|},$$

it follows therefore that

$$(4) \qquad p \nmid |T_G(b) : T_G(b) \cap N_G(L^{w_{j_o}})|.$$

As $N_G(L^{w_{j_o}}) = KC_G(x^{w_{j_o}})$ by Lemma 6.7, (4) implies that $KC_G(x^{w_{j_o}})$ contains a p-Sylow subgroup of $T_G(b)$, and hence

$$(5) \qquad \nu(|T_G(b)|) \le \nu(|KC_G(x^{w_{j_o}})|).$$

Since $x \in K_1$ and $\delta(B_1) \underset{G}{=} \delta(K_1)$ by Lemma 6.9, it follows that $\delta(B_1)$ is conjugate under G to a p-Sylow subgroup of $C_G(x^{w_{j_o}})$. Hence

$$(6) \qquad \nu(|KC_G(x^{w_{j_o}})|) = \nu(|K\delta(B_1)|).$$

Therefore statement e) of Theorem 6.11 follows at once from (1), (5) and (6).

(c) By (2) $\mu^{z_{i_o}}(\ell) = \mu(\ell^{z_{i_o}^{-1}}) \ne 0$ for some $i_o \in \{1,2,\ldots,m\}$. Clearly $\ell^{z_{i_o}^{-1}}$ is the class sum of $L_1 = (x^{z_{i_o}^{-1}})^K$. As $K_1 = (x^{z_{i_o}^{-1}})^G$ we may assume that $\delta(B_1)$ is a p-Sylow subgroup of $C_G(x^{z_{i_o}^{-1}})$. As $K \cap C_G(x^{z_{i_o}^{-1}}) = C_K(x^{z_{i_o}^{-1}})$

is a normal subgroup of $C_G(x^{z_{i_o}^{-1}})$, Sylow's theorem implies that $\delta(B_1) \cap K$ is a p-Sylow subgroup of $C_K(x^{z_{i_o}^{-1}})$. Hence $\delta(B_1) \cap K = D_1$ is a defect group of the conjugacy class $L_1 = (x^{z_{i_o}^{-1}})^K$ of K. Since $\mu(\ell^{z_{i_o}^{-1}}) \neq 0$, Lemma 3.7 implies that $\delta(b) \underset{H}{\leq} D_1 = \delta(B_1) \cap K \underset{H}{=} \delta(L_1)$.

Since $\sum_{j=1}^{t} f_j = \sum_{s=1}^{k} q_s c_s$ by Lemma 6.9,

$$\sum_{j=1}^{t} f_j = q_1 \sum_{i=1}^{m} \ell^{z_i^{-1}} + \sum_{s=2}^{k} q_s c_s,$$

Hence $\ell^{z_{i_o}^{-1}} \in \operatorname{Sup} f_{j_o}$ for some $j_o \in \{1,2,\ldots,t\}$. Therefore $\delta(L_1) \underset{H}{\leq} \delta(f_{j_o})$ by Theorem 3.8. As $f_{j_o} = f^{y_{j_o}}$, the defect groups $\delta(b)$ and $\delta(b_{j_o})$ have the same order, by Lemma 6.10. Thus $|\delta(L_1)| \leq |\delta(b)|$, and

$$\delta(b) \underset{H}{=} \delta(B_1) \cap K$$

which finishes the proof of Theorem 6.10.

§7. Blocks with normal defect groups

In this section we study the blocks of the group algebra FG whose defect groups are normal in G, and we obtain a generalization of Brauer's first main theorem on blocks (see [6], 5C).

The results of this section are based on the following

Definition 7.1. Let U be a subgroup of G, and let $b \leftrightarrow f \leftrightarrow \mu$ be a block of the group algebra FU. For every conjugacy class K_i, $i = 1,2,\ldots,k$, of G let

$$\mu^G\left(\sum_{g\in K_i} g\right) = \mu\left(\sum_{g\in K_i\cap U} g\right) .$$

If there is a block $B \leftrightarrow e \leftrightarrow \lambda$ of FG such that μ^G and λ are in the same equivalence class of linear characters of the center Z(FG) of FG, then b^G is defined, and $B = b^G$.

In case b^G is defined, the map sending b to b^G is called the Brauer correspondence.

If D is a p-subgroup and $H = N_G(D)$, then the Brauer correspondence induces the inverse mapping of the Brauer homomorphism σ mapping injectively the blocks $B \leftrightarrow e \leftrightarrow \lambda$ of FG with defect group $\delta(B) \underset{G}{\equiv} D$ onto the blocks $b \leftrightarrow f \leftrightarrow \mu$ of H with defect group D.

Proposition 7.2. Let D be a normal p-subgroup of G and let K be a normal subgroup of G containing $C_G(D)$. Then:

a) b^G exists for each block $b \leftrightarrow f \leftrightarrow \mu$ of FK, and b^G is the unique block $B \leftrightarrow e \leftrightarrow \lambda$ of FG covering b.

b) For each defect group $\delta(b)$ of $b \leftrightarrow f \leftrightarrow \mu$ there exists a defect group $\delta(B)$ of $B \leftrightarrow e \leftrightarrow \lambda$ such that $\delta(B) \cap K = \delta(b)$.

c) If F is a splitting field for K, then $|\delta(B) : \delta(b)| = p^r$, where $r = \nu(|T_G(b) : K|)$.

Proof ([18] and [30]). a) By Lemma 6.2 b is covered by at least one block $B \leftrightarrow e \leftrightarrow \lambda$ of FG. Since D is a normal p-subgroup of G every defect group $\delta(B)$ of B contains D by Corollary 4.7. Hence

$$C_G(\delta(B)) \le C_G(D) \le K .$$

Let K_i be a conjugacy class of G such that $K_i \not\subseteq K$. Then there is an element $x \in K_i$ with $x \notin K$. If (K_i) is a p-Sylow subgroup of $C_G(x)$ such that $\delta(B)^g \le \delta(K_i)$ for some $g \in G$, then

$$x \in C_G(\delta(B)^g) \le C_G(D) \le K,$$

a contradiction. Thus $\delta(B) \not\le_G \delta(K_i)$. Hence $\lambda(c_i) = 0$ by Lemma 4.9.

If the conjugacy class K_i of G is contained in K,

then $c_i = \sum_{g \in K_i} g \in T = ZFG \cap ZFK$. By Theorem 6.6 there exists an F-algebra isomorphism π from $\mu(T)$ onto $\lambda(T)$ such that

$$\pi\mu(a) = \lambda(a) \quad \text{for all} \quad a \in T,$$

because $B \leftrightarrow e \leftrightarrow \lambda$ covers $b \leftrightarrow f \leftrightarrow \mu$. Hence $\pi\mu(c_i) = \lambda(c_i)$ for every $K_i \leq K$. Since $\lambda(c_i) = 0$ for every $K_i \nleq K$, it follows that μ^G defined as in Definition 7.1 and λ belong to the same equivalence class of linear characters of ZFG. Thus μ^G exists, and b^G is defined. Furthermore, $B = b^G$ by Proposition 2.1.

b) By Theorem 6.11 for every defect group $\delta(b)$ of $b \leftrightarrow f \leftrightarrow \mu$ there is a defect group $\delta(B)$ of $B \leftrightarrow e \leftrightarrow \lambda$ such that $\delta(b) = \delta(B) \cap K$.

If F is a splitting field for K, then Theorem 6.11 e) asserts that

$$\nu(|\delta(B):\delta(b)| = \nu\left(\frac{|\delta(B)|}{|\delta(B)\cap K|}\right) = \nu\left(\frac{|K\delta(B)|}{|K|}\right) = \nu(|T_G(b):K|).$$

This completes the proof of Proposition 7.2.

<u>Corollary 7.3</u>. Suppose that D is a normal p-subgroup of G such that $C_G(D) \leq D$. Then FG has only one block.

<u>Proof</u>. In Proposition 7.2 choose $K = D$. Thus FG has as many blocks as FD, namely one, because FD is a local ring by Lemma 4.3.

Another corollary of Proposition 7.2 is the following generalization of the first main theorem on blocks which was proved by R. Brauer ([6], Theorem (5C)) assuming that F be a splitting field for G.

Theorem 7.4. Let F be a field of characteristic $p > 0$ dividing the order of the finite group G. Let D be a p-subgroup of G and let K be a normal subgroup of $H = N_G(D)$ containing $C_G(D)$.

Then there exists a one-to-one correspondence ψ between the blocks $B \leftrightarrow e \leftrightarrow \lambda$ of FG with defect groups $\delta(B) \underset{G}{=} D$ and the representatives $b \leftrightarrow f \leftrightarrow \mu$ of H-conjugacy classes of blocks of FK having the property that $b \leftrightarrow f \leftrightarrow \mu$ is only covered by blocks of FH with defect $d = \nu(|D|)$.

The correspondence ψ is given by $B = b^G$.

Proof ([18] and [30]). By the first main theorem on blocks (see Theorem 4.12) we may assume that D is normal in G, i.e. $G = H$.

Let $B \leftrightarrow e \leftrightarrow \lambda$ be a block of FG. Then by Lemma 6.8 there exists exactly one conjugacy class of blocks $b \leftrightarrow f \leftrightarrow \mu$ of FK covered by B. If b is a representative of such a G-conjugacy class of blocks of FK, then $B = b^G$ by Proposition 7.2. Thus $b \leftrightarrow f \leftrightarrow \mu$ is only covered by blocks of FH with defect $\nu(|D|)$.

Suppose that $b \leftrightarrow f \leftrightarrow \mu$ is a block of FK which is only covered by blocks of FG with defect $\nu(|D|)$. Then Proposition 7.2 asserts that $B = b^G$ is the only block of FG covering b. As D is normal in G any defect group $\delta(B)$ of B contains D by Corollary 4.7. Thus $\delta(B) = D$.

Remark. More precisely, we have to say that R. Brauer's Theorem 5C of [3] follows at once from Theorem 7.4 and the following lemma.

Lemma 7.5. Let D be a p-subgroup of G and let K be a normal subgroup of $H = N_G(D)$ containing $C_G(D)$ such that F is a splitting field for K. Then the following properties of the block $b \leftrightarrow f \leftrightarrow \mu$ of FK are equivalent:

(1) $b \leftrightarrow f \leftrightarrow \mu$ is only covered by blocks of FH with defect $\nu(|D|)$.

(2) $\delta(b) =_H D \cap K$ and $\nu(|DK|) \geq \nu(|T_H(b)|)$.

Proof. Again we may assume that D is normal in G. Clearly (1) implies (2) by Theorem 6.11 a) and e).

Suppose that (2) holds. By Proposition 7.2 a) the block $b \leftrightarrow f \leftrightarrow \mu$ of FK is only covered by the block $B = b^G$ of FG. If $\delta(b)$ is a defect group of b, then by Proposition 7.2 b) there is a defect group $\delta(B)$ of B such that $\delta(b) = \delta(B) \cap K = D \cap K$. Since $\nu(|DK|) \geq \nu(|T_H(b)|)$ Theorem 6.11 a) and d) implies that $\nu(|DK|) = \nu(|\delta(B)K|) = \nu(|T_G(b)|)$. Hence $\delta(B) = D$, because $D \leq \delta(B)$ by Corollary 4.7.

Remark. As the proof of Lemma 7.5 shows the implication (2) $\Rightarrow$ (1) even holds if I is not a splitting field for K.

§8. Brauer's main theorem on blocks with normal defect groups

If D is a p-subgroup of the finite group G, then the main result of this section asserts that there is a one-to-one correspondence between the blocks of FG with D as one of their defect groups and the $N_G(D)/D$-conjugacy classes of blocks of $F[C_G(D)D/D]$ with defect zero satisfying a certain condition. In view of Brauer's first main theorem, the problem of counting the blocks of FG with defect group D is therefore reduced to the knowledge of the blocks of defect zero of $F[C_G(D)D/D]$ and a knowledge of how $N_G(D)/D$ acts on these blocks by conjugation.

The proof of our main result is preceded by several lemmas.

<u>Lemma 8.1</u>. Let D be a normal p-subgroup of G, $K = DC_G(D)$, $\bar{K} = K/D$, and $\bar{G} = G/D$.

If τ is the canonical group algebra homomorphism from FG onto $F\bar{G}$, then:

a) τ induces a one-to-one correspondence between the class sums of the p-regular conjugacy classes of K and the class sums of the p-regular conjugacy classes of $\bar{K}$.

b) Each p-regular conjugacy class C of K is contained in $C_G(D)$.

c) $\tau(C)$ has defect group $\delta(C)/D$.

<u>Proof</u> [4]. Let $|G| = p^a q$, where $(p,q) = 1$. Let K_j be a p-regular conjugacy class of K, and let $u_j \in K_j$. As u_j is

p-regular and as

$$u_j = cd = dc \in C_G(D) \cdot D = K, \; c \in C_G(D), \quad d \in D,$$

Lemma 5.1 implies that $u_j \in C_G(D)$. Hence b) holds.

Because of b) $D \leq C_K(u_j)$. Hence $C_K(u_j)/D \leq C_{\overline{K}}(u_jD)$.

Let $xD \in C_{\overline{K}}(u_jD)$. Then $x^{-1}u_jx = u_jd = du_j$ for some $d \in D$. Since u_j is p-regular and $d \in D$, it follows that

$$1 = (x^{-1}u_jx)^k = d^k$$

and $d = 1$, if k is the order of u_j. Thus $x \in C_K(u_j)$, and

$$(*) \qquad C_{\overline{K}}(u_jD) = C_K(u_j)/D.$$

Since D is a normal subgroup of $C_K(u_j)$, D is contained in any defect group $\delta(K_j)$ of K_j. Therefore (*) implies that c) holds.

If $\overline{K}_j = (u_jD)^{\overline{K}}$, then $\tau(K_j) = \overline{K}_j$, because

$$|\overline{K} : C_{\overline{K}}(u_jD)| = |K : C_K(u_j)|$$

by (*).

Let K_i be a p-regular conjugacy class of K such that $\overline{K}_i = \overline{K}_j$. Let $u_i \in K_i$. Then

$$(u_iD)^{\overline{K}} = (u_jD)^{\overline{K}},$$

and $u_i = xu_jx^{-1}d$, for some $x \in K$ and $d \in D$. Since

$D \leq \delta(K_j)$, xu_jx^{-1} and d commute. As u_i and xu_jx^{-1} are p-regular, Lemma 5.1 implies that $u_i = xu_jx^{-1}$. Thus $K_i = K_j$, and τ is injective.

Let $\bar{K}_j$ be a p-regular conjugacy class of $\bar{K}$, and let $\bar{u}_j = u_jD \in \bar{K}_j \leq \bar{K} = C_G(D)D/D$. Then we may assume that $u_j \in C_G(D)$. If K is the order of $\bar{u}_j$, then $u_j^k \in C_G(D)$. Since $(p,k) = 1$, there are integers x and y such that

$$xp^a + yk = 1.$$

Hence

$$u_j = u_j^{xp^a} \cdot u_j^{yk} \in u_j^{xp^a}D.$$

As p^a is the highest power of p dividing the order of G, the element $u_j^{xp^a}$ is p-regular. If $v_j = u_j^{p^a x}$, then $K_j = v_j^K$ is a p-regular conjugacy class of K such that $\tau(K_j) = \bar{K}_j$. Hence a) holds.

Lemma 8.2. With the notation of Lemma 8.1 the following assertions hold:

a) τ induces a one-to-one correspondence between the blocks $b \leftrightarrow f \leftrightarrow \mu$ of FK and the blocks $\bar{b} \leftrightarrow \bar{f} \leftrightarrow \bar{\mu}$ of $F\bar{K}$ such that

α) $\bar{f} = \tau(f)$,

β) $\delta(\bar{b}) \underset{K}{=} \delta(b)/D$.

b) Let $b_i \leftrightarrow f_i \leftrightarrow \mu_i$ and $b_j \leftrightarrow f_j \leftrightarrow \mu_j$ be blocks of FK, and let x be an element of G. Then $\bar{x}^{-1}\bar{f}_i\bar{x} = \bar{f}_j$ if and only if $x^{-1}f_ix = f_j$.

c) $$T_{\bar{G}}(\bar{b}) = T_G(b)/D.$$

Proof. a) ([5], 2G) Since $\ker \tau$ is a nilpotent ideal of FG by Lemma 4.5, $\tau(f) \neq 0$ for every block $b \leftrightarrow f \leftrightarrow \mu$ of FK. If $\tau(f)$ were not a primitive idempotent of $ZF\bar{K}$, then there would be at least two different blocks $\bar{b}_1 \leftrightarrow \bar{f}_1 \leftrightarrow \bar{\mu}_1$ and $\bar{b}_2 \leftrightarrow \bar{f}_2 \leftrightarrow \bar{\mu}_2$ of $F\bar{K}$ satisfying

(*) $$\bar{\mu}_1(\tau(f)) = 1 = \bar{\mu}_2(\tau(f)).$$

Clearly $\bar{\mu}_1\tau$ and $\bar{\mu}_2\tau$ are linear characters of FK. Since by (*) $\mu(f) = 1 = \bar{\mu}_i\tau(f)$, Proposition 2.2 implies that $\bar{\mu}_1\tau$ and $\bar{\mu}_2\tau$ are in the same equivalence class of linear characters as μ. Hence there is an F-algebra isomorphism π from $\bar{\mu}_1\tau(ZFK)$ onto $\bar{\mu}_2\tau(ZFK)$ such that

$$\pi\bar{\mu}_1\tau(c_i) = \bar{\mu}_2\tau(c_i)$$

for every conjugacy class K_i of K with class sum c_i. By Lemma 8.2 a) this implies that

(**) $$\pi\bar{\mu}_1(\bar{c}_i) = \bar{\mu}_2(\bar{c}_i)$$

for every p-regular conjugacy class $\bar{K}_i$ of $\bar{K}$ with class sum $\bar{c}_i$. Since $\bar{f}_1$ and $\bar{f}_2$ are both a linear combination of p-regular class sums of $\bar{K}$ by Theorem 5.3, we obtain from (**) that

$$1 = \pi\bar{\mu}_1(f_1) = \bar{\mu}_2(f_1) = 0.$$

This contradiction proves that $\tau(f)$ is a primitive idempotent of $Z(F\bar{K})$.

Now let $f_1, f_2, \ldots, f_n$ be the centrally primitive idempotents of FK. Then

$$1 = f_1 + f_2 + \ldots + f_n .$$

Since τ is an algebra epimorphism of FK onto $F\bar{K}$, we have

$$\tau(1) = \bar{1} = \tau(f_1) + \tau(f_2) + \ldots + \tau(f_n).$$

Since $\tau(f_i)$, $i = 1,2,\ldots,n$ are centrally primitive idempotents, a) α) holds .

The assertion a) β) follows immediately from Theorem 5.3, Lemma 8.1 c) and the definition of a defect group of a block.

b) It is clear that $\bar{x}^{-1}\bar{f}_i\bar{x} = \bar{f}_j$ follows from $x^{-1}f_i x = f_j$. Conversely, if $\bar{x}^{-1}\bar{f}_i\bar{x} = \bar{f}_j$, then

$$x^{-1}f_i x - f_j \in \ker \tau \cap Z(FK).$$

Since $\ker \tau$ is nilpotent by Lemma 4.5, there is an integer $c \geq 0$ such that

$$(x^{-1}f_i x - f_j)^{p^c} = 0.$$

Hence $x^{-1}f_i x = f_j$, because $x^{-1}f_i x$ and f_1 are block idempotents of FK.

c) follows immediately from b).

We now can prove Brauer's main theorem on blocks with normal defect groups which in the generality presented here is due to W. Hamernik and the author [18]. Ring theoretical proofs of special cases of this theorem were given before by Y. Kawada [22] and W. F. Reynolds [32].

Theorem 8.3. Let F be a field of characteristic $p > 0$ dividing the order of the finite group G. Let D be a p-subgroup of G, $H = N_G(D)$, $K = C_G(D)D$, $\bar{H} = H/D$ and $\bar{K} = K/D$.

Then there is a one-to-one correspondence $\phi = \tau\sigma$ between the blocks $B \leftrightarrow e \leftrightarrow \lambda$ of FG having defect group $\delta(B) \underset{G}{=} D$ and the $\bar{H}$-conjugacy classes of blocks $\bar{b} \leftrightarrow \bar{f} \leftrightarrow \mu$ of $F\bar{K}$ having defect zero and the property that $\bar{b} \leftrightarrow \bar{f} \leftrightarrow \bar{\mu}$ is only covered by blocks of $F\bar{H}$ with defect zero.

Proof. By Theorem 4.12 we may assume that D is a normal subgroup of G. Since the normal subgroup $K = C_G(D)D$ of G satisfies the hypothesis of Theorem 7.4 there is a one-to-one correspondence between the blocks $B \leftrightarrow e \leftrightarrow \lambda$ of FG with defect group D and the representatives $b \leftrightarrow f \leftrightarrow \mu$ of G-conjugacy classes of blocks of FK having defect group D and the property that $b \leftrightarrow f \leftrightarrow \mu$ is only covered by blocks of FG with defect $d = \nu(|D|)$.

Proposition 8.4. Let D be a normal p-subgroup of G, $K = DC_G(D)$, $\bar{K} = K/D$, and $\bar{G} = G/D$.

Then the canonical group algebra homomorphism τ from FG onto $F\overline{G}$ induces a one-to-one correspondence between the representatives $b \leftrightarrow f \leftrightarrow \mu$ of G-conjugacy classes of blocks of FK and the representatives $\bar{b} \leftrightarrow \bar{f} \leftrightarrow \bar{\mu}$ of $\overline{G}$-conjugacy classes of blocks of $F\overline{K}$. Furthermore,

a) $$\delta(\tau(f)) =_{\overline{K}} \delta(b)/D$$

b) $$\nu(|T_{\overline{G}}(\bar{b}) : \overline{K}|) = \nu(|T_G(b) : K|).$$

c) $b \leftrightarrow f \leftrightarrow \mu$ is only covered by blocks of FG with defect $\nu(|D|)$ if and only if $\bar{b} \leftrightarrow \bar{f} \leftrightarrow \bar{\mu}$ is only covered by blocks of $F\overline{G}$ with defect zero.

Proof. By Lemma 8.2 it remains to show the assertion c).

Suppose that $b \leftrightarrow f \leftrightarrow \mu$ is a block of FK with the property that $b \leftrightarrow f \leftrightarrow \mu$ is only covered by blocks $B \leftrightarrow e \leftrightarrow \lambda$ of FG with defect $d = \nu(|D|)$. As D is normal in G Corollary 4.7 implies that $\delta(B) = D$. Furthermore, $B = b^G$ by Proposition 7.2 a). If $\{g_j \in G \mid j=1,2,..,t\}$ is a transversal of the inertia subgroup $T_G(b)$ in G, then $e = \sum_{j=1}^{t} f^{g_j}$. Let $\bar{g}_j = g_jD$ for $j = 1,2,..,t$. Then Lemma 8.2 implies

(*) $$\tau(e) = \sum_{j=1}^{t} \tau(f)^{\bar{g}_j}.$$

If $\overline{U}$ denotes the F-subalgebra of $\overline{T} = ZF\overline{K} \cap ZF\overline{G}$ generated by the class sums of all p-regular conjugacy classes $\overline{C}_i \leq \overline{K}$, $i = 1,2,..,s$, of $\overline{G}$ with defect zero, then $\tau(e)$ is a primitive idempotent of $\overline{T}$ contained in $\overline{U}$ by

Proposition 9.3. Thus $\tau(e) \in \overline{V}$ where $\overline{V}$ is the F-subspace of $ZF\overline{G}$ generated by all class sums of the conjugacy classes $\overline{C}$ of $\overline{G}$ with defect zero. If $\tau(e) = \bar{e}_1 + \bar{e}_2 + .. + \bar{e}_r$, where the $\bar{e}_i$ are centrally primitive idempotents of $F\overline{G}$, then $\bar{e}_i = \tau(e)\bar{e}_i \in \overline{V}$, because $\overline{V}$ is an ideal of $ZF\overline{G}$ by Lemma 3.6. Thus (*) asserts that $\bar{b} \leftrightarrow \tau(f) \leftrightarrow \bar{\mu}$ is only covered by blocks of $F\overline{G}$ with defect zero.

Conversely, suppose that the block $\bar{b} \leftrightarrow \bar{f} \leftrightarrow \bar{\mu}$ of $F\overline{K}$ is only covered by blocks of $F\overline{G}$ with defect zero. Then by a) and Lemma 8.2 c) there exists exactly one block $b \leftrightarrow f \leftrightarrow \mu$ of FK such that $\bar{f} = \tau(f)$ and $T_{\overline{G}}(\bar{b}) = T_G(b)/D$. Therefore $\bar{g}_j = g_jD$, $j = 1,2,..,t$, is a transversal of $T_{\overline{G}}(\bar{b})$ whenever $\{g_j \in G \mid j = 1,2,..,t\}$ is a transversal of $T_G(b)$, and

$$(**) \qquad \tau\Big(\sum_{j=1}^{t} f^{g_j}\Big) = \sum_{j=1}^{t} \tau(f)^{\bar{g}_j} \in \overline{T} = ZF\overline{K} \cap ZF\overline{G}.$$

By Proposition 7.2 $e = \sum_{j=1}^{t} f^{g_j}$ is a centrally primitive idempotent of FG with defect group $\delta(e) \geq D$. Hence it remains to show that $\delta(e) \leq D$.

By (**) $\tau(e) = \sum_{j=1}^{t} \bar{f}^{\bar{g}_j} \in \overline{V} \cap \overline{T}$, because $\bar{b} \leftrightarrow \bar{f} \leftrightarrow \bar{\mu}$ is only covered by blocks of $F\overline{G}$ with defect zero. Since $\overline{U} = \overline{V} \cap \overline{T}$, $\tau(e) \in \overline{U}$. Therefore $\tau(e)$ is a linear combination of class sums $\bar{c}_i$ of p-regular conjugacy classes $\overline{C}_i \leq \overline{K}$ of $\overline{G}$ with defect zero, $i = 1,2,..,s$. Thus

$$\tau(e) = \sum_{i=1}^{s} r_i\bar{c}_i \ , \quad \text{some } r_i \in F.$$

By Lemma 9.1 for every $\overline{C}_i$, $i = 1,2,..,s$, there is exactly one p-regular conjugacy class C_i of G with defect group D such that $\tau(c_i) = \overline{c}_i$, where c_i is the class sum of C_i for $i = 1,2,..,s$. Hence $\tau(e) = \sum_{i=1}^{s} r_i\, \tau(c_i)$, and

$$e = \sum_{i=1}^{s} r_i c_i + v \quad \text{for some} \quad v \in \ker \tau.$$

By Lemma 4.7 the ideal $\ker \tau$ of FG is nilpotent. Since $\sum_{i=1}^{s} r_i c_i \in ZFG$, it follows that

$$e = e^{p^h} = (\sum_{i=1}^{s} r_i c_i)^{p^h}$$

for a suitable integer $h \geq 1$. Clearly $\sum_{i=1}^{s} r_i c_i \in J_D$. As J_D is an ideal of ZFG, it follows that

$$e = (\sum_{i=1}^{s} r_i c_i)^{p^h} \in J_D.$$

Hence $\delta(e) \leq D$ by Theorem 3.8. This completes the proof of Proposition 8.4.

§9. Conjugacy classes and blocks

By Osima's theorem (see Theorem 5.3) it is clear that there are close relations between the p-regular conjugacy classes of the finite group G and the blocks of the group algebra FG. In this section we give upper bounds for the number of blocks $B \leftrightarrow e \leftrightarrow \lambda$ with a given defect group D of G by using the number of p-regular conjugacy classes of G with defect group D.

Lemma 9.1. With the notation of Lemma 8.1 the following statement holds:

τ induces a one-to-one correspondence between the p-regular conjugacy classes C of G with defect groups $\delta(C) \underset{G}{\equiv} D$ and the p-regular conjugacy classes $\bar{C}$ of $\bar{G} = G/D$ contained in $\bar{K}$ with defect zero.

Proof. [24]. Let C be a p-regular conjugacy class of G with defect group $\delta(C) = D$. Then $C = g^G$ for some $g \in C$, and D is the p-Sylow subgroup of $C_G(g)$. Hence $g \in C_G(D)$, and $C \leq K = DC_G(D)$, because K is a normal subgroup of G. Thus $\tau(C) \leq \bar{K}$, and C is the disjoint union of finitely many p-regular conjugacy classes C_i of K with defect group $\delta(C_i) = D$, $i = 1,2,\ldots,s$. Therefore Lemma 8.1 b) implies that $C \leq C_G(D)$. Hence $D \leq C_G(x)$ for every $x \in C$. Thus $C_G(x)/D \leq C_{\bar{G}}(xD)$ for every $x \in C$. If $\bar{y} = yD \in C_{\bar{G}}(\bar{x})$, then $y^{-1}xy = xd = dx$ for some $d \in D$. As x is p-regular, Lemma 5.1

implies that $d = 1$. Hence $y \in C_G(x)$, and $C_{\bar{G}}(\bar{x}) = C_G(x)/D$ for every $x \in C$. Furthermore, $\tau(C) = \bar{x}^{\bar{G}}$ is a p-regular conjugacy class of $\bar{G}$ contained in $\bar{K}$ with defect zero.

If, conversely, $\bar{C}$ is a p-regular conjugacy class of $\bar{G}$ contained in $\bar{K}$ with defect zero, then $\bar{C}$ is a disjoint union of finitely many p-regular conjugacy classes $\bar{C}_i$ of $\bar{K}$ with defect zero for $i = 1,2,\ldots,s$. By Lemma 8.1 there exists for each $i \in \{1,2,\ldots,s\}$ an element $t_i \in C_G(D)$ such that $C_i = t_i^K$ is a p-regular conjugacy class of K with defect group $\delta(C_i) = D$. Furthermore, $C_i \neq C_j$ if $i \neq j$. Let $C = t_1^G$. Then $D \leq C_G(t_1)$, and $\bar{C} = \bar{t}_1^{\bar{G}}$, where $t_1 = \bar{t}_1 D$. Since t_1 is p-regular, it follows again that $C_G(T_1)/D = C_{\bar{G}}(\bar{t}_1)$. As $\bar{C} = \bar{t}_1^{\bar{G}}$ has defect zero, and D is normal in G, the p-regular conjugacy class C of G has defect group $\delta(C) = D$, and $\tau(C) = \bar{C}$.

Lemma 9.2 [22]. If D is a normal p-subgroup of G, and if e is a centrally primitive idempotent of FG, then

$$e = \sigma(e) \in Z(F(C_G(D)\cdot D) \cap Z(FC_G(D)),$$

where σ is the Brauer homomorphism from ZFG into $ZF(N_G(D))$.

Proof. As D is a normal p-subgroup of G $\ker \sigma$ is a nilpotent ideal of ZFG by Theorem 4.2 and Proposition 4.6. Thus $\sigma(e_i) \neq 0$ for every centrally primitive idempotent e_i of FG, $i = 1,2,\ldots,n$. Since $G = N_G(D)$ and since σ is an F-algebra homomorphism from ZFG into $ZFN_G(D)$ by Theorem

4.2, it follows that

$$1 = \sigma(1) = \sum_{i=1}^{n} e_i = \sum_{i=1}^{n} \sigma(e_i).$$

Hence every $\sigma(e_i)$ is a centrally primitive idempotent of FG. Therefore there is a permutation π of $\{1,2,\dots,n\}$ such that $e_i = \sigma(e_{\pi(i)})$ for $i = 1,2,\dots,n$. If $e_{\pi(i)} = \sum_{j=1}^{k} r_j c_j$, where $r_j \in F$, and c_j is the class sum of the conjugacy class K_j of G, then

$$e_i = \sigma(e_{\pi(i)}) = \sum_{j=1}^{k} \sigma(r_j)\,\sigma(c_j) = \sum_{j=1}^{k} r_j\,\sigma(c_j).$$

Thus $\sigma(c_j \neq 0$ for at least one $j \in \{1,2,\dots,q\}$, if we assume that $\operatorname{Sup} e_{\pi(i)} = \{c_j \mid j = 1,2,\dots,q\}$. Therefore $C_G(D) \cap K_j \neq \phi$. As $C_G(D)$ is a normal subgroup of G, it follows that $K_j \leq C_G(D)$. Hence $\sigma(c_j) = c_j$ whenever $c_j \in \operatorname{Sup} e_{\pi(i)}$. Thus

$$\sigma(e_i) = \sum_{j=1}^{k} r_j'(c_j) = \sum_{j=1}^{k} r_j c_j = e_i \quad \text{for } i = 1,2,\dots,n.$$

Furthermore, Theorem 4.2 implies that

$$e_i = \sigma(e_i) \in ZFDC_G(D) \cap ZFC_G(D).$$

<u>Proposition 9.3</u> [24]. Let F be a field of characteristic $p > 0$, and let G be a finite group. Let D be a normal p-subgroup of G, $K = DC_G(D)$, $\overline{G} = G/D$ and $\overline{K} = K/D$. Let $\overline{U}$ be the F-subalgebra of $\overline{T} = ZF\overline{K} \cap ZF\overline{G}$ generated by the class sums of all p-regular conjugacy classes $\overline{C} \leq \overline{K}$ of $\overline{G}$

with defect zero.

If d is the number of blocks $B_s \leftrightarrow e_s \leftrightarrow \lambda_s$ of FG with defect group $\delta(B_s) = D$, and if c is the number of the p-regular conjugacy classes C of G with defect group $\delta(C) = D$, then

$$c = \sum_{s=1}^{d} \dim_F \tau(e_s)\overline{T}/\tau(e_s)J(\overline{T}) + \dim_F J(\overline{U}),$$

where τ is the canonical ring epimorphism from FG onto $F\overline{G}$ and where $J(A)$ denotes the Jacobson radical of the ring A.

Furthermore, the idempotents $\tau(e_s)$, $s = 1,2,\ldots,d$, are the primitive idempotents of $\overline{T}$ contained in $\overline{U}$.

Proof. By Lemma 9.1 there is a one-to-one correspondence between the p-regular conjugacy classes C_i of $\overline{G}$ with defect group $\delta(C_i) = D$, $i = 1,2,\ldots,c$, and the p-regular conjugacy classes $\overline{C}_i \leq \overline{K}$ of $\overline{G}$ with defect zero. Let $\overline{C}_1, \overline{C}_2,\ldots,\overline{C}_c$, $\overline{C}_{c+1},\ldots,\overline{C}_t$ be all the conjugacy classes of $\overline{G}$ with defect zero, and let $\bar{c}_j$ be the class sum of $\overline{C}_j$ for $j = 1,2,\ldots,t$. If $\overline{V}$ is the F-subalgebra of $ZF\overline{G}$ generated by $\{\bar{c}_j \mid j = 1,2,\ldots,t\}$, then $\overline{V}$ is an ideal of $ZF\overline{G}$ by Lemma 3.6. Hence

$$\overline{U} = \overline{V} \cap ZF\overline{K} \text{ is an ideal of } \overline{T} = ZF\overline{K} \cap ZF\overline{G}.$$

Since $\{\bar{c}_i \mid i = 1,2,\ldots,c\}$ is a basis of the F-vector space $\overline{U}$, it follows that $c = \dim_F\overline{U}$.

Let $B_s \leftrightarrow e_s \leftrightarrow \lambda_s$, $s = 1,2,\ldots,d$, be the blocks of FG with defect group D. Then $e_s \in ZFK \cap ZFC_G(D)$ by Lemma 9.2, because D is a normal p-subgroup of G. Therefore e_s is a linear combination of class sums of p-regular conjugacy classes C_{sj}, $j = 1,2,\dots,k_s$, with defect group $\delta(C_{is}) = D$ as is easily seen by means of Theorem 5.3. By Lemma 4.5 the kernel $\ker \tau$ of the canonical group algebra homomorphism τ from FG onto $F\bar{G}$ is nilpotent. Hence, if c_{sj} is the class sum of C_{sj}, then

$$(*) \qquad 0 \neq \tau(e_s) = \sum_{j=1}^{k_s} h_{sj}\tau(c_{sj}) \in \bar{U} \leq ZF\bar{K} \cap ZF\bar{G} = \bar{T},$$

where $h_{sj} \in F$, because by Lemma 9.1 $\tau(C_{sj})$ is a p-regular conjugacy class of $\bar{G}$ contained in $\bar{K}$ having defect zero.

As D is a normal p-subgroup of G and as $K = DC_G(D)$ satisfies the hypothesis of Proposition 7.2 there is for every block $B_s \leftrightarrow e_s \leftrightarrow \lambda_s$, $s = 1,2,\dots,d$, a uniquely determined block $b_s \leftrightarrow f_s \leftrightarrow \mu_s$ of FK such that

$$e_s = \sum_{i=1}^{t_s} f_s^{g_{si}},$$

where $\{g_{si} \mid i = 1,2,\dots,t_s\}$ is a transversal of the inertia subgroup $T_G(b_s)$ in G. Thus

$$\tau(e_s) = \sum_{i=1}^{t_s} \tau(f_s)^{\bar{g}_{si}}, \quad \text{where} \quad \bar{g}_{si} = g_{si}D.$$

Let $\gamma_s = \{\tau(f_s)^{\bar{g}_{si}} \mid i = 1,2,\dots,t_s\}$. Then Proposition 8.4 implies that $\gamma_1,\gamma_2,\dots,\gamma_d$ are exactly all the $\bar{G}$-conjugacy classes of blocks of $F\bar{K}$ having defect zero. Therefore Lemma 6.2 a) implies that $\tau(e_s)$ is a primitive idempotent of $\bar{T}$.

As $\tau(e_s) \in \overline{U}$ by (*), and as $\overline{U}$ is an ideal of $\overline{T}$, $\tau(e_s)$ is a primitive idempotent of $\overline{U}$ for every $s = 1,2,\ldots,d$.

Conversely, if u is a primitive idempotent of $\overline{U}$, then u is also a primitive idempotent of $\overline{T}$ by Lemma 1.4. Thus by Lemma 6.2 b) there is a block $\bar{b} \leftrightarrow \bar{f} \leftrightarrow \bar{\mu}$ of $F\overline{K}$ such that

$$u = \sum_{j=1}^{t} \bar{f}^{\bar{g}_j}$$

where $\{\bar{g}_j \mid j = 1,2,\ldots,t\}$ is a transversal of $T_{\overline{G}}(\bar{b})$ in $\overline{G}$. Since $\gamma = \{\bar{f}^{\bar{g}_j} \mid j = 1,2,\ldots,t\}$ is a $\overline{G}$-conjugacy class of primitive idempotents of $F\overline{K}$, and since $u \in \overline{U} \leq \overline{U}_o$, where $\overline{U}_o$ is the ideal of $ZF\overline{K}$ generated by all class sums of $\overline{K}$ of defect zero, it follows that

$$\bar{f}^{\bar{g}_j} = u\bar{f}^{\bar{g}_j} \in \overline{U}_o \quad \text{for } j = 1,2,\ldots,t.$$

Thus γ is a $\overline{G}$-conjugacy class of blocks of $F\overline{K}$ having defect zero. Hence $\gamma \in \{\gamma_s \mid s = 1,2,\ldots,d\}$, and

$$u \in \{\tau(e_s) \mid s = 1,2,\ldots,d\}.$$

Therefore $\{\tau(e_s) \mid s = 1,2,\ldots,d\}$ is the set of all primitive idempotents of $\overline{U}$.

Since $\overline{U}$ is an ideal of $\overline{T}$, $\tau(e_s)\overline{T} \leq \overline{U}$ for every $s = 1,2,\ldots,d$. Let $w = \sum_{s=1}^{d} \tau(e_s)$. Then

$$\overline{U} = w\overline{T} \oplus (1-w)\overline{U}.$$

As $\overline{U}$ is a finite-dimensional vector space over F, the ring

$(1-w)\overline{U}$ satisfies the minimum condition and the maximum condition on ideals. If $(1-w)\overline{U}$ is not a nil ring, then $(1-w)\overline{U}$ contains a primitive idempotent u of $\overline{U}$ by Lemma 4.10. But $u \notin \{\tau(e_s) \mid s = 1,2,\ldots,d\}$. This contradiction shows that $(1-w)\overline{U}$ is a nil ring. As $(1-w)\overline{U}$ is a commutative ring with maximum condition on ideals, $(1-w)\overline{U}$ is nilpotent. Thus $(1-w)\overline{U}$ is contained in the Jacobson radical $J(\overline{U})$ of $\overline{U}$.

Since $\tau(e_s)$ is a primitive idempotent of the finite-dimensional F-algebra $\overline{T} = ZF\overline{K} \cap ZF\overline{G}$ with identity element, each ideal $\tau(e_s)\overline{T}$ of $\overline{T}$ is a finite-dimensional local commutative algebra over the field F with characteristic $p > 0$. Therefore by Theorem 6.4 there exists a subfield $\overline{W}_s$ of $\tau(e_s)\overline{T}$ such that

(1) $\tau(e_s)\overline{T} = \overline{W}_s + \tau(e_s)\,J(\overline{T})$, and

(2) $\overline{W}_s \cap \tau(e_s)\,J(\overline{T}) = 0$

for $s = 1,2,\ldots,d$, where $J(\overline{T})$ denotes the Jacobson radical of $\overline{T}$. Thus

$$\tau(e_s)\overline{T} \,/\, \tau(e_s)\,J(\overline{T}) \cong \overline{W}_s, \quad \text{and}$$

$$\overline{U} = w\overline{T} \oplus (1-w)\overline{U} = \sum_{s=1}^{d} \oplus\, \overline{W}_s + \Big(\sum_{s=1}^{d} \tau(e_s)J(\overline{T}) + (1-w)\overline{U}\Big).$$

Clearly, $J(\overline{U}) = \sum_{s=1}^{d} \tau(e_s)J(\overline{T}) + (1-w)\overline{U}$, and

$$J(\overline{U}) \cap \sum_{s=1}^{d} \oplus\, W_s = 0.$$

Hence $c = \dim_F\overline{U} = \sum_{s=1}^{d} \dim_F\tau(e_s)\overline{T}/\tau(e_s)J(\overline{T}) + \dim_F J(\overline{U})$

This completes the proof of Proposition 9.3.

By means of Proposition 9.3 it is now easy to prove the main result of this section which is due to the author [24].

Theorem 9.4. Let F be a field of characteristic $p > 0$, and let G be a finite group. Let D be a p-subgroup of G, $H = N_G(D)$, $K = DC_G(D)$, $\bar{H} = H/D$ and $\bar{K} = K/D$. Let $\bar{U}$ be the F-subalgebra of the center $ZF\bar{H}$ of $F\bar{H}$ generated by the class sums of all p-regular conjugacy classes of $\bar{H}$ contained in $\bar{K}$ with defect zero, and let $\bar{T} = ZF\bar{H} \cap ZF\bar{K}$.

If d is the number of blocks $B_s \leftrightarrow e_s \leftrightarrow \lambda_s$ of FG with defect groups $\delta(B_s) \underset{G}{=} D$, and if c is the number of p-regular conjugacy classes C of G with defect groups $\delta(C) \underset{G}{=} D$, then:

a) the set $\{\tau\sigma(e_s) \mid s = 1,2,\ldots,d\}$ consists of all primitive idempotents of $\bar{T}$ contained in $\bar{U}$,

b) if $\bar{\lambda}_s$ is the linear character of $\bar{T}$ belonging to the primitive idempotent $\tau\sigma(e_s)$ of $\bar{T}$ for $s = 1,2,\ldots,d$, then

$$c = \sum_{s=1}^{d} \dim_F \bar{\lambda}_s(\bar{T}) + \dim_F J(\bar{U})$$

where $J(\bar{U})$ is the Jacobson radical of $\bar{U}$, σ is the Brauer homomorphism belonging to D, and where τ is the canonical homomorphism from the group algebra FH onto $F\bar{H}$.

Proof. By Lemma 4.8 the Brauer homomorphism σ induces a one-to-one correspondence between the conjugacy classes C_i of G with defect group $\delta(C_i) \underset{G}{=} D$ and the conjugacy classes $C_i{}'$

of $H = N_G(D)$ with defect group $\delta(C_i') = D$. Clearly, if C_i is p-regular, then also $C_i' = \sigma(C_i)$ is p-regular, and vice versa. Since, by the first main theorem on blocks (see Theorem 4.12) the Brauer homomorphism σ also induces a one-to-one correspondence between the blocks $B_s \leftrightarrow e_s \leftrightarrow \lambda_s$ of FG with defect groups $\delta(B_s) \underset{G}{=} D$ and the blocks $b_s \leftrightarrow f_s \leftrightarrow \mu_s$ of FH with defect group D, Theorem 9.4 now easily follows from Proposition 9.3.

Theorem 9.4 has several applications which will be given now. In order to state them we need the following

Definition. Let F be a field of characteristic $p > 0$. Then the conjugacy class C of the finite group G is called F-potent, if its class sum c is not a nilpotent element of the center ZFG of FG. If C is not F-potent, C is F-nilpotent.

Corollary 9.5. The number d of blocks $B_s \leftrightarrow e_s \leftrightarrow \lambda_s$ of the group algebra FG with defect groups $\sigma(B_s) \underset{G}{=} D$ is less or equal to the number c_F of p-regular F-potent conjugacy classes C of G with defect groups $\delta(C) \underset{G}{=} D$.

Proof. Let c be the number of conjugacy classes C of G with defect groups $\delta(C) \underset{G}{=} D$. Then with the notations of Theorem 9.4 we have

$$c = \sum_{s=1}^{d} \dim_F \overline{\lambda}_s(T) + \dim_F J(\overline{U}) .$$

Furthermore, $\tau\sigma(g) \in J(\bar{U})$ for the class sum q of every F-nilpotent conjugacy class Q with defect groups $\delta(Q) \underset{G}{=} D$, because τ and σ are F-algebra homomorphisms. Hence $\dim_F J(\bar{U}) \geq c - c_F$, which implies that

$$c_F \geq \sum_{s=1}^{d} \dim_F \bar{\lambda}_s(\bar{T}) \geq d.$$

Remark. It is clear that Corollary 9.5 contains Y. Kawada's theorem asserting that $d \leq c$ (see [22], Theorem 4.5).

Furthermore, the well-known result stating that the number of all the blocks of the group algebra FG is less or equal to the number of all F-potent p-regular conjugacy classes of G follows also at once from Corollary 9.5.

For the following applications of Theorem 9.4 we need Maschke's theorem. It follows immediately from the following auxiliary result which we shall also apply in §11.

Lemma 9.6. Let G be a group and let U be a subgroup of finite order $n = |G : U|$. If n is a unit in the ring R, then an exact sequence

$$0 \to M' \to M \xrightarrow{\alpha} M'' \to 0$$

of RG-modules splits if it splits as a sequence of RU-modules.

Proof. (Bass [2], p.559). Let $\gamma : M'' \to M$ be an RU-module homomorphism such that $\alpha\gamma m'' = m''$ for every $m'' \in M''$. Let $\{g_i \in G \mid i = 1,2,\ldots,n\}$ be a right transversal of U in G, and set $\gamma'(m) = \sum_{i=1}^{n} \gamma(mg_i^{-1})g_i \quad m \in M''$. If $g \in G$, then

$$\gamma'(mg) = \sum_{i=1}^{m} \gamma(mgg_i^{-1})g_i = \sum_{i=1}^{n} \gamma(mgg_i^{-1})(g_i g^{-1})g$$

$$= [\sum_{i=1}^{n} \gamma(mgg_i^{-1})g_i g^{-1}]g = \gamma(m)g,$$

because $g_i g^{-1}$ is also a right transversal of U in G, and γ' is independent of the particular choice of a right transversal of U in G, as γ is RU-linear. Thus γ' is RG-linear, and

$$\alpha\gamma'(m) = \alpha(\sum_{i=1}^{n} \gamma(mg_i^{-1})g_i) = \sum_{i=1}^{n} \alpha\gamma(mg_i^{-1})g_i = n\cdot m$$

for every $m \in M''$. Thus $\frac{1}{n}\gamma'$ is an RG-linear inverse for α.

<u>Theorem 9.7</u> (Maschke). Let F be a field and G a finite group with order $|G| = n$. Then the group algebra FG is a semi-simple Artinian ring if and only if n is a unit in F.

<u>Proof</u>. If n is a unit in F, then FG is semi-simple Artinian by Lemma 9.6, because every F-module is projective.

If n is divisible by the characteristic p of F, then $y = \sum_{g \in G} g$ is a central nilpotent non-zero element of FG, and FG is not semi-simple Artinian.

<u>Corollary 9.8</u>. With the notation of Theorem 9.4 the following statement holds:

If p does not divide the index of D in $K = DC_G(D)$, then the number c of p-regular conjugacy classes C of G

with defect groups $\delta(C) \underset{G}{\equiv} D$ and the number d of blocks $B_s \leftrightarrow e_s \leftrightarrow \lambda_s$ of FG with defect groups $\delta(B_s) \underset{G}{\equiv} D$ are related by

$$c = \sum_{s=1}^{d} \dim_F \overline{\lambda}_s(\overline{T}).$$

Proof. By Theorem 9.4 we have

$$c = \sum_{s=1}^{d} \dim_F \overline{\lambda}_s(\overline{T}) + \dim_F J(\overline{U}).$$

Since $\overline{U}$ is the F-subalgebra of $\overline{T} = Z F\overline{K} \cap Z F\overline{H}$ generated by the class sums of all p-regular conjugacy classes of $\overline{H}$ contained in $\overline{K}$ with defect zero, $J(\overline{U})$ is a nilpotent ideal of the commutative F-algebra $\overline{T}$. Hence $J(\overline{U})$ consists of nilpotent elements of the center of the group algebra $F\overline{K}$. As p does not divide the order $|\overline{K}| = |DC_G(D)/D|$ of $\overline{K}$, the group algebra $F\overline{K}$ is semi-simple by Theorem 9.7. Hence $J(\overline{U}) = 0$.

Corollary 9.9. Let F be a field of characteristic $p > 0$, and let G be a finite group. If D is a p-subgroup of G such that p does not divide the index of D in $K = DC_G(D)$, and if F is a splitting field for the group $\overline{K} = K/D$, then the number d of blocks $B_s \leftrightarrow e_s \leftrightarrow \lambda_s$ of FG with defect groups $\delta(B_s) \underset{G}{\equiv} D$ equals the number c of p-regular conjugacy classes C of G with defect groups $\delta(C) \underset{G}{\equiv} D$.

Proof. From Corollary 9.8 follows that

$$c = \sum_{s=1}^{d} \dim_F \tau\sigma\lambda_s(\overline{T}).$$

Since $\tau\sigma\lambda_s(\overline{T})$ is a finite extension field of F which by Theorem 6.4 may be assumed to be contained in $T = ZF\overline{K} \cap ZF\overline{H}$, $F = \tau\sigma\lambda_s(\overline{T})$ for $s = 1,2,\ldots,d$, because F is a splitting field for the group $K = DC_G(D)/D$. Hence $c = d$.

Corollary 9.10. a) If d_a is the number of blocks of highest defect of the group algebra FG, and if c_a is the number of p-regular conjugacy classes of G of highest defect, then

$$d_a \leq c_a.$$

b) If D is a p-Sylow subgroup of G, and if F is a splitting field for the group $\overline{K} = DC_G(D)/D$, then $c_a = d_a$.

Proof. Since $p \nmid |K| = |DC_G(D) : D|$ for every p-Sylow subgroup D of G, the assertion a) follows at once from Corollary 9.8.

Clearly Corollary 9.9 implies b).

Remark. Obviously Corollary 9.10 contains the Theorem of Brauer-Nesbitt asserting that the number d_a of blocks of FG of highest defect equals the number c_a of p-regular conjugacy classes of G of highest defect, if F is a splitting field for G.

Since in Theorem 9.4 the case $D = 1$ is not excluded, it

immediately also contains the following result which coincides with Theorem 13 A of R. Brauer [4], if F is a splitting field for G.

Corollary 9.11. Let F be a field of characteristic $p \geq 0$, and let G be a finite group.

If U is the F-subalgebra of the center ZFG of the group algebra FG generated by the class sums of the c_o conjugacy classes C of G with defect zero, and if $B_s \leftrightarrow e_s \leftrightarrow \lambda_s$ are the d_o blocks of FG with defect zero, then

$$c_o = \sum_{s=1}^{d_o} \dim_F \lambda_s(ZFG) + \dim J(U),$$

where $J(U)$ denotes the Jacobson radical of U.

§10. Conjugacy classes and simple modules

From Maschke's theorem (see Theorem 9.7) and Corollary 9.10 we immediately obtain the well-known theorem asserting that the number s of non isomorphic simple modules of the group algebra FG equals the number of conjugacy classes of G, if F is a splitting field for G whose characteristic p does not divide the order $|G|$ of G. In this section we present for the convenience of the reader the well-known proof of R. Brauer's more general theorem stating that in case p divides $|G|$ the number c of p-regular conjugacy classes equals s.

Lemma 10.1 (R. Brauer). Let F be a field of characteristic $p > 0$, and let R be a finite dimensional F-algebra with Jacobson radical $J(R)$. Suppose that $R/J(R)$ is a direct sum of full rings of matrices over F (i.e. F is a splitting field for R). Let S be the F-subspace of R generated by all commutators $ab - ba$, $a,b \in R$. If $T(R) = \{a \in R \mid a^{p^m} \in S$ for some integer $m\}$, then:

a) $T(R)$ is an F-subspace of R containing S.

b) $J(R) \subseteq T(R)$.

c) $\dim_F (R/T(R))$ equals the number s of non isomorphic simple FG-modules.

Proof. a) Let $a,b \in R$. Then by Lemma 5.2

$$(ab-ba)^p \equiv (ab)^p - (ba)^p \equiv a(ba)^{p-1}b) - (ba)^{p-1}b)a$$
$$\equiv 0 \bmod S.$$

Hence $S \le T(R)$. That $T(R)$ is a vector space over F follows at once from Lemma 5.2.

b) As $J(R)$ is nilpotent, $J(R) \subseteq T(R)$.

c) From a) and b) follows at once that

$$T(R) \,/\, J(R) = T(R/J(R)).$$

Since F is a splitting field for $R/J(R)$, the ring $R/J \cong \sum_{i=1}^{s} \oplus (F)_{n_i}$, where $(F)_{n_i}$ denotes the ring of all $n_i \times n_i$ matrices over F, and s is the number of simple (non isomorphic) R-modules. Hence

$$T(R/J(R)) = \sum_{i=1}^{s} \oplus T[(F)_{n_i}].$$

Since $R/T(R) \cong R/J(R)/T(R)/J(R)$, it suffices to show that $\dim_F T[(F)_{n_i}] = n_i^2 - 1$, in order to prove that $s = \dim_F(R/T(R))$. Hence we may assume that R is a ring of $n \times n$ matrices of F. Let $\{e_{ij} \mid i,j = 1,2,\ldots,n\}$ be a full set of matrix units of R. Then $e_{i1}e_{ij} - e_{1j}e_{i1} \in S$ for $i \neq j$, and $0 \neq e_{ii} - e_{11} = e_{i1}e_{1i} - e_{1i}e_{i1} \in S$ for $i \neq 1$.

Hence $n^2 - 1 \le \dim_F S \le \dim_F T(R) \le n^2$. As $e_{11}^{p^m} = e_{11}$ for all m, $e_{11} \notin T(R)$. Thus $\dim_F T(R) \le n^2 - 1$ and

$$S = T, \text{ and } \dim_F T(R) = n^2 - 1,$$

which completes the proof of Lemma 10.1.

Theorem 10.2 (R. Brauer). Let G be a finite group and G be a field of characteristic $p > 0$ such that F is a splitting field for the group algebra FG.

Then the number s of isomorphism classes of simple FG-modules equals the number c of p-regular conjugacy classes of G.

Proof. By Lemma 5.1 there exists for every element $g \in G$ exactly one p-regular element $g_{p'}$ and exactly one p-singular element g_p such that $g = g_p g_{p'} = g_{p'} g_p$. Let p^m be the order of g_p and let $n \geq 0$ be any natural number. Then by Lemma 5.2

$$(g-g_{p'})^{p^{m+n}} \equiv (g_p g_{p'})^{p^{m+n}} - g_{p'}^{p^{m+n}} \equiv 0 \bmod S,$$

where S is the subspace of FG generated by all commutators $ab - ba$, $a,b \in R = FG$. Thus $g \equiv g_{p'} \bmod T(R)$ by Lemma 10.1, where $T(R) = \{a \in R \mid a^{p^t} \in S$ for some $t \geq 0\}$. For every pair $h,k \in G$ the element

$$k-k^h = h(h^{-1}k) - (h^{-1}k)h \in S \subseteq T(R) \cdot$$

Therefore, if $r_1, r_2, \ldots, r_c$ is a system of representatives of the c p-regular conjugacy classes $K_i = r_i^G$, $i = 1,2,\ldots,c$, of G, then $\{r_i + T(R) \mid i = 1,2,\ldots,c\}$ is a system of generators of the F-vector space $R/T(R)$. Hence it suffices to show that they are independent over F. Otherwise, there are $a_i \in F$ such that

$$0 \neq a = \sum_{i=1}^{c} a_i r_i \in T(R).$$

Thus $a^{p^w} \in S$ for some integer w. Suppose that the order of G is $|G| = p^a q$ with $(p,q) = 1$. Then (see proof of Theorem 5.3) there is an integer x such that $p^x \equiv 1 \bmod q$. If $t \geq \max\{w,x\}$, then by Lemma 5.2

$$(*)\ 0 \equiv a^{p^t} \equiv \left(\sum_{i=1}^{c} a_i r_i\right)^{p^t} \equiv \sum_{i=1}^{c} a_i^{p^t} r_i^{p^t} \equiv \sum_{i=1}^{c} a_i^{p^t} r_i \bmod S .$$

Clearly the F-vector space S is generated by all commutators $hk - kh$, where $h,k \in G$. Thus, if $\sum_{g\in G} r_g g \in S$, then $\sum_{g\in K_i} r_g = 0$, where K_i is a conjugacy class of G. Since the r_i, $i = 1,2,\ldots,c$, are representatives of all p-regular conjugacy classes K_i of G, it follows from (*) that $a_i^{p^t} = 0$, and so $a_i = 0$ for every i. This completes the proof of Theorem 10.2.

§11. Frobenius reciprocity theorem and Clifford's theorem

In order to keep these lectures self-contained Clifford's theorem and the Frobenius reciprocity theorem are proved in this section. As an application we obtain here the theorem of Green, Stonehewer and Villamayor on the Jacobson radical of a group ring.

In this section the group G is not necessarily finite, and F may be any commutative ring unless stated otherwise. If S is a subring of the ring R and M is a right R-module, then M is also a right S-module which is denoted by M_S.

Theorem 11.1 (Nakayama). Let F be a commutative ring and U a subgroup of the group G. If M is a right FU-module, and N is a right FG-module, then:

a) $\mathrm{Hom}_{FG}(M \otimes_{FU} FG, N) \cong_F \mathrm{Hom}_{FU}(M, N_{FU})$

b) $\mathrm{Hom}_{FG}(N, \mathrm{Hom}_{FU}(FG, M)) \cong_F \mathrm{Hom}_{FU}(N_{FU}, M)$

Proof. (H. N. Ward [44]). a) For every $\lambda \in \mathrm{Hom}_{FU}(M, N_{FU})$ define $\lambda^* \in \mathrm{Hom}_{FG}(M \otimes_{FU} FG, N)$ by

$$\lambda^*(m \otimes a) = (\lambda m)a \quad m \in M, \quad a \in FG.$$

Then $\lambda \to \lambda^*$ is obviously a K-linear map. If $\lambda^* = 0$, then $\lambda^*(m \otimes 1) = \lambda(m) = 0$ for every $m \in M$, and $\lambda = 0$. If $\tau \in \mathrm{Hom}_{FG}(M \otimes_{FU} FG, N)$, then let $\lambda = \tau|M$, where $\tau|M$ denotes the restriction of τ to $M \otimes 1 \cong M$. Thus $\tau = \lambda^*$, and the map $\lambda \to \lambda^*$ is the required isomorphism.

b) For every $\lambda \in \mathrm{Hom}_{FG}(N, \mathrm{Hom}_{FU}(FG,M))$ define $\lambda^* \in \mathrm{Hom}_{FU}(M, N_{FU})$ by $\lambda^*(n) = [\lambda(n)](1)$, $n \in N$. Then clearly $\lambda \to \lambda^*$ is an injective K-linear map. If $\alpha \in \mathrm{Hom}_{FU}(M, N_{FU})$, then the map λ defined by $[\lambda(n)]a = \alpha(n)a$, $n \in N$, $a \in FG$ belongs to $\mathrm{Hom}_{FG}(N, \mathrm{Hom}_{FU}(FG,M))$ and is uniquely determined by α. Furthermore, $\lambda^* = \alpha$. Thus $\lambda \to \lambda^*$ is the required isomorphism of b).

<u>Lemma 11.2</u>. Let F be a commutative ring and let U be a subgroup of finite index in the group G. If N is a right FG-module and M a right FU-module, then

$$\mathrm{Hom}_{FU}(FG,M) \cong_{FG} M \otimes_{FU} FG.$$

<u>Proof</u>. Let $\{g_i \in G \mid i = 1,2,\ldots,n\}$ be a left transversal of U in G. For every $\phi \in \mathrm{Hom}_{FU}(FG,M)$ let

$$\tau(\phi) = \sum_{i=1}^{n} (g_i^{-1}) \otimes g_i \in M \otimes_{FU} FG.$$

Then τ is an injective F-homomorphism. Since FG is a free right FU-module, for every set $\{m_i \in M \mid i = 1,2,\ldots,n\}$ there is a $\phi \in \mathrm{Hom}_{FU}(FG,M)$ such that $\phi(g_i^{-1}) = m_i$, $i = 1,2,\ldots,n$. Hence τ is an F-module isomorphism.

For every $g \in G$ there is a $u_i \in U$ and a permutation $i \to \pi(i)$ of the index set $\{1,2,\ldots,n\}$ such that $gg_i^{-1} = g_{\pi(i)}^{-1}u_i$. Hence

$$\tau(\phi g) = \sum_{i=1}^{n} (\phi g)g_i^{-1} \otimes g_i = \sum_{i=1}^{n} \phi(gg_i^{-1}) \otimes g_i$$

$$= \sum_{i=1}^{n} \phi(g_{\pi(i)}^{-1}) \otimes g_i = \sum_{i=1}^{n} \phi(g_{\pi(i)}^{-1}) \otimes u_i g_i$$

$$= \sum_{i=1}^{n} \phi(g_{\pi(i)}^{-1}) \otimes g_{\pi(i)}g = \tau(\phi)g.$$

Therefore τ is an FG-module isomorphism, and Lemma 11.2 holds.

Definition. Let R be a ring and M a right R-module. Then the socle of M is the sum of all simple R-submodules of M.

As is well known (see Lambek [23], p.59) the socle of M is the direct sum of a subfamily of the family of all simple submodules of M. Hence it is completely reducible. Thus the socle is the largest completely reducible submodule of M. The co-socle of M is the largest completely reducible factor module of the right R-module M, if M is a right R-module with minimum condition on R-submodules.

If N is a simple R-submodule of the completely reducible R-module M, then the homogeneous N-component of M is the sum of all simple R-submodules of M which are isomorphic to N. If M is Artinian or Noetherian, then the N-component of M is a direct sum of finitely many simple R-submodules which are isomorphic to N. The following result is taken from Huppert ([19], p.557).

Lemma 11.3. Let R be a right Artinian ring, and let M be a finitely generated right R-module and N a simple right R-module.

a) If the homogeneous N-component M/M' of the co-socle of M is isomorphic to the direct sum of k copies of N, then

$$\mathrm{Hom}_R(M,N) \cong_R \mathrm{Hom}_R(N,N)^k .$$

b) If the homogeneous N-component M'' of the socle of M is isomorphic to the direct sum of s copies of N, then

$$\mathrm{Hom}_R(N,M) \cong_R \mathrm{Hom}_R(N,N)^s .$$

Proof. a) For every $0 \neq \alpha \in \mathrm{Hom}_R(M,N)$ the kernel $\ker \alpha \geq M'$ by the definition of M/M', because M/M' has a finite composition length. Thus $\mathrm{Hom}_R(M,N) \cong \mathrm{Hom}_R(M/M',N) \cong \mathrm{Hom}_R(N^k,N)$, and a) holds.

b) Let $0 \neq \alpha \in \mathrm{Hom}_R(N,M)$. If $\alpha N \nleq M''$, then $M'' \oplus \alpha N > M''$, and M'' is not the homogeneous N-component of M. Thus $\alpha N \leq M''$ for every $0 \neq \alpha \in \mathrm{Hom}_R(N,M)$. Therefore $\mathrm{Hom}_R(N,M) \cong \mathrm{Hom}_R(N,M'') \cong \mathrm{Hom}_R(N,N^s)$, and b) holds.

Theorem 11.4 (Frobenius reciprocity theorem). Let G be a finite group and F a field which is a splitting field for G. Let U be a subgroup of G, M an irreducible FU-module and N an irreducible FG-module.

a) If the homogeneous N-component of the co-socle of $M \otimes_{FU} FG$ is isomorphic to the direct sum of k copies of N, and if the homogeneous M-component of the socle of N_{FU} is isomorphic to the direct sum of s copies of M, then $s = k$.

b) If the homogeneous N-component of the socle of $M \otimes_{FU} FG$

is isomorphic to the direct sum of q copies of N, and if the homogeneous M-component of the co-socle of N_{FU} is isomorphic to the direct sum of t copies of M, then $t = q$.

Proof. Since $\mathrm{Hom}_{FG}(N,N) = F = \mathrm{Hom}_{FU}(M,M)$, the proof follows at once from the above results of §11.

Theorem 11.5 (Clifford). Let R be a ring and N a normal subgroup of finite index of the group G. Let M be a simple RG-module containing the simple RN-module $W \neq 0$.

If $\{g_j \in G \mid j = 1,2,\ldots,t \leq |G:N|\}$ is a transversal of the inertia subgroup

$$T_G(W) = \{g \in G \mid W \cong_{RN} W_g\}$$

of W in G, then there is an integer z such that

$$M_{RN} \cong_{RN} (\sum_{j=1}^{t} \oplus Wg_j)^z.$$

In particular, M_{RN} is a completely reducible RN-module.

Proof. Let $\{h_i \in G \mid i = 1,2,\ldots,n = |G:N|\}$ be a transversal of N in G, and let $W = wRN$. Then

$$(*) \qquad M = wRG = \sum_{i=1}^{n} wRNh_i = \sum_{i=1}^{n} Wh_i.$$

Therefore M_{RN} is a completely reducible RN-module, because Wh_i is a simple RN-module for $i = 1,2,\ldots,n$. Clearly $N \leq T_G(W) = \{g \in G \mid W \cong_{RN} Wg\}$. Thus $t = |G:T_G(W)| \leq |G:N| = n$. Let $\{g_j \in G \mid j = 1,2,\ldots,t\}$ be a transversal of $T_G(W)$ in G. If M_j is the Wg_j-component of the completely reducible RN-module M_{RN}, then (*) implies by application of Lambek [23], p.62, Proposition 3 that

$$(**) \qquad M = \sum_{j=1}^{t} \oplus M_j$$

If M_1 is the direct sum of z simple submodules of M which are RN-isomorphic to W, then also the length of every composition series of the RN-module M_j is z for $j = 1, 2, \ldots, t$. Hence (**) implies

$$M_{RN} \cong_{RN} \left(\sum_{j=1}^{t} \oplus Wg_j\right)^z,$$

which completes the proof of Theorem 11.5.

In the following part of this section we derive the theorem of Green, Stonehewer and Villamayor (see [15] and [39]) from Clifford's theorem and Lemma 9.6. It yields a useful relation between the Jacobson radical $J(RG)$ of the group ring RG of the group G over the ring R and the Jacobson radical $J(RN)$ of the group ring RN of the normal subgroup N of finite index $|G:N| = n < \infty$ of G.

<u>Lemma 11.6</u>. If R is a ring and N a normal subgroup of finite index of the group G, then

$$J(RN) \leq J(RG).$$

<u>Proof</u> [39]. It suffices to show that $Mx = 0$ for every $x \in J(RN)$

and every simple FG-module M. Let $\{g_i \in G \mid i = 1,2,\ldots,n = |G:N|\}$ be a transversal of N in G and let $M = mFG$, then

$$M_{RN} = \sum_{i=1}^{n} mg_i\, RN.$$

Hence M_{RN} is a finitely generated RN-module. By Zorn's Lemma it therefore contains a maximal RN-submodule T. Thus Tg_i is a maximal RN-submodule of M_{RN} for $i = 1,2,\ldots,n$. Therefore $(M/Tg_i)x = 0$ for every $x \in J(RN)$, and

$$Mx \subseteq \bigcap_{i=1}^{n} Tg_i = S \quad \text{for every} \quad x \in J(RN).$$

Clearly S is an FN-submodule of M which is properly contained in M, because $M \supsetneqq T \geq S$. For every $s \in S$ there are $t_i \in T$ such that $s = t_i g_i$ for $i = 1,2,\ldots,n$. Let g_j be any element of the transversal $\{g_i \in G \mid i = 1,2,\ldots,n\}$ of N in G, which we chose above. Then

$$sg_j = t_i g_i g_j = t_i h_{ij} g_{\pi(i)} \in Tg_{\pi(i)} \quad \text{for} \quad i = 1,2,\ldots,n,$$

where $g_i g_j = h_{ij} g_{\pi(i)}$, $h_{ij} \in N$ and π is a permutation of $\{1,2,\ldots,n\}$. Thus $sg_j \in \bigcap_{i=1}^{n} Tg_{\pi(i)} = \bigcap_{i=1}^{n} Tg_i = S$, and S is a proper RG-submodule of M. Therefore $S = 0$, and Lemma 11.6 holds.

<u>Lemma 11.7</u>. Let N be a normal subgroup of the group G such that $n = |G:N| < \infty$ is a unit in the ring R.

If M is a simple RN-module, then $M \otimes_{RN} RG$ is a completely reducible RG-module.

Proof. Clearly $V = M \otimes_{RN} RG$ is a finitely generated RG containing (up to isomorphism) the simple RN-module M. Therefore V_{RN} is completely reducible by Theorem 11.5. Let U be an RG-submodule of V, and let α be the natural homomorphism $V \to V/U$. Then the exact sequence

$$0 \to U \to V \to V/U \to 0$$

splits as a sequence of RN-modules. Hence by Lemma 9.6 it splits as a sequence of RG-modules. Therefore U is a direct summand of the FG-module V, and V is a completely reducible FG-module (see Lambek [23], p.61).

Theorem 11.8 (Green Stonehewer, Villamayor). If N is a normal subgroup of finite index $n = |G:N|$ of the group G such that n is a unit in the ring R, then

$$J(RG) = J(RN)RG.$$

Proof. By Lemma 11.6 we know that $J(RN)RG \leq J(RG)$.

Let M be a simple RN-module. Then $V = M \otimes_{RN} RG$ is a completely reducible RG-module by Lemma 11.7. Hence $Vx = 0$ for every $x \in J(RG)$. Let $\{g_i \in G \mid i = 1,2,\ldots,n\}$ be a transversal of N in G. Then $x = \sum_{i=1}^{n} r_i g_i$, where the elements $r_i \in RN$ are uniquely determined by x. Thus

$$0 = (M \otimes 1)x = \sum_{i=1}^{n} \oplus\, Mr_i \otimes g_i,$$

and $Mr_i = 0$ for $i = 1,2,\ldots,n$. Since M was an arbitrary simple FN-module, it follows that $r_i \in J(RN)$ for $i = 1, 2,\ldots,n$. Thus $x \in J(RN)RG$, and Theorem 11.8 holds.

§12. The blocks of a p-nilpotent group

As an application of the block theory developed so far we determine in this section the structure of the block ideals of the group algebra FG of the finite p-nilpotent group G over the field F with characteristic $p > 0$ following the author's paper [25].

The group G is called p-nilpotent, if G contains a normal subgroup N whose order is not divisible by p such that G/N is a p-group. As is well known N is uniquely determined by G and called the p-complement of G. Furthermore, $G = NP$ and $N \cap P = 1$ for every p-Sylow subgroup P of G. (See Huppert [19], p.427).

Lemma 12.1. Let F be a field of characteristic $p > 0$ and let N be the p-complement of the finite p-nilpotent group G. Then the following statements hold:

a) Every block $b \leftrightarrow f \leftrightarrow \mu$ of FN is covered by a unique block $B \leftrightarrow e \leftrightarrow \lambda$ of FG.

b) If $T_G(b)$ is the inertia subgroup of $b \leftrightarrow f \leftrightarrow \mu$ in G, and if a is a primitive idempotent of $b = fFN$, then

$$T_G(aFN) = \{g \in G \mid aFN \cong_{FN} aFNg\} = T_G(b).$$

Proof. a) Let $\{g_j \in G \mid j = 1,2,\dots,t\}$ be a transversal of $T_G(f) = \{g \in G \mid f^g = f\}$ in G. Then

$$x = \sum_{j=1}^{t} f^{g_j} = e_1 + e_2 + \dots + e_r$$

for uniquely determined centrally primitive idempotents $e_i \in FG$, $i = 1,2,\dots,r$. By Theorem 5.3 each of these idempotents e_i of FG is a linear combination of p-regular conjugacy class sums of G. Since G is p-nilpotent, every p-regular element $g \in G$ is contained in N. Thus

$$e_i \in T = ZFN \cap ZFG \quad \text{for} \quad i = 1,2,\dots,r.$$

By Lemma 6.2 the idempotent x of T is primitive in T. Hence $r = 1$, and $B_1 \leftrightarrow e_1 \leftrightarrow \lambda_1$ is the unique block of FG covering the block $b \leftrightarrow f \leftrightarrow \mu$ of FN.

b) Since p does not divide the order of N, the block ideal $b = fFN$ of FN is a simple Artinian ring by Theorem 9.7. If a is a primitive idempotent of b, then $V = afFN = aFN$ is a minimal right ideal of FN. Suppose that $g \in T_G(b)$. Then $Vg = faFNg = agfg^{-1}FNg = agfFN$. Define $\phi \in \mathrm{Hom}_{FN}(agfFN, g^{-1}agfFN)$ by $\phi(agr) = g^{-1}agr$ for every $r \in fFN$. Then ϕ is an isomorphism, and the FN-modules $Vg = agfFN \cong g^{-1}agfFN \cong afFN = V$ are all isomorphic, because gag^{-1} is also a primitive idempotent of the simple Artinian ring b. Therefore $g \in T_G(V)$, and $T_G(b) \le T_G(V)$.

If $T_G(V) > T_G(b)$, then there would be an element $h \in T_H(V)$ with $h \notin T_G(b)$, and an FN-module isomorphism $\phi_h \in \mathrm{Hom}_{FN}(V,Vh)$. Since $a \in V = aFN$, $\phi_h(a) = ar_h h$ for some $r_h \in FN$. Because of $a = a^2$ it follows that $ar_h h = \phi(a) = \phi(a^2) = \phi(a)a = ar_h ha$. Hence

$$ar_h = ar_h hah^{-1}.$$

Clearly $hah^{-1} = hfh^{-1}hah^{-1}$, because $a = fa$. Thus

$$ar_h = (ar_h)f(hfh^{-1})(hah^{-1})$$

as $f \in ZFN$. Since $h \notin T_G(f)$, the block idempotents f and hfh^{-1} of FN are different. Therefore their product is zero. Hence $ar_h = 0$, and $\phi_h(a) = 0$. This contradiction proves that $T_G(V) = T_G(b)$.

Lemma 12.2. Let F be a field of characteristic $p > 0$ and let G be a finite p-nilpotent group with p-complement N.

Then every block $B \leftrightarrow e \leftrightarrow \lambda$ of FG with defect zero is a simple Artinian ring.

Furthermore, if $b \leftrightarrow f \leftrightarrow \mu$ is a block of FN covered by $B \leftrightarrow e \leftrightarrow \lambda$, and if M is (up to isomorphisms) the unique simple FN-module of b, then $T = M \otimes_{FN} FG$ is (up to isomorphisms) the unique simple FG-module of B.

Proof. By hypothesis there is a p-Sylow subgroup P of G such that $G = NP$ and $N \cap P = 1$. Let $B \leftrightarrow e \leftrightarrow \lambda$ be a block of defect zero of FG. By Lemma 6.8 B covers a unique G-conjugacy class of blocks $b_j \leftrightarrow f_j \leftrightarrow \mu_j$, $j = 1,2,\dots,t$, of FN, where $t = |G:T_G(f)|$ for $f = f_1$.

Since the order of N is not divisible by p and $\delta(B) = 1$,

$$\nu(|T_G(f)|) = \nu(|N\delta(B)|) = 0$$

by Theorem 6.11. Hence $T_G(f) = N$ because $N \leq T_G(f) \leq G$ and

$(G/N) = |P|$.

By Lemma 12.1 a) the block $b \leftrightarrow f \leftrightarrow \mu$ of FN is covered by a unique block $B \leftrightarrow e \leftrightarrow \mu$ of FG. Therefore

$$(*) \qquad e = \sum_{d \in P} f^d .$$

By Theorem 9.7 the ring fFN is a simple Artinian ring. If a is a primitive idempotent of $b = fFN$, and if $V = aFN$, then $T_G(f) = T_G(V)$ by Lemma 12.1 b). Let $M = aFG = \sum_{d \in P} \oplus\, aFNd$. If $W \neq 0$ is a simple FG-module contained in M, then Clifford's theorem (see Theorem 11.5) implies that

$$W_{FN} \cong_{FN} \sum_{d \in P} \oplus\, aFNd,$$

because $T_G(V) = T_G(f) = N$. Therefore $M = W$ is a simple right ideal of the block ideal B generated by the idempotent a. From Theorem 1.7 follows now immediately that B is a simple Artinian ring.

For the sake of completeness we give a ring theoretical proof of the following well-known result which usually is proved using characters.

<u>Lemma 12.3</u>. Let F be a field of characteristic $p > 0$ which is a splitting field for the finite group G. If p does not divide the order of G, then p also does not divide the vector space dimension $\dim_F V$ of every simple FG-module V.

<u>Proof</u>. By Theorem 9.7 the group algebra FG is semi-simple. Thus we may assume that $V = aFG$, where a is a primitive

idempotent of FG. Let $n = \dim_F V$. Since F is a splitting field for G the theorem of Artin-Wedderburn implies that $S = \mathrm{End}_{FG} V = F$. Let F_n denote the ring of all $n \times n$ matrices over F. Fix a basis $\{v_i \mid i = 1,2,\ldots,n\}$ of the left vector space V over F. Thus $v_i g = \sum_{j=1}^{n} f_{ij}(g) v_j$, $f_{ij}(g) \in F$. The map τ defined by

$$\tau(g) = (f_{ij}(g)) \in F_n$$

is a ring epimorphism from FG onto F_n. Let K_i be any conjugacy class of G with class sum c_i and $|K_i| = h_i$. Then $\tau(c_i)$ is a diagonal matrix (a_{ij}), where $a_{ii} = f$ for some $f \in F$, because c_i is in the center of FG. Hence its trace $\mathrm{tr}\,\tau(c_i) = nf$. Now, if n were divisible by p, then

$$\mathrm{tr}\,\tau(c_i) = nf = 0.$$

Since the trace is a class function,

$$\mathrm{tr}\,\tau(c_i) = h_i \mathrm{tr}\,\tau(g_i) \quad \text{for every} \quad g_i \in K_i.$$

Now h_i is a unit in F, because

$$h_i = |G : C_G(g_i)|, \quad \text{and} \quad p \nmid |G|.$$

Hence $\mathrm{tr}\,\tau(g) = 0$ for every $g \in G$, as every $g \in G$ lies in at least one conjugacy class K_i of G. Therefore the trace of every $n \times n$ matrix of F_n is zero, because $\tau(FG) = F_n$. This contradiction proves that $p \nmid n$.

Lemma 12.4. Let F be a field of characteristic $p > 0$ which is a splitting field for the finite p-nilpotent group G.

Then every block $B \leftrightarrow e \leftrightarrow \lambda$ of FG contains (up to isomorphism) a unique simple FG-module.

Proof. By Corollary 1.8 B contains at least one simple module. Hence it suffices to show that the number s of isomorphism classes of simple FG-modules equals the number b of blocks $B_i \leftrightarrow e_i \leftrightarrow \lambda_i$ of FG, $i = 1,2,\ldots,b$. By Theorem 10.2 s is the number of p-regular conjugacy classes K_i of G. As G is p-nilpotent every p-regular conjugacy class K_i of G, $i = 1,2,\ldots,s$, is contained in N. Hence their class sums c_i form an F-vector space basis of $T = ZFN \cap ZFG$, because N consists of p-regular elements. From Lemma 12.1 and Lemma 6.2 it follows now immediately that $\{e_j \mid j = 1,2,\ldots,b\}$ is the set of primitive idempotents of T. As T is an s-dimensional semi-simple algebra over the splitting field F, $s = \dim_F T = b$.

Lemma 12.5. Let F be a field of characteristic $p > 0$ and G be a finite group of order $|G| = p^a q$, where $(p,q) = 1$. If U is a finitely generated indecomposable projective right FG-module, then $p^a \mid \dim_F U$.

Proof. Let P be a p-Sylow subgroup of G and let $\{g_j \mid j = 1,2,\ldots,t\}$ be a left transversal of P in G. Then

$$(*) \qquad FG = \sum_{j=1}^{t} \oplus\, g_j FP,$$

where FP is the group algebra of the p-group P. By Lemma 4.3 the ring FP is local (i.e. its Jacobson radical is its unique maximal right ideal).

As FG is Artinian, we may assume that $U = aFG$, where a is a primitive idempotent of FG. Hence $FG = aFG \oplus (1-a)FG$, and by (*) U is a finitely generated projective right FP-module. Since FP is local, U is a finitely generated free FP-module by a well-known theorem of Kaplansky [21]. Thus $\dim_F U = np^a$, where n is the number of basis elements of the free right FP-module U.

<u>Lemma 12.6</u>. Let F be a field of characteristic $p > 0$, and let N be the p-complement of the finite p-nilpotent group G. Then the following statements hold:

a) Every block $B \leftrightarrow e \leftrightarrow \lambda$ of FG covers a unique G-conjugacy class of blocks $b \leftrightarrow f \leftrightarrow \mu$ of FN, and

$$e = \sum_{j=1}^{t} f^{g_j},$$

where $\{g_j \mid j = 1,2,\ldots,t\}$ is a transversal of $T_G(f) = \{g \in G \mid f = f^g\}$ in G.

b) $|G : N| \geq t|\delta(B)|$, where $\delta(B)$ is any defect group of $B \leftrightarrow e \leftrightarrow \lambda$. Equality holds, if F is a splitting field for N.

<u>Proof</u>. a) Since G is p-nilpotent, there is a p-Sylow subgroup P such that $G = NP$ and $N \cap P = 1$. By Lemma 6.8 $B \leftrightarrow e \leftrightarrow \lambda$ covers exactly one G-conjugacy class of blocks

$b_j \leftrightarrow f_j \leftrightarrow \mu_j$, $j = 1,2,\ldots,t$ of FN. Let $f = f_1$ and $\{g_j \in G \mid j = 1,2,\ldots,t\}$ be a transversal of $T_G(f)$ in G. Then Lemma 12.1 implies

$$e = \sum_{j=1}^{t} f^{g_j}.$$

b) By Theorem 6.11 we have

$$\nu(|T_G(f)|) \geq \nu(|N\delta(B)|) = |\delta(B)|$$

for every defect group $\delta(B)$ of $B \leftrightarrow e \leftrightarrow \lambda$, because N consists of p-regular elements. As $N \subseteq T_G(f) \subseteq G$ and $|G/N| = |P|$, it follows that

$$|P| = |G:N| \geq t|\delta(B)|.$$

Lemma 12.7. Let F be a field of characteristic $p > 0$ which is a splitting field for the finite p-nilpotent group G.

Then the block $B \leftrightarrow e \leftrightarrow \lambda$ of FG has defect zero if and only if B is a simple Artinian ring.

Proof. If $B \leftrightarrow e \leftrightarrow \lambda$ has defect zero, then B is a simple Artinian ring by Lemma 12.2.

Conversely, let $B \leftrightarrow e \leftrightarrow \lambda$ be a block of FG such that B is a simple Artinian ring. Then by Lemma 12.6 there is a block $b \leftrightarrow e \leftrightarrow \mu$ of the group algebra FN of the p-complement N of G such that

(1) $$e = \sum_{j=1}^{t} f^{g_j},$$

where $\{g_j \mid j = 1,2,\ldots,t\}$ is a transversal of the inertia subgroup $T_G(f)$ of $b \leftrightarrow f \leftrightarrow \mu$ in G. Furthermore, Lemma 12.6 b) implies that

$$(2) \qquad |G:N| = |\delta(B)|t,$$

where $\delta(B)$ is any defect group of $B \leftrightarrow e \leftrightarrow \lambda$. Because of (1), we have

$$(3) \qquad eFG = \sum_{d\in P} \oplus\, eFNd = \sum_{j=1}^{t} \sum_{d\in P} \oplus\, f^{g_i}FNd,$$

where P is a p-Sylow subgroup of G.

By Theorem 9.7 fFN is a simple finite dimensional algebra over the splitting field F. Therefore $\dim_F fFN = q^2$ for some integer q, which by the theorem of Artin-Wedderburn is the vector space dimension over F of the unique (up to isomorphism) simple right FN-module contained in fFN. As $B = eFG$ is a simple Artinian ring, a similar argument shows that $\dim_F eFG = v^2$ where v is the F-vector space dimension of the unique (up to isomorphism) simple eFG-module M. Because of (3) we have then

$$(4) \qquad v^2 = |P|tq^2.$$

Since $B = eFG$ is simple Artinian, M is a finitely generated projective indecomposable FG-module, thus $|P| \mid \dim_F M = v$ by Lemma 12.5. Hence there is an integer w such that $v = |P|w$, and (4) implies

$$|P|^2 w^2 = |P|tq^2.$$

As $p \nmid q$ by Lemma 12.4, we obtain $t = |P|$. Hence $\delta(B) = 1$ by (2), which completes the proof of Lemma 12.7.

Proposition 12.8. Let F be a field of characteristic $p > 0$ which is a splitting field for the finite p-nilpotent group G with p-complement N. Then the following assertions hold:

a) Every block $B \leftrightarrow e \leftrightarrow \lambda$ of FG covers a unique G-conjugacy class of blocks $b \leftrightarrow f \leftrightarrow \mu$ of FN, and

$$e = \sum_{i=1}^{t} f^{g_i},$$

where $\{g_j \mid j = 1,2,\ldots,t\}$ is a transversal of the inertia subgroup $T_G(f)$ in G.

b) If M is the unique simple FN-module of b, then $T = M \otimes_{FN} FG$ is the unique indecomposable projective right FG-module of B and the unique simple right FG-module W of $B \leftrightarrow e \leftrightarrow \lambda$ is isomorphic as an FN-module to $\sum_{j=1}^{t} \oplus M \otimes g_j$.

c) If $d(B)$ denotes the defect of the block $B \leftrightarrow e \leftrightarrow \lambda$ and if $J(B)$ is the Jacobson radical of B then

$$\dim_F J(B) = (p^{d(B)} - 1)(\dim_F W)^2.$$

Proof. a) follows at once from Lemma 12.6.

b) As fFN is a simple Artinian ring by Theorem 9.7, we may assume that $M = afFN$, where a is a primitive idempotent of FN. Thus

$$T = M \otimes_{FN} FG = aFG \cong \sum_{d \in P} \oplus\, aFNd,$$

where P is a p-Sylow subgroup of G. Let $|P| = p^a$, and let Q be an indecomposable finitely generated direct summand of T. Thus Q is also a projective right FG-module, and $p^a \mid \dim_F Q$ by Lemma 12.5. As F is a splitting field for G, $B = eFG$ contains only one simple FG-module W by Lemma 12.4. Therefore Corollary 1.7 implies that Q is the unique indecomposable finitely generated projective FG-module contained in $B = eFG$. Since FG is a quasi-Frobenius ring (see [9], p.420), we may assume that W is the unique simple FG-submodule contained in Q. Thus Clifford's theorem (see Theorem 11.5) implies that the FN-module W is isomorphic as an FN-module to

$$W_{FN} \cong_{FN} \left(\sum_{j=1}^{t} \oplus aFNg_j\right)^z,$$

for some integer z, because $T_G(f) = T_G(M) = \{g \in G \mid M \cong_{FN} Mg\}$ by Lemma 12.1 b). Since therefore W_{FN} is completely reducible as an FN-module, the Frobenius reciprocity theorem (see Theorem 11.4) asserts that W occurs z-times as a direct summand in the socle of $T = aFG = \sum_{d \in P} \oplus aFNd$. Hence

$$T = Q^z,$$

because FG is a Quasi-Frobenius ring. Let m be the length of a composition series of Q. Then $\dim_F Q = m \dim_F W$, because W is the unique simple FG-module contained in B. Thus

$$\dim_F T = p^a \dim_F(aFN) = z^2 mt \dim_F(aFN),$$

and $p^a = mz^2t$. Because of Lemma 12.6 it follows that

$$m\, z^2 = p^a t^{-1} = \left|\frac{T_G(f)}{N}\right| = |\delta(B)|,$$

where $\delta(B)$ is any defect group of $B \leftrightarrow e \leftrightarrow \lambda$. If $m = 1$, then Q is a simple right ideal of B, and B is a simple Artinian ring. Hence $\delta(B) = 1$ by Lemma 12.7, and b) holds by Lemma 12.2. Thus we may assume that $m \neq 1$. Because of $p^a / \dim_F Q$, we find an integer w such that $\dim_F Q = wp^a$. Hence $\dim_F T = p^a zw = p^a \dim_F aFN$. From $m \neq 1$ and $mz^2 = |\delta(B)|$ and the assumption that $z \neq 1$ we obtain that $p / \dim_F aFN$, which contradicts Lemma 12.3. Thus $z = 1$, and b) holds.

c) follows now easily from a) and b). Decompose the idempotent f into orthogonal primitive idempotents

$$f = a_1 + a_2 + \ldots + a_n$$

of fFN. Since by b) then also $(a_1)^{g_j}$ is a primitive idempotent of eFG, the set $\{a_i^{g_j} | 1 \quad 1,2,\ldots,n;\ j = 1,2,\ldots,t\}$ consists of nt orthogonal primitive idempotents of eFG such that

$$e = \sum_{i=1}^{n} \sum_{j=1}^{t} (a_i)^{g_j}.$$

As a is a primitive idempotent of fFN and fFN is a simple algebra over the splitting field F, clearly $n = \dim_F afFN = \dim_F aFN$. Hence $nt = \dim_F W = \dim_F(\sum_{j=1}^{t} \oplus\, aFNg_j)$. By Lemma 12.6 b) $|\delta(B)| = d(B) = \frac{|P|}{t}$. Hence the unique maximal FG-submodule $aJ(FG)$ of aFG has a composition series consisting of $p^{d(B)} - 1$ composition factors which are all isomorphic to W. Since

$$B = eFG = \sum_{i=1}^{n} \sum_{j=1}^{t} \oplus (a_i)^{g_j} FG,$$

the Jacobson radical

$$\dim_F J(B) = \sum_{i=1}^{n} \sum_{j=1}^{t} \dim_F[(a_i)^{g_j} J(FG)] = (p^{d(B)}-1)(\dim_F W)^2,$$

which completes the proof of Proposition 12.8.

The following results prepare the generalization of Proposition 12.8 for group algebras of p-nilpotent groups over arbitrary fields of characteristic $p > 0$. We start with a lemma which is due to Y. Kawada [22].

<u>Lemma 12.9</u>. Let F be a field of characteristic $p > 0$ and E an extension field of F. If $B \leftrightarrow e \leftrightarrow \lambda$ is a block of the group algebra FG of the finite group G with defect groups $\delta(B) \underset{G}{=} D$, then there are finitely many blocks $B_i^* \leftrightarrow e_i^* \leftrightarrow \lambda_i^*$ of EG with defect groups $\delta(B_i^*) \underset{G}{=} D$ such that

$$e = e_1^* + e_2^* + \ldots + e_n^*.$$

<u>Proof</u>. As $e \in ZFG$ and $EG \cong E \otimes_F FG$, $e \in ZEG$. Thus there are block idempotents $e_i^* \in EG$ such that

$$e = e_1^* + e_2^* + \ldots + e_n^*.$$

Since $\delta(B) \underset{G}{=} D$, $e \in I_D$ but $e \notin \hat{I}_D$ by Theorem 3.8. Let

$I_D^* = \{z \in ZEG \mid c_i \in \text{Sup } z \text{ implies } \delta(K_i) \underset{G}{\leq} D\}$, and

$\hat{I}_D^* = \{z \in ZEG \mid c_i \in \text{Sup } z \text{ implies } \delta(K_i) \underset{G}{<} D\}$. By Lemma 3.6 $e \in I_D^*$, and I_D^* and $\hat{I}_D^*$ are ideals of the center ZEG of EG. Hence $e_i^* = e_i^* e \in I_D^*$ for $i = 1,2,\ldots,n$. If $e_i^* \in \hat{I}_D^*$ for some $i \in \{1,2,\ldots,n\}$, then $e_i^* = e_i^* e \in \hat{I}_D^* e$. As $e \notin \hat{I}_D$, Lemma 3.7 implies that $\hat{I}_D e$ is a nilpotent ideal of ZFG. Hence $\hat{I}_D^* e = E \otimes_F \hat{I}_D e$ is a nilpotent ideal of EG. But $0 \neq e_i^*$ is idempotent, and so $e_i^* \notin \hat{I}_D^*$ for $i = 1,2,\ldots,n$. Therefore Theorem 3.8 implies

$$\delta(e_i^*) \underset{G}{=} D \quad \text{for} \quad i = 1,2,\ldots,n.$$

Lemma 12.10. Let F be a field of characteristic $p > 0$ and E an extension field of F. Let N be the p-complement of the finite p-nilpotent group C, and let E be a splitting field for N.

If $b \leftrightarrow f \leftrightarrow \mu$ is a block of FN which is covered by the block $B \leftrightarrow e \leftrightarrow \lambda$ of FG with defect groups $\delta(B) \underset{G}{=} D$, and if $T_G(b)$ is the inertia subgroup of $b \leftrightarrow f \leftrightarrow \mu$ in G, then there are finitely many centrally primitive idempotents f_i^* of EN such that the following assertions hold:

a) $$f = f_1^* + f_2^* + \ldots + f_n^* .$$

b) $|T_G(f_i^*)| = p^{a-d}$ for $i=1,2,\ldots,n$, where $|D|=p^d$ and $|G/N|=p^a$.

c) If $\{g_{ij} \in G \mid j = 1,2,\ldots,t\}$ is a transversal of $T_G(f_i^*)$ in G, then

$$e_i^* = \sum_{j=1}^{p^{a-d}} (f_i^*)^{g_{ij}}$$

is a centrally primitive idempotent of EG.

d) There exists $1 \le k \le n$ such that $e = e_1^* + e_2^* + \dots + e_k^*$ and $kp^{a-d} = nt$.

<u>Proof</u>. By Lemma 12.6 $e = \sum_{j=1}^{n} f^{g_j}$, and from Lemma 12.9 follows that there are primitive idempotents f_i^*, $i = 1,2,\dots,n$, of the center ZEN of EN such that

$$f = f_1^* + f_2^* + \dots + f_n^*.$$

Because of Lemma 12.1 every block $b_i^* \leftrightarrow f_i^* \leftrightarrow \mu_i^*$ of EN is covered by a unique block $B_i^* \leftrightarrow e_i^* \leftrightarrow \lambda_i^*$ of EG. If $\{g_{ij} \in G \mid j = 1,2,\dots,t_i\}$ is a transversal of $T_G(f_i^*)$ in G, then

$$e_i^* = \sum_{j=1}^{t_i} (f_i^*)^{g_{ij}}$$

by Lemma 12.6 a). Therefore $e = \sum_{i=1}^{n} \sum_{j=1}^{t} (f_i^*)^{g_j}$ implies that $e = e_1^* + e_2^* + \dots + e_k^*$ for some $1 \le k \le n$. Since each e_i^* is a block idempotent of EG, this decomposition is unique. Hence, if $\delta(B_i^*)$ is a defect group of the block $B_i^* \leftrightarrow e_i^* \leftrightarrow \lambda_i^*$ of EG, then $\delta(B_i^*) \underset{G}{\equiv} \delta(B)$ for $i = 1,2,\dots,n$ by Lemma 12.9. Hence Lemma 12.6 b) implies

$$t_i = |G : T_G(f_i^*)| = \frac{|G:N|}{|\delta(B_i^*)|} = p^{a-d}.$$

Therefore $p^{a-d} = |T_G(f_i^*)|$ for $i = 1,2,\dots,n$, which completes the proof of Lemma 12.10.

As is usual the algebra R over the field F is called <u>separable</u>, if $R \otimes_F E$ is semi-simple and Artinian for every extension field E of F. By Maschke's theorem the group algebra FG of the finite group G over the field F of characteristic $p \nmid |G|$ is a separable algebra over F. But more generally, the following well-known result holds.

Proposition 12.11. If J(FG) is the Jacobson radical of the group algebra FG of the finite group G over the field F of characteristic $p > 0$, then FG/J(FG) is a separable algebra over F.

Proof. Let A be the prime field of F. Then AG/J(AG) is a finite dimensional semi-simple algebra over the perfect field A. By [9], Theorem 69.11 there is a finite separable extension field E of A which is a splitting field for G. Hence $E \otimes_A AG/J(AG)$ is a semi-simple finite-dimensional algebra over the splitting field E by [9], Theorem 69.4. Therefore the theorem of Artin-Wedderburn implies that

$$(1) \qquad E \otimes_A AG/J(AG) \cong \sum_{i=1}^{k} \oplus E_{n_i},$$

where E_{n_i} denotes the ring of all $n_i \times n_i$ matrices over E. Furthermore,

$$(2) \qquad J(EG) = E \otimes_A J(AG).$$

If K is an extension field of F, and if H is the algebraic closure of K, then $E \le H$, because $A \le K$. As E_{n_i} is a central simple algebra over E, Theorem 68.1 of Curtis-Reiner [9] implies

$$(3) \qquad H \otimes_E E_{n_i} = H_{n_i}.$$

Hence it follows from (1) that

$$(4) \qquad H \otimes_E E \otimes_A AG/J(AG) = H \otimes_A AG/J(AG) = \sum_{i=1}^{k} \oplus H_{n_i}.$$

Therefore $HG/H \otimes J(AG)$ is semi-simple, and

$$(5) \qquad J(HG) = H \otimes_A J(AG).$$

Since $J(AG) \subseteq J(FG) \subseteq J(KG) \subseteq J(HG)$, (5) implies that

$$(6) \qquad J(KG) = K \otimes_F J(FG) = K \otimes_F F \otimes_A J(AG) = K \otimes_A J(AG).$$

Thus $FG/J(FG)$ is a separable algebra over F.

Corollary 12.12. Let F be a field of characteristic $p > 0$ and let E be an extension field of F.

If $B \leftrightarrow e \leftrightarrow \lambda$ is a block of the group algebra FG of the finite group G over the field F, then

$$J(E \otimes_F B) = E \otimes_F J(B),$$

where $J(A)$ denotes the Jacobson radical of the ring A.

Proof. Let B_i, $i = 1,2,\ldots,s$, be the block ideals of FG. If $J(B_i)$ denotes the Jacobson radical of the ring B_i, then

$$J(FG) = \sum_{i=1}^{s} \oplus J(B_i) .$$

Since $E \otimes_F J(B_i) \leq J(E \otimes_F B_i)$ for every i, it suffices to show that

$$\dim_E J(E \otimes_F B_i) = \dim_E E \otimes_F J(B_i) = \dim_F J(B_i).$$

From Proposition 12.11 we obtain

$$\dim_E J(E \otimes_F FG) = \dim_E J(EG) = \dim_F J(FG) = \sum_{i=1}^{s} \dim_F J(B_i).$$

Hence $J(E \otimes_F B_i) = E \otimes_F J(B_i)$ for $i = 1,2,\ldots,s$.

Corollary 12.13. If FG is the group algebra of the finite group G over the field F of characteristic $p > 0$, then every block ideal B of the group algebra FG splits over its Jacobson radical $J(B)$.

Proof follows at once from Corollary 12.12 and the theorem of Wedderburn and Malcev ([9], p.491).

With these auxiliary results we now can determine the dimension of the Jacobson radical $J(B)$ of every block ideal B of the group algebra FG of the finite p-nilpotent group G over any field F of characteristic $p > 0$.

Theorem 12.14. Let F be a field of characteristic $p > 0$, and let N be the p-complement of the finite p-nilpotent group G of order $|G| = p^a q$, where $(p, q) = 1$. Then the following assertions hold:

(a) Every block $B \leftrightarrow e \leftrightarrow \lambda$ of FG covers a unique G-conjugacy class of blocks $b \leftrightarrow f \leftrightarrow \mu$ of the group algebra FN, and

$$e = \sum_{j=1}^{t} f^{g_j} ,$$

where $\{g_j \in G \mid j = 1,2,\ldots,t\}$ is a transversal of $T_G(b)$ in G.

(b) If d is the defect of $B \leftrightarrow e \leftrightarrow \lambda$ and if $J(B)$ is the Jacobson radical of the block ideal B, then

$$\dim_F J(B) = (p^d - 1)p^{a-d}\, t \dim_F b.$$

Proof. (a) follows immediately from Lemma 12.6 a).

(b) Let E be any extension field of F which is a splitting field for G, e.g. the algebraic closure of F. By Lemma 12.10 there are finitely many centrally primitive idempotents $f_i{*}$ of the group algebra EN such that:

(1) $f = f_1{*} + f_2{*} + .. + f_n{*}$

(2) $|T_G(f_i{*})| = p^{a-d}$ for $i = 1,2,..,n$.

(3) If $\{g_{ij} \in G \mid j = 1,2,...,t_i\}$ is a transversal of $T_G(f_i{*})$ in G, then $e_i{*} = \sum_{j=1}^{t_i} (f_i{*})^{g_{ij}}$ is a centrally primitive idempotent of EG.

(4) $e = e_1{*} + e_2{*} + .. + e_k{*}$ for some $1 \le k \le n$ satisfying $kp^{a-d} = nt$.

Hence if $b_i{*}$ is the block ideal of $f_i{*}$ in EN and $B_i{*}$ is the block ideal of $e_i{*}$ in EG for $i = 1,2,..,k$, then Proposition 12.8 asserts that

$$\dim_E J(B_i{*}) = (p^{d(B_i{*})}-1)p^{2(a-d(B_i{*}))}\dim_E b_i{*},$$

where $d(B_i{*})$ is the defect of the block $B_i{*} \leftrightarrow e_i{*} \leftrightarrow \lambda_i{*}$ of the group algebra EG. By Lemma 12.9 we obtain from (2) and (3) that

$$d(B_i{*}) = d = d(B) \quad \text{for } i = 1,2,..,k$$

Hence

$$\dim_E J(B_i^*) = (p^d-1)p^{2(a-d)} \dim_E b_i^*.$$

Thus another application of (4) yields

$$\dim_E J(eEG) = (p^d-1)p^{2(a-d)}(\sum_{i=1}^{k} \dim_E b_i^*).$$

As (1) and (4) imply

$$\dim_E(eEG) = p^{2a-d}(\sum_{i=1}^{k} \dim_E b_i^*) = p^a t \dim_E(E \otimes_F b),$$

it follows

$$(5) \qquad \dim_E J(eEG) = (p^d-1)p^{a-d} t \dim_F b.$$

Since Corollary 12.12 asserts that

$$J(E \otimes_F B) = J(eEG) = E \otimes_F J(B)$$

we obtain from (5)

$$\dim_F J(B) = (p^d-1)\, p^{a-d} t \dim_F b.$$

Remark 12.15. Actually Theorem 12.14 allows us to determine the F-vector space dimension of the Jacobson radical $J(FG)$ of group algebras FG of a larger class of finite groups G than the p-nilpotent groups G. For, let H be an extension of a finite p-nilpotent group G by a p-regular group B. Then the theorem of Stonehewer-Green and O. E. Villamayor (Theorem 11.8) asserts that

$$\dim_F J(FH) = |B| \dim_F J(FG) .$$

Thus using Theorem 12.14 it is easy to compute $\dim_F J(FH)$.

As a corollary of Proposition 12.8 we now easily obtain the following result which recently was proved by W. F. Reynolds ([34], Corollary 4) using a generalization of the Witt-Berman induction theorem for group algebras FG of finite groups G over arbitrary fields F of characteristic $p > 0$.

<u>Corollary 12.16</u>. Every block of the group algebra FG of the finite p-nilpotent group G contains a unique simple FG-module.

<u>Proof</u>. Let $B \leftrightarrow e \leftrightarrow \lambda$ be a block of FG with defect d having more than one simple FG-module. Then $B/J(B)$ is not a simple Artinian ring, where $J(B)$ is the Jacobson radical of B. Since B splits over $J(B)$ by Corollary 12.14, there exists a subring $S \cong B/J(B)$ of B such that $B = S + J(B)$, and $S \cap J(B) = 0$. As F has characteristic $p > 0$, the idempotent e lies in S. Thus $e = e_1 + e_2$, $0 \neq e_i \in S$, where e_1 and e_2 are orthogonal central idempotents of S.

Let E be the algebraic closure of F. Then by Lemma 12.9 there are finitely many blocks $B_i^* \leftrightarrow e_i^* \leftrightarrow \lambda_i^*$ of EG with defect $|\delta(B_i)| = d$ such that

(1) $$e = e_1^* + e_2^* + \ldots + e_n^*.$$

Because of Corollary 12.13 $S \cong B/J(B)$ is a separable F-algebra, and

a) $$E \otimes_F B = E \otimes_F S + E \otimes_F J(B),$$

b) $$E \otimes_F S \cap E \otimes_F J(B) = 0,$$

c) $$E \otimes_F J(B) = J(E \otimes_F B) .$$

Hence (1) implies that

d) $$B_i^* = e_i^*(E \otimes_F B) = e_i^*(E \otimes_F S) + J(B_i^*), \text{ and}$$

e) $$e_i^*(E \otimes_F S) \cap J(B_i^*) = 0.$$

Since E is a splitting field for G Proposition 12.8 implies that B_i^* is a quasi-local ring. Hence $(E \otimes_F S)e_i^*$ is a simple Artinian ring with identity element e_i^*. Therefore (1), d) and e) imply that the set $\{e_i^* \mid i = 1,2,\ldots,n\}$ consists of all centrally primitive idempotents of $E \otimes_F S$ and of all centrally primitive idempotents of $E \otimes_F B$. As e_1 and e_2 are orthogonal central idempotents of S, they are orthogonal central idempotents of $E \otimes_F S$. Hence we may assume that

$$e_1 = e_1^* + e_2^* + \ldots + e_k^* \quad \text{and} \quad e_2 = e_{k+1}^* + e_{k+2}^* + \ldots + e_n^*$$

for some integer k with $1 \leq k < n$. Therefore e_1 and e_2 lie in the center of $EG = E \otimes_F FG$. In particualr, e_1 and e_2 are central idempotents of FG. Thus $e = e_1 + e_2$ is not a centrally primitive idempotent of FG. This contradiction proves Corollary 12.16.

By means of Corollary 12.16 we now can prove also assertion b) of Proposition 12.8 for <u>arbitrary</u> fields F of characteristic $p > 0$.

<u>**Theorem 12.17.**</u> Let F be a field of characteristic $p > 0$ and G a finite p-nilpotent group with p-complement N having index p^a. Then the following assertions hold:

a) Every block $B \leftrightarrow e \leftrightarrow \lambda$ of FG with defect d covers a unique G-conjugacy class of blocks $b \leftrightarrow f \leftrightarrow \mu$ of FN, and

$$e = \sum_{j=1}^{t} f^{g_j}$$

where $\{g_j \in G \mid j = 1,2,\ldots,t\}$ is a transversal of the inertia group $T_G(f)$ in G.

b) B contains a unique (up to isomorphisms) indecomposable projective FG-module Q and a unique simple FG-module W.

c) Every composition series of the FG-module Q has length $m = p^d$.

d) There are integers r and z such that if M is the unique simple FN-module of b, then:

1) $T = M \otimes_{FN} FG \cong_{FG} Q^r$,

2) $W_{FN} \cong_{FN} (\sum_{j=1}^{t} \oplus M \otimes g_j)^z$,

3) $rtz = p^{a-d}$.

e) The Jacobson radical $J(B)$ of the block ideal B has F-vector space dimension

$$\dim_F J(B) = (p^d-1)p^{a-d}\, t \dim_F b.$$

<u>Proof.</u> The assertions a) and e) hold by Theorem 12.14. Since by Corollary 12.16 $B = eFG$ has only one simple FG-module W, it also has (up to isomorphisms) only one indecomposable

projective right FG-module Q by Corollary 1.8. Thus it suffices to show c) and d).

As fFN is a simple Artinian ring by Theorem 9.7 we may assume that $M = aFN$ where a is a primitive idempotent of $b = fFN$. Thus

$$(1) \qquad T = M \otimes_{FN} FG \cong aFG = \sum_{d \in P} \oplus\, aFNd,$$

where P is a p-Sylow subgroup of G with order $|P| = p^a$. As T is a finitely generated projective right FG-module, $T \cong Q^r$ for some integer $r \geq 1$ by b). Furthermore, we may assume that W is the socle of Q, because FG is a quasi-Frobenius ring. Hence $W \leq T$, and Clifford's theorem implies that there is an integer $z \geq 1$ such that

$$(2) \qquad W_{FN} \cong {}_{FN} \left(\sum_{j=1}^{t} \oplus\, aFNg_j \right)^z,$$

because $T_G(f) = T_G(M) = \{g \in G \mid M \cong_{FN} Mg\}$ by Lemma 12.1 b). Because of e) we know

$$(3) \qquad \dim_F J(B) = (p^d - 1)p^{a-d} t \dim_F b.$$

Now Lemma 12.10 implies

$$(4) \qquad \dim_F B = p^a t \dim_F b.$$

As Q is the unique indecomposable projective FG-module of B there is an integer $s \geq 1$ such that $B \cong Q^s$. Therefore $\dim_F W = \dim_F Q - \frac{1}{s}(p^d - 1)p^{a-d} t \dim_F b$, because B is a quasi-local Artinian ring. Hence

(5) $$\dim_F Q = \frac{1}{s} p^a t \dim_F b$$

implies

(6) $$\dim_F W = \frac{1}{s} p^{a-d} t \dim_F b .$$

If m is the length of a composition series of Q, then (5) and (6) imply

(7) $$m = p^d$$

which proves c). By (2) we know that

(8) $$\dim_F W = tz \dim_F aFN$$

and (1) implies

(9) $$\dim_F T = p^a \dim_F aFN .$$

As $T \cong Q^r$ we obtain from (7) and (8) that

(10) $$\dim_F T = rp^d tz \dim_F aFN.$$

Hence (9) and (10) imply $rtz = p^{a-d}$ which completes the proof of Theorem 12.17.

Remark 12.18. a) In particular Theorem 12.17 contains Proposition 12.8 because if F is a splitting field for N then Lemma 12.6 b) asserts that $t = p^{a-d}$. Hence $rz = 1$ by Theorem 12.17 d), and $r = 1 = z$.

b) Without any restrictions on the field F the full assertion of Proposition 12.8 holds for the blocks

$B \leftrightarrow e \leftrightarrow \lambda$ of the group algebra FG with highest defect $d(B) = a$, because Theorem 12.17 d) then implies

$$rtz = p^{a-d} = p^0 = 1, \quad \text{and} \quad r = 1 = z.$$

c) If q is the length of any composition series of the simple Artinian ring $b = aFN$ then we also have shown in the proof of Theorem 12.17 that the unique indecomposable projective FG-module Q of B occurs $s = rtq$ times as a direct summand of B. Therefore by Theorem 12.17 the Cartan matrix of the block $B \leftrightarrow e \leftrightarrow \lambda$ of FG is an $s \times s$ matrix $C = (c_{ij})$ where $c_{ij} = 0$ for $i \neq j$ and $c_{ij} = p^d$ for $i = 1,2,\ldots, s = rtq$.

<u>Remark 12.19</u>. The full assertion of Proposition 12.8 does <u>not</u> hold for <u>arbitrary</u> fields F and every block $B \leftrightarrow e \leftrightarrow \lambda$ of the group algebra FG as the following example shows. Let F be the field with two elements and let G be the dihedral group of order 6. Then G is 2-nilpotent. Its p-complement $N = \{1,a,a^2\}$ is a cyclic group of order 3. FN has two block idempotents $e_1 = 1 + a + a^2$ and $e_2 = a + a^2$ which are also the only block idempotents of FG. The block $B_2 \leftrightarrow e_2 \leftrightarrow \lambda_2$ of FG has defect zero and the block $B_1 \leftrightarrow e_1 \leftrightarrow \lambda_1$ of FG has highest defect.

$M = e_2FN$ is the only simple FN-module of the block $b = e_2FN$ of FN. But

$$T = e_2FN \otimes_{FN} FG = e_2FG$$

is <u>not</u> indecomposable because the idempotent e_2 of T splits, i.e. $e_2 = f_1 + f_2$, $f_i f_j = \delta_{ij}$, where $f_1 = e_2\,(1 + a + b + a^2 b)$ and $f_2 = e_2\,(1 + a^2 + a^2 b + b)$. Thus Theorem 12.17 d) cannot be strengthened.

§13. Group algebras with central radicals

Using the results of the previous sections it is now easy to present H. Spiegel's ring theoretical proof [37] for D. A. R. Wallace's theorem [42] determinating the structure of all finite groups G whose group algebras FG have the property that their Jacobson radical J(FG) is contained in the center ZFG of FG, where F is a field of characteristic $p > 0$.

Since clearly the Jacobson radical J(FG) of the group algebra of any abelian group G lies in the center of FG, we assume in this section without further mention that the finite group G is not abelian.

Lemma 13.1. Let F be a field of characteristic $p > 0$ and G a finite group of order $|G| = p^a q$, where $(p,q) = 1$. If the Jacobson radical J(FG) of the group algebra FG satisfies $J(FG)^2 = 0$, then $p^a = 2$.

Proof. Let U_1 be the indecomposable projective FG-module such that $U_1/U_1J(FG)$ is the simple FG-module belonging to the 1-representation of G. Then $\dim_F[U_1/U_1J(FG)] = 1$. Since $[U_1J(FG)]J(FG) = U_1[J(FG)]^2 = 0$, and since FG is a quasi-Frobenius ring, it follows that $\dim_F(U_1J(FG)) = 1$. Hence $\dim_F U_1 = 2$. But $p^a | \dim_F U_1$ by Lemma 12.5. Thus $p^a = 2$.

Lemma 13.2. The element w of the group algebra FG annihilates the augmentation ideal ωFG of FG if and only if

$$w = k \sum_{g \in G} g \quad \text{for some} \quad k \in F.$$

The proof is trivial.

Lemma 13.3 (Wallace [42]). The following properties of the ideal I of the group algebra FG are equivalent:

(1) $$I \leq ZFG.$$

(2) $$(1-x)I = 0 \quad \text{for every} \quad x \in G'$$

(3) $$I \leq sFG, \quad \text{where} \quad s = \sum_{y \in G'} y.$$

Proof. (1) => (2): Let $w \in I$ and $a,b \in G$. Then $(ab)w = a(bw) = (bw)a = b(wa) = (ba)w$, and $(ab-ba)w = 0$. Hence $(1 - b^{-1}a^{-1}ba)w = 0$. Since every element $x \in G'$ is a finite product of commutators $b^{-1}a^{-1}ba$ of G an easy inductive argument now shows that $(1 - x)w = 0$, because if $x = c_1c_2..c_s$, where each c_i is a commutator of G, then

$$\begin{aligned}(1-x)w &= (1-c_1c_2...c_s)w = w - c_sw + c_sw - c_1c_2...c_sw \\ &= (1-c_s)w + (1-c_1c_2...c_{s-1})c_sw \\ &= (1-c_1c_2...c_{s-1})w' \quad \text{for some} \quad w' \in I.\end{aligned}$$

(2) => (3). Let $\{g_i \in G \mid i = 1,2,...,n\}$ be a transversal of G' in G. Then $w = \sum_{i=1}^{n} r_ig_i$ for uniquely determined $r_i \in FG'$. Since $(1-x)w = 0$, it follows that $(1-x)r_i = 0$ for every $x \in G'$ and $i = 1,2,...,n$. Hence Lemma 13.2 asserts that $r_i = k_is$ for some $k_i \in F$, where $s = \sum_{y \in G'} y$. Thus $w = \sum_{i=1}^{n} r_ig_i \in sFG$.

(3) => (1) Clearly $s \in ZFG$. Since sFG is generated by the elements s_u, $u \in G$, as an F-vector space, and since for every $v \in G$ we have

$$\begin{aligned}(su)v &= s(uv) = (uv)s \\ &= (vuu^{-1}v^{-1})(uv)s = vu(u^{-1}v^{-1}uv)s \\ &= vus = v(su),\end{aligned}$$

it follows that $sFG \leq ZFG$.

Lemma 13.4. If $J(FG) \leq ZFG$, then the following are equivalent:

(1) $J(FG) \neq sFG$, where $s = \sum_{y \in G'} y$.

(2) $p \nmid |G'|$.

Proof. By Lemma 13.3 $J(FG) \leq sFG$. If (1) holds, and if $p \mid |G'|$, then $s^2 = |G'|s = 0$, and $J(FG) = sFG$, a contradiction. Hence (2) follows from (1).

Suppose that (2) holds and $J(FG) = sFG$. By Lemma 11.6 we have

$$J(FG') \subseteq J(FG) \cap FG' \subseteq J(FG').$$

Hence $JFG' = sFG' \neq 0$, and $p \mid |G'|$ by Theorem 9.7, a contradiction. Therefore $J(FG) \neq sFG$.

Lemma 13.5. If $J(FG) \leq ZFG$, then $p \mid |G'|$.

Proof. Otherwise $J(FG) = sFG$ by Lemma 13.4, where $s = \sum_{y \in G'} y$. Furthermore $s^2 = 0$, and $J(FG)^2 = 0$. Therefore $p^a = 2$ by Lemma 13.1. As $p \mid |G'|$, it follows that $|G'| = 2n$, where

n is odd. By Huppert [19], p.30, Satz 6.7 G' contains a normal subgroup N of index 2. Thus $G'' \leq N < G'$. Let P be a 2-Sylow subgroup of G'. Then P is also a 2-Sylow subgroup of G. Since $G'' \cap P = 1$, the group $H = G''P$ is a characteristic subgroup of G' whose index is not divisible by 2. Therefore H is normal in G, and Theorem 11.8 implies

$$1 = \dim_F J(FG') = |G':H| \dim_F J(FH).$$

Hence $H = G''P = G'$, because $\dim_F J(FH) \geq 1$. Therefore $K = G'/G''$ is a normal subgroup of order 2 in the group $S = G/G''$ and $|S:K|$ is odd. By the Theorem of Zassenhaus ([19], p.126) there is then a subgroup U of S such that $S = KU$ and $K \cap U = 1$. Since U has index 2, it is normal in S, and $S = K \times U$ is abelian, because $U \cong G/G'$. Hence $G' = G'' \neq G'$, a contradiction. This completes the proof of Lemma 13.5.

Following Huppert([19], p.497)the finite group G is called a <u>Frobenius group for its p-Sylow subgroup P</u>, if G is p-nilpotent, and $n \neq qnq^{-1}$ for every $1 \neq q \in P$ and every $1 \neq n \in N$, where N is the p-complement of G.

<u>Lemma 13.6</u>. Let F be a field of characteristic $p > 0$ and G a finite p-nilpotent group of order $|G| = p^a q$, where $(p,q) = 1$. Then the following statements are equivalent:

(1) $$\dim_F J(FG) = p^a - 1,$$

(2) G is a Frobenius group for any of its p-Sylow subgroups P.

Proof. By Corollary 12.12 we may assume that F is a splitting field for G. Let N be the p-complement of G, and let P be any p-Sylow subgroup of G. Then $e = \frac{1}{|N|} \cdot \sum_{n \in N} n$ is a centrally primitive idempotent of G, because

$$(*) \qquad FG = \sum_{n \in N} (1-n)FG \oplus eFG \quad \text{and} \quad eFG \cong FP.$$

Suppose that the group G has minimal a among the p-nilpotent groups G for which (1) does not imply (2). Then there exists an element $1 \neq x \in P$ of order p and an element $1 \neq n \in N$ such that $n^x = n$. If $a \neq 1$, then $\langle N,x \rangle$ is contained in a maximal normal subgroup G_1 of G with index $|G:G_1| = p$. As G_1 is p-nilpotent, $G_1 = NP_1$ for every p-Sylow subgroup P_1 of G_1 containing $\langle x \rangle$, and

$$(**) \qquad FG_1 \cong \sum_{n \in N} (1-n)FG_1 \oplus FP_1.$$

Thus $\dim_F J(FG_1) \geq p^{a-1}-1$, and using (*), (**) and Lemma 11.6 it is easy to see that

$$p^a-1 = \dim_F J(FG) \geq p \dim_F J(FG_1) + p - 1.$$

Hence $\dim_F J(FG_1) = p^{a-1}-1$, and G_1 is a Frobenius group for P_1, which contradicts $n^x = n$, $x \in \langle x \rangle \leq P_1$. Thus $a = 1$. Therefore, $J(FG) = J(eFG)$, because $\dim_F J(FG) = p-1$. Since 1 is in the central support of e the block $B = eFG \leftrightarrow e \leftrightarrow \lambda$ has defect groups $\delta(B) =_G P$. Therefore B is the only block ideal of FG having positive defect by Theorem 12.17 and Lemma 12.9. Hence Corollary 9.10 b) asserts that 1 is the unique p-regular conjugacy class of G with positive defect, because $a = 1$. Thus $n^x \neq n$, a contradiction! Therefore, (1) implies (2)

Conversely, if (2) holds, then every p-regular conjugacy class $C \neq 1$ of G has defect zero. Therefore Corollary 9.5 asserts that every block $B' \leftrightarrow e' \leftrightarrow \lambda'$ of FG different from $B \leftrightarrow e \leftrightarrow \lambda$ has defect zero. Hence another application of Theorem 12.17 yields $\dim_F J(FG) = p^a - 1$, which completes the proof of Lemma 13.6.

The proof of the next result consists of an easy modification of a similar proof due to Brauer and Nesbitt [3] in case F is algebraically closed.

Lemma 13.7. Let F be a field of characteristic $p > 0$ and G a finite group of order $|G| = p^a q$, where $(p,q) = 1$. If $J(FG)$ is the Jacobson radical of the group algebra FG, then $\dim_F J(FG) \geq p^a - 1$.

Proof.[24]. Since $R = FG$ is an Artinian ring, the identity element 1 of R is a sum of finitely many orthogonal primitive idempotents $e_{ij_i} \in R$, where $1 \leq i \leq m$, $1 \leq j_i \leq k_i$, and $e_{ij_i}R \cong e_{ts_t}R$ if and only if $i = t$ and $1 \leq j_i, s_t \leq k_i$. Furthermore, let $e_i = e_{i1}$ for $i = 1,2,\ldots,m$, and assume that the idempotents e_i are ordered so that $e_1R/e_1J(R)$ is the irreducible FG-module belonging to the 1-representation of G. Let $f_i = \dim_F e_iR/e_iJ(R)$ and $u_i = \dim_F e_iR$ for $i = 1, 2,\ldots,m$. If $j = \dim_F J(R)$, then $g-j = \sum_{i=1}^{m} k_i f_i$, because R/J is a semi-simple Artinian ring. Since $g = \sum_{i=1}^{m} k_i u_i$, it follows that

$$j = \sum_{i=1}^{m} k_i(u_i - f_i) \geq u_1 - 1,$$

because $k_i > 0$, $u_i \geq f_i$ for $i = 1,2,\ldots,m$, and $f_1 = 1$. By Lemma 13.5 we know that $p^a | u_1$. Hence $j \geq p^a - 1$.

The following theorem was proved by D. A. R. Wallace in [42] using the results of modular representation theory of finite groups for group algebras over algebraically closed fields. Its generalization and the proof given here are due to H. Spiegel [37].

Theorem 13.8. Let F be a field of characteristic $p > 0$ and G a finite non-abelian group of order $|G| = p^a q$, where $(p,q) = 1$.

Then the following statements are equivalent:

(1) The Jacobson radical $J(FG)$ of the group algebra FG is contained in the center ZFG of FG.

(2) The order $|G'|$ of the commutator subgroup G' of G is not divisible by p, and for every p-Sylow subgroup P of G $\dim_F J(FG'P) = p^a - 1$.

(3) For every p-Sylow subgroup P of G the group $H = G'P$ is a Frobenius group for its p-Sylow subgroup P.

Proof. $H = G'P$ is a p-nilpotent group, if (2) holds. Thus (2) and (3) are equivalent by Lemma 13.6.

Suppose that (1) holds. Then Lemma 13.5 asserts that $p \nmid |G'|$. Therefore $J(FG) \subsetneqq sFG$, where $s = \sum_{g \in G'} g$, by Lemma 13.4 and Lemma 13.3. Clearly $H = G'P$ is a normal subgroup of G whose index $|G:H|$ is not divisible by p. Hence Theorem 11.8 asserts that

(*) $$\dim_F J(FG) = |G:G'P| \dim_F J(FG'P).$$

Since $J(FG) < sFG$, $\dim_F J(FG) < \dim_F sFG$. Thus $\dim_F J(FG) < |G:G'|$, and therefore (*) implies that

$$\dim_F J(FG'P) < p^a.$$

By Lemma 13.7 $\dim_F J(FG'P) \geq p^a - 1$. Hence $\dim_F J(FG'P) = p^a - 1$, and (2) follows from (1).

Conversely, assume that (2) holds. Let $s = \sum_{y \in G'} y$, and $e = \frac{1}{|G'|} s$. Since $\dim_F J(FG'P) = p^a - 1$, the Jacobson radical

$$J(FG'P) = J(B),$$

where $B = eFG'P$, because e is a central idempotent of $FG'P$ and $B \cong F(P)$. Thus $J(FG'P) \leq sFG'P$. Clearly $H = G'P$ is a normal subgroup of G whose index $|G:H|$ is not divisible by p. Therefore Theorem 11.8 asserts that

$$J(FG) = J(FG'P) \leq sFG.$$

Hence $J(FG) \leq ZFG$ by Lemma 13.3, and Theorem 13.8 holds.

§14. REFERENCES

[1] E. Artin, C. Nesbitt and R. M. Thrall, Rings with minimum condition. Ann Arbor, 1948.

[2] H. Bass, Algebraic K-Theory. New York, Amsterdam, 1968.

[3] R. Brauer and C. Nesbitt, On the modular characters of groups. Annals of Mathematics (2) 42 (1941), 556-590.

[4] R. Brauer, Zur Darstellungstheorie der Gruppen endlicher Ordnung I. Math. Z. 63 (1956), 406-444.

[5] R. Brauer, Zur Darstellungstheorie der Gruppen endlicher Ordnung II. Math. Z. 72 (1959), 25-46.

[6] R. Brauer, On blocks and sections in finite groups I. Amer. J. Math. 89 (1967), 1115-1136.

[7] R. Brauer, On the first main theorem on blocks of characters of finite groups. Illinois J. Math. 14 (1970), 183-187.

[8] E. Cline, Modular representation theory. Lecture Notes, California Institute of Technology, Pasadena, 1966.

[9] C. W. Curtis and I. Reiner, Representation theory of finite groups and associative algebras. New York, London, 1962.

[10] W. Feit, Modular representation theory. Lecture Notes Yale University, New Haven, 1970.

[11] P. Fong, On the characters of p-solvable groups. Transact. Amer. Math. Soc. 98 (1961), 263-284.

[12] A. Geddes, A short proof of the existence of coefficient fields fcr complete, equicharacteristic local rings. J. London Math. Soc. 29 (1954), 334-341.

[13] J. A. Green, Blocks of modular representations. Math. Z. 79 (1962), 100-115.

[14] J. A. Green, Some remarks on defect groups. Math. Z. 107 (1968), 133-150.

[15] J. A. Green and S. E. Stonehewer, The radicals of some group algebras. J. Algebra 13 (1969), 137-142.

[16] J. A. Green, Axiomatic representation theory for finite groups. J. Pure and Applied Algebra 1 (1971), 41-77.

[17] M. Hall, The theory of finite groups. New York, 1959.

[18] W. Hamernik and G. Michler, On Brauer's main theorem on blocks with normal defect groups. J. Algebra (to appear).

[19] B. Huppert, Endliche Gruppen I. Berlin, Heidelberg, New York, 1967.

[20] B. Huppert, Modulare Darstellungstheorie. Vorlesungsausarbeitung, Mainz, 1968.

[21] I. Kaplansky, Projective modules. Annals of Math. 68 (1958), 372-377.

[22] Y. Kawada, On blocks of group algebras of finite groups. Science reports of the Tokyo Kyoiku Daigaku A 9 (1966), 87-110.

[23] J. Lambek, Lectures on rings and modules. Waltham (Mass.), Toronto, London, 1966.

[24] G. Michler, Conjugacy classes and blocks of group algebras. Symposium on "Associative algebras", Rome, November 23-26, 1970.

[25] G. Michler, The blocks of p-nilpotent groups over arbitrary fields.

[26] M. Nagata, Local rings. New York, 1962.

[27] M. Osima, Notes on blocks of group characters. Math. J. Okayama University 4 (1955), 175-188.

[28] M. Osima, On block idempotents of modular group rings. Nagoya Math. J. 28 (1966), 429-433.

[29] D. S. Passman, Central idempotents in group rings. Proc. Amer. Math. Soc. 22 (1969), 555-556.

[30] D. S. Passman, Blocks and normal subgroups. J. Algebra 12 (1969), 569-575.

[31] W. F. Reynolds, Blocks and normal subgroups of finite groups. Nagoya Math. J. 22 (1963), 15-32.

[32] W. F. Reynolds, Block idempotents and normal p-subgroups. Nagoya Math. J. 28 (1966), 1-13.

[33] W. F. Reynolds, Twisted group algebras over arbitrary fields. Illinois J. Math. 15 (1971), 91-103.

[34] W. F. Reynolds, Blocks and F-class algebras of finite groups. To appear.

[35] W. F. Reynolds. Sections and ideals of centers of group algebras. To appear.

[36] A. Rosenberg, Blocks and centers of group algebras. Math. Z. 76 (1961), 209-216.

[37] H. Spiegel, Das Radikal der modularen Gruppenalgebra einer endlichen Gruppe. Diplomarbeit, Tübingen, 1970.

[38] O. E. Villamayor, On the semi-simplicity of group algebras I. Proc. Amer. Math. Soc. 9 (1958), 621-627.

[39] O. E. Villamayor, On the semi-simplicity of group algebras II. Proc. Amer. Math. Soc. 10 (1959), 27-31.

[40] D. A. R. Wallace, Note on the radical of a group algebra. Proc. Cambridge Phil. Soc. 54 (1958), 128-130.

[41] D. A. R. Wallace, On the radical of a group algebra. Proc. Amer. Math. Soc. 12 (1961), 133-137.

[42] D. A. R. Wallace, Group algebras with central radicals. Proc. Glasgow Math. Assoc. 5 (1962), 103-108.

[43] D. A. R. Wallace, Lower bounds for the radical of the group algebra of a finite p-soluble group. Proc. Edinburgh Math. Soc. (2) 16 (1968), 127-134.

[44] H. N. Ward, The analysis of representations induced from a normal subgroup. Michigan Math. J. 15 (1968), 417-428.

[45] H. Zassenhaus, The theory of groups. Göttingen, 1956.

CLOSURE SPACES WITH APPLICATIONS TO RING THEORY

by

R. S. Pierce*

Department of Mathematics
University of Hawaii
Honolulu, Hawaii

1. Introduction. An appropriate subtitle for these notes would be "how to make good things happen simultaneously." In order to understand what this means, consider the following situation. Suppose that we want to construct an object with two properties A and B. We find that there is a way to build something with property A, but not B. However, by enlarging our object, it can be made to have property B, though in the process, A is lost. But then it turns out that another enlargement will restore A while losing B. This would seem to be a hopeless pursuit--one that could go on forever, producing ever larger objects which have one, but not the other of our desired properties. However, in mathematics (and perhaps only in mathematics), such an infinite process may lead somewhere. By making infinitely many adjustments, we may sometimes get an object with exactly the properties A and B that are desired. Roughly speaking, we push the difficulties off to infinity, where they disappear.

In 1958, Kaplansky published a remarkable paper [7] in which he applied this general technique to settle some problems concerning projective modules. He showed that many questions concerning infinite modules could be reduced to the countably infinite case. Then if countability can be exploited to solve the problem--something that occurs rather often--the matter under consideration is completely settled.

---

*Supported in part by National Science Foundation Grant GP-29248.

The general technique that Kaplansky used so effectively has been picked up, modified, and improved by many authors. However, Kaplansky's paper [7] merits much of the credit for teaching a generation of algebraists the power of pure set theory. Among ring theorists, it has become customary to associate Kaplansky's name with this method of making infinitely many adjustments. For example, in [9], Osofsky refers to the "snaking argument of Kaplansky" (see p. 634), and in [5], Griffith speaks of "Kaplansky's infinite juggling." However, the technique was certainly born before 1958. In fact, the process of making an infinite sequence of corrections is the general idea behind the familiar proof of the Schroeder-Bernstein theorem, which dates back to 1897.

The purpose of these notes is to arrange the ideas involved in "Kaplansky's method" and its various generalizations, so that they will constitute a systematic theory. Our hope is that this effort will yield a tool that can be used routinely by all kinds of mathematicians, more or less in the way that Zorn's lemma is now used. Nothing that we have to say is profound, and, with the possible exception of Theorem 8.7, nothing in the notes is original.

Our formulation of the theory will be in terms of closure operators on a set, and some closely related concepts. The theory could be developed in the context of lattice theory. However, it is the author's feeling that this abstract approach would have many disadvantages and almost no virtues. The main argument in favor of a concrete formulation is that the resulting theory lies closer to the applications than does its abstract counterpart. We have intentionally introduced terminology that conforms with module theory. As a result, it isn't necessary to reinterpret all of our notions when they are applied in the last two sections. Hopefully, this procedure also motivates our ideas, at least to ring theorists.

These notes are self-contained in the sense that all statements are proved, modulo filling in easy computations. Omitting the proofs of routine exercises means that many results are just presented as assertions. When this happens, our meaning is that the lemma or proposition under consideration can easily be obtained from appropriate definitions and earlier results. The last two sections presuppose a little knowledge of module theory. However, even here, only the most basic notions of the subject are encountered.

Notational conventions. We will use the more or less standard notation of set theory. Upper case Latin letters usually denote sets, while lower case Latin letters stand for elements. The inclusion of sets is indicated by writing $X \subseteq Y$. The expression $X \subset Y$ means that $X \subseteq Y$ and $X \neq Y$. We use the symbol $\emptyset$ for the empty set. The difference between sets X and Y is denoted by $X - Y$. That is, $X - Y = \{x \in X : x \notin Y\}$. It will be convenient to write $X \cup x$ instead of the more usual $X \cup \{x\}$. Mappings will be written on the left, so that $\varphi\psi$ means: apply ψ before φ. If $\varphi : X \to Y$ is a mapping, then $\operatorname{Im} \varphi$ denotes the image of φ, that is, $\operatorname{Im} \varphi = \{\varphi(x) : x \in X\} = \varphi X$. We will frequently consider <u>directed</u> collections of sets $\{X_i : i \in I\}$. By this we mean that $\{X_i : i \in I\}$ is a <u>non-empty</u> set of sets such that for each i and j in I, there exists $k \in I$ such that $X_i \cup X_j \subseteq X_k$.

The cardinal number of a set X is denoted by $|X|$. The lower case Greek letters α and β always stand for infinite, regular cardinal numbers. Thus, when we use an expression such as "for all α," what we really mean is "for all infinite, regular cardinal numbers α." The property of regularity of the cardinal number α is important in our theory. What regularity means is that if $\{X_i : i \in I\}$ is a collection of sets, with $|X_i| < \alpha$ for all $i \in I$, and if $|I| < \alpha$, then $|\cup_{i\in I} X_i| < \alpha$. Of course, $\aleph_0$, $\aleph_1$, $\aleph_2$, and in fact all $\aleph_\xi$ for which ξ is not a limit ordinal are regular. We will follow the convention of considering cardinal numbers as particular ordinal numbers. The ordinal numbers will also be denoted by lower case Greek letters, particularly λ, ξ, η, and ζ. As usual, ω is the first infinite ordinal. The expression $P(M)$ will denote the power set M, that is, the set of all subsets of M. It is also convenient to introduce the notation $P_\alpha(M) = \{X \in P(M) : |X| < \alpha\}$, and $F(M) = P_{\aleph_0}(M) - \{\emptyset\}$. Thus, $F(M)$ is the set of all finite, non-empty subsets of M.

Our notation for the module theoretic part of the notes is also fairly standard. If A and B stand for modules, then the expression $A < B$ represents the statement that A is a submodule of B. The zero module will be denoted by 0 as usual. Of course, 0 will also stand for the zero element of rings and modules. The plus sign $+$ will designate either the sum of two elements of a module or ring, or the lattice sum of two submodules of a larger module. The symbol $\oplus$ represents direct sums, either inner or outer. It will also appear briefly as the symmetric difference in a Boolean algebra. We will use $\oplus$

both as a binary operation symbol (as in $A \oplus B$), and as an infinitary operation symbol (as in $\oplus_{i \in I} A_i$). If φ is a homomorphism from a module M to a module N, then the kernel of φ is designated by $\operatorname{Ker} \varphi$. That is, $\operatorname{Ker} \varphi = \{x \in M : \varphi(x) = 0\} = \varphi^{-1} 0$. The ring of all endomorphisms of a module M is denoted by $E(M)$. Finally, the projective dimension of a module M is abbreviated by $\operatorname{Pd} M$. We will briefly explain this notion when it occurs in Section 8.

2. Closure operators. In this section, we collect some well-known properties of closure operators. Actually, we will be concerned with closure operators of a special kind, generally called algebraic closure operators. For simplicity the adjective "algebraic" will be suppressed.

2.1. Definition. Let M be any set. A pre-closure operator on M is a mapping $\Gamma : P(M) \to P(M)$ such that for all X and Y in $P(M)$:

(a) $X \subseteq \Gamma(X)$;

(b) if $\{X_i : i \in I\} \subseteq P(M)$ is directed, then $\Gamma(\cup_{i \in I} X_i) = \cup_{i \in I} \Gamma(X_i)$.

If Γ also satisfies

(c) $\Gamma(\Gamma(X)) = \Gamma(X)$,

then Γ is called a closure operator on M.

Note that (b) implies: if $X \subseteq Y$, then $\Gamma(X) \subseteq \Gamma(Y)$.

2.2. Examples. (a) For $X \in P(M)$, define $I(X) = X$. Then I is a closure operator, called the identity operator.

(b) Let A be a fixed subset of M. Define $\Gamma(X) = X \cup A$ for all $X \in P(M)$. Then Γ is a closure operator. Note that if $A = M$, then $\Gamma(X) = M$ for all $X \in P(M)$.

(c) Let $\varphi : M \to M$ be any mapping. For $X \in P(M)$, define $\Gamma(X) = X \cup \varphi(X)$. Then Γ is a pre-closure operator on M, but generally not a closure operator. Note that $\Gamma(X) = X$ if and only if $\varphi(X) \subseteq X$.

(d) Let M be a universal algebra with any number of finitary operations. For any set X in $P(M)$, denote by $\Sigma(X)$ the subalgebra of M that is generated by X. Then Σ is a closure operator on M. This particular closure operator (especially in the case that M is a module) motivates most of the theory that we will develop, and it will be useful to check the meanings of the various concepts that are introduced in the context of this example. Actually, this example is very general. In fact, it follows from the proof of a theorem of Birkhoff and Frink in [3] that every closure operator on a set M can be realized as such a subalgebra closure for a suitably defined algebra structure on M.

Notation. We will denote the set of all pre-closure operators on a set M by $\mathcal{Q}(M)$, or just by $\mathcal{Q}$ when there is no danger of confusion.

2.3. <u>Lemma</u>. Let Γ and Δ be pre-closure operators on M. Define $(\Gamma\Delta)(X) = \Gamma(\Delta(X))$ for all $X \in P(M)$. Then $\Gamma\Delta$ is a pre-closure operator on M.

Note that even when Γ and Δ are closure operators, the product $\Gamma\Delta$ may not be a closure operator.

2.4. Proposition. For Γ and Δ in $\mathcal{Q}$, define $\Gamma \leq \Delta$ if $\Gamma(X) \subseteq \Delta(X)$ for all $X \in P(M)$. Then $\mathcal{Q}$ becomes a partially ordered semigroup with identity I (defined in 2.2 (a)), and I is the smallest preclosure operator in the ordering of $\mathcal{Q}$. If $\Gamma \in \mathcal{Q}$, then Γ is a closure operator if and only if Γ is idempotent.

2.5. Corollary. If Γ and Δ are in $\mathcal{Q}$, then $\Gamma, \Delta \leq \Gamma\Delta$.

2.6. Corollary. If Γ and Δ are in $\mathcal{Q}$, and Γ is a closure operator, then $\Delta \leq \Gamma$ if and only if $\Gamma\Delta = \Delta\Gamma = \Gamma$.

Proof. $\Delta\Gamma = \Gamma$ implies $\Delta \leq \Gamma$ by 2.5. If $\Delta \leq \Gamma$, then $\Gamma \leq \Delta\Gamma \leq \Gamma^2 = \Gamma$. Similarly, $\Gamma\Delta = \Gamma$.

2.7. Lemma. (a) Let $\emptyset \neq \{\Gamma_i : i \in I\} \subseteq \mathcal{Q}$. Then $\cup_{i\in I}\Gamma_i \in \mathcal{Q}$, where $(\cup_{i\in I}\Gamma_i)(X) = \cup_{i\in I}\Gamma_i(X)$ for all $X \in P(M)$.

(b) Let Γ and Δ be in $\mathcal{Q}$. Then $\Gamma \cap \Delta \in \mathcal{Q}$, where $(\Gamma \cap \Delta)(X) = \Gamma(X) \cap \Delta(X)$ for all $X \in P(M)$.

Proof. (a) is routine, and 2.1 (a) clearly holds for $\Gamma \cap \Delta$. Let $\{X_i : i \in I\}$ be directed in $P(M)$. Then $(\Gamma \cap \Delta)(\cup_i X_i) = \cup_{i,j}\Gamma(X_i) \cap \Delta(X_j) = \cup_i \Gamma(X_i) \cap \Delta(X_i)$, since $\{X_i : i \in I\}$ is directed. Thus, 2.1 (b) holds for $\Gamma \cap \Delta$.

2.8. Lemma. (a) If $\emptyset \neq \{\Gamma_i : i \in I\} \subseteq \mathcal{Q}$, and if $\Delta \in \mathcal{Q}$, then $(\cup_{i\in I}\Gamma_i)\Delta = \cup_{i\in I}(\Gamma_i\Delta)$.

(b) If $\{\Gamma_i : i \in I\}$ is directed in $\mathcal{Q}$, and if $\Delta \in \mathcal{Q}$, then $\Delta(\cup_{i\in I}\Gamma_i) = \cup_{i\in I}(\Delta\Gamma_i)$.

Proof. (b) If $X \in P(M)$, then $\{\Gamma_i(X) : i \in I\}$ is directed since $\{\Gamma_i : i \in I\}$ is directed. Thus, $\Delta(\cup_i\Gamma_i)(X) = \Delta(\cup_i\Gamma_i X) = \cup_i\Delta\Gamma_i(X) = (\cup_i\Delta\Gamma_i)(X)$.

The next result can be considered as one formulation of the "method" that was described very roughly in the first paragraph of the introduction.

2.9. Proposition. Let $\Gamma \in \mathcal{Q}$, and define

$$\overline{\Gamma} = \cup_{n<\omega}\Gamma^n.$$

Then $\overline{\Gamma}$ is the smallest closure operator in $\mathcal{Q}$ that contains Γ.

Proof. By 2.5, $I \leq \Gamma \leq \Gamma^2 \leq \Gamma^3 \leq \ldots$. Hence, by 2.8, $\Gamma\,\overline{\Gamma} = \cup_{n<\omega}\Gamma^{n+1} = \overline{\Gamma}$, and by induction, $\Gamma^n\overline{\Gamma} = \overline{\Gamma}$. Thus, $\overline{\Gamma}^2 = (\cup_{n<\omega}\Gamma^n)\overline{\Gamma} = \cup_{n<\omega}\Gamma^n\overline{\Gamma} = \overline{\Gamma}$. Therefore, $\overline{\Gamma}$ is a closure operator. Clearly, $\Gamma \leq \overline{\Gamma}$. If Δ is a closure operator and $\Gamma \leq \Delta$, then $\Gamma^n \leq \Delta^n = \Delta$ for all n. Thus, $\overline{\Gamma} \leq \Delta$.

Notation. Henceforth, if $\Gamma \in \mathcal{Q}$, then $\overline{\Gamma}$ denotes the smallest closure operator containing Γ. Of course, $\overline{\Gamma}$ is explicitly constructed by the result of 2.9. We call $\overline{\Gamma}$ the closure of Γ.

If Γ is any pre-closure operator on M, then a set $X \in P(M)$ is said to be closed with respect to Γ (or simply Γ-closed) if $\Gamma X = X$. It is clear that X is Γ-closed if and only if X is $\overline{\Gamma}$-closed.

There are some simple properties of closure operators that are used repeatedly. We collect these identities in the following lemma.

2.10. Lemma. Let Γ be a closure operator on M. Then:

(a) $\Gamma(\Gamma X \cup Y) = \Gamma(X \cup Y)$;

(b) $\Gamma(\cup_{i\in I}\Gamma X_i) = \Gamma(\cup_{i\in I}X_i)$;

(c) $\Gamma(\cap_{i\in I}\Gamma X_i) = \cap_{i\in I}\Gamma X_i$.

Proof. (a) $\Gamma(\Gamma X \cup Y) \subseteq \Gamma^2(X \cup Y) = \Gamma(X \cup Y) \subseteq \Gamma(\Gamma X \cup Y)$. The proof of (b) is similar. (c) For all $j \in I$, we have $\Gamma X_j \supseteq \cap_{i\in I}\Gamma X_i$. Thus, $\Gamma X_j = \Gamma^2 X_j \supseteq \Gamma(\cap_i \Gamma X_i)$. Hence, finally, $\cap_i \Gamma X_i \supseteq \Gamma(\cap_i \Gamma X_i) \supseteq \cap_i \Gamma X_i$.

2.11. Definition. Let $\varphi : M \to N$ be a mapping. For $\Gamma \in \mathcal{Q}(N)$, define $\Gamma_\varphi : P(M) \to P(M)$ by

$$\Gamma_\varphi X = \varphi^{-1}(\Gamma\varphi(X)).$$

For $\Delta \in \mathcal{Q}(M)$, define $\Delta^\varphi : P(N) \to P(N)$ by

$$\Delta^\varphi X = \varphi(\Delta\varphi^{-1}(X)).$$

2.12. Lemma. (a) If $\varphi : M \to N$ is any mapping and $\Gamma \in \mathcal{Q}(N)$, then $\Gamma_\varphi \in \mathcal{Q}(M)$. Moreover, $(\Gamma\Delta)_\varphi \geq \Gamma_\varphi\Delta_\varphi$, with equality if φ is surjective. Finally, $X \in P(M)$ is Γ_φ-closed if and only if $\varphi^{-1}\varphi X = X$ and $\Gamma\varphi X \cap \operatorname{Im}\varphi = \varphi X$.

(b) If $\varphi : M \to N$ is a surjective mapping and $\Delta \in \mathcal{Q}(M)$, then $\Delta^\varphi \in \mathcal{Q}(N)$. Moreover, $(\Gamma\Delta)^\varphi \geq \Gamma^\varphi\Delta^\varphi$. Finally, $Y \in P(N)$ is Δ^φ-closed if and only if $\Delta\varphi^{-1}Y = \varphi^{-1}Y$.

(c) If $\varphi : M \to N$ is a surjective map, then $(\Gamma_\varphi)^\varphi = \Gamma$ for all $\Gamma \in \mathcal{Q}(N)$ and $(\Delta^\varphi)_\varphi \geq \Delta$ for all $\Delta \in \mathcal{Q}(M)$.

Proof. Suppose that $\Gamma_\varphi X = X$, that is, $\varphi^{-1}\Gamma\varphi X = X$. Then $\varphi X = \varphi\varphi^{-1}\Gamma\varphi X = \Gamma\varphi X \cap \operatorname{Im}\varphi$, and $\varphi^{-1}\varphi X = \varphi^{-1}\varphi\varphi^{-1}\Gamma\varphi X = \varphi^{-1}\Gamma\varphi X = X$. Conversely, if $\varphi^{-1}\varphi X = X$ and $\Gamma\varphi X \cap \operatorname{Im}\varphi = \varphi X$, then $X = \varphi^{-1}\varphi X = \varphi^{-1}(\Gamma\varphi X \cap \operatorname{Im}\varphi) = \varphi^{-1}\Gamma\varphi X = \Gamma_\varphi X$. The rest of the proof is routine.

2.13. Corollary. Let $\varphi : M \to N$ be a mapping. If Γ is a closure operator on N, then Γ_φ is a closure operator on M. If Δ is a closure operator on N, and if φ is surjective, then Δ^φ is a closure operator on N.

2.14. Examples. (a) Let $\varphi : M \to N$ be an arbitrary mapping, and let I be the identity operator on N. Then $I_\varphi(X) = \varphi^{-1}\varphi X$ for all $X \in P(M)$. If φ is surjective, and I is the identity operator on M, then I^φ is the identity operator on N.

(b) Let $M \subseteq N$, and suppose that φ is the inclusion mapping. In this particular case, we will write Γ_M instead of Γ_φ, and call Γ_M the restriction of Γ to M. Thus, for all $X \in P(M)$,

$$\Gamma_M(X) = \Gamma X \cap M.$$

Note that if M is Γ-closed, then $\Gamma_M X = \Gamma X$ for all $X \in P(M)$. Thus, in this case, Γ_M is the restriction of Γ in the usual sense.

3. Closure spaces. We turn now to the class of all pairs (M, Σ), where M is a set and Σ is a closure operation on M. Such objects will be called closure spaces. We will impose a category structure on this class.

3.1. Definition. Let (M, Σ) and (N, Θ) be closure spaces. A morphism of (M, Σ) to (N, Θ) is a mapping $\Phi : P(M) \to P(N)$ such that:

(a) if $\{X_i : i \in I\} \subseteq P(M)$ is directed, then $\Phi(\cup_{i \in I} X_i) = \cup_{i \in I} \Phi(X_i)$;

(b) $\Phi(\Sigma X) = \Theta(\Phi X) = \Phi X$ for all $X \in P(M)$.

Note. It follows from (a) that $X \subseteq Y$ implies $\Phi(X) \subseteq \Phi(Y)$.

3.2. Lemma. If $\Phi : (M, \Sigma) \to (N, \Theta)$, and $\Psi : (N, \Theta) \to (P, \Xi)$ are morphisms, then $\Psi\Phi : (M, \Sigma) \to (P, \Xi)$ is a morphism.

As a consequence of this lemma, we see that the collection of closure spaces and their morphisms defined in 3.1 forms a category which we refer to as the category of all closure spaces. The identity morphism of an object (M, Σ) is the closure operator Σ (which is plainly a morphism of (M, Σ) to itself).

The partial ordering that was defined for pre-closure operators has an analogue for morphisms of closure spaces.

3.3. Definition. Let Φ and Ψ be morphisms from the closure space (M, Σ) to the closure space (N, Θ). Define $\Phi \leq \Psi$ if $\Phi(X) \subseteq \Psi(X)$ for all $X \in P(M)$.

3.4. Lemma. If $\Phi \leq \Psi$, then $\Omega\Phi \leq \Omega\Psi$ and $\Phi T \leq \Psi T$ (assuming that the compositions are defined).

3.5. Lemma. (a) Let $\{\Phi_i : i \in I\}$ be a directed set of morphisms from (M, Σ) to (N, Θ). For $X \in P(M)$, define:

$$(\cup_{i\in I}\Phi_i)(X) = \cup_{i\in I}\Phi_i(X).$$

Then $\cup_{i\in I}\Phi_i$ is a morphism from (M, Σ) to (N, Θ). Moreover, $\Omega(\cup_{i\in I}\Phi_i) = \cup_{i\in I}\Omega\Phi_i$ and $(\cup_{i\in I}\Phi_i)T = \cup_{i\in I}\Phi_i T$ (assuming that the compositions are defined).

(b) Let Φ and Ψ be morphisms from (M, Σ) to (N, Θ). For $X \in P(M)$, define:

$$(\Phi \cap \Psi)(X) = \Phi(X) \cap \Psi(X).$$

Then $\Phi \cap \Psi$ is a morphism from (M, Σ) to (N, Θ).

Proof. The proof of (a) is straightforward. To see that $\Phi \cap \Psi$ satisfies 3.1 (a), just copy the argument used to establish 2.7 (b). The identity $(\Phi \cap \Psi)\Sigma = \Phi \cap \Psi$ follows directly from definitions, while $\Theta(\Phi \cap \Psi)(X) = \Theta(\Phi X \cap \Psi X) = \Theta(\Theta\Phi X \cap \Theta\Psi X) = \Theta\Phi X \cap \Theta\Psi X = \Phi X \cap \Psi X = (\Phi \cap \Psi)(X)$ by 2.10 (c).

The following property of morphisms will be important for some of our applications.

3.6. Definition. A morphism $\Phi : (M, \Sigma) \to (N, \Theta)$ of closure spaces is called unital if $\Phi(M) = N$.

Obviously, if $\Phi \leq \Psi$, and if Φ is unital, then Ψ is unital. Also, if Φ and Ψ are both unital, then so is $\Phi \cap \Psi$.

We turn now to some important examples of morphisms of closure spaces.

3.7. Lemma. Let (M, Σ) and (N, Θ) be closure spaces. For any mapping $\varphi : M \to N$, define $\varphi_* : P(M) \to P(N)$ and $\varphi^* : P(N) \to P(M)$ by

$$\varphi_*(X) = \Theta\varphi(\Sigma X) \quad \text{and} \quad \varphi^*(Y) = \Sigma\varphi^{-1}(\Theta Y).$$

Then φ_* and φ^* are morphisms of closure spaces. In all cases, φ^* is unital; φ_* is unital if and only if $\Theta(\varphi M) = N$ (in particular, if φ is surjective).

In general, the morphisms defined in 3.7 do not have useful properties. However, if φ satisfies $\varphi\Sigma = \Theta\varphi$ $(\varphi^{-1}\Theta = \Sigma\varphi^{-1})$, then $\varphi_*X = \varphi\Sigma X = \Theta\varphi X$ (respectively, $\varphi^*Y = \varphi^{-1}\Theta Y = \Sigma\varphi^{-1}Y$), and the morphisms φ_* and φ^* are more tractable in these cases.

3.8. Example. Let M and N be universal algebras of the same type, with all operations of M and N assumed to be finitary. Let Σ and Θ be the closure operations on M and N respectively that map a set into the subalgebra that it generates, as in Example 2.2 (d).

If φ is any homomorphism of M to N, then $\varphi\Sigma = \Theta\varphi$, so that $\varphi_*X = \varphi\Sigma X = \Theta\varphi X$ for all $X \in P(M)$. It is not true that $\varphi^{-1}\Theta = \Sigma\varphi^{-1}$, but we do have $\Sigma\varphi^{-1}\Theta = \varphi^{-1}\Theta$; thus, $\varphi^*Y = \varphi^{-1}\Theta Y$ for all $Y \in P(N)$.

3.9. Lemma. Let (M, Σ) and (N, Θ) be closure spaces. Suppose that $\varphi : M \to N$ is a mapping such that

(a) φM is Θ-closed, and

(b) $\Theta_\varphi \geq \Sigma$.

Then $\varphi_*X = \Theta\varphi X$ for all $X \in P(M)$, and $\varphi^*Y = \varphi^{-1}\Theta Y$ for all $Y \in P(N)$.

Proof. By 2.6, $\Theta_\varphi\Sigma = \Sigma\Theta_\varphi = \Theta_\varphi$. Thus, $\Theta\varphi X = \Theta\varphi X \cap \text{Im }\varphi = \varphi\varphi^{-1}\Theta\varphi X = \varphi\Theta_\varphi X = \varphi\Theta_\varphi\Sigma X = \Theta\varphi\Sigma X = \varphi_*X$, and $\varphi^{-1}\Theta Y = \varphi^{-1}(\Theta Y \cap \text{Im }\varphi) = \varphi^{-1}\Theta(\varphi\varphi^{-1}\Theta Y) = \Theta_\varphi\varphi^{-1}\Theta Y = \Sigma\Theta_\varphi\varphi^{-1}\Theta Y = \Sigma\varphi^{-1}\Theta Y = \varphi^*Y$.

Remarks. (1) The conditions (a) and (b) of 3.9 are clearly satisfied if $\varphi\Sigma = \Theta\varphi$. Thus, 3.9 can be considered as a generalization of Example 3.8.

(2) An important particular case of 3.9 occurs when M is a Θ-closed subset of N, $\Sigma = \Theta_M$, and φ is the inclusion mapping of M into N. We then have $\varphi_*X = \Theta X$ for $X \in P(M)$, and $\varphi^*Y = \Theta Y \cap M$ for $Y \in P(N)$.

From our standpoint, the most important endomorphisms of closure spaces are the ones that are closure operators.

3.10. Definition. Let (M, Σ) be a closure space. A pre-Σ operator on M is a pre-closure operator Γ on M, such that

$$\Gamma\Sigma = \Sigma\Gamma = \Gamma.$$

If also $\Gamma^2 = \Gamma$, then Γ is called a Σ-operator.

Thus, the pre-Σ operators on M are exactly the endomorphisms Γ of (M, Σ) that satisfy $\Gamma(X) \supseteq X$ for all $X \in P(M)$, i.e., $\Gamma \geq I$.

3.11. Proposition. The set $\mathcal{Q}(M, \Sigma)$ of all pre-Σ operators on M is a sub-semigroup of $\mathcal{Q}(M)$. The identity element of $\mathcal{Q}(M, \Sigma)$ is Σ, and Σ is also the smallest element in the ordering of $\mathcal{Q}(M, \Sigma)$. Finally, $\mathcal{Q}(M, \Sigma)$ is closed under the formation of directed unions.

3.12. Corollary. If $\Gamma \in \mathcal{Q}(M, \Sigma)$, then $\overline{\Gamma} \in \mathcal{Q}(M, \Sigma)$.

3.13. Corollary. If the hypotheses of 3.9 are satisfied, and if $\Gamma \in \mathcal{Q}(N, \Theta)$, then $\Gamma_\varphi = \varphi^*\Gamma\varphi_* \in \mathcal{Q}(M, \Sigma)$.

4. Additive morphisms. There is a special class of morphisms that is important enough to merit special attention.

4.1. Definition. Let $\Phi : (M, \Sigma) \to (N, \Theta)$ be a morphism of closure spaces. Then Φ is additive if

(a) $\Phi\emptyset = \Theta\emptyset$, and

(b) $\Phi(X \cup Y) = \Theta(\Phi X \cup \Phi Y)$ for all X and Y in $P(M)$.

Remarks. (1) Every morphism $\Phi : (M, \Sigma) \to (N, \Theta)$ satisfies $\Phi\emptyset \supseteq \Theta\emptyset$ and $\Phi(X \cup Y) \supseteq \Theta(\Phi X \cup \Phi Y)$.

(2) In general, if φ is a mapping of M to N, then the induced morphisms φ_* and φ^* need not be additive. However, if the conditions of 3.9 are satisfied, then it is easy to see that φ_* is additive.

4.2. Lemma. If $\Phi : (M, \Sigma) \to (N, \Theta)$ and $\Psi : (N, \Theta) \to (P, \Xi)$ are additive morphisms, then $\Psi\Phi$ is additive.

Note that Σ is an additive endomorphism of (M, Σ) by 2.10 (b). Thus, the class of additive morphisms is a subcategory of the category of all morphisms of closure spaces.

4.3. Lemma. If $\{\Phi_i : i \in I\}$ is a directed set of additive morphisms of (M, Σ) to (N, Θ), then $\cup_{i\in I}\Phi_i$ is additive.

4.4. Corollary. If Γ is an additive pre-Σ operator, then $\overline{\Gamma}$ is additive.

4.5. Proposition. An additive morphism $\Phi : (M, \Sigma) \to (N, \Theta)$ satisfies

$$(*) \qquad \Phi(\cup_{i\in I}X_i) = \Theta(\cup_{i\in I}\Phi X_i)$$

for every family $\{X_i : i \in I\} \subseteq P(M)$.

Proof. For $|I| = 0$, 1, and 2, (*) follows from 4.1 (a), 3.1 (b), and 4.1 (b) respectively. For $I = \{1, 2, \ldots, n\}$, $n \geq 3$, the identity (*) is obtained by induction on n, using 4.1 (b) and 2.10 (a): $\Phi(X_1 \cup X_2 \cup \ldots \cup X_n) = \Theta(\Phi X_1 \cup \Phi(X_2 \cup \ldots \cup X_n)) = \Theta(\Phi X_1 \cup \Theta(\Phi X_2 \cup \ldots \cup \Phi X_n)) = \Theta(\Phi X_1 \cup \Phi X_2 \cup \ldots \cup \Phi X_n)$. Now suppose that I is infinite. For $F \in F(I)$, write $X_F = \cup_{i\in F}X_i$. Then $\Phi(\cup_i X_i) = \Phi(\cup_{F\in F(I)}X_F) = \cup_F \Phi X_F$, because $\{X_F : F \in F(I)\}$ is directed. Hence, using the finite case, $\Phi(\cup_i X_i) = \cup_F \Theta(\cup_{i\in F}\Phi X_i) = \Theta(\cup_F \cup_{i\in F}\Phi X_i) = \Theta(\cup_i \Phi X_i)$.

In particular, $\Phi X = \Theta(\bigcup_{x \in X} \Phi x)$ for every $X \in P(M)$. Thus, Φ is uniquely determined by its values on singletons. This observation is made precise as follows.

4.6. Proposition. Let (M, Σ) and (N, Θ) be closure spaces, and suppose that $\Phi : M \to P(N)$ is a mapping such that

(a) $y \in \Sigma(\{x_1, \ldots, x_n\})$ implies $\Phi y \subseteq \Theta(\Phi x_1 \cup \ldots \cup \Phi x_n)$, and

(b) $\Theta\Phi x = \Phi x$ for all $x \in M$.

Then there is a unique extension of Φ to an additive morphism of (M, Σ) to (N, Θ) defined by $\Phi X = \Theta(\bigcup_{x \in X} \Phi x)$.

Proof. The uniqueness has been noted, and it is clear that 3.1 (a) holds. Also, $\Theta\Phi = \Phi$ by definition. If $X \subseteq M$ is finite, then $\Phi\Sigma X = \Phi X$ by (a). The condition $\Phi\Sigma = \Phi$ follows easily from 3.1 (a). Thus, Φ is a morphism. By 2.10 (b), $\Phi(X \cup Y) = \Theta(\bigcup_{x \in X \cup Y} \Phi x) = \Theta((\bigcup_{x \in X} \Phi x) \cup (\bigcup_{y \in Y} \Phi y)) = \Theta(\Theta(\bigcup_{x \in X} \Phi x) \cup \Theta(\bigcup_{y \in Y} \Phi y)) = \Theta(\Phi X \cup \Phi Y)$.

Additive pre-Σ operators can be constructed using 4.6. However, it is necessary to supplement the conditions (a) and (b).

4.7. Corollary. Let (M, Σ) be a closure space. Suppose that $\Gamma : M \to P(M)$ is a mapping such that

(a) $y \in \Sigma(\{x_1, \ldots, x_n\})$ implies $\Gamma y \subseteq \Sigma(\Gamma x_1 \cup \ldots \cup \Gamma x_n)$,

(b) $\Sigma\Gamma x = \Gamma x$ for all $x \in M$, and

(c) $x \in \Gamma x$ for all $x \in M$.

Then there is a unique extension of Γ to an additive pre-Σ operator on M. If Γ also satisfies

(d) $y \in \Gamma x$ implies $\Gamma y \subseteq \Gamma x$,

then the extension of Γ is a closure operator.

It is convenient to let the same symbol denote a mapping $M \to P(M)$ and its additive extension, as given by 4.7.

We use 4.7 to construct two particular pre-closure operators on modules. These are actually additive pre-Σ operators, where Σ is the closure operator defined in 2.2 (d), that is, $\Sigma(X)$ = submodule of M generated by X. In both of the following examples, and in all future discussions of closure operators on modules, Σ will be this closure operator.

4.8. Example. Let R be a ring, M an R-module, and suppose that $M = \oplus_{i \in I} M_i$ is an internal direct sum decomposition of M. Let π_i denote the projection endomorphism of M onto M_i associated with this decomposition. For $x \in M$, define

$$Kx = \oplus_{i \in F} M_i, \text{ where } F = \{i \in I : \pi_i x \neq 0\}.$$

Then K extends uniquely to an additive Σ-operator. A subset A of M is K-closed if and only if A is a submodule of M such that $A = \oplus_{i \in J} M_i$ for some subset J of I.

Proof. Plainly, Kx is a submodule of M, that is, $\Sigma Kx = Kx$. If $y \in \Sigma\{x_1, \ldots, x_n\} = Rx_1 + \ldots + Rx_n$, then $\pi_i y \neq 0$ implies $\pi_i x_k \neq 0$ for some k. Thus, $Ky \subseteq Kx_1 + \ldots + Kx_n = \Sigma(Kx_1 \cup \ldots \cup Kx_n)$. Clearly, $x \in Kx$ for all $x \in M$. If $y \in Kx = \oplus_{i \in F} M_i$, then $\pi_i y = 0$ for all $i \in I - F$. Hence, $Ky \subseteq Kx$. By 4.7, there is a unique extension of K to an additive Σ-operator on M. It is obvious from the definitions that the K-closed subsets of M are exactly the sums $\oplus_{i\in J}M_i$, where J is a subset of I.

4.9. Example. Let R be a ring, M an R-module, and $\mathcal{D}$ a subset of $E(M)$, such that $1_M \in \mathcal{D}$. For $x \in M$, denote $\mathcal{D}x = \{\varphi x : \varphi \in \mathcal{D}\}$. Then there is a unique additive, pre-Σ operator Λ such that

$$\Lambda x = \Sigma(\mathcal{D}x)$$

for all $x \in M$. A subset A of M is Λ-closed if and only if A is a $\mathcal{D}$-invariant submodule of M.

Proof. If $y \in \Sigma\{x_1, \ldots, x_n\} = Rx_1 + \ldots + Rx_n$, and if $\varphi \in \mathcal{D}$, then $\varphi y \in R\varphi x_1 + \ldots + R\varphi x_n \subseteq \Sigma\mathcal{D}x_1 + \ldots + \Sigma\mathcal{D}x_n = \Sigma(\Lambda x_1 \cup \ldots \cup \Lambda x_n)$. Thus, 4.7 (a) is satisfied. Plainly, $\Sigma\Lambda x = \Lambda x$ for $x \in M$, and $x \in \Lambda x$ since $1_M \in \mathcal{D}$. Thus, the existence and uniqueness of Λ follows from 4.7. If $\Lambda A = A$, then $\Sigma A = A$, so that A is a submodule of M. Moreover, if $x \in A$ and $\varphi \in \mathcal{D}$, then $\varphi x \in \Lambda A = A$. Hence, $\mathcal{D}A \subseteq A$. Conversely, if $\mathcal{D}A \subseteq A$ and A is a submodule of M, then $\Sigma(\mathcal{D}x) \subseteq A$ for all $x \in A$. Hence, $A \subseteq \Lambda A \subseteq A$.

5. Bounded morphisms. We now begin the development of a rather specialized aspect of the theory of closure spaces. This section and the following one contain the principal results of these notes.

Recall our standing assumption that α always denotes an infinite, regular cardinal number. The deeper theorems also require the hypothesis $\alpha > \aleph_0$, but we won't impose this inequality as a blanket hypothesis.

5.1. Definition. Let (M, Σ) be a closure space. An ordered pair Y/X of subsets of M is called an α-extension in (M, Σ) if $Y \supseteq X$ and there exists $W \in P_\alpha(M)$ such that $\Sigma Y = \Sigma(X \cup W)$. If $X/\emptyset$ is an α-extension, then X is said to be α-generated.

Remarks. (1) If $X \subseteq Y$ and $|Y - X| < \alpha$, then Y/X is an α-extension.

(2) By 2.10 and our definition, Y/X is an α-extension if and only if $Y \supseteq X$ and $\Sigma Y/\Sigma X$ is an α-extension. For the most part, 5.1 will be applied only to Σ-closed subsets of M.

5.2. Lemma. (a) If Z/Y and Y/X are α-extensions, then Z/X is an α-extension.

(b) If Z/X is an α-extension, and $X \subseteq Y \subseteq Z$, then Z/Y is an α-extension.

Proof. (a) By definition, $\Sigma Z = \Sigma(Y \cup W_1)$ and $\Sigma Y = \Sigma(X \cup W_2)$ for some W_1 and W_2 in $P_\alpha(M)$. By 2.10, $\Sigma Z = \Sigma(\Sigma Y \cup W_1) = \Sigma(\Sigma(X \cup W_2) \cup W_1) = \Sigma(X \cup (W_1 \cup W_2))$, and $W_1 \cup W_2 \in P_\alpha(M)$.

(b) $\Sigma Z = \Sigma(X \cup W) \subseteq \Sigma(Y \cup W) \subseteq \Sigma Z$ for some $W \in P_\alpha(M)$.

5.3. Lemma. If Y/X is an α-extension and $Z \in P(M)$, then $(Y \cup Z)/(X \cup Z)$ is an α-extension.

Proof. Let $\Sigma Y = \Sigma(X \cup W)$, $W \in P_\alpha(M)$. Then $\Sigma(Y \cup Z) = \Sigma(\Sigma Y \cup Z) = \Sigma(\Sigma(X \cup W) \cup Z) = \Sigma((X \cup Z) \cup W)$.

5.4. Lemma. If $\{Z_i : i \in I\} \subseteq P(M)$, and Z_i/X is an α-extension for all $i \in I$, where $|I| < \alpha$, then $(\bigcup_{i \in I} Z_i)/X$ is an α-extension.

Proof. By definition, there exist $W_i \in P_\alpha(M)$ such that $\Sigma Z_i = \Sigma(X \cup W_i)$. Thus, by 2.10 (b), $\Sigma(\bigcup_{i \in I} Z_i) = \Sigma(\bigcup_i \Sigma Z_i) = \Sigma(\bigcup_i \Sigma(X \cup W_i)) = \Sigma(\bigcup_i (X \cup W_i)) = \Sigma(X \cup \bigcup_i W_i)$. By the regularity of α, $|\bigcup_i W_i| < \alpha$.

5.5. Corollary. Let λ be an ordinal number with $\lambda < \alpha$. Suppose that $\{X_\xi : \xi \leq \lambda\} \subseteq P(M)$ satisfies:

(a) $\xi < \eta \leq \lambda$ implies $X_\xi \subseteq X_\eta$;

(b) if η is a limit ordinal, then $X_\eta = \bigcup_{\xi < \eta} X_\xi$;

(c) $X_{\xi+1}/X_\xi$ is an α-extension for all $\xi < \lambda$.

Then X_λ/X_0 is an α-extension.

Proof. This follows by transfinite induction on λ, using 5.2 and 5.4 for the non-limit and limit steps respectively.

5.6. Definition. Let $\Phi : (M, \Sigma) \to (N, \Theta)$ be a morphism of closure spaces. Then Φ is α-bounded if

(a) $\Phi\emptyset = \Theta\emptyset$, and

(b) if Y/X is an α-extension, then $\Phi Y/\Phi X$ is an α-extension.

In more detail, 5.6 (b) means that if $\Sigma Y = \Sigma(X \cup Z)$ for some $Z \in P_\alpha(M)$, then $\Phi Y = \Theta(\Phi X \cup W)$ for some $W \in P_\alpha(N)$.

Of course, 5.6 defines α-boundedness for pre-Σ operators as a special case of α-boundedness for morphisms.

5.7. Lemma. Let $\Phi : (M, \Sigma) \to (N, \Theta)$ and $\Psi : (N, \Theta) \to (P, \Xi)$ be α-bounded morphisms. Then $\Psi\Phi$ is α-bounded.

By the remark (2) following 5.1, it follows that Σ (considered as an endomorphism of (M, Σ)) is α-bounded. Thus, the class of α-bounded morphisms is a subcategory of the class of all morphisms of closure spaces.

5.8. Lemma. Let $\{\Phi_i : i \in I\}$ be a directed family of α-bounded morphisms of (M, Σ) to (N, Θ). Suppose also that $|I| < \alpha$. Then $\cup_{i \in I}\Phi_i$ is α-bounded.

Proof. Clearly, 5.6 (a) holds for $\cup_i\Phi_i$. Assume that Y/X is an α-extension. Then so is $\Phi_i Y/\Phi_i X$ for all $i \in I$. Let $Z = \cup_{i \in I}\Phi_i X$. By 5.3, $\Phi_i Y \cup Z/Z$ is an α-extension for all $i \in I$, so that by 5.5, $(\cup_i\Phi_i)Y/(\cup_i\Phi_i)X = (\cup_i\Phi_i Y) \cup Z/Z$ is an α-extension.

5.9. Corollary. Let $\alpha > \aleph_0$. If Γ is an α-bounded pre-Σ operator, then $\overline{\Gamma}$ is α-bounded.

It is sometimes useful to have an alternative characterization of α-bounded morphisms.

5.10. Proposition. Let $\Phi : (M, \Sigma) \to (N, \Theta)$ be a morphism of closure spaces such that $\Phi\emptyset = \Theta\emptyset$. Then Φ is α-bounded if and only if $\Phi(X \cup x)/\Phi X$ is an α-extension for all $x \in M$ and all $X \in P(M)$.

Proof. The condition is plainly necessary. Conversely, suppose that every $\Phi(X \cup x)/\Phi X$ is an α-extension. Let Y/X be any α-extension. Then $\Sigma Y = \Sigma(X \cup W)$, where $W \in P_\alpha(M)$. By an induction, using 5.2, $\Phi(X \cup F)/\Phi X$ is an α-extension for all $F \in F(W)$. Since the family $\{X \cup F : F \in F(W)\}$ is directed and has cardinality less than α, we conclude by 5.4 that $\Phi Y/\Phi X = \Phi(X \cup W)\Phi X = (\cup_{F \in F(M)}\Phi(X \cup F))/\Phi X$ is an α-extension.

5.11. Corollary. If β is a regular cardinal number greater than α, and if Φ is an α-bounded morphism of closure spaces, then Φ is β-bounded.

5.12. Corollary. Let (M, Σ) and (N, Θ) be closure spaces, and suppose that $\varphi : M \to N$ is a mapping such that φM is Θ-closed and $\Theta_\varphi \geq \Sigma$. Then φ_* is α-bounded for all α.

Proof. By 3.9, $\varphi_* X = \Theta\varphi X$ for all $X \in P(M)$. Hence, $\varphi_*\emptyset = \Theta\emptyset$, and $\varphi_*(X \cup x) = \Theta\varphi(X \cup x) = \Theta(\varphi X \cup \varphi x) = \Theta(\Theta\varphi X \cup \varphi x) = \Theta(\varphi_* X \cup \varphi x)$.

5.13. Corollary. If $\Phi : (M, \Sigma) \to (N, \Theta)$ is an additive morphism, then Φ is α-bounded if and only if Φx is α-generated for all $x \in M$.

Proof. If Φ is α-bounded, then $\Phi x/\Theta\emptyset$ is an α-extension, that is, Φx is α-generated. Conversely, assume that Φx is α-generated for all x. If $X \in P(M)$, then $\Phi(X \cup x) = \Theta(\Phi X \cup \Phi x) = \Theta(\Phi X \cup \Theta W) = \Theta(\Phi X \cup W)$ for some $W \in P_\alpha(M)$. Thus, $\Phi(X \cup x)/\Phi X$ is an α-extension.

5.14. Example. Let M be an R-module, and let K be the pre-Σ operator defined in 4.8. It is clear from 5.13 that K is

α-bounded if and only if each of the summands M_i is α-generated.

5.15. Example. Let M be an R-module, and let Λ be the pre-Σ operator defined in 4.9. It follows from 5.13 that Λ is α-bounded if and only if $\Sigma(\mathcal{D}x)$ is α-generated for all $x \in M$. Naturally, this condition is satisfied if $|\mathcal{D}| < \alpha$. However, there is another important case in which it holds. Indeed, suppose that M can be written as an internal direct sum $\oplus_{i \in I} M_i$ with associated projection endomorphisms π_i. Let $\mathcal{D} = \{\pi_i : i \in I\}$. Since $\{i \in I : \pi_i x \neq 0\}$ is finite for all $x \in M$, it is clear that $\Sigma(\mathcal{D}x)$ is finitely generated. Thus, Λ is α-bounded for all α. If Λ is defined in this way, corresponding to a direct decomposition $M = \oplus_{i \in I} M_i$, then a submodule A of M is Λ-closed if and only if $A = \oplus_{i \in I}(A \cap M_i)$. In fact, we saw in 4.9 that A is Λ-closed if and only if A is a submodule such that $\mathcal{D}x \subseteq A$ for all $x \in A$. In our case, this means that $\pi_i A \subseteq A$ for all $i \in I$. Assuming that this holds, then A is clearly the direct sum of all its submodules $A \cap M_i$. Conversely, if $A = \oplus_{i \in I}(A \cap M_i)$, then we must have $\pi_i x \in A \cap M_i \subseteq A$ for all $i \in I$, and for all $x \in A$. Thus, A is Λ-closed.

It is possible to improve 5.10 in an interesting way. The result will not be used later, so that its proof could be omitted without loss of continuity.

5.16. Proposition. Let $\Phi : (M, \Sigma) \to (N, \Theta)$ be a morphism of closure spaces such that $\Phi\emptyset = \Theta\emptyset$. Then Φ is α-bounded if and only if $\Phi(X \cup x)/\Phi X$ is an α-extension for all $x \in M$ and all $X \in P(M)$ satisfying $|X| \leq \alpha$.

Proof. Assume that $\Phi(X \cup x)/\Phi X$ is an α-extension whenever $|X| \leq \alpha$. We will work for a contradiction to the assumption that Φ is not α-bounded. By 5.10, this assumption entails the existence of a set $Y \subseteq M$ and $y \in M$ such that

(1) $\Phi(Y \cup y) \supset \Theta(\Phi Y \cup W)$

for all $W \subseteq \Phi(Y \cup y)$ such that $|W| < \alpha$. By a suitable application of Zorn's lemma, there is a maximal well-ordered collection $\{Z_\xi : \xi < \lambda\}$ of subsets of Y with the properties:

(2) $\xi < \eta < \lambda$ implies $Z_\xi \subset Z_\eta$;

(3) $|Z_\xi| < \alpha$ for all $\xi < \lambda$;

(4) $\Phi(Z_{\xi+1} \cup y) \not\subseteq \Theta(\Phi(Z_\xi \cup y) \cup \Phi Y)$ if $\xi + 1 < \lambda$.

It is evident from (2) and (3) that $\lambda \leq \alpha$. Let $X = \cup_{\xi<\lambda} Z_\xi$. Then $|X| \leq \alpha$ by (3). We contend that $|X| = \alpha$, and therefore, since α is regular, $\lambda = \alpha$ also. Assume that $|X| < \alpha$. Our initial hypothesis implies that $\Phi(X \cup y) = \Theta(\Phi X \cup W)$ for some $W \in P_\alpha(N)$. Consequently, $\Theta(\Phi(X \cup y) \cup \Phi Y) = \Theta(\Phi X \cup W \cup \Phi Y) = \Theta(\Phi Y \cup W) \subset \Phi(Y \cup y)$ by (1). By 3.1 (a), $\Phi(Y \cup y) = \cup\{\Phi(Z \cup y) : X \subseteq Z \subseteq Y, |Z| < \alpha\}$. Thus, there exists $Z_\lambda \subseteq Y$ with $X \subset Z_\lambda$, $|Z_\lambda| < \alpha$, and $\Phi(Z_\lambda \cup y) \not\subseteq \Theta(\Phi(X \cup y) \cup \Phi Y)$. This contradicts the maximality of $\{Z_\xi : \xi < \lambda\}$. Therefore, $|X| = \alpha$ and $\lambda = \alpha$, Using our hypothesis again, we can find $V \in P_\alpha(N)$ such that $V \subseteq \Theta(\Phi X \cup V) = \Phi(X \cup y) = \cup_{\xi<\alpha}\Phi(Z_\xi \cup y)$. Since $|V| < \alpha$ and α is regular, there exists $\xi < \alpha$ such that $V \subseteq \Phi(Z_\xi \cup y)$. It then follows that $\Theta(\Phi(Z_\xi \cup y) \cup \Phi Y) \supseteq \Theta(V \cup \Phi X) = \Phi(X \cup y) \supseteq \Phi(Z_{\xi+1} \cup y)$, which contradicts (4). Thus, we are forced to the conclusion that Φ is α-bounded.

We conclude this section with an important simultaneity theorem for α-bounded morphisms. Two useful lemmas precede the main result.

5.17. Lemma. Let $\Phi : (M, \Sigma) \to (N, \Theta)$ be a unital morphism. Suppose that Z/X is an α-extension in (M, Σ), and $W/\Phi X$ is an α-extension in (N, Θ), where $\Phi Z \subseteq W$. Then there exists $Y \in P(M)$ such that Y/X is an α-extension, $Z \subseteq Y$, and $\Phi Y \supseteq W$.

Proof. Let $\Theta W = \Theta(\Phi X \cup V)$, $V \in P_\alpha(N)$. Since Φ is unital, $V \subseteq \Phi M = \cup\{\Phi(Z \cup F) : F \in F(M)\}$. Hence, for each $v \in V$, there exists $F_v \in F(M)$ such that $v \in \Phi(Z \cup F_v)$. Let $Y = \cup_{v\in V}(Z \cup F_v)$. Then $Z \subseteq Y$, $|Y - Z| < \alpha$, and therefore Y/X is an α-extension by 5.2. Moreover, $\Phi X \cup V \subseteq \cup_{v\in V}\Phi(Z \cup F_v) \subseteq \Phi Y$. Hence, $W \subseteq \Theta W = \Theta(\Phi X \cup V) \subseteq \Theta\Phi Y = \Phi Y$.

5.18. Lemma. Assume that $\alpha > \aleph_0$. Let Φ and Ψ be unital morphisms from (M, Σ) to (N, Θ), such that $\Phi \leq \Psi$, and Ψ is α-bounded. Suppose that $X \in P(M)$ is such that $\Phi X = \Psi X$. Let Z/X be an α-extension in (M, Σ). Then there is an α-extension Y/X in (M, Σ) such that $Z \subseteq Y$ and $\Phi Y = \Psi Y$.

Proof. Since Ψ is α-bounded, $\Psi Z/\Psi X = \Psi Z/\Phi X$ is an α-extension. By 5.17, there exists $Y_1 \supseteq Z$ such that Y_1/X is an α-extension and $\Phi Y_1 \supseteq \Psi Z$. Continuing this process, we get sets $X \subseteq Z = Y_0 \subseteq Y_1 \subseteq Y_2 \subseteq \dots \subseteq Y_n \subseteq \dots$, such that each Y_n/X is an α-extension and $\Phi(Y_{n+1}) \supseteq \Psi(Y_n)$. Let $Y = \cup_{n<\omega} Y_n$. Then Y/X is an α-extension by 5.5. Moreover, $Z \subseteq Y$, and $\Phi Y \subseteq \Psi Y = \cup_{n<\omega}\Psi Y_n \subseteq \cup_{n<\omega}\Phi Y_{n+1} = \Phi Y$.

5.19. Proposition. Assume that $\alpha > \aleph_0$. Let Φ and Ψ be α-bounded, unital morphisms from (M, Σ) to (N, Θ). Suppose that $X \in P(M)$ is such that $\Phi X = \Psi X$. If Z/X is an α-extension in (M, Σ), then there is an α-extension Y/X in (M, Σ) such that $Z \subseteq Y$ and $\Phi Y = \Psi Y$.

Proof. By 3.5 (b), $\Phi \cap \Psi$ is a unital morphism which is $\leq \Phi$ and $\leq \Psi$. Thus, we can apply 5.18 successively to get a chain of sets

$$X \subseteq Z \subseteq U_0 \subseteq V_0 \subseteq U_1 \subseteq V_1 \subseteq \dots \subseteq U_n \subseteq V_n \subseteq \dots$$

such that all U_n/X and V_n/X are α-extensions, and $(\Phi \cap \Psi)U_n = \Phi U_n$, $(\Phi \cap \Psi)V_n = \Psi V_n$ for all n. Let $Y = \bigcup_{n<\omega} U_n = \bigcup_{n<\omega} V_n$. By 5.4, Y/X is an α-extension, and $\Phi Y = (\Phi \cap \Psi)Y = \Psi Y$ is a consequence of 3.1 (a).

6. Inductive families. It is well known that if Γ is a closure operator on the set M, then the collection of all Γ-closed subsets of M forms a complete lattice. In this section, we will investigate related, but somewhat more general families of sets.

6.1. Definition. (1) Let (M, Σ) be a closure space. An inductive family on (M, Σ) is a subset H of $P(M)$ satisfying

(a) if $X \in H$, then $\Sigma X = X$;

(b) if $\{X_i : i \in I\}$ is a directed subset of H, then $\bigcup_{i \in I} X_i \in H$, and

(c) $M \in H$.

(2) If the inductive family H satisfies the condition

(b') if $\{X_i : i \in I\}$ is any subset of H, then $\Sigma(\bigcup_{i \in I} X_i) \in H$,

we will call H additive.

(3) If the inductive family H satisfies the conditions

(d) $\Sigma\emptyset \in H$, and

(e) if $X \in H$ and Z/X is an α-extension, then there is an α-extension Y/X such that $Z \subseteq Y \in H$,

we will say that H is α-bounded.

Remarks. (1) The condition (b') is plainly stronger than (b). Note that (b') implies $\Sigma\emptyset \in H$. Moreover, conditions (b), (d), and (e) imply condition (c). In fact, it follows from (b) and (d) that H contains a maximal element which, by (e), is necessarily M.

(2) $\aleph_1$-bounded, additive families of "nice" subgroups of primary abelian groups were introduced by Paul Hill in [6] for the proof of his generalized Ulm theorem for totally projective groups. Griffith in [5] gives a proof of Hill's theorem, using only the existence of $\aleph_1$-bounded, inductive families of "nice" subgroups. (An error in the proof of Theorem 67 in [5] leaves open the possibility that the class of groups to which Griffith's proof applies may properly include the class of totally projective groups.)

It is convenient to shorten some of the terminology introduced in 6.1. Thus, we will sometimes drop the adjective "inductive," and simply speak of a family. Also, α-bounded families will occasionally be referred to as α-families.

The connection between closure operators and inductive families is a special case of a more general connection between families and morphisms.

6.2. Proposition. Let Φ and Ψ be unital morphisms of (M, Σ) to (N, Θ). Define

$$H = \{X \in P(M) : \Sigma X = X \text{ and } \Phi X = \Psi X\}.$$

Then

(a) H is an inductive family on (M, Σ);

(b) if Φ and Ψ are additive, then H is additive;

(c) if $\alpha > \aleph_0$, and if Φ and Ψ are α-bounded, then H is α-bounded.

Proof. It is routine to verify 6.1 (a) and (b); (c) follows from the hypothesis that Φ and Ψ are unital. If Φ and Ψ are additive, then 6.1 (b') follows from 4.5. Suppose that Φ and Ψ are α-bounded. Then $\Phi\Sigma\emptyset = \Phi\emptyset = \Theta\emptyset = \Psi\emptyset = \Psi\Sigma\emptyset$. Thus, 6.1 (d) is satisfied. The property 6.1 (e) is just a restatement of 5.19.

6.3. Corollary. Let Γ be a pre-Σ operator on M. Then $H = \{X \in P(M) : \Gamma X = X\}$ is an inductive family on (M, Σ). If Γ is additive, then so is H. If $\alpha > \aleph_0$ and if Γ is α-bounded, then H is α-bounded.

Not every inductive family is obtained by the construction in 6.3. It follows from 2.10 and the observation that X is Γ-closed if and only if X is $\overline{\Gamma}$-closed that the family H defined in 6.3 is closed under intersection. The converse is true and well known: if H is an inductive family that is closed under intersection, then there exists a Σ-operator Γ on M such that $H = \{X \in P(M) : \Gamma X = X\}$. Indeed, Γ is defined by $\Gamma X = \bigcap\{Y \in H : X \subseteq Y\}$. This result won't be used, so we omit its easy proof.

We now prove a simultaneity theorem for inductive families. The result is analogous to 5.19.

6.4. Proposition. Let H and J be inductive families on the closure space (M, Σ). Then:

(a) $H \cap J$ is an inductive family;

(b) if H and J are additive, then $H \cap J$ is additive;

(c) if $\alpha > \aleph_0$, and if H and J are α-bounded, then $H \cap J$ is α-bounded.

Proof. (a) and (b) follow routinely from the definitions. Assume that H and J are α-bounded. Let $X \in H \cap J$, and suppose that Z/X is an α-extension. Repeated application of the definition of α-boundedness gives a chain of sets

$$X \subseteq Z \subseteq U_0 \subseteq V_0 \subseteq U_1 \subseteq V_1 \subseteq \dots \subseteq U_n \subseteq V_n \subseteq \dots,$$

such that $U_n \in H$, $V_n \in J$, and U_n/X and V_n/X are α-extensions for all $n < \omega$. Let $Y = \bigcup_{n<\omega} U_n = \bigcup_{n<\omega} V_n$. Then $Y \in H \cap J$, and Y/X is an α-extension by 5.4.

6.5. Corollary. Suppose that $\alpha > \aleph_0$. Let H be an α-family on the closure space (M, Σ). If Φ and Ψ are α-bounded, unital morphisms from (M, Σ) to (N, Θ), then there is an α-family $\mathcal{J} \subseteq \mathcal{H}$ such that $\Phi X = \Psi X$ for all $X \in \mathcal{J}$. In particular, if Γ is an α-bounded, pre-Σ operator on M, then there is an α-family $\mathcal{J} \subseteq \mathcal{H}$ such that $\Gamma X = X$ for all $X \in \mathcal{J}$.

It is possible to weaken the condition 6.1 (e) for α-families. The result is analogous to 5.10, and the proof is similar, so that we omit it.

6.6. Proposition. Let $\mathcal{H}$ be an inductive family on (M, Σ), with $\Sigma\emptyset \in \mathcal{H}$. Then $\mathcal{H}$ is α-bounded if and only if it satisfies:

(e') if $X \in \mathcal{H}$ and $x \in M$, then there is an α-extension Y/X with $Y \in \mathcal{H}$.

6.7. Corollary. If the inductive family $\mathcal{H}$ is α-bounded, then $\mathcal{H}$ is β-bounded for all regular cardinals $\beta \geq \alpha$.

6.8 Corollary. An additive family $\mathcal{H}$ is α-bounded if and only if for each $x \in M$, there is an α-generated $Y \in \mathcal{H}$ with $x \in Y$.

Proof. Let $X \in \mathcal{H}$, $x \in M$. By hypothesis, $x \in Y \in \mathcal{H}$ for some α-generated Y. By 5.3, $(X \cup Y)/X$ is α-generated, and therefore so is $\Sigma(X \cup Y)/X$. Since $\mathcal{H}$ is additive, $\Sigma(X \cup Y) \in \mathcal{H}$.

We now consider the question: is the image of an inductive family under a morphism again an inductive family?

6.9. Proposition. Let $\Phi : (M, \Sigma) \to (N, \Theta)$ be an additive, unital morphism of closure spaces. Suppose that $\mathcal{H}$ is an additive family on (M, Σ). Then ΦH is an additive family on (N, Θ). If Φ and H are α-bounded, then ΦH is α-bounded.

Proof. Clearly, $\Theta\Phi X = \Phi X$ for all $X \in H$. Let $\{X_i : i \in I\} \subseteq H$. Then $\Theta(\cup_i \Phi X_i) = \Phi(\cup_i X_i) = \Phi\Sigma(\cup_i X_i) \in \Phi H$ by the additivity of Φ and H. Finally, $N = \Phi M \in H$, since Φ is unital. Thus, ΦH is an additive family. Suppose that Φ and H are α-bounded. Let $X \in H$, and let $W/\Phi X$ be an α-extension in (N, Θ). By 5.17 there is an α-extension Z/X in (M, Σ) such that $\Phi Z \supseteq W$. Since H is α-bounded, there exists $Y \in H$ with $Z \subseteq Y$ and Y/X an α-extension. Finally, since Φ is α-bounded, $\Phi Y/\Phi X$ is an α-extension in (N, Θ), and $W \subseteq \Phi Y \in \Phi H$.

6.10. Example. Let (N, Θ) be a closure space, and suppose that $M \subseteq N$ satisfies $\Theta M = N$. Define $\Phi : (M, I) \to (N, \Theta)$ by $\Phi X = \Theta X$ for $X \in P(M)$. Then Φ is an additive morphism that is α-bounded for all α. (Actually, $\Phi = \varphi_*$, where φ is the inclusion mapping of M to N.) Since $\Theta M = N$, Φ is unital. By 6.9,

$$\omega_M = \{\Theta X : X \in P(M)\}$$

is an additive, α-family on (N, Θ) for all α.

Without additivity, the verification of 6.1 (b) in ΦH runs into trouble. However, there is one case in which the difficulty can be overcome.

6.11. Proposition. Let $\Phi : (M, \Sigma) \to (N, \Theta)$ be a unital morphism of closure spaces that satisfies

(*) $\Phi X \subseteq \Phi Y$ implies $\Sigma X \subseteq \Sigma Y$.

Let H be an inductive family on (M, Σ). Then ΦH is an inductive family on (N, Θ). If Φ and H are α-bounded, then ΦH is α-bounded.

Proof. If $\{\Phi X_i : i \in I\}$ is a directed subset of ΦH, with $X_i \in H$ for all i, then by (*), $\{X_i : i \in I\}$ is a directed subset of H. Therefore, $\cup_i \Phi X_i = \Phi \cup_i X_i \in \Phi H$. The remaining parts of the proof of 6.9 are still valid.

There is a particular type of inductive family that is stable under arbitrary morphisms. Except in trivial cases, these families are not α-bounded. Nevertheless, they are useful in some applications. For instance, in [1], Auslander uses such families to prove a basic theorem on global dimension.

6.12. Definition. Let (M, Σ) be a closure space. A filtration of (M, Σ) is a well-ordered collection $F = \{X_\xi : \xi \leq \lambda\} \subseteq P(M)$ satisfying:

(a) $\Sigma X_\xi = X_\xi$ for all $\xi \leq \lambda$;

(b) $\xi < \eta \leq \lambda$ implies $X_\xi \subseteq X_\eta$;

(c) if $\eta \leq \lambda$ is a limit ordinal, then $X_\eta = \cup_{\xi<\eta} X_\xi$;

(d) $X_0 = \Sigma\emptyset$ and $X_\lambda = M$.

If F also satisfies

(e) $X_{\xi+1}/X_\xi$ is an α-extension for all $\xi < \lambda$,

then F is called an α-filtration.

Plainly, any filtration is an additive family. Moreover, it is evident from the definitions that unital morphisms carry filters to filters.

6.13. Proposition. Let F be a filtration of the closure space (M, Σ), and suppose that $\Phi : (M, \Sigma) \to (N, \Theta)$ is a unital morphism. Then ΦF is a filtration of (N, Θ). If F is an α-filtration and Φ is α-bounded, then ΦF is an α-filtration.

We conclude this section by showing that α-filtrations can be constructed within any α-family.

6.14. Proposition. Let H be an α-family on the closure space (M, Σ). Then there is an α-filtration $F \subseteq H$.

Proof. Construct X_η by transfinite induction on η. For $\eta = 0$, let $X_0 = \Sigma\emptyset$. Note that $\Sigma\emptyset \in H$ by 6.1 (d). Assume that $X_\xi \in H$ have been defined for all $\xi < \eta$ such that 6.12 (b), (c), and (e) hold. If η is a limit ordinal, let $X_\eta = \cup_{\xi<\lambda} X_\xi$. Then $X_\eta \in H$ by 6.1 (b). Suppose that $\eta = \xi + 1$. If $X_\xi = M$, terminate the process, taking $\lambda = \xi$. Assume that $X_\xi \subset M$. Choose $x \in M - X_\xi$. By 6.1 (e) there exists $X_{\xi+1} \in H$ such that $X_{\xi+1} \supseteq X_\xi \cup x$ and $X_{\xi+1}/X_\xi$ is an α-extension. This completes the inductive construction, which must terminate with an α-filter $F \subset H$.

7. Applications to modules. Throughout this section and the next one, we will consider closure spaces of the form (M, Σ), where M is the set of elements of a left R-module, and Σ is the closure operator on M that maps a set X to the submodule generated by X. Plainly, $\Sigma(X \cup Y) = \Sigma X + \Sigma Y$, and $\Sigma\emptyset = 0$. Also, $\Sigma A = A$ if and only if A is a submodule of M. The same symbol Σ will be used to denote this closure operator on every module that we consider. This abuse of notation should cause no confusion. Reference to the closure operator Σ will usually be suppressed in the terminology introduced in the last section. For instance, we will speak of a family or a filtration on M, rather than on (M, Σ).

For the most part, the terminology of Section 5 was chosen so that it conforms with the corresponding concepts in module theory. However, there is some ambiquity in using the symbolism B/A to denote either an extension or a quotient module. Whenever this expression occurs, its meaning should be clear from the context. It is worth noting that B/A is an α-extension if and only if the quotient module B/A is α-generated.

The following property of module filtrations will be used several times.

7.1. Lemma. Let $\{A_\xi : \xi \leq \lambda\}$ be a filtration of M such that A_ξ is a direct summand of $A_{\xi+1}$ for all $\xi < \lambda$. Then M is isomorphic to $\oplus_{\xi<\lambda}(A_{\xi+1}/A_\xi)$.

Proof. Let $A_{\xi+1} = A_\xi \oplus B_\xi$, so that $B_\xi \cong A_{\xi+1}/A_\xi$. We prove by induction on η that $A_\eta = \oplus_{\xi<\eta} B_\xi$. For $\eta = 0$, the desired conclusion is a restatement of the first part of 6.12 (d). If η is a limit ordinal greater than 0, then by 6.12 (c), $A_\eta = \cup_{\xi<\eta} A_\xi = \cup_{\xi<\eta}(\oplus_{\zeta<\xi} B_\zeta) = \oplus_{\zeta<\eta} B_\zeta$. If $\eta = \xi + 1$, then $A_\eta = A_\xi \oplus B_\xi = \oplus_{\zeta<\eta} B_\zeta$.

Our first application of closure operators ia a slight generalization* of Kaplansky's main theorem in [7].

7.2. Theorem. Let $M = P \oplus Q = \oplus_{i\in I} M_i$, where each M_i is α-generated. Assume that $\alpha > \aleph_0$. Then $P = \oplus_{j\in J} N_j$, where each N_j is α-generated.

Proof. Let K be the Σ-operator associated with the decomposition $M = \oplus_{i\in I} M_i$ in Example 4.8. Let Λ be the pre-Σ operator associated with the decomposition $M = P \oplus Q$ by the construction of 4.9 and 5.15. By 5.14 and 5.15, K and Λ are α-bounded. Let Γ be the closure of the product $K\Lambda$. By 2.5, 2.9, 3.11, 3.12, and 5.9, Γ is an α-bounded Σ-operator with $\Gamma \geq K$ and $\Gamma \geq \Lambda$. By 6.3 and 6.14, there is an α'-filtration $\{A_\xi : \xi \leq \lambda\}$ of M such that $\Gamma A_\xi = A_\xi$ for all $\xi \leq \lambda$. Thus, by 2.6, $KA_\xi = A_\xi$ and $\Lambda A_\xi = A_\xi$, so that by 4.8 and 5.15,

$$A_{\xi+1} = A_\xi \oplus K_\xi \quad (K_\xi = \text{a sum of } M_i\text{'s}), \text{ and}$$
$$A_\xi = (A_\xi \cap P) \oplus (A_\xi \cap Q).$$

---

*Kaplansky's statement is for $\alpha = \aleph_1$. It is likely that the generalization to arbitrary α was apparent to many persons. The author first heard the extension explicitly mentioned in 1963 by Carol Walker.

By the modular law,

$$A_{\xi+1} \cap P = (A_\xi \cap P) + N_\xi, \text{ where } N_\xi = (A_\xi \cap Q + K_\xi) \cap P.$$

Using the modular law once more, we obtain

$$\begin{aligned} A_\xi \cap P \cap N_\xi &= A_\xi \cap (A_\xi \cap Q + K_\xi) \cap P = (A_\xi \cap Q + A_\xi \cap K_\xi) \cap P \\ &= A_\xi \cap Q \cap P = 0. \end{aligned}$$

Thus, $A_\xi \cap P$ is a direct summand of $A_{\xi+1} \cap P$. Moreover,

$$A_{\xi+1}/A_\xi \cong (A_{\xi+1} \cap P)/(A_\xi \cap P) \oplus (A_{\xi+1} \cap Q)/(A_\xi \cap Q).$$

Since $A_{\xi+1}/A_\xi$ is α-generated by 6.12 (e), it follows that $(A_{\xi+1} \cap P)/(A_\xi \cap P)$ is α-generated. Therefore, $\{A_\xi \cap P : \xi \leq \lambda\}$ is an α-filtration of P. By virtue of 7.1, the proof is complete.

One of the principal applications of 7.2 in [7] is the result that every projective module is a direct sum of countably generated projective modules. The following variant of that theorem seems to be quite useful.

7.3. Proposition. Let P be a projective module. Then there is an $\aleph_1$-bounded Σ-operator Π on P such that if $A < P$ satisfies $\Pi A = A$, then P/A and A are projective modules.

Proof. Recall that a module is projective if and only if it is a direct summand of a free module. Thus, we can write $M = P \oplus Q$, where $M = \oplus_{i \in I} M_i$, with $M_i \cong R$ as a left R-module. Let Γ be the operator on M, defined in the proof of 7.2. Since each M_i is $\aleph_1$-generated, it follows that Γ is $\aleph_1$-bounded. Let $\Pi = \Gamma_P$ be the restriction of Γ to P, that is, $\Pi A = \Gamma A \cap P$ for $A < P$. Note that by 4.8 and 5.15,

(1) $A < M$ implies ΓA and $M/\Gamma A$ are free, and

(2) $A < M$ implies $\Gamma A = (\Gamma A \cap P) \oplus (\Gamma A \cap Q)$.

Thus, if $A < B < M$, then by (2),

(3) $\Gamma B/\Gamma A \cong (\Gamma B \cap P)/(\Gamma A \cap P) \oplus (\Gamma B \cap Q)/(\Gamma A \cap Q)$.

If B/A is $\aleph_1$-generated, so is $\Gamma B/\Gamma A$, because Γ is $\aleph_1$-bounded; therefore, if $A < B < P$ and B/A is $\aleph_1$-generated, then $\Pi B/\Pi A$ is also $\aleph_1$-generated by (3). This proves that Π is $\aleph_1$-bounded. If $\Pi A = A$, then A is projective by (1) and (2), and P/A is projective by (1) and (3).

7.4. Corollary. Let P be a projective module that is generated by a set Y. Then there is an $\aleph_1$-family $\mathcal{J}$ on P such that

(a) if $A \in \mathcal{J}$, then A and P/A are projective, and

(b) if $A \in \mathcal{J}$, then $A = \Sigma X$ for some set $X \subseteq Y$.

Proof. Apply 6.5 with $\mathcal{H} = \mathcal{W}_Y$ (defined in 6.10) and $\Gamma = \Pi$ (as in 7.3).

As an application of this corollary, we will prove a theorem due to G. M. Bergman (see [2]). Bergman's proof uses sheaf-theoretic ideas. Recall that a ring R is left hereditary if every left ideal of R is a projective R-module.

7.5. Theorem. Let R be a left hereditary ring. Let S be a sub-Boolean ring of R. That is, 0 and 1 are in S, $e \in S$ implies $e^2 = e$, and if e and f are in S, then $ef = fe \in S$ and $e \oplus f = e + f - 2ef \in S$. Then S is hereditary.

Remark. It is well known and easy to check that S is a Boolean ring under multiplication and symmetric difference $\oplus$. Of course, S is generally not a subring of R.

Proof. Let J be an ideal of S. We must prove that J is projective as an S-module. Let

$$RJ = \{re : r \in R, \ e \in J\}.$$

Then RJ is a left ideal of R, since $r_1e_1 + r_2e_2 = (r_1e_1 + r_2e_2)(e_1 + e_2 - e_1e_2)$. Since R is hereditary, it follows that RJ is a projective R-module. By 7.4 and 6.14, there is a filtration $\{A_\xi : \xi \leq \lambda\}$ of RJ such that for all $\xi \leq \lambda$,

(1) $A_\xi = RJ_\xi$, where $J_\xi = A_\xi \cap J$,

(2) RJ/A_ξ is projective as an R-module,

(3) $A_{\xi+1}/A_\xi$ is countably generated as an R-module $(\xi < \lambda)$.

It follows from (2) and the exact sequence

$$0 \to A_{\xi+1}/A_\xi \to RJ/A_\xi \to RJ/A_{\xi+1} \to 0$$

that

(4) $A_{\xi+1}/A_\xi$ is a projective R-module.

Thus, A_ξ is a direct summand of $A_{\xi+1}$, say

(5) $A_{\xi+1} = A_\xi \oplus B_\xi$.

We now show

(6) J_ξ is a direct summand of $J_{\xi+1}$, and

(7) $J_{\xi+1}/J_\xi$ is countably generated.

For the proof of (6), define $K_\xi = \{h \in J_{\xi+1} : hJ_\xi = 0\}$. Then K_ξ is an ideal of S, and $K_\xi \cap J_\xi = 0$. It suffices to show that $J_{\xi+1} = J_\xi + K_\xi$. Let $e \in J_{\xi+1}$. By (1) and (5), there exists $f \in J_\xi$, $r \in R$, and $b \in B_\xi$ such that $e = rf + b$. If $g \in J_\xi$, then $gb = ge - grf = eg - grf \in A_\xi \cap B_\xi = 0$. Hence, $ge(1 - f) = 0$. This shows that $e(1 - f) \in K_\xi$, so that $e = ef + e(1 - f) \in J_\xi + K_\xi$. To prove (7), note that by (1) and (3), there is a countable set $E = \{e_0, e_1, e_2, \ldots\} \subseteq J_{\xi+1}$ such that $A_{\xi+1} = A_\xi + Re_0 + Re_1 + Re_2 + \ldots$. If $e \in J_{\xi+1}$, then we can write $e = rf + s_0e_0 + s_1e_1 + \ldots + s_{n-1}e_{n-1}$ for suitable r and s_i in R, and $f \in J_\xi$. Consequently, $e = e(f \cup e_0 \cup e_1 \cup \ldots \cup e_{n-1})$ lies in the ideal of S generated by E and J_ξ. (Note: $f \cup e_0 \cup e_1 \cup \ldots \cup e_{n-1} = 1 - (1 - f)(1 - e_0)(1 - e_1) \ldots (1 - e_{n-1}) \in Sf + Se_0 + Se_1 + \ldots + Se_{n-1}$.) Thus, $J_{\xi+1}/J_\xi$ is generated by the cosets of elements in E, which proves (7). The proof of 7.5 is now easily completed. It is clear from (1) and (7) that $\{J_\xi : \xi \leq \lambda\}$ is a filtration of J. By 7.1 and (6), J is a direct sum of countably generated ideals of S. However, every countably generated ideal in a Boolean ring is projective. Indeed, if $L = Se_0 + Se_1 + Se_2 + \ldots$, then $L = Se_0 \oplus Se_1(1 - e_0) \oplus Se_2(1 - e_0)(1 - e_1) \oplus \ldots$, and each term in this direct sum is a summand of S, hence projective. Thus, we conclude that J is projective.

8. An application to projective dimension. Our objective in this section is to generalize Kaplansky's theorem that every projective module is a direct sum of countably generated projective modules. Can anything similar to this result be proved for modules of projective dimension greater than zero? We cannot generally expect to get direct sum decompositions for such modules. It is known, for instance, that there are indecomposable abelian groups of very large cardinality (see [4]). However, a rephrasing of Kaplansky's theorem suggests a less ambitious conjecture. Our restatement is this: if P is a projective R-module, then there is an $\aleph_1$-filtration $\{A_\xi : \xi \leq \lambda\}$ of P such that P/A_ξ is projective for all $\xi \leq \lambda$. Plainly, this assertion is implied by Kaplansky's theorem, and conversely, Kaplansky's theorem follows easily from the existence of such a filtration, using 7.1 and the argument that led to (4) in the proof of 7.5. We therefore propose the following conjecture.

8.1. Conjecture. Let M be an R-module of finite projective dimension n. Then there is an $\aleph_1$-filtration $\{A_\xi : \xi \leq \lambda\}$ of M such that $\mathrm{Pd}\,(M/A_\xi) \leq n$ for all $\xi \leq \lambda$.

In a sense, 8.1 is the converse of a well-known result due to Auslander (see [1]). Auslander shows that if there is a filtration $\{A_\xi : \xi \leq \lambda\}$ of the module M with $\mathrm{Pd}\ A_{\xi+1}/A_\xi \leq n$ for all $\xi < \lambda$, then $\mathrm{Pd}\ M \leq n$. It is not hard to see (using Theorem 2, p. 169 in [8] for instance) that if $\mathrm{Pd}\ M/A_\xi \leq n$ for all $\xi \leq \lambda$, then $\mathrm{Pd}\ A_{\xi+1}/A_\xi \leq n$ for all $\xi < \lambda$.

We will show that 8.1 is valid if R satisfies a certain chain condition. It would be interesting to know whether or not this chain condition is essential for the validity of 8.1, but this point is left open.

8.2. Definition. Let α be an infinite, regular cardinal number. A ring R is left α-Noetherian if every left ideal of R is α-generated.

Thus, $\aleph_0$-Noetherian is equivalent to the usual Noetherian property.

The property of being α-Noetherian can be formuated as an ascending chain condition.

8.3. Lemma. A ring R is left α-Noetherian if and only if R satisfies the ascending α-chain condition on left ideals: if $\{J_\xi : \xi < \alpha\}$ is a well ordered, ascending chain of left ideals of R, then there is some $\eta < \alpha$ such that $J_\xi = J_\eta$ for all $\xi \geq \eta$.

The proof of this result is just a minor variant of the familiar argument for the usual Noetherian case. Since 8.3 won't be needed, we omit its proof. However, we will need the generalization of another standard result concerning Noetherian rings.

8.4. Lemma. Let R be a left α-Noetherian ring, and suppose that M is an α-generated left R-module. Then every submodule of M is α-generated.

Proof. Let $N < M$. Our objective is to show that N is α-generated. First consider the special case in which M is cyclic. Thus, there is an R-module homomorphism φ of R onto M. Since $\varphi^{-1}N$ is a submodule of R, that is, a left ideal, and R is left α-Noetherian, it follows that $\varphi^{-1}N$ is α-generated. Hence, $N = \varphi\varphi^{-1}N$ is α-generated. Now consider the general case in which $M = \Sigma Y$, where $|Y| < \alpha$. Well order Y, say $Y = \{y_\xi : \xi < \lambda\}$, where $\lambda < \alpha$. Let $A_\eta = \Sigma\{y_\xi : \xi < \eta\}$. Then $\{A_\eta : \eta \leq \lambda\}$ is a filtration of M, and for all $\xi < \lambda$, the quotient module $A_{\xi+1}/A_\xi$ is generated by the coset of y_ξ, that is, $A_{\xi+1}/A_\xi$ is cyclic. It follows that $\{A_\eta \cap N : \eta \leq \lambda\}$ is a filtration of N, and for all $\xi < \lambda$, $(A_{\xi+1} \cap N)/(A_\xi \cap N)$ is isomorphic to the submodule $(A_{\xi+1} \cap N + A_\xi)/A_\xi$ of $A_{\xi+1}/A_\xi$. By the special case that we initially considered, $(A_{\xi+1} \cap N)/(A_\xi \cap N)$ is α-generated. By 5.5, N is α-generated.

The importance of the α-Noetherian property for our work is that it permits us to conclude that the restriction morphism from a module to one of its submodules (considered as closure spaces) is α-bounded.

8.5. <u>Corollary</u>. Let R be a left α-Noetherian ring. Let $\varphi : M \to N$ be an injective homomorphism of left R-modules. Then the morphism $\varphi^* : (N, \Sigma) \to (M, \Sigma)$ of closure spaces is α-bounded.

<u>Proof</u>. Note that $\varphi^*X = \Sigma\varphi^{-1}\Sigma X = \varphi^{-1}\Sigma X$ for all $X \in P(N)$. In particular, $\varphi^*\emptyset = \varphi^{-1}0 = \ker \varphi = 0 = \Sigma\emptyset$, since φ is injective. Moreover, if Y/X is an α-extension, then the quotient module $\Sigma Y/\Sigma X$ is α-generated, and $\varphi^*Y/\varphi^*X = \varphi^{-1}\Sigma Y/\varphi^{-1}\Sigma X \cong (\Sigma Y \cap \operatorname{Im} \varphi)/(\Sigma X \cap \operatorname{Im} \varphi) \cong (\Sigma Y \cap \operatorname{Im} \varphi + \Sigma X)/\Sigma X$ is α-generated by 8.4. Thus, φ^* is α-bounded.

8.6. <u>Corollary</u>. Let R be a left α-Noetherian ring, and let P, F, and A be left R-modules, with $A < F$. Suppose that $\Phi : (P, \Sigma) \to (F, \Sigma)$ is an α-bounded morphism of closure spaces. For $X \subseteq P$, define

$$\Phi_A X = \Phi X \cap A.$$

Then Φ_A is an α-bounded morphism.

<u>Proof</u>. Let $\varphi : A \to F$ be the inclusion homomorphism. Then $\varphi^*\Phi X = \varphi^{-1}\Sigma\Phi X = \varphi^{-1}\Phi X = \Phi X \cap A = \Phi_A X$. Thus, 8.6 follows from 8.5 and 5.7.

We can now formulate the main result of this section.

8.7. <u>Theorem</u>. Let R be a left α-Noetherian ring, where $\alpha > \aleph_0$. Let M be a left R-module with $\operatorname{Pd} M = n < \infty$. Then there is an α-filtration $\{A_\xi : \xi \leq \lambda\}$ of M such that $\text{Pd}\, M/A_\xi \leq n$ for all $\xi \leq \lambda$.

Thus, the conjecture 8.1 is true for modules over rings in which every left ideal is countably generated, in particular, Noetherian rings.

The proof of 8.7 requires two more preliminary results, the first of which is a well-known "diagram lemma." No hypotheses on R are needed for these preliminary steps.

8.8. Lemma. Hypotheses.

(a) $0 \to A_0 \to A_1 \to \dots \to A_n \to A_{n+1} \to 0$ is an exact sequence of left R-modules, with the homomorphisms $\varphi_k : A_k \to A_{k+1}$ for $k \leq n$.

(b) For $k \leq n+1$, B_k is a submodule of A_k such that if $k < n$, then $\varphi_k B_k = B_{k+1} \cap \text{Ker}\, \varphi_{k+1}$.

(c) $B_{n+1} = \varphi_n B_n$.

Conclusion. There is an exact sequence

$$0 \to A_0/B_0 \to A_1/B_1 \to \dots \to A_n/B_n \to A_{n+1}/B_{n+1} \to 0.$$

Proof. We argue by induction on n. In the case $n = 1$, we have the exact sequence $0 \to A_0 \to A_1 \to A_2 \to 0$, $B_k < A_k$ for $k = 0, 1, 2$, $\varphi_0 B_0 = B_1 \cap \text{Ker}\, \varphi_1$, and $B_2 = \varphi_1 B_1$. Let $C = \text{Ker}\, \varphi_1 = \text{Im}\, \varphi_0$. Then $A_2/B_2 \cong (A_1/C)/((B_1 + C)/C) \cong A_1/(B_1 + C) \cong (A_1/B_1)/((B_1 + C)/B_1)$. Moreover, $(B_1 + C)/B_1 \cong C/(C \cap B_1) = \varphi_0 A_0/\varphi_0 B_0 \cong A_0/B_0$, since φ_0 is injective. Hence, the required exact sequence $0 \to A_0/B_0 \to A_1/B_1 \to A_2/B_2 \to 0$

exists. Assume now that $n > 1$, and the lemma holds for $n - 1$. Define $\bar{A}_n = \operatorname{Im} \varphi_{n-1} = \operatorname{Ker} \varphi_n$ and $\bar{B}_n = \varphi_{n-1}B_{n-1}$. We then obtain two exact sequences.

$$0 \to A_0 \to A_1 \to \dots \to A_{n-1} \to \bar{A}_n \to 0,$$
$$0 \to \bar{A}_n \to A_n \to A_{n+1} \to 0,$$

where $\bar{A}_n \to A_n$ is inclusion. Note that

$$\bar{A}_n \cap B_n = \operatorname{Im} \varphi_{n-1} \cap B_n = \varphi_{n-1}B_{n-1} = \bar{B}_n.$$

Thus, by the induction hypothesis and the case $n = 1$, we get the exact sequences:

$$0 \to A_0/B_0 \to A_1/B_1 \to \dots \to A_{n-1}/B_{n-1} \to \bar{A}_n/\bar{B}_n \to 0,$$
$$0 \to \bar{A}_n/\bar{B}_n \to A_n/B_n \to A_{n+1}/B_{n+1} \to 0,$$

which can be pasted together to get the required exact sequence.

8.9. Lemma. Let M be a left R-module. Then there is a free left R-module F and a surjective homomorphism $\varphi : F \to M$ such that

$$K(\operatorname{Ker} \varphi) = F,$$

where K is the Σ-operator on F defined in 4.8 with respect to a suitable decomposition $F = \oplus_{i \in I} F_i$ of F as a direct sum of copies of R.

Proof. Let $X = \{x_a : a \in M\}$, $Y = \{y_a : a \in M\}$ be disjoint sets, each of which is indexed by M. Let F be the free module on $X \cup Y$, so that F has the direct sum decomposition

$$F = (\oplus_{a \in M} Rx_a) \oplus (\oplus_{a \in M} Ry_a),$$

with each factor isomorphic to R. Let the Σ-operator K be associated with this decomposition. Define the homomorphism $\varphi : F \to M$ by the conditions

$$\varphi(x_a) = \varphi(y_a) = a.$$

Plainly, φ is surjective, and $x_a - y_a \in \operatorname{Ker} \varphi$ for all $a \in M$. Note that by the definition in 4.8, $K(x_a - y_a) = Rx_a \oplus Ry_a$. Thus, $K(\operatorname{Ker} \varphi) = F$.

Proof of 8.7. By repeated use of 8.9, we can construct an exact sequence

(1) $0 \to P \to F_1 \to \ldots \to F_n \to M \to 0,$

with homomorphisms $\varphi_n : F_n \to M$, $\varphi_k : F_k \to F_{k+1}$ for $1 \leq k < n$, and with φ_0 taken to be the inclusion mapping of $P = \operatorname{Ker} \varphi_1$ into F_1. By virtue of the construction, each F_k is free, and there is a Σ-operator K_k on F_k (associated as in 4.8 with a direct decomposition of F_k

into copies of R), such that

(2) $K_k(\text{Ker } \varphi_k) = F_k$, and

(3) $K_k(B) = B$ implies $B < F_k$ and F_k/B is free.

It follows from (1) that P is projective, because $\text{Pd } M = n$. In fact, one way to define projective dimension is just this: $\text{Pd } M \leq n$ if and only if every sequence of the form (1) begins with a projective module by P. By Schanuel's lemma ([8], p. 167), if one sequence of the form (1) has P projective, then P will be projective in every such sequence. It follows from 7.3 that there is an $\aleph_1$-bounded Σ-operator Π on P such that

(4) $\Pi(B) = B$ implies $B < P$ and P/B is projective.

Since $\alpha > \aleph_0$, 5.11 implies that Π is α-bounded.

By recursion, define morphisms $\Phi_k : (P, \Sigma) \to (F_k, \Sigma)$ for $1 \leq k \leq n$ according to the schema

$$\Phi_1 = K_k(\varphi_0)_*, \quad \Phi_k = K_k(\varphi_{k-1})_*\Phi_{k-1} \quad \text{for} \quad k > 1.$$

Note that Φ_k is α-bounded by 5.12 and 5.14. Also, Φ_k is unital. In fact, $\Phi_0 P = K_1\varphi_0 P = K_1(\text{Ker } \varphi_1) = F_1$ by (2), and, assuming inductively that $\Phi_{k-1}P = F_{k-1}$, we get $\Phi_k P = K_k\varphi_{k-1}F_{k-1} = K_k(\text{Ker } \varphi_k) = F_k$ by (2) again. Since K_k is a closure operator, it is clear that

(5) $K_k\Phi_k = \Phi_k$.

For $1 \leq k \leq n$, denote for convenience

(6) $C_k = \text{Ker } \varphi_k = \text{Im } \varphi_{k-1}$.

Define morphisms $\Psi_k : (P, \Sigma) \to (C_k, \Sigma)$ as in 8.6 by

(7) $\Psi_k X = \Phi_k X \cap C_k$.

Since R is α-Noetherian, Ψ_k is α-bounded (by 8.6). Also, Ψ_k is unital, because, as was shown above, Φ_k is unital.

Finally, define morphisms $\Omega_k : (P, \Sigma) \to (C_k, \Sigma)$ by

$$\Omega_1 = (\varphi_0)_*, \quad \Omega_k = (\varphi_{k-1})_*\Phi_{k-1} \quad \text{for} \quad 1 < k \leq n,$$

where the φ_{k-1} are considered here as homomorphisms of F_{k-1} onto C_k. It is also useful to define $\Omega = \Omega_{n+1} : (P, \Sigma) \to (M, \Sigma)$ by $\Omega_{n+1} = (\varphi_n)_*\Phi_n$. By 5.7 and 5.12, all Ω_k are α-bounded and unital. Note that by 3.8,

(8) $\Omega_k B = \varphi_{k-1}(\Phi_{k-1}B)$ for all $B < P$.

By applying 6.2, 6.3, and 6.4 to the Σ-operator Π, and the various pairs of morphisms (Ψ_k, Ω_k), we find that there is an α-family H on P such that for all $B \in H$,

(9) $\Pi B = B$, and

(10) $\Omega_k B = \Psi_k B$ for $1 \leq k \leq n$.

By (7) and (8), equation (10) can be rewritten as

(11) $\varphi_{k-1}(\Phi_{k-1}B) = \Phi_k B \cap \operatorname{Ker} \varphi_k$ for $1 \leq k \leq n$.

To this we can add

(12) $\Omega(B) = \varphi_n(\Phi_n B)$.

Using 8.8 together with (11) and (12), it is now easy to show that

(13) $B \in H$ implies $\operatorname{Pd} M/\Omega B \leq n$.

Indeed, if we write $B_k = \Phi_k B$ for $1 \leq k \leq n$, then by (5) and (3), each F_k/B_k is free. Also, P/B is projective by (9) and (4). Thus, by (11), (12), and 8.8, there is an exact sequence

$$0 \to P/B \to F_1/B_1 \to \ldots \to F_n/B_n \to M/\Omega B \to 0.$$

Consequently, $\operatorname{Pd} M/\Omega B \leq n$, as claimed in (13).

To complete the proof, choose an α-filtration $F \subseteq H$. This is possible by 6.14. Let $\{A_\xi : \xi \leq \lambda\} = \Omega F$. By 6.13, $\{A_\xi : \xi \leq \lambda\}$ is an α-filtration of M, and by (13), $\operatorname{Pd} M/A_\xi \leq n$ for all $\xi \leq \lambda$.

REFERENCES

1. Auslander, M., On the dimension of modules and algebras, III. Nagoya Math. J. 9 (1955), 67-77.

2. Bergman, G. M., Hereditary commutative rings, and centers of hereditary rings, To appear.

3. Birkhoff, G. and O. Frink, Representations of lattices by sets, Trans. Amer. Math. Soc. 64 (1948), 299-316.

4. Corner, A. L. S., Large torsion-free abelian groups with prescribed topological endomorphism rings, To appear.

5. Griffith, P. A., "Infinite Abelian Group Theory," Chicago, 1970.

6. Hill, Paul, On the classification of abelian groups, To appear.

7. Kaplansky, I., Projective modules, Ann. of Math. 68 (1958), 372-377.

8. Kaplansky, I., "Fields and Rings," Chicago, 1969.

9. Osofsky, B. L., Homological dimension and cardinality, Trans. Amer. Math. Soc. 151 (1970), 641-650.

ON GOLDMAN'S PRIMARY DECOMPOSITION

by

Hans H. Storrer

Research Institute for Mathematics
Swiss Federal Institute of Technology
Zurich, Switzerland

In a recent paper [12] Goldman introduced a primary decomposition for Noetherian modules over an arbitrary ring, generalizing the classical Lasker-Noether theory for finitely generated modules over a commutative Noetherian ring.

In Goldman's theory, the role of prime ideals is played by the prime kernel functors and a good deal of his paper is devoted to the development of the general theory of kernel functors, modules of quotients and related concepts.

In this paper, we present another approach, which we think requires less background and which moreover follows very closely the treatment of the commutative case as presented e.g. in Bourbaki's chapter on primary decomposition.

The first section of this article contains some material on rational extensions of modules. Most of these results are well known [6,8], but for the convenience of the reader we have included proofs, some of which may be shorter than the existing ones.

The second section motivates and presents the definitions of the objects which will play the role of prime ideals, the atoms: A module is called atomic if it is rationally complete and a rational extension of every non-zero submodule and an atom is an isomorphism class of atomic submodules. Thus an atom is in a way an analogue of an isomorphism class of indecomposable injective modules. For a commutative ring, the former are in a one-one correspondence with the prime ideals, whereas the latter are not. This is one of the reasons why we work with atoms.

Section 3 introduces the concept of associated atom: An atom represented by the atomic module A is associated to a module M is a non-zero submodule of A is isomorphic to a submodule of M. $N \subseteq M$ is a primary submodule if M/N is co-primary, i.e. has only one associated atom. With these definitions, the well-known results on existence and uniqueness of a primary decomposition carry over from the case of a finitely generated module over a commutative Noetherian ring to the case of a Noetherian module over an arbitrary ring.

In section 4, two generalizations of the Krull-Akizuki Theorem are presented. In the next section, we prove the equivalence of the present theory with the one by Goldman. More precisely it is shown, that there is a one-one correspondence between atoms and prime kernel functors (this has recently been announced by Hudry [14,15]) and that (at least for Noetherian modules) Goldman's primary modules are the same as our co-primary modules.

In section 6, we show how to obtain the Lesieur-Croisot tertiary decomposition. This is done by introducing an equivalence relation on the set of atoms. The last section contains some results on the connection between atomic modules and the compressible and quasi-simple modules recently introduced by Goldie [11] and Koh [16].

The author wishes to thank Tulane University for its hospitality, John Dauns for many helpful conversations and the "Schweizerischer Nationalfonds zur Förderung der wissenschaftlichen Forschung" for financial support.

1. Rational Extensions of Modules

Here and throughout this paper, we shall use the following conventions: All rings have a unit element and "module" means "right R-module". The injective hull of a module M is denoted by E(M) and if $N \subseteq M$, $x \in M$, then $x^{-1}N = \{r \in R \mid xr \in N\}$. Ann x denotes the annihilator of x.

A module M is said to be a <u>rational extension</u> of a submodule N if for every module A such that $N \subseteq A \subseteq M$ and every homomorphism $\phi: A \to M$, $N \subseteq \operatorname{Ker} \phi$ implies $\phi = 0$ [8].

(1.1) <u>Lemma</u>. If $N \subseteq M$, then the following statements are equivalent:

(a) $N \subseteq M$ is rational,

(b) $\operatorname{Hom}(M/N, E(M)) = 0$,

(c) For all $m_1, m_2 \in M$, $m_2 \neq 0$, there exists an $r \in R$ such that $m_1 r \in N$, $m_2 r \neq 0$.

Proof. The equivalence of (a) and (b) is clear. If (c) is not satisfied, there exist elements $m_1, m_2 \neq 0$ in M such that $m_1 r \in N$ implies $m_2 r = 0$ for all $r \in R$. Then $\phi: N + m_1 R \to M$ given by $(n + m_1 r) = m_2 r$ is a well-defined non-zero homomorphism with $N \subseteq \operatorname{Ker} \phi$, hence $N \subseteq M$ is not rational. If (a) is false, then there is a non-zero $\psi: A \to M$ with $N \subseteq \operatorname{Ker} \psi$, thus $\psi(m_1) = m_2 \neq 0$ for some $m_1, m_2 \in M$, and since then $m_1 r \in N$ implies $m_2 r = 0$ for all $r \in R$, (c) is not satisfied.

By taking $m_1 = m_2$ in (c), one sees, that a rational extension is essential. The converse is not true in general, e.g. $\mathbb{Z}/(2) \subseteq \mathbb{Z}/(4)$ is essential, but not rational. However, if the singular submodule $Z(N)$ of N is zero, then every essential extension M of N is rational. Indeed, for any $m_1, m_2 \in M$, $m_2 \neq 0$, $m_1^{-1}N$ is an essential right ideal and cannot be contained in $\operatorname{Ann} m_2$ (for otherwise $m_2 \in Z(M)$, and $Z(M) = 0$ since $N \subseteq M$ is essential and $Z(N) = 0$). This means, that there exists an $r \in m_1^{-1}N$, $r \notin \operatorname{Ann} m_2$.

A module is said to be <u>rationally complete</u>, if it has no proper rational extensions. In particular, any injective module is rationally complete. The following lemma shows, that rationally complete modules satisfy a weak form of injectivity.

(1.2) <u>Lemma</u>. A module M is rationally complete if and only if the following condition holds:

(*) If B is any module and if C is a submodule of B such that $\operatorname{Hom}(B/C, E(M)) = 0$, then any homomorphism from C into M can be extended to a homomorphism from B into M.

Proof. Let M satisfy (*) and suppose $M \subseteq P$ is rational. Since $E(M) \subseteq E(P)$, we have $\operatorname{Hom}(P/M, E(M)) = 0$ by (1.1,b). Thus the identity map $M \to M$ lifts to a homomorphism $\phi : P \to M$, which is mono because $M \subseteq P$ is essential. If $p \in P$, then $\phi(p) = m = \phi(m)$ for some $m \in M$, hence $P = M$ and M is rationally complete.

The following proof of the converse is due to P. M. Cohn. Assume, that M is rationally complete and assume further, that $\mathrm{Hom}(B/C,E(M)) = 0$. Any $\phi: C \to M$ extends to a $\psi: B \to E(M)$ and we have to show, that $\psi(B) \subseteq M$. ψ induces an isomorphism $B/D \to \psi(B)/\psi(B) \cap M$, where $D = \{b \in B \mid \psi(b) \in M\}$ and by the isomorphism theorem, $B/D \cong M + \psi(B)/M$. Since $C \subseteq D$, there is an epimorphism $B/C \to B/D$ and therefore $\mathrm{Hom}(B/D,E(M)) = 0$. Observing that $E(M + \psi(B)) = E(M)$, we see, that $\mathrm{Hom}(M + \psi(B)/M, E(M + \psi(B)) = 0$, thus $M \subseteq M + \psi(B)$ is rational by (1.1,b). Since M is assumed to have no proper rational extensions $\psi(B) \subseteq M$.

Next, we show, that every module M has a rational extension which is rationally complete. This extension is called the rational completion $\bar{M}$ and is unique up to isomorphism over M. We can construct $\bar{M}$ in the following way: Let

$$\bar{M} = \{x \in E(M) \mid \forall\ y \neq 0,\ y \in E(M)\ \exists\ r \in R,\ xr \in M,\ yr \neq 0\}.$$

The crucial point is to show, that the set $\bar{M}$ is a submodule of $E(M)$. The given condition is equivalent to the following: If $y \neq 0$, then $y(x^{-1}M) \neq 0$. Suppose now, that $x_1, x_2 \in \bar{M}$, but $z = x_1 - x_2 \notin \bar{M}$. Then $y(z^{-1}M) = 0$ for some $y \neq 0$ and $\phi: M + zR \to E(M)$ given by $m + zr \longmapsto yr$ is a well-defined homomorphism and can be extended to $\psi: E(M) \to E(M)$. $\psi(z) = y \neq 0$, so either $\psi(x_1) = m_1$ or $\psi(x_2) = m_2$ is non-zero. But this contradicts the assumption, that

$y_i(x_i^{-1}M) \neq 0$ for $i = 1,2$, hence $x_1 - x_2 \in \overline{M}$. A similar argument shows, that $x \in \overline{M}$ implies $xr \in \overline{M}$ for all $r \in R$.

By construction, $M \subseteq \overline{M}$ is rational and it remains to show, that $\overline{M}$ is rationally complete. If $\overline{M} \subseteq T$ is rational, then we can assume, that $\overline{M} \subseteq T \subseteq E(M)$ since there exists a monomorphism $T \to E(M)$ extending the identity on $\overline{M}$. It is easily seen, that the concept of rational extension is transitive, hence $M \subseteq T$ is rational. Let $t \in T$, $y \in E(M)$, $y \neq 0$. Then there is an $r \in R$ such that $0 \neq yr \in T$ and furthermore, there exists an $s \in R$ such that $trs \in M$, $yrs \neq 0$, whence $t \in \overline{M}$.

The proof shows, that $\overline{M}$ contains every rational extension of M in $E(M)$. Since the injective hull is unique up to isomorphism over M, so is $\overline{M}$.

Another characterization of $\overline{M}$ is as follows.

(1.3) Lemma. $\overline{M}$ is the intersection of the kernels of all endomorphisms of $E(M)$ which annihilate M.

Proof. Let L be the intersection described above. Let $x \in \overline{M}$ and let $\phi: E(M) \to E(M)$ with $M \subseteq \mathrm{Ker}\,\phi$. Suppose $y = \phi(x) \neq 0$. Then there exists an $r \in R$ such that $xr \in M$, $yr \neq 0$, a contradiction. Thus $\phi(x) = 0$ and $x \in L$. On the other hand, if $x \notin \overline{M}$, then there exist a $y \in E(M)$, $y \neq 0$, such that $xr \in M$ implies $yr = 0$. Then $\psi: xR + M \to E(M)$ given by $xr + m \longmapsto yr$ extends to a non-zero endomorphism of $E(M)$ annihilating M.

The following result will be used later.

(1.4) <u>Lemma</u>. Let R be commutative and let $R \to S$ be a surjective ring homomorphism. Then every rationally complete S-module N is rationally complete as an R-module.

Proof (Bland [2]). Let $N \subseteq M$ be a rational extension of R-modules and let I be the kernel of $R \to S$. For any $r \in I$, multiplication by r is an R-homomorphism $M \to M$ which annihilates N. Since $N \subseteq M$ is rational, this must be the zero map, i.e. $rM = 0$ for every $r \in I$. This means, that M can be considered as an S-module and $N \subseteq M$ will be a rational extension of S-modules, whence $N = M$.

Recall, that a module is said to be uniform, if it is an essential extension of every non-zero submodule. We shall say, that a module is <u>strongly uniform</u>, if it is non-zero and a rational extension of every non-zero submodule. It is easy to see, that submodules and rational extensions of strongly uniform modules are again strongly uniform. Examples of strongly uniform modules are the simple modules, the uniform modules with singular submodule zero, and, most important for us, the modules of the form R/P, where P is a prime ideal of the commutative ring R:

(1.5) <u>Lemma</u>. If P is an ideal of the commutative ring R, then R/P is strongly uniform if and only if P is prime.

Proof. Let R/P be strongly uniform and let $a,b \in R$, $a,b \notin P$. Then R/P is a rational extension of $bR + P/P$, thus there exists an $r \in R$ such that $(a+P)r \neq 0$, $(1+P)r \in bR + P/P$, i.e. $r = bs + p$ $(s \in R, p \in P)$ and $ar \notin P$. Since $ar = abs + ap \notin P$, it follows, that $ab \notin P$, thus P is prime.

Conversely, let P be prime. We wish to show, that $bR + P/P \subseteq R/P$ is rational for any $b \notin P$. For this let $a \notin P$, $c \in R$. Then $(a+P)b \neq 0$, $(c+P)b \in bR + P/P$ and we are done.

A module will be called atomic, if it is strongly uniform and rationally complete, i.e. if it is a rational extension of every non-zero submodule and has no proper rational extension. Replacing "rational" by "essential", one sees, that an atomic module is in some sense analogous to an indecomposable injective module.

As an illustration, we describe the atomic R-modules, where R is commutative. We claim, that they are (up to isomorphism) just the quotient fields of the integral domains R/P, viewed as R-modules.

In fact an atomic module is the rational completion of each of its cyclic submodules, and the latter are of the form R/P, P a prime, by (1.5). Now the quotient field K of R/P is the injective hull of R/P as R-module and $R/P \subseteq K$ is a rational extension of R/P-modules hence of R-modules. Thus it remains to show, that K is rationally complete as an R-module, and this follows immediately from (1.4), since K is injective as an R/P-module.

2. Atoms

Before proceeding, we wish to motivate the definitions in this section. Our aim is to generalize the notion of "prime ideal" to the noncommutative case in such a way, that we obtain a theory of "associated primes" and "primary decomposition" paralleling the classical theory for commutative Noetherian rings.

One possibility is to consider two-sided prime ideals and one is then led to the "tertiary decomposition" introduced by Lesieur and Croisot [21], which is in a certain sense the only possibility [24].

Here we do not insist, that our "primes" be two-sided ideals. Instead, we wish a definition in terms of right ideals and right modules, which, incidentally, would allow a generalization to suitable categories.

Now Lemma (1.5) is just what we need: If R is commutative, then an ideal P is prime if and only if R/P is strongly uniform. (It should be mentioned, however, that there are other characterizations; see section 7).

Thus, for an arbitrary ring, we are led to consider right ideals P such that R/P is strongly uniform. These right ideals have been called <u>critical</u> by Lambek and Michler [18], where they occur under a different, but equivalent definition. We shall adhere to their terminology.

In the commutative case, a prime ideal P is associated to a module M, if R/P is isomorphic to a submodule of M or, equivalently, if P is the annihilator of a non-zero element of M.

If we generalize this definition to the case of an arbitrary ring, replacing "prime" by "critical", it is immediate, that two critical right ideals P and Q such that $R/P \cong R/Q$ should be considered as equivalent (at least for the purpose of defining associated "primes"). Thus it seems, that a suitable generalization of "prime" would be "equivalence class of critical right ideals" (under the equivalence relation just defined).

However, this is not yet quite satisfactory, for as in the commutative case, the module R/P (P critical) should have only one associated "prime" namely P itself (or rather the equivalence class of P).

The following example shows, that this will not be the case in general.

(2.1) Example. Let R be the ring of upper triangular 2×2 matrices over a field and let

$$P = \{\begin{pmatrix} 0 & 0 \\ 0 & \gamma \end{pmatrix}\}, \quad Q = \{\begin{pmatrix} \alpha & \beta \\ 0 & 0 \end{pmatrix}\}.$$

Both P and Q are critical right ideals, Q is maximal and P is not, thus R/Q is not isomorphic to R/P, yet R/Q is isomorphic to a submodule of R/P.

This example together with the remark preceding it shows, that two critical right ideals P and Q should be considered equivalent not only if $R/P \cong R/Q$, but also if a non-zero submodule of R/P is isomorphic to a submodule of R/Q. Now it is easy to see, that this happens if and only if $\overline{R/P} \cong \overline{R/Q}$. Since the modules of the form $\overline{R/P}$ (P critical) are just the atomic modules, we see that the new equivalence classes of critical right ideals are in a one-one correspondence to the isomorphism classes of atomic modules. Hence we finally have obtained our desired generalization of "prime", which we will call "atom" (following a suggestion by J. Lambek): An isomorphism class of atomic modules.

Let us repeat the new definitions made in the preceding remarks: A right ideal P is called critical if R/P is strongly uniform. Examples of critical ideals are:
(i) Every maximal right ideal is critical, (ii) if R is commutative, then the critical ideals are exactly the prime ideals and more generally (iii) a two-sided ideal P of an arbitrary ring R is critical as a right ideal if and only if R/P is a right Ore domain [15]. (iii) can be proved as follows:

Assume that R/P is a right Ore domain. We have to show, that $aR + P/P \subseteq R/P$ is rational for every $a \notin P$. Let $b,c \in R$, $c \notin P$. If $b \in P$, then $(b+P)1 \in aR + P/P$, $(c+P)1 \neq 0$. If $b \notin P$, then $bc \notin P$ since R/P is a domain and because of the Ore condition there exist $x,y \in R$ such that $bcx - ay \in P$ and $bcx \notin P$. Thus $r = cx \notin P$, hence $(b+P)r \in aR + P/P$ and $(c+P)r \neq 0$. Therefore (1.1,c) is satisfied for all $(b+P), (c+P) \in R/P$, $(c+P) \neq 0$.

Conversely, let P be critical. Let $a,b \in R$, $a,b \notin P$, then R/P is a rational extension of $bR + P/P$ and there exists an $r \in R$ such that $(a+P)r \neq 0$ and $(1+P)r \in bR+P/P$

i.e. $r = bs + p$, $s \in R$, $p \in P$. Then we have $ar = abs + ap \notin P$ which implies $ab \notin P$, hence R/P is a domain. To prove the Ore condition, given a,b as before, we choose an $r \in R$ such that $0 \neq (a+P)r \in bR + P/P$. Then $ar \notin P$ and $ar - bs \in P$ for some $s \in R$.

(2.2) Lemma. P is a critical right ideal if and only if $\mathrm{Hom}(R/I, E(R/P)) = 0$ for every right ideal I properly containing P.

Proof. This is just a restatement of the definition in view of (1.1).

We shall say, that two right ideals I and J are related [18] if $a^{-1}I = b^{-1}J$ for some $a \notin I$, $b \notin J$.

(2.3) Proposition. The following are equivalent for two critical right ideals P and Q.

(a) P and Q are related,

(b) A non-zero submodule of R/P is isomorphic to a submodule of R/Q,

(c) $\overline{R/P} \cong \overline{R/Q}$.

Proof. (a) $\Rightarrow$ (b). The map sending $1 + a^{-1}P$ into $a + P$ is an isomorphism of $R/a^{-1}P$ onto a submodule of R/P; similarly, $R/b^{-1}Q$ is isomorphic to a submodule of R/Q and since $a^{-1}P = b^{-1}Q$, (b) follows.

(b) $\Rightarrow$ (c). Since R/P and R/Q are both strongly uniform with isomorphic submodules M and M', we have $\overline{R/P} = \overline{M} \cong \overline{M'} = \overline{R/Q}$.

(c) $\Rightarrow$ (a). Let $\phi : \overline{R/P} \to \overline{R/Q}$ be an isomorphism. If $\phi(1+P) = x \in \overline{R/Q}$ then there exists an $a \in R$ such that $0 \neq xa = b+Q \in R/Q$. Then $a \notin P$, $b \notin Q$ and since ϕ is mono $ar \notin P \iff br \in Q$ for all $r \in R$, i.e. P and Q are related.

If R is commutative, two prime ideals are related if and only if they are equal. Thus every cyclic submodule of a fixed atomic module is isomorphic to R/P, for a fixed prime ideal P.

We have already defined an <u>atom</u> to be an isomorphism class of atomic modules. In the light of (2.3) an atom can also be viewed as an equivalence class of related critical right ideals.

We have also remarked, that "atomic module" is an analogue of "indecomposable injective module." We shall now discuss this analogy in more detail. Recall that a right ideal I is called <u>irreducible</u>, if I is not the intersection of any two properly larger ideals, or equivalently, if R/I is a uniform module.

The following result was proved by Dlab [5]. Another proof can be given along the lines of (2.3).

(2.4) <u>Proposition</u>. The following are equivalent for two irreducible right ideals I and J.

(a) I and J are related,

(b) A non-zero submodule of R/I is isomorphic to a submodule of R/J,

(c) $E(R/I) \cong E(R/J)$.

(2.5) Corollary. The assignment $A \longmapsto E(A)$ induces a one-one map from the set of atoms into the set of isomorphism classes of indecomposable injectives.

To describe this connection more precisely, we need another definition, due to Lesieur and Croisot [20]: The core (or heart) $C(E)$ of an indecomposable injective module E is the intersection of the kernels of all endomorphism of E with non-zero kernel.

We need two preliminary results, the first of which has a useful corollary.

(2.6) Proposition (Lambek/Michler [18,2.7]). Let E be an indecomposable injective module and suppose, that P is maximal among the annihilators of non-zero elements of E. Then P is critical.

Proof. By assumption, an isomorphic copy of R/P is contained in E, hence $E \cong E(R/P)$. Let I properly contain P. We claim, that $\mathrm{Hom}(R/I,E) = 0$, which by (2.2) will prove the proposition. Thus suppose, that $\phi : R/I \to E$ is non-zero with kernel K/I. Then ϕ induces a monomorphism $R/K \to E$; thus by maximality $P = I = K$.

(2.7) Corollary. If R is right Noetherian, then every module M contains a strongly uniform submodule.

Proof. By a result of Matlis [22] every injective module is a direct sum of indecomposable injectives, and each of these contains a strongly uniform submodule R/P by (2.6).

Thus $E(M)$ contains a strongly uniform submodule and hence so does M, since $M \subseteq E(M)$ is essential. This result will also follow from (3.3).

(2.8) Lemma. Let U be a strongly uniform submodule of the indecomposable injective E. If $0 \neq V \subseteq U$, then $\phi(V) = 0$ implies $\phi(U) = 0$ for all $\phi : E \to E$.

Proof. Suppose $\phi(u) \neq 0$, $u \in U$. Since $U \subseteq E$ is essential, there is an $r \in R$ such that $0 \neq \phi(u)r \in U$ and since $V \subseteq U$ is rational, there is an $s \in R$ such that $urs \in V$, $\phi(urs) \neq 0$, a contradiction.

(2.9) Proposition. Let E be an indecomposable injective module. Then the following subsets of E are equal:

(a) C_1 = sum of all strongly uniform submodules of E (the empty sum is the zero submodule by definition).

(b) $C_2 = \{x \in E \mid \text{Ann } x \neq R \text{ is critical}\} \cup \{0\}$.

(c) $C_3 = C(E)$, the core of E.

(d) $C_4 = \{x \in E \mid \text{Ann } x$ is maximal among the $\text{Ann } y$, $0 \neq y \in E\} \cup \{0\}$.

Proof. Clearly $C_2 \subseteq C_1$, and if E has no strongly uniform submodules, then $C_1 = C_2$. Let now U and V be arbitrary non-zero strongly uniform submodules, and put $U \cap V = W \neq 0$. Any map $\phi: E \to E$ annihilating U annihilates W and hence V by (2.8). Thus $\overline{U} = \{x \in E \mid \phi(U) = 0 \Rightarrow \phi(x) = 0$ for all $\phi: E \to E\} = \overline{V}$ (1.3). This module contains every strongly uniform module, thus equals C_1 and is itself strongly uniform, thus $C_1 \subseteq C_2$.

Any $\phi: E \to E$ with non-zero kernel annihilates a submodule of C_2, hence all of C_2 by (2.8), whence $C_2 \subseteq C_3$.

The following proof for $C_3 \subseteq C_4$ is taken from [20,3.1]. Let $0 \neq x \in C_3$. Suppose $\text{Ann } x$ is not maximal, then $\text{Ann } x \subsetneqq \text{Ann } y$, $0 \neq y \in E$, thus $x \longmapsto y$ induces a well-defined map $xR \to yR$, which can be extended to a $\phi: E \to E$. Since $\text{Ann } x \subsetneqq \text{Ann } y$ is proper, there is an $r \in R$ such that $xr \neq 0$, $yr = 0$, whence $xr \in \text{Ker } \phi \neq 0$ and $x \notin \text{Ker } \phi$, contradicting the assumption $x \in C_3$.

Finally, $C_4 \subseteq C_2$ follows from (2.6).

Thus an indecomposable injective module E contains a strongly uniform submodule if and only if $C(E) \neq 0$ and in this case $C(E)$ is the unique atomic submodule of E. It follows from (2.7) that $C(E) \neq 0$ for every indecomposable injective E if R is right Noetherian. We obtain the following improvement of (2.5).

(2.10) <u>Proposition</u>. Let R be right Noetherian. Then the assignment $A \longmapsto E(A)$ induces a bijection between the set of atoms and the set of isomorphism classes of indecomposable injectives, with inverse induced by $E \longmapsto C(E)$.

It remains to give an example of an indecomposable injective E which does not contain any strongly uniform submodule. In the commutative case, this means, that no submodule of E is isomorphic to R/P, P a prime.

(2.11) Example. Let K be a field and let R be the ring of polynomials in countably many commuting variables $x_1, x_2, \ldots,$ over K, subject to the relations

(*) $x_1^2 = 0$ and $x_n^2 = x_{n-1}$ for $n \geq 2$.

This ring appears in [19] in another context.

R is clearly local with maximal ideal J generated by the x_i. J is nil but not nilpotent (in fact $J^2 = J$). Any element r of R can be written as a polynomial in a single variable, $r = a_0 + a_1x_n + \ldots + a_kx_n^k$ say, and by factoring out the smallest power of x_n with non-zero coefficient, x_n^j say, we get $r = x_n^j(a_j + a_{j+1}x_n + \ldots. + a_kx_n^{k-j}) = x_n^ju$, where u is a unit. We can now establish the following properties of R.

(i) Since the unique maximal ideal J is nil, it is also the unique prime ideal of R.

(ii) Because of the representation $r = x_n^ju$ and the relations (*), it is easily seen, that no element of R annihilates J, in other words, R does not contain minimal ideals.

(iii) R is uniform as a right R-module. It suffices to show, that the intersection of two non-zero principal ideals rR and sR is non-zero. This follows immediately from the representations $r = x_m^iu$, $s = x_n^jv$ and the fact, that because of the relations (*) x_m is a power of x_n if $m \leq n$.

Let now E = E(R). This is an indecomposable injective by (iii), it does not contain any simple submodule by (ii), yet the only module of the form R/P, P a prime is simple by (i). Thus E is the desired example.

3. The primary decomposition

Recall, that we defined an atom $\mathcal{A}$ to be an isomorphism class of atomic modules, or equivalently, an equivalence class of critical right ideals. We often shall write $\mathcal{A} = [A]$, A denoting an atomic module in $\mathcal{A}$.

As already mentioned in the introduction to section 2, an atom $\mathcal{A}$ is associated to a module M, if there is an $x \in M$, whose annihilator is a critical ideal belonging to $\mathcal{A}$. However, we shall work with the following equivalent definition: $\mathcal{A} = [A]$ is associated to M if a non-zero submodule of A is isomorphic to a submodule of M. It is readily verified, that this definition agrees with the usual one for commutative rings [3,ch.IV,§1].

One might think, that $\mathcal{A} = [A]$ is associated to M implies A is isomorphic to a submodule of $\bar{M}$. This is not so in general, as shown by the following example: Let $R = \mathbb{Z}$, $A = \mathbb{Q}$ and $M = \mathbb{Z} \oplus \sum_p \mathbb{Z}/p\mathbb{Z}$, where p runs through all primes. It can be shown, that M is rationally complete. Obviously, $[A]$ is associated to M, yet A does not embed into M. This also shows, that $N \subseteq M$ does not imply $\bar{N} \subseteq \bar{M}$ in general, although it does imply $E(N) \subseteq E(M)$.

The set of associated atoms of M will be denoted by Ass M. We are now going to prove the main properties of Ass M. The commutative analogue can be found in Bourbaki's chapter on primary decomposition [3] together with other results, which also would generalize.

(3.1) Proposition.

(a) If M is the union of submodules M_i, then $\text{Ass } M = \bigcup_i \text{Ass } M_i$.

(b) If A is atomic and $\mathcal{A} = [A]$, then for every non-zero submodule M of A, $\text{Ass } M = \{\mathcal{A}\}$.

(c) If $N \subseteq M$, then $\text{Ass } N \subseteq \text{Ass } M \subseteq \text{Ass } N \cup \text{Ass } M/N$.

(d) If M is the direct sum of submodules M_i, then $\text{Ass } M = \bigcup_i \text{Ass } M_i$.

Proof. (a) Clearly $\bigcup_i \text{Ass } M_i \subseteq \text{Ass } M$. If $\mathcal{A} \in \text{Ass } M$, then a critical right ideal belongs to $\mathcal{A}$ if it is the annihilator of an element of M, hence of an element of some M_i.

(b) By definition $\mathcal{A} \in \text{Ass } M$. Suppose $\mathcal{B} = [B] \in \text{Ass } M$, then a submodule B' of B is isomorphic to $A' \subseteq A$. Thus $A = \overline{A'} \cong \overline{B'} = B$.

(c) Clearly $\text{Ass } N \subseteq \text{Ass } M$. Suppose $[A] \in \text{Ass } M$. Let $0 \neq M' \subseteq M$ be isomorphic to $A' \subseteq A$, and let $L = M' \cap N$. If $L = 0$, then M' is isomorphic to a submodule of M/N, hence $[A] \in \text{Ass } M/N$. If $L \neq 0$, $\text{Ass } L = \{[A]\}$ by (b), and $[A] \in \text{Ass } L \subseteq \text{Ass } N$.

(d) Since M is the union of all the finite sums of the M_i, it suffices to prove the result for finite sums by (a) and by induction, one reduces the proof to the case of two summands, which is immediate by (b).

An important property of modules over a commutative Noetherian ring is that Ass $M = \emptyset$ if and only if $M = 0$ [3,ch.IV,§1]. We cannot omit "Noetherian", as the module R of Example (2.11) shows.

In the non-commutative case, the result is still true for modules over a right Noetherian ring by (2.7). We shall now prove a more general result and we will need the following Lemma, due to Goldie [10, 1.2; 11, 1.06].

(3.2) Lemma. Every Noetherian module contains a non-zero uniform submodule.

Proof. Suppose M does not contain any uniform submodule. Then there exist two non-zero submodules N_1, N_1' with zero intersection. None of them is uniform, hence N_1' contains again two non-zero submodules N_2, N_2' with zero intersection and so on. In this way we obtain a proper ascending chain of direct sums $N_1 \subseteq N_1 + N_2 \subseteq N_1 + N_2 + N_3 \subseteq \ldots\ldots$.

(3.3) Proposition. If M is a Noetherian module over the arbitrary ring R, then Ass $M \neq \emptyset$.

Proof. By (3.2) M contains a uniform submodule $N \neq 0$, which is also Noetherian. Let A be maximal among all submodules of N such that $\mathrm{Hom}(N/A, E(N)) \neq 0$, put $B = N/A$ and consider a non-zero map $\phi : B \to E(N)$. If K/A is the kernel of ϕ, then ϕ induces a monomorphism $N/K \cong (N/A)/(K/A) \to E(N)$, hence $A = K$ by maximality and ϕ is a monomorphism. Since $E(N)$ is uniform, $E(N) \cong E(B)$

and $\phi(B)$ has non-zero intersection with N. Thus it remains to show that B is strongly uniform. If $B' = N'/A$ is a non-zero submodule of B, then $B/B' \cong N/N'$, hence $\mathrm{Hom}(B/B', E(B)) = 0$ by maximality of A, and B is strongly uniform by (1.1, b).

We are now ready to establish the primary decomposition for an arbitrary Noetherian module. First two more definitions.

We will say, that a module M is <u>co-primary</u>, if $\mathrm{Ass}\, M$ consists of a single element. A submodule Q of M is a <u>primary submodule</u> if M/Q is co-primary. More precisely, Q is said to be $\mathcal{A}$-primary in M, if $\mathrm{Ass}\, M/Q = \{\mathcal{A}\}$.

(3.4) <u>Lemma</u>. A Noetherian uniform module U is co-primary.

Proof. By (3.3), $\mathrm{Ass}\, U \neq \emptyset$. Suppose $\mathcal{A} = [A]$ and $\mathcal{B} = [B]$ are both associated to U. Then there are submodules A', B' in U, isomorphic to submodules of A, B, respectively. Since $A' \cap B' = C \neq 0$ is isomorphic to a submodule of A and to a submodule of B, we get $A \cong \overline{C} \cong B$, whence $\mathcal{A} = \mathcal{B}$.

For the rest of this section M always denotes a Noetherian module over an arbitrary ring.

A <u>primary decomposition</u> of a submodule N of M is a representation of N as a finite intersection of primary submodules. By passing to M/N it is sufficient to investigate the primary decomposition of the zero submodule.

(3.5) <u>Proposition</u>. The zero submodule of M possesses a primary decomposition.

Proof. Since M is Noetherian, every submodule of M is a finite intersection of irreducible submodules. The proof of this is standard [27,p.208]. Thus, in particular, $0 = Q_1 \cap \ldots \cap Q_n$, where the Q_i are irreducible. Therefore the modules M/Q_i are uniform and Noetherian, hence co-primary by (3.4).

A somewhat different procedure is used in Bourbaki [3,ch.IV,§2]. One first proves, that $\mathrm{Ass}\, M$ is finite and then uses the following: If S is any subset of $\mathrm{Ass}\, M$, there exists an $N \subseteq M$ such that $\mathrm{Ass}\, M/N = S$ and $\mathrm{Ass}\, N = \mathrm{Ass}\, M \setminus S$, which provides another way of producing primary submodules. It is not hard to see that these arguments also work in the non-commutative case.

A primary decomposition $0 = Q_1 \cap Q_2 \cap \ldots \cap Q_n$ (Q_i A_i - primary) is said to be <u>reduced</u> if

(1) There exists no i such that $\bigcap_{i \neq j} Q_j \subseteq Q_i$.

(2) The A_i are all distinct.

Given any primary decomposition of 0, we can always find one satisfying (1) by omitting certain Q_i. Then (2) can be achieved using the following lemma:

(3.6) <u>Lemma</u>. If $Q_1, \ldots, Q_k \subseteq M$ are all A-primary, then $Q = \bigcap_{i=1}^{k} Q_i$ is also A-primary.

Proof. M/Q is isomorphic to a submodule of the direct sum $\sum_{i=1}^{k} M/Q_i$. By (3.1) $\mathrm{Ass}\, M/Q \subseteq \mathrm{Ass}\, \Sigma\, M/Q_i = \{A\}$, and by (3.3) $\mathrm{Ass}\, M/Q \neq \emptyset$.

Finally we deduce the usual uniqueness property.

(3.7) <u>Proposition</u>. If $0 = Q_1 \cap \dots \cap Q_n$ is a reduced primary decomposition, then $\bigcup_{i=1}^{n} \mathrm{Ass}\, M/Q_i = \mathrm{Ass}\, M$ and the union is disjoint.

Proof. Since M is isomorphic to a submodule of $\sum_{i=1}^{n} M/Q_i$, $\mathrm{Ass}\, M \subseteq \bigcup_{i=1}^{n} \mathrm{Ass}\, M/Q_i$ by (3.1). Conversely, if A_i is associated to M/Q_i, let $P_i = \bigcap_{i \neq j} Q_j$. Then $P_i \cong P_i + Q_i/Q_i \subseteq M/Q_i$ and since the decomposition is reduced $P_i \neq 0$, hence $\mathrm{Ass}\, P_i = \{A_i\}$ by (3.3), thus $A_i \in \mathrm{Ass}\, M$. The union is disjoint because the decomposition satisfies condition (2).

We now discuss briefly the commutative case. By [3,ch.IV,§2] one sees immediately, that for a commutative Noetherian ring R our primary decomposition reduces to the classical one. In particular, an ideal of R is primary in the present sense if and only if it is primary in the classical sense.

For an arbitrary commutative ring, Zariski and Samuel [27,p.252] have a definition of "primary submodule", which can be restated as follows: Define the <u>primary radical</u> Rad M of M to be the set of all $r \in R$, such that $Mr^n = 0$ for some natural number n. M is said to be ZS-co-primary if it satisfies (*): If $r \in R$ annihilates a non-zero element of M, then $r \in \mathrm{Rad}\, M$. In this case Rad M is a prime ideal.

Indeed, if $rs \in \text{Rad } M$, $r \notin M$, then $Mr^n \neq 0$, $Mr^n s^n = 0$ for some n. Therefore s annihilates a non-zero element of M and $s \in \text{Rad } M$ by (*).

(3.8) Proposition. Let M be a Noetherian module over the arbitrary ring R. Then M is co-primary if and only if it is ZS-co-primary. In this case $\text{Rad } M$ is the only prime ideal associated to M.

Proof. First assume, that (*) holds and let the prime ideals P and Q be associated to M, that is $P = \text{Ann } x$, $Q = \text{Ann } y$ for some $x, y \in M$. If $p \in P$, then $Mp^n = 0$ by (*), whence $yp^n = 0$, thus $p^n \in Q$ and $p \in Q$ since Q is prime. Here we did not use that M is Noetherian.

However, this assumption is used to prove the converse. Thus suppose M is Noetherian and has only one associated prime P. Then by (3.3) every non-zero submodule of M contains an element whose annihilator is P. Let now $xr = 0$ for some $0 \neq x \in M$. Then $xRr = 0$ and since some $xs \in xR$ has annihilator P, $r \in P$. If we set $N_i = \{m \in M \mid mr^i = 0\}$ we get an ascending chain of submodules which must stop, $N_n = N_{n+1}$ say. We claim that $Mr^n = 0$. Suppose not and let $yr^n \neq 0$ $(y \in M)$. Then yr^nR contains a non-zero element $0 \neq yr^nt$ $(t \in R)$ with annihilator P. Since $r \in P$, we have $yr^ntr = 0$, i.e. $yt \in N_{n+1} = N_n$ and consequently $ytr^n = 0$. This contradiction shows, that $r \in \text{Rad } M$, i.e. (*) is satisfied. From (*) it immediately follows, that $P \subseteq \text{Rad } M$. To show the reverse inclusion, let $r \in \text{Rad } M$.

Then some power of r annihilates M, hence r annihilates a non-zero submodule $N \subseteq M$. Since N contains an element whose annihilator is P, $r \in P$.

It follows from (3.8) that for a Noetherian module over a commutative ring our primary decomposition is the same as the one studied by Fisher [7, 9.3].

We conclude this section with the following remark: If R is right Noetherian, then the term "atom" could have been replaced throughout by "isomorphism class of indecomposable injective" because of (2.10). Furthermore, one obtains the same sets Ass M using either concept. In this case our primary decomposition reduces to a special case of the one considered by Gabriel [9, p.389] for locally Noetherian categories.

4. A characterization of certain Artinian rings

The Krull-Akizuki Theorem asserts, that a commutative Noetherian ring is Artinian if and only if every prime ideal is maximal [27, p.203].

We shall present two generalizations.

An atom [A] is said to be <u>maximal</u> if A contains a simple submodule.

(4.1) <u>Proposition</u>. A right Noetherian ring is right Artinian if and only if every atom is maximal.

Proof. If R is a right Artinian, then every module contains a simple submodule, hence every atom is maximal. Conversely, if R is right Noetherian and if every atomic module has a simple submodule, then every module contains a simple submodule by (2.7). Since R has no infinite set of orthogonal idempotents, R is left perfect by Theorem P of Bass [1]. Since a right Noetherian, left perfect ring is right Artinian, the result follows.

Michler [23] has proved a similar result using Goldman's equivalent notion of prime kernel functor [12].

(4.2) <u>Lemma</u>. Let R be right Artinian. Then every critical right ideal is maximal if and only if every simple module is rationally complete.

Proof. The condition is necessary, because R/P (P critical) has a simple submodule S and if S is rationally complete, then S = R/P, whence P is maximal.

To prove sufficiency, suppose S is simple but not rationally complete, and let $s \in \overline{S}$, $s \notin S$. If P = Ann s, then R/P is strongly uniform but not simple.

(4.3) Proposition. Let R be right Noetherian. Then every critical right ideal is maximal if and only if R is a finite direct product of primary right Artinian rings.

Proof. By a result of Courter [4], a finite direct product of primary right Artinian rings has the property, that every simple module is rationally complete and (4.2) applies.

Conversely, if R is right Noetherian and if every critical ideal is maximal, then R is right Artinian by (4.1) and every simple module is rationally complete by (4.2). It then follows from [25, 2.10], then R is a finite direct product of primary right Artinian rings.

For still another generalization see [18, 3.6].

5. Goldman's theory

The only purpose of this section is to establish the
equivalence between Goldman's theory and the present one.
For the convenience of the reader, we briefly recall the
relevant definitions; for more details, the reader should
consult Goldman's paper [12] or one of the many others
dealing with torsion theories, modules of quotients and
related concepts. (See [17] for a survey and a bibliography.)

An idempotent kernel functor σ : Mod R $\to$ Mod R is a
left exact subfunctor of the identity functor on Mod R such
that $\sigma(M/\sigma(M)) = 0$ for all M $\in$ Mod R. These functors are
partially ordered in the obvious way ($\sigma \leq \tau$ if $\sigma(M) \subseteq \tau(M)$
for all M). Given any module S, there exists a unique
largest idempotent kernel functor σ such that $\sigma(S) = 0$.
σ is denoted by τ_S and it is given by the formula

$$\tau_S(M) = \{m \in M \mid \phi(m) = 0 \text{ for all } \phi : M \to E(S)\}.$$

From this it follows, that $S' \subseteq S$ implies $\tau_S \leq \tau_{S'}$ and
that $\tau_S = \tau_{S'}$ if $S' \subseteq S$ is essential.

An idempotent kernel functor μ is called a prime, if
$\mu = \tau_S$, where S is τ_S-supporting, i.e. S/S' is
τ_S-torsion for every non-zero submodule S' of S. This
last condition can be written as $\text{Hom}(S/S', E(S)) = 0$. By
(1.1), this means, that S is τ_S-supporting if and only if
S is strongly uniform. Note, that $\tau_S = \tau_{\overline{S}}$.

(5.1) Proposition. There is a one-one correspondence between atoms and prime idempotent kernel functors given by $[A] \longmapsto \tau_A$.

Proof. By what has been said above, the primes are exactly the idempotent kernel functors of the form $\tau_{\overline{S}}$, where S is strongly uniform, or, equivalently, where $\overline{S}$ is atomic. It remains to show, that $[A] \to \tau_A$ is one-one. Thus let A, B be atomic and suppose $\tau_A = \tau_B = \tau$. Since $\tau(A) = 0$ $\mathrm{Hom}(A, E(B)) \neq 0$, thus there is a non-zero map $\phi : A' \to B$, where $A' \subseteq A$. This map must be a monomorphism by [12, 6.2], hence A' is isomorphic to $B' \subseteq B$ and $A = \overline{A'} \cong \overline{B'} = B$.

Goldman defines the set $\mathbb{P}(M)$ of associated primes of a module M to be the set of all those τ_N $(N \subseteq M)$, which happen to be primes.

(5.2) Proposition. For any module M, there is a bijection between $\mathbb{P}(M)$ and Ass M.

Proof. If $[A] \in \mathrm{Ass}\, M$, then some non-zero $A' \subseteq A$ is isomorphic to a submodule M' of M'. Then $\tau_{M'} = \tau_{A'} = \tau_A$ is a prime and $\tau_A \in \mathbb{P}(M)$.

Conversely, suppose $\tau_A \in \mathbb{P}(M)$, i.e. $\tau_A = \tau_N$ $(0 \neq N \subseteq M$, A atomic). As in the proof of (5.1) there is a non-zero map $\phi : A' \to N$, $A' \subseteq A$ which must be a monomorphism. Thus $[A] \in \mathrm{Ass}\, N \subseteq \mathrm{Ass}\, M$.

A module Q is Goldman primary, if τ_Q is a prime and if Q is stable, which means, that $\tau_Q = \tau_{Q'}$ for every $0 \neq Q' \subseteq Q$.

(5.3) Lemma. A Goldman primary module is co-primary.

Proof. The only associated atom of Q is the one corresponding to the prime τ_Q.

(5.4) Proposition. A co-primary module Q is Goldman primary if and only if $\operatorname{Ass} Q' \neq \emptyset$ for every non-zero submodule Q' of Q.

Proof. The condition is necessary, because $\tau_Q = \tau_{Q'}$ is a prime by assumption, whence Q' contains a non-zero submodule of the corresponding atomic module by the proof of (5.2).

To prove sufficiency, we first note, that by assumption there is a non-zero map from every non-zero submodule of Q into $E(A)$, where the atomic module A represents the only element of $\operatorname{Ass} Q$. In particular, for each $q \in Q$ there is a non-zero map $qR \to E(A)$ and from this it follows, that Q embeds into a direct product E of copies of $E(A)$. By assumption, every $Q' \subseteq Q$ $(Q' \neq 0)$ contains a non-zero submodule isomorphic to $A' \subseteq A$. Thus we get the following chain of inequalities: $\tau_A = \tau_{E(A)} = \tau_E \leq \tau_Q \leq \tau_{Q'} \leq \tau_{A'} = \tau_A$. Hence $\tau_{Q'} = \tau_A$ for every $0 \neq Q' \subseteq Q$ and Q is Goldman primary.

There are co-primary modules which are not Goldman primary, e.g. the module $R \oplus R/J$ with R and J as in (2.11), but for Noetherian modules, the two concepts coincide by (3.3). Thus the primary decomposition for Noetherian modules obtained by Goldman is the same as ours.

The concept of "prime kernel functor" is more useful than that of "atom", when it comes to forming rings and modules of quotients [12]. Indeed, if we wish to define these concepts for an atom [A], we have to pass to the associated kernel functor τ_A or to the associated idempotent filter $\mathcal{F}_A = \{I \subseteq R \mid \mathrm{Hom}(R/I, E(A)) = 0\}$ of right ideals, according to taste. See [17] for the various aspects of these constructions and some history.

6. The tertiary decomposition

In this section, we briefly study the relation between our primary decomposition and the tertiary decomposition introduced by Lesieur and Croisot [20] for finitely generated modules over a right Noetherian ring. We will generalize this slightly to Noetherian modules over an arbitrary ring. The reader is also referred to a paper by Fisher [7], where a more general class of modules appears. We also wish to mention the paper by Michler [23], where the tertiary and Goldman's primary decomposition are compared.

We recall the relevant definitions, see [7, 13, 21, 24]: Let M be a module. An element r of R is said to be an *annihilating element* for M if $Nr = 0$ for some non-zero submodule of M. The set $T(M)$ of annihilating elements of M is a two-sided ideal. The *tertiary radical* $\text{Ter}\, M$ is the set of all elements of R annihilating an essential submodule of M, or, equivalently, the set of all elements of R, which are annihilating elements for all non-zero submodules of M. A module M is *co-tertiary*, if $T(M) \leq \text{Ter}\, M$, and $N \leq M$ is a *tertiary submodule* if M/N is co-tertiary.

We now introduce an equivalence relation in the set of atoms as follows: $[A] \sim [B]$ if $T(A) = T(B)$. We do not resist the temptation to use the name "*molecule*" for an equivalence class of atoms. A molecule is associated to M if one of the atoms it contains is associated to M in our previous sense.

(6.1) Proposition. A Noetherian module is co-tertiary if and only if it has only one associated molecule.

Proof. Suppose M is co-tertiary and let $[A],[B] \in \mathrm{Ass}\, M$. We have to show, that $[A] \sim [B]$. Thus let $A'r = 0$ for some non-zero submodule A' of A. Since $[A] \in \mathrm{Ass}\, M$, a non-zero submodule A'' of A' is isomorphic to a submodule of M, thus $r \in T(M)$, and $T(M) \subseteq \mathrm{Ter}\, M$, since M is co-tertiary. Hence r annihilates an essential submodule L of M and $L \cap D \neq 0$, where $D \subseteq M$ is isomorphic to a submodule of B. Therefore $r \in T(B)$. It follows, that $T(A) = T(B)$, whence $[A] \sim [B]$.

Conversely, assume that all the associated atoms of M are equivalent. Let $r \in T(M)$. We have to show, that r annihilates an essential submodule of M. It suffices to show, that every non-zero submodule U of M contains a non-zero submodule U^* annihilated by r. By assumption $Nr = 0$ for some $0 \neq N \subseteq M$ and since N is Noetherian, it contains a non-zero submodule isomorphic to a submodule of A, where $[A] \in \mathrm{Ass}\, M$. Thus $r \in T(A)$, hence $r \in T(B)$ for all $[B] \in \mathrm{Ass}\, M$. Let now $U \neq 0$ be any submodule of M. Then $0 \neq U' \subseteq U$ is isomorphic to $B' \subseteq B$, for some $[B] \in \mathrm{Ass}\, M$. Since $T(A) = T(B)$ $B''r = 0$ for some $0 \neq B'' \subseteq B$ and if $0 \neq U^* \subseteq U$ is isomorphic to $B' \cap B''$, then $U^*r = 0$.

(6.2) Corollary. A Noetherian co-primary module is co-tertiary.

From this, it is clear, that the primary decomposition established in (3.6) is also a tertiary decomposition. The former is finer than the latter in the sense, that more primary submodules are needed to yield an irredundant representation of the zero submodule.

The existence of a reduced tertiary decomposition and the uniqueness of the associated molecules also follow immediately. Indeed, all the relevant properties of associated atoms also hold for associated molecules.

If R is right Noetherian, there is a bijection between molecules and two-sided prime ideals of R as follows.

If A is an atomic module, then A is uniform, hence $T(A) = \text{Ter } A$ and the latter is known to be a prime ideal [13]. Another way of obtaining $\text{Ter } A$ is to consider the (necessarily unique) maximal ideal among the annihilators of non-zero submodules of A [21]. If $[A] \sim [B]$, then $T(A) = T(B)$ by definition, hence $\text{Ter } A = \text{Ter } B$ is the same prime ideal.

Conversely, if P is a two-sided prime ideal, then $E(R/P)$ is isotypical, i.e. a direct sum of n copies of the indecomposable injective E [9, p.421]. The core $C(E)$ is atomic by (2.9) and P corresponds to the molecule containing $[C(E)]$.

7. Compressible and quasi-simple modules

In Section 2 we generalized the notion of prime ideal to the non-commutative case using the fact, that P is prime if and only if R/P is strongly uniform. There is at least one other characterization, equivalent to the first one in the commutative case but not in general.

Following Goldie [11], we say, that a module M is <u>compressible</u>, if every non-zero submodule of M contains an isomorphic copy of M.

(7.1) <u>Lemma</u>. If R is commutative, then an ideal P is prime if and only if R/P is compressible.

Proof. If P is prime and if $a + P$ is a non-zero element of R/P, then the map sending $1 + P$ to $a + P$ is a well-defined monomorphism $R/P \to aR + P/P$.

Conversely, if R/P is compressible, and $a \notin P$, then the monomorphism $R/P \to aR + P/P$ sends $1 + P$ into $ar + P \neq 0$ for some $r \in R$. If $b \notin P$, then $(1 + P)b \neq 0$, whence $(ar + P)b \neq 0$, i.e. $ab \notin P$.

In the non-commutative case, the concepts of "compressible" and "strongly uniform" are no longer equivalent, even for cyclic modules. Indeed, if R is an integral domain which does not satisfy the right Ore condition, then R is compressible, but not strongly uniform, whereas the module R/P of Example 2.1 is strongly uniform but not compressible.

However, we have the following result:

(7.2) Proposition. If R is right Noetherian, then every compressible module M is strongly uniform.

Proof. By compressibility, M is isomorphic to a submodule of each if its finitely generated submodules, hence M is itself finitely generated, since R is Noetherian. Thus by (2.7) M contains a strongly uniform submodule and must therefore be strongly uniform itself, again by compressibility.

Recall that a module M is said to be quasi-injective, if every map from a submodule of M into M can be extended to an endomorphism of M. M is quasi-injective if and only if $\Lambda M = M$, where Λ denotes the endomorphism ring of $E(M)$.

The quasi-injective hull $\tilde{M}$ of M is in a sense the smallest quasi-injective module containing M; it is given by $\tilde{M} = \Lambda M$. See [6] for proofs.

(7.3) Lemma. An atomic module A is quasi-injective.

Proof. Let $A' \leq A$ (we can assume $A' \neq 0$) and let $\phi: A' \to A$. By (1.1) $\mathrm{Hom}(A/A', E(A)) = 0$, hence by (1.2) ϕ can be extended to a map $A \to A$.

(7.4) Lemma. Let Q be a quasi-injective strongly uniform module. Then the endomorphism ring of Q is a division ring.

Proof. Let $\phi: Q \to Q$ be a non-zero endomorphism with kernel K. Suppose $K \neq 0$ and let $q \notin K$, $q \in Q$. Since $K \leq Q$ is rational, there is an $r \in R$ such that $qr \in K$

$\phi(q)r \neq 0$, a contradiction. Thus every endomorphism is a monomorphism. Let now $N = \phi(M)$, then the map $\psi: N \to M$ given by $\psi\phi(m) = m$ can be extended to $\overline{\psi}: M \to M$ by quasi-injectivity. If $x \in M$, then $\overline{\psi}(x) = m = \psi\phi(m) = \overline{\psi}\phi(m)$ for some $m \in M$. Since $\overline{\psi}$ is mono, $x = \phi(m)$, thus ϕ is onto and hence an automorphism.

In particular it follows, that the endomorphism ring of an atom is a division ring. (See also [12,p.43]). Wong [26] has a related result in view of (2.9,d).

(7.5) Proposition. A module is strongly uniform if and only if the endomorphism ring of its quasi-injective hull is a division ring.

Proof. If M is strongly uniform, then $\overline{M}$ is atomic, thus quasi-injective and we have that $\Lambda\overline{M} = \overline{M}$, where Λ is the endomorphism ring of $E(M) = E(\overline{M})$. Therefore $\tilde{M} = \Lambda M \subseteq \Lambda\overline{M} = \overline{M}$ and (5.2) implies, that End $\tilde{M}$ is a division ring.

Conversely, let End $\tilde{M}$ be a division ring and let $0 \neq N \subseteq A \subseteq M$. Then any $\phi: A \to M$ is either zero or can be extended to an automorphism of M, hence $\phi(N) = 0$ implies $\phi = 0$, i.e. $N \subseteq M$ is rational.

Let us recall now the following definitions due to Koh [16]: A module M is quasi-simple if

(i) End $\tilde{M}$ is a division ring.

(ii) M is compressible.

A right ideal I is <u>almost maximal</u>, if R/I is quasi-simple.

Note that under the presence of (i) every map from M into a submodule of M is a monomorphism. Thus (i), (ii) can be replaced by (i), (iii), where

(iii) For every non-zero submodule N of M, there is a non-zero map $M \to N$.

By the results derived above, the following proposition easily follows.

(7.6) <u>Proposition</u>.

(a) A module is quasi-simple if and only if it is strongly uniform and compressible.

(b) If R is right Noetherian, then a module is quasi-simple if and only if it is compressible.

Similar statements hold for almost maximal right ideals.

(7.7) <u>Proposition</u>. A two-sided ideal I is almost maximal as a right ideal if and only if it is critical as a right ideal.

Proof. By (7.6,a) any almost maximal right ideal is critical. Let now I be critical, then it is sufficient to show, that (iii) holds for $M = R/I$ and $N = aR + I/I$, $a \notin I$. This is immediate, since $1 + I \longmapsto a + I$ is a well defined non-zero map because I is two-sided.

In view of the result preceding (2.2), we obtain the following corollary: A two-sided ideal I is almost maximal as a right ideal if and only if R/I is a right Ore domain.

We conclude these notes with a question: Does every atomic module contain a quasi-simple submodule? If this were true, we would obtain a one-one correspondence between atoms and classes of related almost maximal right ideals. What can be said if R is right Noetherian?

References

1. Bass, H.: Finitistic dimension and a homological generalization of semi-primary rings. Trans. Amer. Math. Soc., 95, 466-488 (1960).

2. Bland, P. E.: On rational and quasi-rational extensions of modules. Thesis. University of South Carolina 1969.

3. Bourbaki, N.: "Algèbra commutative". Chap. III et IV. Paris: Hermann 1961.

4. Courter, R.: Finite direct sums of complete matrix rings over perfect completely primary rings. Canad. J. Math. 21, 430-446 (1969).

5. Dlab, V.: Rank theory of modules. Fund. Math. 64, 313-324 (1969).

6. Faith, C.: "Lectures on injective modules and quotient rings." Lecture Notes in Mathematics Nr. 49. Berlin, Heidelberg, New York: Springer 1967.

7. Fisher, J. W.: Decomposition theories for modules. Trans. Amer. Math. Soc. 145, 241-269 (1969).

8. Findlay, G. D. and J. Lambek: A generalized ring of quotients I, II. Canad. Math. Bull. 1, 77-85, 155-167 (1958).

9. Gabriel, P.: Des catégories abéliennes. Bull. Soc. Math. France 90, 323-448 (1962).

10. Goldie, A. W.: The structure of prime rings under ascending chain conditions. Proc. London Math. Soc., 8, 589-608, (1958).

11. Goldie, A. W.: The structure of Noetherian rings. Tulane Ring Year Lecture Notes.

12. Goldman, O.: Rings and modules of quotients. J. Algebra 13, 10-47 (1969).

13. Herstein, I. N.: Topics in Ring Theory. Chicago: University of Chicago Press 1969.

14. Hudry, A.: Sur la localisation dans une catégorie de modules. C. R. Acad. Sci. Paris Sér. A-B 270, 925-928 (1970).

15. ________: Sur les anneaux localement homogènes. C. R. Acad. Sci. Paris Sér. A-B 271, 1214-1217 (1970).

16. Koh, K.: Quasi-simple modules and almost maximal one-sided ideals. Tulane Ring Year Lecture Notes.

17. Lambek, J.: "Torsion theories, additive semantics and rings of quotients." Lecture Notes in Mathematics Nr. 177. Berlin, Heidelberg, New York: Springer 1971.

18. Lambek, J. and G. Michler: The torsion theory at a prime ideal of a right Noetherian ring. To appear.

19. Lazard, D. and P. Huet: Dominions des anneaux commutatifs. Bull. Sci. Math. 94, 193-199 (1970).

20. Lesieur, L. and R. Croisot: Coeur d'un module. Journ. Math. Pures Appl. 42, 367-406 (1963).

21. ____________________: Algèbre noethérienne non commutative, Mémor. Sci. Math. Fasc. CLIV, 1963.

22. Matlis, E.: Injective modules over Noetherian rings, Pacific J. Math. 8, 511-528 (1958).

23. Michler, G.: Goldman's primary decomposition and the tertiary decomposition. J. Algebra 16, 129-137 (1970).

24. Riley, J. A.: Axiomatic primary and tertiary decomposition theory, Trans. Amer. Math. Soc. 105, 177-201 (1962).

25. Storrer, H. H.: Rational extensions of modules. Pacific J. Math. To appear.

26. Wong, E. T.: Endomorphisms of the quasi-injective hull of a module. Canad. Math. Bull. 13, 149-150 (1970).

27. Zariski O. and P. Samuel: Commutative Algebra, Vol. I, Princeton, N. J.: Van Nostrand 1958.

Lecture Notes in Mathematics

Comprehensive leaflet on request

Vol. 74: A. Fröhlich, Formal Groups. IV, 140 pages. 1968. DM 16,-

Vol. 75: G. Lumer, Algèbres de fonctions et espaces de Hardy. VI, 80 pages. 1968. DM 16,-

Vol. 76: R. G. Swan, Algebraic K-Theory. IV, 262 pages. 1968. DM 18,-

Vol. 77: P.-A. Meyer, Processus de Markov: la frontière de Martin. IV, 123 pages. 1968. DM 16,-

Vol. 78: H. Herrlich, Topologische Reflexionen und Coreflexionen. XVI, 166 Seiten. 1968. DM 16,-

Vol. 79: A. Grothendieck, Catégories Cofibrées Additives et Complexe Cotangent Relatif. IV, 167 pages. 1968. DM 16,-

Vol. 80: Seminar on Triples and Categorical Homology Theory. Edited by B. Eckmann. IV, 398 pages. 1969. DM 20,-

Vol. 81: J.-P. Eckmann et M. Guenin, Méthodes Algébriques en Mécanique Statistique. VI, 131 pages. 1969. DM 16,-

Vol. 82: J. Wloka, Grundräume und verallgemeinerte Funktionen. VIII, 131 Seiten. 1969. DM 16,-

Vol. 83: O. Zariski, An Introduction to the Theory of Algebraic Surfaces. IV, 100 pages. 1969. DM 16,-

Vol. 84: H. Lüneburg, Transitive Erweiterungen endlicher Permutationsgruppen. IV, 119 Seiten. 1969. DM 16,-

Vol. 85: P. Cartier et D. Foata, Problèmes combinatoires de commutation et réarrangements. IV, 88 pages. 1969. DM 16,-

Vol. 86: Category Theory, Homology Theory and their Applications I. Edited by P. Hilton. VI, 216 pages. 1969. DM 16,-

Vol. 87: M. Tierney, Categorical Constructions in Stable Homotopy Theory. IV, 65 pages. 1969. DM 16,-

Vol. 88: Séminaire de Probabilités III. IV, 229 pages. 1969. DM 18,-

Vol. 89: Probability and Information Theory. Edited by M. Behara, K. Krickeberg and J. Wolfowitz. IV, 256 pages. 1969. DM 18,-

Vol. 90: N. P. Bhatia and O. Hajek, Local Semi-Dynamical Systems. II, 157 pages. 1969. DM 16,-

Vol. 91: N. N. Janenko, Die Zwischenschrittmethode zur Lösung mehrdimensionaler Probleme der mathematischen Physik. VIII, 194 Seiten. 1969. DM 16,80

Vol. 92: Category Theory, Homology Theory and their Applications II. Edited by P. Hilton. V, 308 pages. 1969. DM 20,-

Vol. 93: K. R. Parthasarathy, Multipliers on Locally Compact Groups. III, 54 pages. 1969. DM 16,-

Vol. 94: M. Machover and J. Hirschfeld, Lectures on Non-Standard Analysis. VI, 79 pages. 1969. DM 16,-

Vol. 95: A. S. Troelstra, Principles of Intuitionism. II, 111 pages. 1969. DM 16,-

Vol. 96: H.-B. Brinkmann und D. Puppe, Abelsche und exakte Kategorien, Korrespondenzen. V, 141 Seiten. 1969. DM 16,-

Vol. 97: S. O. Chase and M. E. Sweedler, Hopf Algebras and Galois theory. II, 133 pages. 1969. DM 16,-

Vol. 98: M. Heins, Hardy Classes on Riemann Surfaces. III, 106 pages. 1969. DM 16,-

Vol. 99: Category Theory, Homology Theory and their Applications III. Edited by P. Hilton. IV, 489 pages. 1969. DM 24,-

Vol. 100: M. Artin and B. Mazur, Etale Homotopy. II, 196 Seiten. 1969. DM 16,-

Vol. 101: G. P. Szegö et G. Treccani, Semigruppi di Trasformazioni Multivoche. VI, 177 pages. 1969. DM 16,-

Vol. 102: F. Stummel, Rand- und Eigenwertaufgaben in Sobolewschen Räumen. VIII, 386 Seiten. 1969. DM 20,-

Vol. 103: Lectures in Modern Analysis and Applications I. Edited by C. T. Taam. VII, 162 pages. 1969. DM 16,-

Vol. 104: G. H. Pimbley, Jr., Eigenfunction Branches of Nonlinear Operators and their Bifurcations. II, 128 pages. 1969. DM 16,-

Vol. 105: R. Larsen, The Multiplier Problem. VII, 284 pages. 1969. DM 18,-

Vol. 106: Reports of the Midwest Category Seminar III. Edited by S. Mac Lane. III, 247 pages. 1969. DM 16,-

Vol. 107: A. Peyerimhoff, Lectures on Summability. III, 111 pages. 1969. DM 16,-

Vol. 108: Algebraic K-Theory and its Geometric Applications. Edited by R. M.F. Moss and C. B. Thomas. IV, 86 pages. 1969. DM 16,-

Vol. 109: Conference on the Numerical Solution of Differential Equations. Edited by J. Ll. Morris. VI, 275 pages. 1969. DM 18,-

Vol. 110: The Many Facets of Graph Theory. Edited by G. Chartrand and S. F. Kapoor. VIII, 290 pages. 1969. DM 18,-

Vol. 111: K. H. Mayer, Relationen zwischen charakteristischen Zahlen. III, 99 Seiten. 1969. DM 16,-

Vol. 112: Colloquium on Methods of Optimization. Edited by N. N. Moiseev. IV, 293 pages. 1970. DM 18,-

Vol. 113: R. Wille, Kongruenzklassengeometrien. III, 99 Seiten. 1970. DM 16,-

Vol. 114: H. Jacquet and R. P. Langlands, Automorphic Forms on GL (2). VII, 548 pages. 1970. DM 24,-

Vol. 115: K. H. Roggenkamp and V. Huber-Dyson, Lattices over Orders I. XIX, 290 pages. 1970. DM 18,-

Vol. 116: Séminaire Pierre Lelong (Analyse) Année 1969. IV, 195 pages. 1970. DM 16,-

Vol. 117: Y. Meyer, Nombres de Pisot, Nombres de Salem et Analyse Harmonique. 63 pages. 1970. DM 16,-

Vol. 118: Proceedings of the 15th Scandinavian Congress, Oslo 1968. Edited by K. E. Aubert and W. Ljunggren. IV, 162 pages. 1970. DM 16,-

Vol. 119: M. Raynaud, Faisceaux amples sur les schémas en groupes et les espaces homogènes. III, 219 pages. 1970. DM 16,-

Vol. 120: D. Siefkes, Büchi's Monadic Second Order Successor Arithmetic. XII, 130 Seiten. 1970. DM 16,-

Vol. 121: H. S. Bear, Lectures on Gleason Parts. III, 47 pages. 1970. DM 16,-

Vol. 122: H. Zieschang, E. Vogt und H. D. Coldewey, Flächen und ebene diskontinuierliche Gruppen. VIII, 203 Seiten. 1970. DM 16,-

Vol. 123: A. V. Jategaonkar, Left Principal Ideal Rings. VI, 145 pages. 1970. DM 16,-

Vol. 124: Séminare de Probabilités IV. Edited by P. A. Meyer. IV, 282 pages. 1970. DM 20,-

Vol. 125: Symposium on Automatic Demonstration. V, 310 pages. 1970. DM 20,-

Vol. 126: P. Schapira, Théorie des Hyperfonctions. XI, 157 pages. 1970. DM 16,-

Vol. 127: I. Stewart, Lie Algebras. IV, 97 pages. 1970. DM 16,-

Vol. 128: M. Takesaki, Tomita's Theory of Modular Hilbert Algebras and its Applications. II, 123 pages. 1970. DM 16,-

Vol. 129: K. H. Hofmann, The Duality of Compact Semigroups and C*-Bigebras. XII, 142 pages. 1970. DM 16,-

Vol. 130: F. Lorenz, Quadratische Formen über Körpern. II, 77 Seiten. 1970. DM 16,-

Vol. 131: A. Borel et al., Seminar on Algebraic Groups and Related Finite Groups. VII, 321 pages. 1970. DM 22,-

Vol. 132: Symposium on Optimization. III, 348 pages. 1970. DM 22,-

Vol. 133: F. Topsøe, Topology and Measure. XIV, 79 pages. 1970. DM 16,-

Vol. 134: L. Smith, Lectures on the Eilenberg-Moore Spectral Sequence. VII, 142 pages. 1970. DM 16,-

Vol. 135: W. Stoll, Value Distribution of Holomorphic Maps into Compact Complex Manifolds. II, 267 pages. 1970. DM 18,-

Vol. 136: M. Karoubi et al., Séminaire Heidelberg-Saarbrücken-Strasbourg sur la K-Théorie. IV, 264 pages. 1970. DM 18,

Vol. 137: Reports of the Midwest Category Seminar IV. Edited by S. MacLane. III, 139 pages. 1970. DM 16,-

Vol. 138: D. Foata et M. Schützenberger, Théorie Géométrique des Polynômes Eulériens. V, 94 pages. 1970. DM 16,-

Vol. 139: A. Badrikian, Séminaire sur les Fonctions Aléatoires Linéaires et les Mesures Cylindriques. VII, 221 pages. 1970. DM 18,-

Vol. 140: Lectures in Modern Analysis and Applications II. Edited by C. T. Taam. VI, 119 pages. 1970. DM 16,-

Vol. 141: G. Jameson, Ordered Linear Spaces. XV, 194 pages. 1970. DM 16,-

Vol. 142: K. W. Roggenkamp, Lattices over Orders II. V, 388 pages. 1970. DM 22,-

Vol. 143: K. W. Gruenberg, Cohomological Topics in Group Theory. XIV, 275 pages. 1970. DM 20,-

Vol. 144: Seminar on Differential Equations and Dynamical Systems, II. Edited by J. A. Yorke. VIII, 268 pages. 1970. DM 20,-

Vol. 145: E. J. Dubuc, Kan Extensions in Enriched Category Theory. XVI, 173 pages. 1970. DM 16,-

Please turn over

Vol. 146: A. B. Altman and S. Kleiman, Introduction to Grothendieck Duality Theory. II, 192 pages. 1970. DM 18,–

Vol. 147: D. E. Dobbs, Cech Cohomological Dimensions for Commutative Rings. VI, 176 pages. 1970. DM 16,–

Vol. 148: R. Azencott, Espaces de Poisson des Groupes Localement Compacts. IX, 141 pages. 1970. DM 16,–

Vol. 149: R. G. Swan and E. G. Evans, K-Theory of Finite Groups and Orders. IV, 237 pages. 1970. DM 20,–

Vol. 150: Heyer, Dualität lokalkompakter Gruppen. XIII, 372 Seiten. 1970. DM 20,–

Vol. 151: M. Demazure et A. Grothendieck, Schémas en Groupes I. (SGA 3). XV, 562 pages. 1970. DM 24,–

Vol. 152: M. Demazure et A. Grothendieck, Schémas en Groupes II. (SGA 3). IX, 654 pages. 1970. DM 24,–

Vol. 153: M. Demazure et A. Grothendieck, Schémas en Groupes III (SGA 3). VIII, 529 pages. 1970. DM 24,–

Vol. 154: A. Lascoux et M. Berger, Variétés Kähleriennes Compactes. VII, 83 pages. 1970. DM 16,–

Vol. 155: Several Complex Variables I, Maryland 1970. Edited by J. Horváth. IV, 214 pages. 1970. DM 18,–

Vol. 156: R. Hartshorne, Ample Subvarieties of Algebraic Varieties. XIV, 256 pages. 1970. DM 20,–

Vol. 157: T. tom Dieck, K. H. Kamps und D. Puppe, Homotopietheorie. VI, 265 Seiten. 1970. DM 20,–

Vol. 158: T. G. Ostrom, Finite Translation Planes. IV. 112 pages. 1970. DM 16,–

Vol. 159: R. Ansorge und R. Hass. Konvergenz von Differenzenverfahren für lineare und nichtlineare Anfangswertaufgaben. VIII, 145 Seiten. 1970. DM 16,–

Vol. 160: L. Sucheston, Constributions to Ergodic Theory and Probability. VII, 277 pages. 1970. DM 20,–

Vol. 161: J. Stasheff, H-Spaces from a Homotopy Point of View. VI, 95 pages. 1970. DM 16,–

Vol. 162: Harish-Chandra and van Dijk, Harmonic Analysis on Reductive p-adic Groups. IV, 125 pages. 1970. DM 16,–

Vol. 163: P. Deligne, Equations Différentielles à Points Singuliers Reguliers. III, 133 pages. 1970. DM 16,–

Vol. 164: J. P. Ferrier, Seminaire sur les Algebres Complétes. II, 69 pages. 1970. DM 16,–

Vol. 165: J. M. Cohen, Stable Homotopy. V, 194 pages. 1970. DM 16,–

Vol. 166: A. J. Silberger, PGL_2 over the p-adics: its Representations, Spherical Functions, and Fourier Analysis. VII, 202 pages. 1970. DM 18,–

Vol. 167: Lavrentiev, Romanov and Vasiliev, Multidimensional Inverse Problems for Differential Equations. V, 59 pages. 1970. DM 16,–

Vol. 168: F. P. Peterson, The Steenrod Algebra and its Applications: A conference to Celebrate N. E. Steenrod's Sixtieth Birthday. VII, 317 pages. 1970. DM 22,–

Vol. 169: M. Raynaud, Anneaux Locaux Henséliens. V, 129 pages. 1970. DM 16,–

Vol. 170: Lectures in Modern Analysis and Applications III. Edited by C. T. Taam. VI, 213 pages. 1970. DM 18,–

Vol. 171: Set-Valued Mappings, Selections and Topological Properties of 2^X. Edited by W. M. Fleischman. X, 110 pages. 1970. DM 16,–

Vol. 172: Y.-T. Siu and G. Trautmann, Gap-Sheaves and Extension of Coherent Analytic Subsheaves. V, 172 pages. 1971. DM 16,–

Vol. 173: J. N. Mordeson and B. Vinograde, Structure of Arbitrary Purely Inseparable Extension Fields. IV, 138 pages. 1970. DM 16,–

Vol. 174: B. Iversen, Linear Determinants with Applications to the Picard Scheme of a Family of Algebraic Curves. VI, 69 pages. 1970. DM 16,–

Vol. 175: M. Brelot, On Topologies and Boundaries in Potential Theory. VI, 176 pages. 1971. DM 18,–

Vol. 176: H. Popp, Fundamentalgruppen algebraischer Mannigfaltigkeiten. IV, 154 Seiten. 1970. DM 16,–

Vol. 177: J. Lambek, Torsion Theories, Additive Semantics and Rings of Quotients. VI, 94 pages. 1971. DM 16,–

Vol. 178: Th. Bröcker und T. tom Dieck, Kobordismentheorie. XVI, 191 Seiten. 1970. DM 18,–

Vol. 179: Seminaire Bourbaki – vol. 1968/69. Exposés 347-363. IV. 295 pages. 1971. DM 22,–

Vol. 180: Séminaire Bourbaki – vol. 1969/70. Exposés 364-381. IV, 310 pages. 1971. DM 22,–

Vol. 181: F. DeMeyer and E. Ingraham, Separable Algebras over Commutative Rings. V, 157 pages. 1971. DM 16,–

Vol. 182: L. D. Baumert. Cyclic Difference Sets. VI, 166 pages. 1971. DM 16,–

Vol. 183: Analytic Theory of Differential Equations. Edited by P. F. Hsieh and A. W. J. Stoddart. VI, 225 pages. 1971. DM 20,–

Vol. 184: Symposium on Several Complex Variables, Park City, Utah, 1970. Edited by R. M. Brooks. V, 234 pages. 1971. DM 20,–

Vol. 185: Several Complex Variables II, Maryland 1970. Edited by J. Horváth. III, 287 pages. 1971. DM 24,–

Vol. 186: Recent Trends in Graph Theory. Edited by M. Capobianco/ J. B. Frechen/M. Krolik. VI, 219 pages. 1971. DM 18,–

Vol. 187: H. S. Shapiro, Topics in Approximation Theory. VIII, 275 pages. 1971. DM 22,–

Vol. 188: Symposium on Semantics of Algorithmic Languages. Edited by E. Engeler. VI, 372 pages. 1971. DM 26,–

Vol. 189: A. Weil, Dirichlet Series and Automorphic Forms. V, 164 pages. 1971. DM 16,–

Vol. 190: Martingales. A Report on a Meeting at Oberwolfach, May 17-23, 1970. Edited by H. Dinges. V, 75 pages. 1971. DM 16,–

Vol. 191: Séminaire de Probabilités V. Edited by P. A. Meyer. IV, 372 pages. 1971. DM 26,–

Vol. 192: Proceedings of Liverpool Singularities – Symposium I. Edited by C. T. C. Wall. V, 319 pages. 1971. DM 24,–

Vol. 193: Symposium on the Theory of Numerical Analysis. Edited by J. Ll. Morris. VI, 152 pages. 1971. DM 16,–

Vol. 194: M. Berger, P. Gauduchon et E. Mazet. Le Spectre d'une Variété Riemannienne. VII, 251 pages. 1971. DM 22,–

Vol. 195: Reports of the Midwest Category Seminar V. Edited by J. W. Gray and S. Mac Lane. III, 255 pages. 1971. DM 22,–

Vol. 196: H-spaces – Neuchâtel (Suisse)- Août 1970. Edited by F. Sigrist, V, 156 pages. 1971. DM 16,–

Vol. 197: Manifolds – Amsterdam 1970. Edited by N. H. Kuiper. V, 231 pages. 1971. DM 20,–

Vol. 198: M. Hervé, Analytic and Plurisubharmonic Functions in Finite and Infinite Dimensional Spaces. VI, 90 pages. 1971. DM 16,–

Vol. 199: Ch. J. Mozzochi, On the Pointwise Convergence of Fourier Series. VII, 87 pages. 1971. DM 16,–

Vol. 200: U. Neri, Singular Integrals. VII, 272 pages. 1971. DM 22,–

Vol. 201: J. H. van Lint, Coding Theory. VII, 136 pages. 1971. DM 16,–

Vol. 202: J. Benedetto, Harmonic Analysis on Totally Disconnected Sets. VIII, 261 pages. 1971. DM 22,–

Vol. 203: D. Knutson, Algebraic Spaces. VI, 261 pages. 1971. DM 22,–

Vol. 204: A. Zygmund, Intégrales Singulières. IV, 53 pages. 1971. DM 16,–

Vol. 205: Séminaire Pierre Lelong (Analyse) Année 1970. VI, 243 pages. 1971. DM 20,–

Vol. 206: Symposium on Differential Equations and Dynamical Systems. Edited by D. Chillingworth. XI, 173 pages. 1971. DM 16,–

Vol. 207: L. Bernstein, The Jacobi-Perron Algorithm – Its Theory and Application. IV, 161 pages. 1971. DM 16,–

Vol. 208: A. Grothendieck and J. P. Murre, The Tame Fundamental Group of a Formal Neighbourhood of a Divisor with Normal Crossings on a Scheme. VIII, 133 pages. 1971. DM 16,–

Vol. 209: Proceedings of Liverpool Singularities Symposium II. Edited by C. T. C. Wall. V, 280 pages. 1971. DM 22,–

Vol. 210: M. Eichler, Projective Varieties and Modular Forms. III, 118 pages. 1971. DM 16,–

Vol. 211: Théorie des Matroïdes. Edité par C. P. Bruter. III, 108 pages. 1971. DM 16,–

Vol. 212: B. Scarpellini, Proof Theory and Intuitionistic Systems. VII, 291 pages. 1971. DM 24,–

Vol. 213: H. Hogbe-Nlend, Théorie des Bornologies et Applications. V, 168 pages. 1971. DM 18,–

Vol. 214: M. Smorodinsky, Ergodic Theory, Entropy. V, 64 pages. 1971. DM 16,–